Goodwyn, Lawrence.

Breaking the
barrier.

$27.95

DATE			

BREAKING THE BARRIER

BREAKING THE BARRIER

The Rise of Solidarity in Poland

LAWRENCE GOODWYN

New York Oxford
OXFORD UNIVERSITY PRESS
1991

Oxford University Press

Oxford New York Toronto
Delhi Bombay Calcutta Madras Karachi
Petaling Jaya Singapore Hong Kong Tokyo
Nairobi Dar es Salaam Cape Town
Melbourne Auckland

and associated companies in
Berlin Ibadan

Published by Oxford University Press, Inc.,
200 Madison Avenue, New York, New York 10016

Oxford is a registered trademark of Oxford University Press

Library of Congress Cataloging-in-Publication Data
Goodwyn, Lawrence.
Breaking the barrier : the rise of Solidarity in Poland /
Lawrence Goodwyn.
p. cm. Includes bibliographical references and index.
ISBN 0-19-506122-5
1. NSZZ "Solidarność" (Labor organization)—History.
2. Poland—Politics and government—1945- I. Title.
HD8537.N783G66 1991
322'.2'09438—dc20 90-39898

9 8 7 6 5 4 3 2 1

Printed in the United States of America
on acid-free paper

In affection and with esteem,
this book is dedicated to
the activists of Solidarność
and to the people of Poland.

$27.95

Acknowledgments

THE STATE OF WAR in Poland transformed normal rules of research and attribution on Solidarność because, among other hazards, General Jaruzelski imposed martial law faster than comprehensive studies could be written and published. The thousands of local, regional, and national activists imprisoned in December 1981, who constituted in the aggregate the key sources to the internal life of the Polish movement, remained in a threatened status under martial law even after they were released from internment a year or more later. Most learned in 1982 that the government had decided to make it easier for them to obtain exit visas than work in Poland. A substantial percentage of the large and reluctant emigration that ensued came to the West through a Solidarność "pipeline" hurriedly established in Paris through the efforts of the government of François Mitterand. I wish to acknowledge the generous cooperation of the Paris Solidarność Committee in facilitating interviews by the author with workers and intellectuals from many parts of Poland who came through this émigré channel. I also wish to thank Anna Husarska of the Paris committee for her extended services as a translator. For the translation of subsequent interviews in Poland and the United States, I thank as well Anna Bohdziewicz, Jacek Jonak, Joanna Williams, Maria Pietras, Magdalena Naylor; and for manuscript translations, Wiktor Osatyński of Warsaw and Professor Witold Kalinowski and Katarzyna Rosner of Warsaw University. I also wish to thank Maria and Ania Jęczmyk and Mieczysław Gos and Polish citizens of Grojec and Gostynin, as well as those in Poznań, Warsaw, and on the Baltic coast. I have benefited greatly from many hours of discussion with my friend and colleague Dirk Philipsen over the theoretical implications of the organic tensions between popular democratic movements and the cultures of politics that have materialized in the West since industrialization. I also wish to thank Tadeusz Szafar of Boston and Ronald Witt and Thomas Naylor of Duke University for comments on the manuscript and Fredric Jameson and Marshall Roderick, also of Duke, for discussions over certain central themes of this inquiry. None of these colleagues, of course, bear responsibility for the final content.

This inquiry would not have been possible without the generous financial support of the Ford Foundation, the National Endowment for the Humanities, the Josiah Trent Foundation, and the Research Council of Duke University. These institutions generously bore the burden of extensive translation and travel expenses over a prolonged period of time. I also thank an individual as well as these funding agencies. Reverend Robert Gamble of Boston has generously underwritten the translation of this work into Polish. I thank him not only for volunteering this exceptional gift, but for the many other forms of assistance and counsel he extended to me in Poland, including a videotaped interview with Regina Matyja, the widow of the leader of the workers' movement at Cegielski-Poznań in 1956.

I warmly thank Dorothy Sapp and Thelma Kithcart of the History Department at Duke University for the care they bestowed upon the manuscript. Sheldon Meyer of Oxford University Press has, as always, been extremely knowledgeable and helpful, as have been Scott Lenz, David Bain, and Karen Wolny. Also, as always, Nell Goodwyn has been imaginative in helping to shape the structure of the interpretation contained in this book.

I have benefited enormously from a prolonged collegial association with Professor Roman Laba of the University of California, Santa Cruz. His knowledge of matters Polish is intricate and sophisticated and his mastery of the sources on the 1970–71 Baltic rising is unsurpassed in the world of Polish scholarship. Professor Laba is a generous colleague and warm friend. His temporary arrest by the Polish security services in 1982 for knowing too much and doing too much to preserve under the pressure of martial law the 1970–71 and 1980–81 archival materials of the Gdańsk Historical Commission of Solidarność provides an appropriately symmetrical commentary on the dynamics of the party's attempt to control social information in Poland and the long and ultimately successful effort of Solidarność to bring such censorship to an end.

All of which is prelude to a delicate topic. I wish to conclude with a personal word to that fraternity of scholars and journalists who specialize in studies of Poland. I could not anticipate in 1982 when I began systematic work on the origins and development of the Polish democratic movement that the literature on the subject would become both large and thematically congruent as to the causal dynamics presumed to be at work in Polish society between 1945 and 1990. I can only say, after eight years of inquiry, that the modes of scholarly inquiry informing the research habits of Poland watchers do not intersect very often with the perspectives on social investigation developed in the past quarter-century by historians of social movements elsewhere. It is interesting and instructive to discover that one's work must stand against such a voluminous literature as now exists on the Solidarność experience. But, then, the good social guardians of the Polish United Workers' party worked zealously and with authoritarian policing power to create a closed system of "source materials" in which the self-activity of the Polish population would systematically remain virtually invisible for thirty-five years.

Since this wall of censorship has been breached, the task now is to recover the history of democratic striving by the Polish population during the long tyranny of the party. This book may perhaps most understandably be viewed, then, as an attempt to pursue what for Poland is a new path of research into the interior of popular insurgency by a proud people who were forever restive under Leninist confinement. From such a perspective, novel for Poland as it is, this inquiry by an American specialist on social movements may seem somewhat less surprising in its evidence and its conclusions. Indeed, now that the written evidence of social protest over four decades is no longer confined solely to police files, I hope this inquiry will be comprehensible as a small harbinger of an avalanche certain to come.

Durham, North Carolina L. G.
August 14, 1990

Contents

Introduction

THROUGHOUT 1989, a fascinated world watched the people of Eastern Europe engage in a series of surprising public acts that effectively released them from the grip of Leninist governance. This did not happen simply because the people of Eastern Europe had suddenly internalized ideas about "freedom." They obviously had possessed such ideas for a very long time. The explanation, equally transparent, is that they found a way to *act* on their ideas.

The ones who led the way were Polish. In some fashion, Poles found a whole series of ways to act that produced one of the largest popular movements in history. Called "Solidarność," it was institutionally grounded in a "free and independent trade union" that in the course of four months in 1980 enrolled most of the adult population of the country. Lurking in this achievement is the central dynamic of history: the process through which private aspirations move to purposeful public activity—through which ideas move to action. Present also is the central task of scholars: to discover the causal connections between idea and action.

In the case of Solidarność, this task has not been performed. Ideas about "freedom" and about "free unions" have been seen to move to action with telepathic speed, literally, as if some peculiar social lightning had "transformed the workers' consciousness" and produced the miracle called Solidarność. It did not happen that way.

Since, as is now clear, the long-term impact of Solidarność was not restricted to Poland, this failure cannot be seen as provincial in scope and meaning. Whatever happened in Poland at the beginning of the 1980s was but the opening phrase of broadly energizing dynamics that continued functioning throughout the first half of the decade in ways that gradually began to transform the politics of the Soviet Union itself. Still functioning after 1985, but with new leading actors on stage in the Soviet Union, these accelerating impulses led to a structural breakthrough in Poland in the spring and summer of 1989. The people who had been thrown in jail in December 1981 formed the new government in Poland, and one of them became prime minister of the state. Thus enriched, the rhythms of popular hope and assertion rippled through Eastern Europe and through the subordinate nationalities of the Soviet Union in the autumn and winter of 1989 and 1990. In the course of this tumult, international politics was fundamentally reshaped, ending the first phase of the post-World War II era, the period known as the Cold War. And yet, the lessons that can be drawn from this remarkable historical change necessarily depend upon how well we understand the dynamics that completely overturned existing structures of governance in the Soviet sphere, broadened previously entrenched assumptions of what was possible, and effectively dismantled so many settled and seemingly reliable certainties.

At first inspection, the Polish movement would appear to be quite open to public view. In sheer size and depth, Solidarność is one of the most elaborate social movements in human history, one that penetrated into every city and hamlet in Poland. This nation stands as a giant laboratory of social life in which occurs a striking transformation from something resembling totalitarianism to something for which no term of traditional political description currently seems adequate. Particularly since there is much general need to learn from the arduous politics through which Solidarność successfully created new democratic forms out of entrenched hierarchy, one would expect that a great deal of effort has been spent to decipher and interpret the internal development of this transforming process.

But this is not the case. While an adequate amount of literary attention has been devoted to Solidarność over the past decade, the energizing dynamics of the Polish movement, strikingly democratic as they were, remain hidden from view. In some ways, part of the responsibility for the conceptual blind spots appears to lie with what are taken to be Polish "authorities," but who are, in fact, partisan participants who, out of a quite understandable self-interest, have encouraged a certain reading of events. But this circumstance routinely surrounds historical material and should present no insuperable hazard to a patient and thorough observer. The real problem, it develops, lies much deeper and does not originate in Poland at all. It has to do, quite basically, with the way contemporary people—all people—have experienced democratic and hierarchical forms in their own social lives and, as a result of these collective experiences, how most modern intellectuals have come to think about politics and society. The issues raised by Solidarność go to the heart of contemporary political theory and thus to the essence of how scholars in the world's universities, and journalists and writers for the world's leading periodicals, have trained themselves to think about social life.

The essential task is to assess the causative roles played by the Communist party in Poland, by the Catholic Church, by the intelligentsia, and by the workers who created the independent trade union they called Solidarność. Additionally, the long reach of the Soviets, poised in increasing anxiety, intruded into and helped shape events from the very outset. This influence reached its apex on the grim December night in 1981 when martial law descended upon Poland. Yet Solidarność did not go away. So organic and energized were the human drives unleashed by the popular movement that Solidarność continued to live—in a thousand underground publications, in networks of people who continued to act in democratic ways, and in an idea about social possibility that had become too real to abandon.

Beneath this energy, fueling it, was the breakdown of the system of production. The crisis was not national, it was structural; it was not Polish, it was Leninist. In attenuated form, the dynamics at work in Poland intruded into the perceptions of the Soviet bureaucracy, creating the political environment in which Mikhail Gorbachev emerged. As it became clear at the end of the 1980s that Gorbachev would not countenance what Brezhnev had demanded—the military crackdown in Poland—Solidarność formally reappeared in society in ways that transformed public life throughout the Eastern bloc. The tumultuous changes of 1989–90 were the result.

This book does not resemble the work I thought I would write when I originally began to research the origins and development of Solidarność. One reason for this is the sheer poverty of information available on social activity in Leninist Poland. I know of no better method of introducing the subject than by detailing in a very personal way the tangled sequence of events through which a historian of America became enmeshed in and then puzzled by the dynamics of the Polish movement and by the way people in the West wrote about Poland.

In 1980 I had no special interest, much less the remotest claims to expertise, in matters Polish. I did not speak the language, had not independently developed a familiarity with English translations of Polish literature, and had no more than what might be called a responsible layman's knowledge of the people of Eastern Europe who had been unwillingly caught up in the Soviet orbit. My area of specialization is U.S. social movements, ranging from the American Revolution to the modern civil-rights era and encompassing a number of agrarian and labor movements in between. The long historical agony of Poland rarely intruded into any of these American happenings. But Solidarność was a social movement too significant for a student of social movements to ignore. I applied for and in November 1981 received research funds to go to Poland. Soon thereafter, on the night of December 12–13, 1981, a coordinated nationwide sweep by the security forces of the state descended upon the leadership of Solidarność throughout the country. The next morning, the world learned that General Jaruzelski had declared a "State of War" in Poland. By the time I arrived in Europe in June 1982, most of the people who had brought Solidarność into being were in prison or in exile. A number of key participants in the Polish democratic movement, both workers and members of the intelligentsia, had been out of the country at the time martial law was declared and had proceeded to establish a variety of support groups in the West. The center of this activity was the Paris Solidarność Committee. Through the generous assistance of the government of French President François Mitterand, the Paris committee had established an efficient émigré pipeline. As the internment camps in Poland began to disgorge hundreds and later thousands of Solidarność activists, a large number were forced into exile and into this pipeline. Through the cooperation of the Paris Solidarność Committee, I was able to interview a selected cross-section of activists. Thus began the search for the origins of Solidarność and the dynamics of its subsequent construction.

It required quite a while—years—to learn with verifiable concreteness how long this process had been unfolding. The first hint came in 1982 from a highly informed Solidarność activist from the Baltic coast. I shall call him Henryk. Some of my respondents talked about the months preceding the Baltic mobilization of 1970, but Henryk went further back. He spoke of a time in the 1960s when as a very young employee of the shipyard he first heard some of the "old militants" talk about independent trade unions. What they had said was: "The party unions are no good, they never defend the workers. They only defend the party. The workers need their own unions."

Late one night in Berlin in the spring of 1983, I talked to another Polish worker in exile. His name was Franek. Some twenty-seven years earlier he had briefly worked in Poznań at an engineering works named Cegielski. A month after he had been hired in May 1956, there had been a "big march" at Cegielski. It had been a very big march, indeed—triggering three days of rioting and armed street warfare known to every Pole as the "Poznań events." In those early days of martial law, I had expended considerable effort in attempting to locate an émigré worker from Poznań who was there in 1956. But in Berlin that night, Franek produced little historically useful information or even clues to possible information. He had been too new to the plant in 1956, too young in age, and too low in the worker hierarchy to have much concrete knowledge of the structure of the organizing work that had preceded the "big march." Most relevant of all, he had left Cegielski soon afterward and thus had no opportunity to learn through postmortems the pertinent details surrounding the effort. Twenty-seven years later, he thought the name of one of the key figures at Cegielski was "Pacher" (it was, in fact, "Taszer"—see pp. 61–65). Franek did have one interesting morsel to contribute. "The workers at Cegielski," he said, "wanted their own union, even back then. They were trying to get their own organization out from under the party. In 1956." The dream of a union free of the party. If what my somewhat uninformed source from Poznań said were true, Solidarność had been a long time coming indeed.

Eventually, the search for origins took me all the way back to 1945–46, to a time when the prewar political parties were being obliterated and when blood flowed on shop floors as socialist and independent trade unionists battled the security police of the Communist party. To understand Solidarność, I would have to endeavor to get a sense of the essential social rhythms of the entire postwar history of Poland! Grappling with that task has been an adventure of eight years' duration.

It was in the course of this long quest that I discovered in concrete terms how the Leninist system crushed the world of the scholar. It became clear early on that foreign studies of Poland were evidentially thin on social activity and that this condition reflected the absence of detailed monographic studies in Polish. Until the emergence of Solidarność dramatically altered the entire trajectory of social evidence, what passed for domestic and foreign sociopolitical scholarship on Poland was largely speculative, inherently descriptive, and, most limiting of all, necessarily reliant on inferences drawn from the public actions of the ruling party. The Western literature on Poland essentially reflected an accumulation of glimpses of the very apex of society—an attempt to make sense of the activities of the party apparat, the Central Committee, and the Politburo, plus whatever social insights could be extracted from the countervailing literature generated by the democratic opposition. The people of Poland in all their rich diversity remained essentially invisible. One result is that the range of opinion generated by Western "Poland watchers" is not grounded upon the kind of evidential density that characterizes debate in a nonpolice state, particularly those advanced industrial societies possessing massive archives and sophisticated systems of classification and retrieval.

The ultimate proof of the intellectual closure achieved in Leninist states is the totality with which access to fundamental social information is systematically foreclosed. One speaks here about evidence of social activity existing in recoverable form. The party ensured that such did not eixst in Poland. For almost two full generations, anything remotely approaching a sociology of the nation remained beyond reach because, quite literally, in-depth sociological studies of the Polish people were impossible for anyone to conduct. No intricate scholarly disputations over the shape and meaning of social life appeared in Polish academic journals because no empirical studies existed to argue about. This condition extended beyond the range of formal state censorship that blanketed social research within the boundaries of the country. In the émigré literature, too, ground-level social research was nonexistent.

It is impossible to overstate the massive impact these circumstances have had on the way Solidarność is understood throughout the world. The dynamics of any society necessarily reflect the organic tension between social hope and popular resignation, between assertion and acceptance, between protest and servility, between movement and stasis. But in Poland, the controlling fact was that social evidence bearing on assertion and protest—the very politics of hope—was locked away in police files. This had the effect of concealing names, dates, and locations of the great majority of the organizers of social protest who had at any time in the postwar period been successful enough in their own locale to attract the attention of the security police. The intelligentsia could generate literary evidence of protest that was subsequently recoverable; indeed, routinely such literature constituted the entirety of the protest itself. Accordingly, it represented the *ideas* of protest—that the censorship should be modified or ended, that a proposed constitution should be amended, that judicial procedures should be altered, and so on. But protest moves from idea to action when it becomes social—that is, when it is organized so that people are acting rather than writing or talking about acting. In People's Poland, it was precisely this kind of evidence that was secreted in the "historical archives" of the Interior Ministry. It was not remotely visible in the sociological or historical literature. What remained outside police files in convenient public view was a party-controlled literature. It described, in very general terms, a society "building socialism" under what the party liked to describe as "the normal rhythms of work." This literature centered on the resolutions of the party while providing no detail at all on actual living Poles who had identities and specific jobs and specific views about life and death or about politics and society. The sheer banality of the great bulk of this party-dominated prose verified its uselessness. From the moment the Polish party consolidated power in 1948, a veil of silence cloaked all social evidence of how the knowledge to create Solidarność was somehow acquired in Poland and who acquired it.

This was the first reality that propelled me along a path distinct from my original intention: written evidence concerning social activity did not exist in Poland beyond the speculative articles that intellectuals generated in the oppositional literature. The necessary bridge of evidence tracing the concept of free social space from private longing to functioning public reality simply did not exist in Leninist Poland. To an extent that was much greater than I had originally imagined, my pursuit of the origins and development of Solidarność would necessarily have to be an undertaking largely grounded in oral research techniques. That specialized work was not, in itself, especially daunting. I had founded the Oral History Program at Duke University in 1971 and had spent a number of years enmeshed in the task of sorting through the thickets of social movements. I had written on the subject extensively and had worked with many graduate students engaged in the same task. Oral evidence, like written evidence, was partisan, incomplete, and often misleading, but if approached with care and poise, it was a useful and often necessary tool.

For Poland, the initial starting point for research was clear. Since Solidarność emerged from an institution on Poland's Baltic coast, the task was to work backward in time from August 1980 to discover the source of each of the organizational and ideological building blocks employed by the shipyard workers in Gdańsk to fashion their new democratic institution. Such sources could obviously exist in any number of places across the country and could emerge from any number of sectors of the working class, the intelligentsia, the Catholic episcopate, or, conceivably, even from forces within the Polish party.

But as this work proceeded over the years, and as the number of articles and books on Solidarność proliferated in the meantime, it gradually became clear that connecting ideas to action was not the central purpose of any of my fellow chroniclers of the Polish movement. The explanations that did emerge were all quite ephemeral on this fundamental question. Moreover, these explanations had begun to build upon one another. A vast and steadily growing literature took form in many languages, not only in Polish and English, but in French, German, Italian, and Spanish. It did not seem to matter whether these advisories about Poland were penned by scholars or by journalists, whether they were constructed out of primary or secondary sources, whether the writer knew Polish or proceeded to work through translators. On all major issues of causality purporting to explain how Solidarność came into being, a worldwide consensus had rather quickly settled into place. This would be fine were it not for the fact that the evidence I kept uncovering fundamentally contradicted the consensus. As this fact became clear, a second unanticipated task came into focus. I no longer could be content merely to pursue causal connections and then present my findings; the findings themselves would contradict a small army of scholars and foreign correspondents. I therefore had to find a way to confront the reigning cultural and intellectual presumptions that guided the way people thought about the creation of democratic forms in modern societies and about the processes of popular politics generally. The stark imperative was to bring these controlling assumptions to high public visibility so they could be forthrightly dissected. This is not easily done. Indeed, to clarify the conceptual dimensions of the problem is a task that will occupy the remainder of this introduction.

At issue are nothing less than the most organic presumptions, beliefs, and expectations that people everywhere bring to the study of "politics" and "history." These are the very beliefs and presumptions that get in the way of understanding the rise of Solidarność. I have brooded about precisely these matters for a number of years and believe that I can now arrange them in an orderly sequence so that their more elusive components can become visible.

The study of Solidarność that follows proceeds from an understanding that I believe is essential in providing the necessary connecting link between idea and action in history—quite simply, that social knowledge is experiential. Humans have a pronounced tendency to behave the way they do because, in the various cultures of the world, they have been taught to conform to certain customs, and, in so doing from very early in life, they experience these forms and come to regard them as "normal." We are blessed by the fact that this process of cultural and ideological instruction is not a finely honed and efficient one, that some people in every culture learn their lessons less fully than others, and that some, while internalizing a requisite number of mannerly ways, also experience a variety of alternative lessons that have not been sanctioned by cultural authority. Social life intermittently contains unanticipated surprises because of this circumstance.

The nature of the surprises, of course, is as varied as social systems, tribal loyalties, religious sects, and the many other influences that in the aggregate cause people to experience and understand society in divergent ways. But while the inevitable disputations that ensue are frequently conducted in a vocabulary of high principle and in terminology arranged to project logical power, the beliefs and customs being projected are passionately held precisely because they have been experientially confirmed in daily life. For better or for worse, social knowledge is experiential, and the beliefs so shaped are taken to be "normal." The standards of social evaluation, including scholarly standards, that are applied to historical events, grow out of this acquired sense of normality.

In considering that vast and varied area of human activity that can loosely be described as "the politics of protest," the first thing that must be acknowledged is that the social activity suggested by the phrase is emphatically not perceived as "normal." The phrase itself refers in some general way to a dim arena of human activity, emotion, perception, and belief that is connected to unusual acts of unsanctioned assertion by previously little-known persons.

In order to grasp the abnormality of protest, however, it is necessary to be clear about those acts that are regarded as politically normal. These common acts of everyday life and politics are performed within a huge array of rules and customs that come down out of the past and are codified in laws and buttressed by enforcing mechanisms, both legal and social. These rules and customs of society can be summarized in a rather simple phrase of description; they constitute the "received culture" within which social and political life takes place. In all its ideological dimensions, secular and religious, this culture determines how the young of the land are raised, how they are instructed in the formulas of making do and getting ahead, which modes of conduct augment one's influence and which modes detract. Though the formulas, variously familial, ethnic, geographical, and sociopolitical, are elaborate and contradictory, they share an educational purpose: to teach the ongoing generations how to live. Here, in highly generalized terms, is the starting point of "politics." Indeed, throughout history, in most societies most of the time, most of what passes as politics takes place—both begins and ends—within the parameters of recieved tradition; the participants may not wholly approve of some or most of their inherited customs, particularly the constraints on human conduct they impose, but people live out their lives under received rules for the elemental reason that they do not know how to go about achieving a significant alteration of the structure in which they are caught. In essence, most "ideas" do not move to "action" because people are afraid to act on their ideas. Quite simply, they do not know what to do that is safe to do. The problem is one of power and the social fear that power can stimulate. Power generates in people experiences that teach the merits of caution.

As an intellectual proposition, this dynamic relationship of the powerful to the powerless is, unfortunately, routinely passed over. Since the relationship itself informs most of daily life, the resulting explanations of "politics" are, accordingly, inappropriately simplified. A rather central element in the language of political analysis becomes apparent; people who object to one or more features of the received culture in which they live—most people—but who also cannot be observed doing anything about it are routinely described as "apathetic." A particular descriptive word applies here, one that will receive enduring attention throughout this book. That word is "consciousness." To become real—that is, to become historical actors—apathetic people, it is presumed, need their consciousness "raised." If one were to follow this tradition, in studying a time of high activity such as the time of Solidarność, one would seek to discover which Poles knew enough to be able to "raise the consciousness" of enough other Poles to make Solidarność possible.

The particular Poles who brought Solidarność onstage—whoever they may turn out to be—obviously were neither apathetic in the face of the one-party state nor hamstrung by low or inadequate consciousness. Somehow, in some manner, they found a way—publicly and in a police state—to perform with sustained purpose against the basic tenets of a Leninist society. By some process, they found a way to act.

In pursuit of how this happened, however, it is timely to issue a precautionary warning as to the cultural power and the political unserviceability of the inherited terms of political description through which social movements such as Solidarność are routinely explained. These terms are so common that they are literally a part of the political premises of the readers of this book; they create inappropriate but highly sanctioned ways of thinking about "protest"—habits of political thought that absolutely stand in the way of understanding Solidarność. To suggest that modern approaches to the analysis of popular (that is, "abnormal") politics are grounded in "inappropriate but highly sanctioned ways of thinking" is obviously to raise the most fundamental questions about contemporary political theory in general and about democratic theory in particular. This is precisely my intent.

The nature of the problem cannot be outlined in a paragraph or two. The source of the problem lies in the nature of the past, the political struggles it enshrines, and the elusive and highly partisan language of social description that has been passed down to us by the victors in those past struggles. Historically, societies are not routinely afflicted with "movements." Things are usually "normal" and people behave in conventional ways. A relatively small number of citizens possessing high sanction move about in an authoritative manner and a much larger number of people without such sanction move about more softly. Some among the multitude may be seen energetically to be doing all they can to acquire a measure of status, but in the meantime, they join their less-energetic neighbors in behaving with conventional deference.

Movements disrupt this normal order. A considerable number of unsanctioned people appear publicly in a new guise; they present petitions or voice demands; they suddenly arrogate to themselves the right to criticize inherited customs and may even issue manifestos proclaiming the precise way they intend to rearrange received habits. Moreover, they have a pronounced tendency to conduct activity out-of-doors where everything is visible. People march, they strike, they demonstrate, and they may even suddenly riot and burn down or otherwise dismantle certain physical signs of established tradition.

Such modes of procedure are rarely viewed with easy tolerance. Indeed, over time, the descriptive terminology applied to this kind of politics has acquired an elaborate metaphorical quality. Images drawn from nature are employed to dramatize the disruptive distinctiveness that movements generate. The historical evidence is conclusive that fire, wind, pestilence, and water provide the favored images. Movements "flare up" and "gather steam." They "boil." They can then "burst into flame" and "burn like a prairie fire" before, in time, "flickering out." A social movement can also be understood as a "gathering storm" that when gathered "sweeps like a cyclone" through vulnerable regions. A kind of sickness may then be detected as prevailing in the afflicted territory. A part of a city or a whole country may be described as "infested with insurgency," traceable to a widespread "virus of discontent." Finally, movements can be understood as liquid. They "wash" against the established order, sometimes as a "tidal wave," before their motive force "drains away."

The common feature of all these terms of description is the point of view of the observer, who stands on the high ground of received historical tradition, from which vantage point prairie fires or tidal waves may be safely monitored. The resulting language may be understood as involving something beyond metaphor. More precisely, it is launched from a perspective located securely outside the social formation that is being characterized. It is a view from afar. A surface appearance is being distantly described, sometimes with partisan hostility, sometimes with affectionate approval. In either case, something approximating a historical result is being spelled out, but not historical causation. So vivid is the image of the tidal wave that its dim and distant origins remain beyond the horizon, as if obscured by the curvature of the earth.[1]

This metaphorical language contains a remarkable feature: though the images are inexorable in their literary force and suggest that a force is being described, the human ingredients that comprise this presence remain invisible. Very few identifiable people can be discerned amidst the floods, fires, and pestilence. Both the metaphorical terminology and the high-ground perspective that is its source conceal historical causation by concealing historical actors. It is hard to resist the thought that such elegant but remote excursions into metaphor do nothing in the end but drain themselves, like yesterday's rain water, into a placid pond of cultural complacency, where they subside into a sheen of motionless surface undisturbed by human presence.

Let us toss a small pebble in the hope we might generate a discernible ripple of causation. Leaving aside all metaphorical display, large-scale democratic movements do not happen in any of these easily characterized ways. Democratic forms are ordered. To function well, they must be experientially tested. Their construction requires overcoming many culturally based hierarchical impediments. They happen, then, when they are organized. They happen in no other way. Democratic movements are built by human actors, and their actions are historically traceable. Let us give a name to this seminal political process. We may call it movement building. So enamored have we become of metaphorical explanations that we know very little about this process. Around the world, "movement building" is not an organic category of political science. While it might be supposed that a key to understanding a large-scale social movement turns in the first instance in locating its origins, the fact is that very few historical studies of insurgency do this. Scholars stop too soon—that is, at a point not far enough back in time to capture the original social formation (necessarily small at the point of origin) from which everything else developed. Research in this dim and uncharted past is difficult because the identity and relationship of the original insurgents is often uncertain and their activities hard to document. In any case, so much becomes known about the movement in its later stages that to most observers these vague beginnings do not seem vital. These "details" can, in any case, be easily handled through metaphorical imagery—even, on occasion, with a single word of description: the movement was "spontaneous."

Indeed, in a Leninist state where social information on protest is locked away in police files, metaphor becomes especially useful, even necessary, to scholarly logic. But even in the more open societies of the West, research into origins of movements acquires a certain verbal efficiency as imagery is employed to overcome all research shortcomings. The custom is more than merely widespread in intellectual circles throughout the world; it effectively governs the habits of research—in the West, in Poland, everywhere.

The dynamics that emerged from a preliminary inquiry into Solidarność brought this whole question to the surface. An organization of Warsaw intellectuals known as "KOR" had been formed in September 1976 in the wake of brutal government repressions of workers in two Polish cities, Radom and Ursus, in June. The activists of KOR persevered from 1976 through 1980, appeared at times to discuss the idea of free trade unions, and launched a newspaper aimed at workers. In 1979, they published a comprehensive "Charter of Workers' Rights" containing ideas that a short eight months later found expression in the twenty-one Gdańsk demands of 1980. In the words of one highly placed source, the newspaper and the charter had "prepared the workers' consciousness for the strikes."[2] My initial inquiries on Solidarność pointed clearly to one fact: KOR presaged Solidarność. Indeed, articles and books had already begun to appear in 1981–82 offering precisely this analysis of the origin and development of the Polish movement.

The first difficulty I experienced with this explanation centered in the June worker protests that gave rise in September to the founding of KOR in 1976. It was not so much what happened in Radom and Ursus that seemed relevant but rather what occurred elsewhere in Poland at the very same time. What happened was both simple and stark. Workers may have protested violently in two cities, but they had also firmly put down their tools and had gone on strike all over Poland. In a great number of cases (how many I had no way of knowing), the specific organizing tactic they used was rather sophisticated. Polish workers employed what is known in America as the sit-down strike and in Poland as the occupation strike. As a mode of assertion, the occupation strike was a tactic that did not fall from the sky. It involved the occupation of a factory rather than picketing outside it and thus required a considerable amount of organized worker cohesion. If, three months before KOR was ever formed, workers across Poland had mounted a series of occupation strikes, they obviously had acquired this knowledge from precincts other than those occupied by Warsaw intellectuals.

But this was not the only historical puzzle. The great wave of strikes on the Baltic coast that gave birth to Solidarność in 1980 embodied a veritable ensemble of elaborate organizing techniques that produced coordinated actions in scores of different cities. The coordination was achieved through a form of social organization that, to my knowledge, was historically unprecedented—an "Interfactory Strike Committee." This expansive structure linked together in a single instrument of assertion hundreds of thousands of workers in many factories in many cities. The institutional edifice was capped by a "Strike Presidium" of worker militants who presided over the entire undertaking. Transparently, the idea of an "interfactory strike committee" was an institutional blueprint for that spectre that had long haunted capitalists—the general strike. In my initial research, I was unable to learn that anyone in KOR had developed technical expertise in the conduct of occupation strikes or the fashioning of interfactory strike committees. Both modes of structural mobilization were necessarily grounded in intense shop-floor organizing activity, and it was difficult to imagine how intellectuals could master, much less impart to workers, such subtle workplace organizing skills. This seemed especially true since the evidence was conclusive, even at an early stage of research, that a number of these skills had become the property of the Polish working class long before KOR was founded.

Isolated here is the impressive raw cultural credibility of views from afar. The interpretation proffered by KOR intellectuals was accepted everywhere—

even though it could not pass the most elementary tests of logic. It was as if KOR's ideas, whatever they might be, were somehow understood to have been projected over an invisible social transom and into the "consciousness" of an unseen group of people, abstractly identified as "the workers." In unspecified ways, this information enabled the workers to make the revolution. Here is the "politics of protest" as seen from afar: KOR generated Solidarność.[3]

This way of understanding popular movements is not some peculiarly Polish malfunction. The same habit is easily visible in the historical literature of the United States. One example will suffice to indicate not only the presence of this controlling assumption but also its prestige at the highest level of academic analysis. Some years ago, while researching the nineteenth-century agrarian revolt in America, I encountered, as all students of American Populism must, a prestigious interpretive work in which the author uncovered for his readers the presence in the U.S. of a pervasive "agrarian myth" about the presumed moral quality of rural life and the hardy "yeoman farmers" it supposedly produced. The author, a gifted stylist, ranged tirelessly across the breadth of Western culture, quoting English poets, French physiocrats, and American Founding Fathers, all of whom were seen to be caught in the grip of romantic notions about "the uniquely productive life" of soil tilling. Having fashioned a telling lens, the author then viewed the protest movement through it, with devastating results. The reformers were seen to be caught in the myth, too. Such was the state of social analysis when the book was written in the 1950s that no one seemed to notice that its conclusions had been achieved without research into the interior of the movement being interpreted. Scarcely an American farmer was visible in this much-admired study. Rather, the "educated classes," in the author's words, had prepared the consciousness of the farmers with a myth, and the farmers had brought forth a shallow and infantile movement based on the myth.[4]

In short, what modern explanations of Solidarność shared with those of an older generation of American intellectual historians was an error in the premise; they endeavored to understand a self-generated social formation without recourse to research inside the formation itself. The need to connect idea to action by research on the causal connections was—quite simply and quite profoundly—not a part of the inherited tradition; the habit of seeing society from afar was.

A final pair of examples is offered to illustrate the sheer scope of this unworkable interpretive habit. In the burgeoning literature assessing the causes of Solidarność, vivid but evidentially empty descriptions appear of a society "transformed" and armed with a "new consciousness" as a result of the 1979 visit to Poland of the new Polish Pope. Conversely, one is told by another author that it was not the Pope of Rome but rather "frustrated nationalism" that was "fundamental" to the movement's emergence in 1980.[5] In such a manner, the bridge of causation from idea to action is crossed not through recourse to evidence of action but by invoking still another idea. No substantiation is offered, because, in views from afar, supporting evidence is not presumed to be needed. The interpretive generalization is enough. One, therefore, looks in vain for evidence of the "transforming" act the Pope performed on the Baltic coast that he inadvertantly failed to perform in all the other parts of Poland where workers somehow could not bring themselves to form interfactory strike committees. Conversely, one is left to wonder what singularly virulent strain of Polish patriotism infested the Lenin Shipyard to produce a "frustrated nationalism" of such transcendent proportions that it burst forth as Solidarność. One poses such questions as a way of dramatizing that such sweeping cultural and psychological categories are so unintentionally all-encompassing as to contain absolutely no causal meaning.

The task, then, went beyond the evidential underpinnings of the historical process through which Solidarność was constructed. Important as this evidence was, there was the more intricate task of recasting prevailing intuitions about historical causality upon which these erroneous interpretations of the Polish movement were based. The need was not only to clear out the space previously occupied by metaphorical imagery and fill it with evidence but also to confront the universe of perception through which contemporary intellectuals formed their own private sense of democratic possibility in the modern world.

The work that follows constitutes a venture into this open terrain, into the world of "the intelligentsia" and the world of the "working class," into that realm where ideas become connected to social action, into that space occupied by formal ideology, into those dark corners of insurgency where actions are concealed beneath the descriptive word "spontaneous."

Since social movements, when they occur, are sequential—beginning simply and growing more intricate as a function of self-generated experience—the evolution of Solidarność necessarily appears in stages. This book therefore may at times seem to take on the appearance of an old-fashioned chronological narrative. But the purpose is less to establish a chronological rhythm than to demonstrate that the evolutionary forms of self-activity that appear in history do so sequentially and as a discernible dialectic of experience. A second purpose is to attempt to rescue the politics of social transformation from allusive abstraction and imbue it with explicit social meaning.

Toward this end, a certain literary strategy is employed throughout this volume. In the first instance, events are described from afar, from outside the social formation that generated the events—in this case, from outside the Polish working class; that is, culturally sanctioned ways of viewing popular movements are employed to establish an introductory perspective. The same historical ground is then retraced, this time from within the popular movement. The underlying causal dynamics that come into view may be seen to alter fundamentally the perception of what has happened. The intention in arraying these two patterns of descriptive analysis alongside one another is to undermine as decisively and as permanently as possible all political and historical views from afar.

As the social evidence of Polish insurgency accumulated, chapter by chapter, a final, all-encompassing cultural assumption, prevailing throughout the Western world and towering over scholarship itself, gradually came into clear focus. It is an assumption so massive in its controlling impact as to edge everything else bearing on this study to the margins—overshadowing the perils embedded in the ingrained habit of "viewing from afar," the paucity of sources on Poland, or the constraints of martial law. I refer to the central understanding throughout the world that social and political "consciousness" is somehow organically an intellectual phenomenon and, as such, is something that essentially happens in the mind. From this premise, it is a fairly simple matter to slide into the assumption that people's consciousness is "raised" by reading manifestos or other kinds of literary exhortations, be they journalistic, political, or, perhaps, theoretical. This intuition is given further impetus by an occupational influence. For most intellectuals, the act of thinking about politics and society inexorably carries with it the impulse to regard thought itself as the energizing component of political activity. The breadth of this assumption explains why so many intellectuals perceive no particular need to uncover an evidential "bridge" from idea to action. The act of writing down an idea—in a manifesto, for example—becomes without further activity the *cause* of any subsequent movement that can be seen in any way to be associated with the idea. The custom of viewing from afar further protects and energizes this illusion.[6]

"Consciousness" is not shaped in such an antiseptically intellectual manner. Rather, as the evidence from Solidarność verifies once again, social knowledge is experiential. It needs to be immediately asserted that people engaged in living through "experience" do so with their minds intact; they can, if they have the will, think about their experience. This book-length body of evidence supports this understanding. As the concluding chapter summarizes, the "transforming" ingredient shaping political and social consciousness consists of intelligence impregnated with social experience—as distinct from conceptual intelligence alone.

In its raw and socially unformed state, insurgency is simply an idea. When it begins its long journey toward ultimate social meaning, it is historically almost invisible. The idea takes the form of a tiny group of people sitting in isolation in a space they have created for themselves—a space to think in. Soon, these people, the human embodiment of "the politics of protest," begin to try to communicate to the larger society that is encased, in varying degrees of discontent, within the received culture. The dissenters may write manifestos or pamphlets or newsletters.

At this juncture (it may last for years or even generations), insurgency is essentially an idea trying to attain a social presence, an idea attempting to find expression in the actions of significant numbers of people. Specified here is the evidential desert in which the students of Solidarność have gotten lost. In literary sources, journalists and scholars have found the idea, "a free and independent trade union," lurking in a subsidiary paragraph of a KOR document written in 1979. Observers have also easily found the idea in action in the Lenin Shipyard in Gdańsk in August 1980. In leaping across this gap between idea and action (it requires only one transitional sentence to complete the passage), they have connected KOR to Solidarność in a causative dynamic.

But the idea of a free and independent trade union in Leninist Poland had been the energizing organizational drive in the Polish working class for decades. Unknown to KOR or anyone else among the Polish intelligentsia or episcopate, this spacious dream had acquired a steadily escalating sequence of organizational expressions over a twenty-five-year period, reaching its essential structural dimension some six years before KOR was formed. The tracks of this sequential experience of movement building are not easily found because, as written evidence, these tracks exist essentially as footprints in Polish police files. But they are recoverable through oral investigation. While the process of discovery cannot accurately be pictured as a protracted "wandering in the desert," it has nevertheless proved to be a very long research journey in a nation socially organized as a police state.

Students of social movements are aware that local-level organizing activity routinely generates documents of written exhortation—leaflets, public demands, organizing appeals, and the like—that can conceivably be recovered, if one knows in which locale to search. But exactly because such assertion in one part of the country is subject to emulation in another, evidence of its existence is socially dangerous to ruling authorities and, in Leninist societies, is consequently suppressed. This, then, was the social vacuum that journalists and scholars faced in attempting to reconstruct the causal dynamics that connected ideas to action in the formation of Solidarność. The fact helped explain the totality of views from afar.

The research task involved in filling this vacuum was, in terms of primary or secondary written sources, rather simpler than one would want. I learned rather quickly that the Polish party's grip on social information left little to draw upon. From the moment the party consolidated power in 1948, a veil of silence cloaked the evidence of social activity. But after thirty-two years, a team of social scientists in the provincial city of Poznań took advantage in 1980–81 of the social opening created by the appearance of Solidarność. They conducted a hurried but often highly germane investigation of the "Poznań events" of 1956. These oral interviews and narratives drawing on relevant housing, income, and other economic data that provided a context for oral evidence were published in 1981, before martial law ended such modes of scholarship later in the year.

A second breakthrough into the closed evidential world of social life in Poland came on the Baltic coast. In the aftermath of the victory in the Lenin Shipyard, the Gdańsk Solidarność created a historical commission to recover all possible material bearing on events before, during, and after the 1970 massacre on the Baltic. The committee uncovered a vast storehouse of materials bearing on coastal organizing in 1970 and 1971 and the extensive demand-gathering campaigns conducted in both years within crafts, shop sections, and enterprises. Many interviews were conducted, and correspondence involving both the party and the worker opposition were retrieved. A member of the group, the American political scientist Roman Laba, managed to get the great bulk of this material out of the country during martial law. Later, in the United States, he collated and computerized the demands. Professor Laba generously shared both his research and his insights with me. This material constitutes an elaborate body of evidence on the coastal working class at an important stage of its organizational and ideological evolution.

Thus, as fortune would have it, though social evidence is desperately thin for the three and a half decades of communist rule that extended to the Polish August of 1980, this condition does not hold for the two most sustained insurgencies that were mounted against the state—at Poznań in 1956–57 and on the Baltic coast in 1970–71. This material served as the evidential base upon which my agenda of oral research for Poland was constructed. I conducted interviews in France, Germany, the United States, and Poland between 1982 and 1989.

What prolonged study of social movements teaches its practitioners is that, in all societies, it is far simpler for activists to visualize an appealing political goal than it is to recruit large numbers of people to perform the public acts necessary to breathe social life into the possibility of achieving that goal. As a veteran Baltic activist said in one of my earliest interviews in Paris in 1982, "the idea of a trade union free of the party comes to you very naturally. That is not the hard thing. The hard thing is finding a way to get it."

What happened in the Lenin Shipyard in the Polish August was organizationally elaborate. The intricate components of this edifice had been learned over time, either directly in an experiential way by the shipyard activists themselves or from prior experiences of other organizers who had passed the knowledge on for further testing by the shipyard militants. For the journalist or scholar, the task is to penetrate to this realm of social activity where movements acquire their structural form and their political meaning. Lurking there are the relevant social facts that organizers of the "politics of protest" learned from their experiences. It is not a body of knowledge that can be acquired in libraries or by developing the skills to write all-encompassing manifestos. One cannot "prepare the consciousness" of others by writing a political charter, for the elementary reason that such literary appeals, in announcing goals, pass over the central problem of what to do to reach the goal. This relationship is a fundamental axiom of democratic politics, one that is, to say the least, not widely grasped in the modern world. Indeed, the myth of KOR would not have attained wide currency had these basic social dynamics of democratic politics been more broadly understood. As this book endeavors to illustrate, mere exhortation passes over a number of the components crucial to the creation of an organized and activated civil society in any nation.[7]

In Poland, "consciousness" was not centered in the knowledge, long internalized by workers, that a free trade union was a worthy objective; the kind of consciousness that produced Solidarność consisted of knowing what acts to perform to recruit enough people to bring that long-cherished goal within practical social reach. This was not an art one learned by working on a newspaper or editing a literary journal or reading for an advanced degree in the social sciences. Political ideology moves from isolated thought to broad social practice only through a social experience of organizing that is intricate, often painfully and even dangerously learned, and is grounded in an intimate grasp of the milieu in which one recruited. It is on this hard-won and sophisticated level of social experience that a given venture into the "politics of protest" stands or falls. This imperative is absolutely the central animating component of popular politics. Indeed, it is precisely because this level of knowledge is so difficult to achieve that helps explain why large-scale democratic movements are so rare in history. Another reason—a hidden truth about modern progressive politics—is that it is hard for democratic advocates to encourage the appearance of social dynamics that they do not themselves understand. To summarize, at the heart of Solidarność was highly relevant social experience, not literary craftsmanship or erudite political analysis.[8]

I hope it is not improper, after immersing myself in the literature of Solidarność, to suggest that a general caution is now in order on the subject of journalistic and scholarly over-identification with the sundry roles played by intellectuals in the politics of protest. Admittedly, the courageous and principled activists of KOR are an interesting and attractive group of people; nevertheless, in the intricate dynamics of movement building, a singularly experiential endeavor above all else, the activating function of protest literature has its limits. If movements are to be historically reconstructed to the extent they can be understood, there remains a professional responsibility to specify the evidential connections between idea and action. To date, the literature on Solidarność has been a signal failure in this respect. The evidential shortfall is pervasive, and it is difficult to resist the speculation that elements of complacency and, in the case of a number of observers, an active capacity for condescension have contributed directly to the tepid results. Nothing is more human than to avoid seeking evidence in places where one unconsciously intuits it cannot exist.

In any case, what has now become the forty-five-year struggle of the Polish citizenry is, in itself, an instructive chapter in the annals of democratic striving among the peoples of the world. As the conclusion of this book suggests, Solidarność is a cumulative political evolution that forces all who study its origins and development to rethink modern politics and contemporary social relations and to rethink as well the narrow circumference of innovation and perception that effectively, if needlessly, confine the people of the present.[9]

The Polish experience vividly confirms that the essential precondition for Solidarność was the breakdown of Leninist production which inexorably led the entire Soviet sphere to economic crisis. Because this breakdown was not exclusively Polish, the potential for change can be retrospectively seen to have existed from East Berlin to Moscow. It happened where it did because people labored in creative ways and with considerable courage to make it happen.

The role of Solidarność in energizing the Soviet sphere was central. The appearance of the Polish movement and its subsequent endurance through the long strains of martial law lifted the pall of resignation that decades of Leninist rule had cast over all of Eastern Europe. The crisis of Leninist production ultimately affected the Soviet bureaucracy itself. A reciprocal relationship exists between the collapse of Leninist economies, the creation of Solidarność, the emergence of Gorbachev, and the sweeping opportunities (and dangers) now visible throughout what so recently had been the exclusive sphere of the Soviets.

But to respond to these dynamics with intelligence, it is necessary to understand their sequential development, and this task, I submit, cannot be achieved by theorizing from afar. The patient, tedious, painful, and often courageous development of democratic forms in Eastern Europe, a process that in places is now just beginning, is in the first instance a Polish story. This book is an effort to explain that development, its present status, and its potential meaning for the years immediately ahead. In 1980, the point of impact for Solidarność seemed Polish; by 1990, it clearly was global. It is possible, if people everywhere can somehow restrain their capacity for condescension, to learn many things from the Polish experience about the dynamics of democratic forms. That we have not yet done so measures the narrowness of contemporary democratic theory and the constricted and complacent angle of vision it provides on the modern world.

BREAKING THE BARRIER

Some historians . . . failing in the simplicity of their hearts to understand the question as to the meaning of power, seem to believe that the combined will of the masses is delegated to historical leaders unconditionally, and therefore, describing any such authority, these historians assume that that authority is the one absolute and real one, and that every other force, opposing that real authority, is not authority, but a violation of authority, and unlawful violence.

Leo Tolstoy
War and Peace

God is in the details.

Mies van der Rohe

1 ▪ An Event of Unknown Origins

> We do not have the full picture and knowledge of the origins of the August events as yet. Many elements are hard to grasp, and I excluded them from my picture.
>
> Andrzej Wajda, in appraising his prize-winning film, *Man of Iron*, in 1981.

THE DATE WAS August 23, 1980; the place, Gdańsk, Poland; the meaning, unclear. Somehow a massive outburst of popular politics had materialized in a police state. Journalists gathered from around the world, expectant. Surely the government would suppress it at any moment; if not, surely then the Russians would come.

It was hard to grasp: a scene dominated by thousands of striking shipyard workers, variously postured in huddled groups, standing, gesturing, sitting apart, moving with purpose, the whole subsiding when the worker-controlled loudspeaker boomed—and periodically, a sudden quiet. So intense was the energy level that the silence seemed to surge: a studied patience that somehow conveyed militancy. It was all very colorful, but still very raw "news" from a puzzled media.

The center of the puzzle was the Lenin Shipyard where sixteen thousand workers had put down their tools. They were evidently supported by thousands of other workers who sent their representatives to sit on a strike committee in the shipyard. The facts took awhile to assimilate. In Washington and Moscow, in the Vatican and East Berlin, people watched with varying loyalties and anxieties as a working class seemed to be rising against a workers' state. The phenomenon was anchored in the northern coastal provinces, from Szczecin on the East German border to Elbląg near the Soviet Union, but it was also reported to be spreading southward to mills and factories in central Poland. It seemed only a question of time before it reached the extreme southwest of the nation containing the strategically vital coal fields of Silesia. If that developed, Communist Poland faced the specter that long had haunted capitalist regimes: a general strike of millions of workers.

It had all happened, it seemed, in the nine days since August 14 when work first stopped and an "occupation strike" had been declared in the sprawling Lenin Shipyard in Gdańsk. A strike committee had been promptly elected and an "Interfactory Strike Committee" had soon materialized with delegates from twenty enterprises along the coast. They formulated an assertive list of demands, the first of which called for free trade unions independent of the Communist party. Such a "free union," if it somehow managed to survive, would represent a fundamental structural change in the workers' relationship to existing power in Poland, a change in power itself. Through some miracle of organizing that no one seemed able to explain, the idled workplaces affiliated with the Interfactory Strike Committee spread their agenda of insurgency in five bewildering days from one shipyard to 20 and then to over 150 enterprises. By the beginning of the tenth day, August 23, the number of affiliated factories exceeded 350. It was almost as if the whole region had decided to stand up to the authorities. The workers announced their intention to negotiate collectively through their Interfactory Strike Committee—a united front of hundreds of industrial enterprises confronting the state.[1]

The historic old port of Gdańsk presented a bizarre spectacle to an increasingly fascinated world. The commercial life of the city seemed to have come to a virtual halt. Government cars and trucks were strangely absent from the streets, idled because workers at gasoline distribution points had struck. By agreement with the Interfactory Strike Committee (MKS in the Polish acronym), bus and tram operators continued to provide service to the population, though transport workers demonstrated their solidarity by staging totally effective one-hour stoppages that brought everything to an absolute standstill. The citizens of Gdańsk appeared gripped by some surpassing emotion, passionately internalized but heretofore publicly invisible.

What *was* visible as a symbol of the strike was the Polish national flag. It was everywhere; Gdańsk was decked in red and white. The social effect was striking. Though the normal life of the city had been stifled, a vivid camaraderie had overtaken the place. Total strangers huddled in solemn, intimate dialogue, occasionally punctuated by sharp laughter. It was if the city had become host to a public conspiracy that extended to the entire population. The stakes, it was evident, were very high.[2]

The pageantry could easily be misread by anyone not intimately acquainted with the complex postwar history of People's Poland. Each flag that waved in the Baltic breeze played a role in an emotional struggle between a powerless population and the ruling party apparatus. Each unfurled pennant was simultaneously a private declaration of resistance and a public affiliation with the cause of the shipyard workers. The red and white flags asserted a shared memory. The cause was historic, cultural, collective, national, tribal; the party was oppressive, anticultural, alien. The contest was about how to live, and the flag was the expression of that fact. The conjunction of red and white was thus revealed as partisan; it belonged to the people and the nation, not to the party.

Gdańsk had been a strategic Baltic port since the Middle Ages when it served as a commercial bulwark of the Hanseatic League. The bay formed a great natural harbor which sheltered the tri-cities of Gdańsk, Gdynia, and Sopot. The dense urban ring that encircled the bay teemed with maritime enterprises. The large shipyard in Gdynia had been on strike since August 15, one day after the standard of revolt had first been raised in Gdańsk. Now Gdynia, too, was decked in red and white, as was Sopot.

Intricate maneuvering by both workers and government authorities had characterized the nine days since August 14. As the party viewed matters, the Lenin Shipyard constituted precisely the wrong symbol, for it had been at the center of protracted turmoil on the Baltic coast in 1970. In light of the volatile mood of the country's working class in the summer of 1980, it seemed imperative to prevent the coast from once again serving as a tinderbox. Accordingly, as soon as it became clear that the sixteen thousand workers in the Lenin Shipyard had gone on strike, the authorities severed all telephone and telex links between Gdańsk and the rest of Poland. All roads leading into the port city were blockaded and a sizable proportion of the nation's police forces, including secret police, were concentrated in the coastal provinces. The government newspaper in Gdańsk, *Głos Wybrzeża (Voice of the Coast)*, had ominously discovered widespread "fear about what is happening around us." The striking workers were depicted as a violent and coercive force and their spokesmen as cunning manipulators possessing concealed agendas.

The newspaper account, in fact, effectively conveyed the public posture that the party wished to show to those citizens who, because they lived in immediate proximity to the crisis, had to be told something:

> The free will of citizens is limited or even brutally violated . . . workers who want to continue serving the population have their tools destroyed and apartments burglarized. Threats of destruction of vital installations are frequent. Cars and trucks are not permitted to transport needed commodities. Gasoline is not sold to the drivers of state enterprises. Individuals try to take over the right to decide what is beneficial and what is not for the daily lives of thousands of people . . . The most important question is this: where do the expectations and postulates of the working people end and political games begin—the political games waged by manipulators who hide themselves behind the backs of the shipyard, port, factory and transport workers? It is no mystery these manipulators are there and that they are active. Worse—it seems that they know no limit in striving toward their goals, toward further tension and the disorganization of daily life.

In this circuitous fashion, the official press betrayed the source of the regime's deepest anxiety: the party's time-honored means of handling Polish workers had somehow proved ineffective. Only one conclusion was publicly admissible—a false worker leadership of political manipulators who "abuse the fact that the authorities . . . reveal a measured attitude which is rarely encountered—even in those cases where a more decisive attitude would be justified."[3]

As every Pole knew, as every Pole had painfully learned in the course of thirty-five years of party control, this was the kind of language that heralded imminent police repression. The threat was directed solely toward the coastal population. The party moved to maintain control over the rest of the country by withholding all essentials about the confrontation on the coast.

Meanwhile, the rank and file of the party also had to be told something. In Warsaw, the party's Central Committee issued an internal letter that spelled out for lower echelons the appropriate political stance: "The anti-socialist elements amongst the Gdańsk shipyard workers made political demands and hostile stipulations in order to seize control of the strike." The demands "put in danger our national survival, our common achievement and our unity, built at such a high price and in such difficult conditions, at the cost of so many sacrifices."[4]

Emboldened by such pronouncements, Jan Szydłak, head of the nation's trade unions and member of the party Politburo, clarified the basic issue: "The authorities do not intend to give up their power or to share it with anyone else." As if to punctuate this judgment, security police began rounding up prominent members of the democratic opposition in Warsaw. The effort to contain unrest also had more subtle dimensions. A special government negotiator, Tadeusz Pyka, was dispatched to the coast. He promptly invoked such routine party methods as "consulting" with workers by bringing in the local troika—the plant manager, the head of the party in the plant, and the head of the plant trade union. These functionaries, all party members, did not have great difficulty grasping the point of view of Pyka, a deputy premier of Poland. Unfortunately, the workers remained on strike despite this meeting of party minds. Pyka then tried for a broader base by having his various troikas bring in some "representative" workers, again selected by the party. Again, agreements; and again the workers remained on strike. When Gdańsk party officials ordered the printing of back-to-work leaflets, printers ran them off, talked it over, and then went out on strike themselves to protest the blatant disinformation contained in the material.[5]

At the source of the conflict, the Gdańsk shipyard itself, the harassed manager became overwhelmed by events. He returned to his office and was scarcely to be seen after the fifth day of the strike. Meanwhile, day after day, worker delegations from other factories arrived to declare their support of the twenty-one Gdańsk demands and to announce that they, too, stood ready to strike or had already done so. Delegates arrived despite the police blockade; they possessed information despite the communication blackout.[6]

In an effort to cope with this avalanche of popular activity, the government notified the shipyard strike committee that a delegation of worker representatives would be received in Warsaw. This gesture failed to elicit a positive response. The strike committee merely reiterated its original demand—that the negotiations should be held in the shipyard itself. Everyone understood the implicit meaning. If the government hoped to co-opt or intimidate the strike committee, it would have to intimidate the entire work force to do so. The authorities tried their sticks and carrots and stood fast. The workers stood fast. And the strike spread, edging toward Silesia.

What was significant about this contest was, first, the elaborate and time-tested array of weapons of containment available to the party and, second, the sophisticated knowledge the workers exhibited in deflecting these tactics. The more one knew of this detailed dance of thrust and counterthrust, the more impressive the politics of the workers appeared. By the same token, the less one knew of the details, the less exceptional the workers' efforts seemed. A simple strike was in progress.

On the ninth day, the government decided to talk to the Interfactory Strike Committee. Mieczysław Jagielski, deputy prime minister of Poland and the party's most prestigious and adroit labor negotiator, arrived on the coast and notified the workers' committee that he would receive them for a preliminary discussion in downtown Gdańsk. The minister went further and agreed to dispatch the Prefect of Gdańsk, Jerzy Kołodziejski, to the shipyard the next day to work out the protocol for a formal negotiating session.

To observers around the world, this decision appeared to mark the first crucial juncture in the tumultuous series of events that would come to be known as "the Polish August." The appearance was not wholly an illusion; a juncture of sorts had indeed been reached. But as will become clear, there was little that was first or even "early" about this moment of time. A rather imposing number of things had to have happened to make this juncture possible. Collectively, these acts comprise not only the immediate foreground of the Polish August, but constitute as well the essential—the absolutely essential—background to the struggle.

It is necessary, then, to specify clearly at the outset a basic causal connection that informed events throughout the entire life of the Polish movement. It is a connection that existed between largely unseen efforts of Polish workers over a prolonged period of time and their highly visible culmination on the Baltic coast in the late summer of 1980. For this long evolution of unseen events explains the appearance of Solidarność as nothing else can.

There is a corollary to this basic causal relationship. The construction of the democratic space that Solidarność occupied was an experience that was deeply Polish, but it would be quite wrong for it to be seen as something marginal, aberrant, or unique. The concrete political development that the Polish experience makes visible cannot be effectively grasped as a sudden sweeping democratization or as some sort of magical coalescing of collective will. Rather, the creation of Solidarność was a protracted development that can most easily be understood as the sequential construction of the building blocks of popular democracy. This is not meant to imply, even for an instant, that the route to Solidarność followed a blueprint that had been conceived by some ingenious worker or perhaps by some creative political thinker among the Polish intelligentsia. Far from it. The ultimate "knowledge" that Solidarność represented was acquired in parcels, sometimes very small parcels, over a prolonged period of time. But these bits and pieces of democratic experience not only can be tracked, they must be tracked if Solidarność is to be understood for what it represented; namely, the achievement of a historically unprecedented plateau of large-scale democratic self-organization. I would suggest that then and only then does it become possible to visualize the deepest implications of the Polish achievement. Solidarność speaks not only to Poles or to other confined populations; the political process generated by Solidarność contains experiential ingredients that are at least arguably available for use in any society whose citizens aspire for more authentically democratic social relations.

It is necessary to pause at this "early" and "crucial" juncture of August 23, 1980, in order to emphasize that the events of that day had their causative meaning in happenings occurring days, months, years, and even decades earlier. To penetrate the fog that covers these years, it is necessary to begin to deal with a sanctioned way of thinking about social movements as an institutional expression of a certain kind of consciousness. Traditionally, one is inclined to believe that the people who brought Solidarność into being were people of relatively "high consciousness." They were not mystified by the self-serving propaganda of the regime; they had come to possess an alternative, autonomous awareness; and this high level of social comprehension and self-perception explains why they were able to do so many surprising things in the course of their protracted encounter with state power.

Though carrying a suggestion of analytical clarity and sometimes even of analytical power, such generalized terms as "high consciousness" and "autonomous awareness" routinely conceal the social realities they are supposed to describe. When a person or group of people is perceived to have advanced from what might be called a normal stage of political awareness to a higher stage, some questions would appear to be in order. What happened to transport the observed person or persons from normal to sophisticated levels of perception? Concretely what? Is it something cerebral—a sudden flash of insight or a leap of the imagination? If so, alterations in consciousness originate in the mind. It seems most intellectuals assume this to be so. Be that as it may, the original question nevertheless persists: what makes these intellectual happenings—these leaps and flashes—occur? Clearly, this is an essential question if human agency in history is to be understood as anything other than a quaint illusion of romantics. Isolated in this manner, the question may fairly be seen for what it actually is—an inquiry about an elementary building block of social information. We must know concretely what happened, originally and subsequently, as the persons involved moved from a routinely compliant state of social conformity to presumably higher realms of insight and subtlety.

To this end, let us shift for a moment to the present tense and take station on the Baltic coast on August 23, 1980. In the early afternoon, the prefect of Gdańsk, Jerzy Kołodziejski, climbs into a government car, is driven to the gates of the shipyard, and immediately bears the brunt of all the tension that has accumulated through the volatile days of confrontation. As he makes his way through a boisterous crowd of strikers at Gate No. 2 of the Lenin Shipyard, Kołodziejski encounters men and women who have been ridiculed as dupes, slandered as coercive hooligans, and accused of disloyalty to the nation. The workers grudgingly make way as Kołodziejski, head down, plods through the parted throng. One of the strikers shouts, "Look straight into the eyes of the workers." A low rumble of assent from the group reinforces the remark.[7]

While Kołodziejski makes his way toward the strike committee, another scene is being played out in the shipyard. The steadily growing mass of delegates from other factories in northern Poland is gathered in an adjoining hall. They are addressed by a new delegate who has managed to get through the police blockade with the aid of some coordinated help from Baltic workers. Delegates nearest the loudspeaker become very quiet when it is understood the newest representative in the hall has traveled from Silesia. He tells the delegates that their fellow workers in Silesia are "completely disoriented."

> Radio and television tell us nothing. We have been cut off from the rest of the country. I was sent here specially by my factory to find out what is really going on. Someone tells us that fifteen factories are on strike, then the radio news says that a similar number have gone back to work. So I am here to observe at first hand what is happening on the Coast. And I believe that Silesia will join in.[8]

A wave of applause engulfs the hall. A member of the workers' presidium then shares with the delegates the "preconditions" for talks that are at that moment being presented to Kołodziejski. The rigorously specific detail mobilized by the strike leaders underscores the intensity of their commitment to these preconditions. It also points back in time to a period before August 23 when workers had performed some activities of which the population of Poland is unaware; the strike committee announces the presence of its couriers, or "helpers." The details of this pronouncement are revealing:

> We categorically demand that all repression of Interfactory Strike Committee helpers cease and that any further repression be prohibited. In the last few days, the militia and security service have repeatedly picked up, interrogated and detained people distributing Strike Committee publications. For instance . . . Andrzej Słominski, Piotr Szczudłowski, Maciej Budkiewicz and Andrzej Madzejski were held for forty-eight hours for distributing Strike Committee leaflets. Słominski and Szczudłowski were brutally beaten while in custody . . . A fifteen year old, Michał Wojciechowski, detained by the security service for some hours for distributing leaflets, was brutally beaten at the militia headquarters in Starogard Gdański . . . Jordan Cezaryk, held by militia and security service in Gdynia for many hours for distributing Strike Committee leaflets, was brutally beaten . . . Faced with such acts of violence by the militia and security service, and the tolerant attitude of the state and administrative authorities towards such practices, working people represented by the Interfactory Strike Committee unequivocally declare such provocative and criminal actions to be in total contravention of the Constitution of the Polish People's Republic and Poland's obligations under the International Conventions of Human Rights. Signed: The Interfactory Strike Committee, August 23, 1980.

Visible here is the public surfacing of a basic ingredient in the successful mobilization of the Baltic workers: the system of couriers who carry the germane details of the Lenin Shipyard strike to other factories along the Baltic coast. Visible also is the party's ruthless attempt to suppress these organizational links through arrests and beatings. Finally, what is visible is the specific level of the workers' tactical sophistication. Having at long last forged a multifactory strike and forced the government into negotiating with their representatives, the workers use their strategic position to protect their tactical achievement; a "precondition" for negotiations must be the elimination of party repression of the couriers. Such is the view from inside the strike. From afar, it merely appears evident that the worker delegates gathered in the shipyard from 370 supporting factories are delighted at the clarity of the strike committee's preconditions.[9]

In the planning session itself, Prefect Kołodziejski is less delighted. But he leaves the problem for his superiors. For the moment, he is content to work out the agenda for a formal meeting at 8 p.m. that same evening between the government commission and the strike committee.

So it was that on the evening of August 23, 1980, a fundamental duality crept into the story of Solidarność, one that was to persist in variously complex dimensions throughout the sixteen months of its life, through the long years of martial law that followed, and down to the present when the history of the period is formally being set in place. In general terms, its components may be characterized as, on the one hand, the surface flow of events, which carry with them their own apparent causation, and on the other hand, the underlying events and *their* causes, which—on August 23 and thereafter—are both determinate and largely unseen. The gap between the two is sufficiently large to cause Solidarność to be fundamentally misunderstood as a large-scale democratic movement. To make this duality visible, it will be necessary from time to time to focus closely on the surface appearance of events and then to retrace the same happenings with the movement's internal needs in mind. The objective is to make as visible as possible the social dynamics that produced Solidarność as a large-scale democratic movement. The ultimate objective is to alter in certain significant ways conventional notions as to what the phrase "democratic politics" means.

Because of the rich complexity of the events of August 23, 1980—that is, the sheer number of dualities between the seen and the unseen that are exposed—this day may be taken as a useful starting point. At 8 p.m. in the evening—in the presence of a crowd of two hundred journalists from around the world, the deputy premier of Poland, Mieczysław Jagielski, led the government commission through Gate No. 2 of the Lenin Shipyard past thousands of striking workers and into a small building where formal negotiations were to take place. The scene, meticulously reported at the time by foreign journalists, reconstructed many times since by sociologists and historians, and quickly integrated into the folklore of the Polish nation, contained all the essential ingredients of high drama. This became starkly evident in the opening moments when the chairman of the Interfactory Strike Committee, a 37-year-old electrician named Lech Wałęsa, formally welcomed the government team "on behalf of 370 enterprises of the Gdańsk region and some from Elbląg and Słupsk."

He quickly emphasized two central ingredients of the occasion—the extent of the social base of the movement and its interior democratic style: "The fact that we represent hundreds of thousands of people makes us feel sure that the cause we are fighting for is just. Coming here may bring home to you what a shipyard is like when the workers are governing themselves. You can see for yourself how orderly it all is." Wałęsa then provided the government team with its first indications that the strike committee did not intend to be stampeded: "The serious matters we must settle require us to act prudently and without haste. We have been waiting patiently for nine days, and we have plenty of patience left."[10]

With an already weakened national economy thrown completely out of joint by the network of work stoppages across the northern provinces, the last thing the government wanted was negotiations conducted "prudently and without haste." So when Wałęsa suggested that the government side offer a general response to the worker proposals, Jagielski jumped at the chance. But the deputy prime minister's rhetorical march through the twenty-one demands does not merit meticulous review because, to be concise, he did not say anything. He emphasized his "concern" and his "sincerity" and assured his listeners of his clear understanding that the times called for "reform" of the existing trade unions; that is to say, he blandly and repeatedly refused to recognize the existence of a demand for unions independent of the party. Jagielski spoke a very long time, but he failed to seize the attention of his audience. They had heard it all before. The minister merely repeated in his own particular style the ancient language of official deception that had characterized the party's public face since it first came to power under the guns of the Red Army in 1945.

The new language of the workers' movement immediately made its appearance in the Lenin Shipyard, conveyed by the chairman of the Strike Presidium.

> Thank you, Mr. Prime Minister. We listened to it very carefully. But I think that we did not say why is it so that every ten years, (let us say, this time it lasted ten years and it may last another ten) we return to the same point, in which we are at present. In order to prevent this repetition, to draw conclusions, one has to know reasons. But you did not tell us the reasons for our nation's vicious circle every ten years. What guarantee do we, the workers, have (and workers work hard and do not love striking at all) that they will be able to demand what they truly deserve? I think that in this coordination, steering, management, checking and control there is something wrong. I am a worker but that is how I see it. We shall never make up for mistakes if we fail to discover their source . . . Mr. Prime Minister: what does the government think, what do they think we should do in order to abandon this vicious circle?

Jagielski responded once again in a time-honored way. His terminology—the officialese of the one-party state—should perhaps be quoted in its entirety at least once, if for no other reason than to convey the flavor of public discourse in People's Poland that the movement sought to bring to an end:

> I can answer that. You said, I have made a note, I did not note everything, that the reason for this development here and for our discussion of them—means among others, that is, that it is linked to something wrong in managing and coordinating. I agree. Something is wrong. Something is wrong. I suggest that we accept the following. The following. Let us make an agreement. Trust me. The coming plenum of the Central Committee will answer this question.

Wałęsa's response set the tone of the Polish August: "We shall prompt them how to answer it. There is only one solution. We suggest what we see, we observe, we notice. Free trade unions! Strong and dynamic, as the workers would like to see them. This is no political matter, this is actual counterbalance and check and control. We shall control ourselves." On this basic point, Wałęsa built a second: identifying and condemning repression by the police as the ultimate expression of state censorship. "We shall see mistakes and prompt solutions, but not with the means which are presently employed, i.e. not by silencing the people who want to voice their opinion, not by arrests, imprisonments, and huge enlargement of the power and violence machinery, i.e. huge development of police forces."[11]

The explosion of applause from the 370 MKS delegates that greeted this speech brought home to government negotiators the depth of the threat to entrenched practices and inherited customs that the existence of the Interfactory Strike Committee represented. Throughout Jagielski's lengthy and aimless review of the twenty-one demands, the range of experience he confronted on the worker presidium was signaled by repeated interruptions as strikers contradicted his clichés. When Jagielski offered perfunctory assurances "that the safety of strike participants and helpers is guaranteed and that they will not suffer any consequences," the strike committee chairman interrupted: "We don't see it quite like this. Plenty of people are sitting in prison and plenty more are beaten up. These are the facts. Since we were to speak frankly, I think this matter should be made known. There is a list of those beaten up, with medical reports. Some of those beaten up were rearrested. So, it is not true that there are no consequences." This sort of candor was not restricted to Wałęsa. Another member of the Strike Presidium, Wojciech Gruszecki, intervened: "A new case has just been reported to us. The mother of Andrzej Słominski, a doctor from Tczew—came here in tears because he has been rearrested and beaten, together with four others, including his brother Wiesław. I tried to telephone the Prefect, Mr. Kołodziejski, but unfortunately could not get through. I request that this matter be cleared up straight away." The Strike Presidium seemed intent on burying government clichés under an avalanche of social facts. Another worker, Florian Wiśniewski, offered a history lesson:

> I have worked in trade unions for many years and know that the problem is really serious. At the moment, it looks difficult from the legal angle because the Labour code is presently so formulated that anyone who goes on strike can get the sack. People come to us from Southern Poland—one brought a paper to show that he lost his job for going on strike. That would be impossible in the present atmosphere on the coast. Things have gone further here . . . This must be put right if people are to be satisfied. We know how strongly people feel on this . . . I think we should decide this on the spot. We are fully representative. Prime Minister, this is not the first time such events have taken place on the coast . . . In the past, the strikers tried hard to make contact, often reaching out directly to the government, but there was no response. Instead, lies were circulated about us and various incidents staged . . .

In response, Jagielski attempted another bureaucratic evasion: "On the basis of information given me before my departure, there are no political prisoners in Poland. That is, persons convicted for their political opinions. I received this statement from the Minister of Justice." But the workers had too much experience. Wałęsa said: "Prime Minister, these cases are known to us and to the public. We know what kind of trials they were. And so, if we are to be honest, I propose that these people should be released while their cases are reexamined. For the truth is we attended these trials. I was there, and others too. I can say straight out, because I am a worker and don't mince words, that they were rigged." The immediate applause from the hall of delegates informed the minister that he was not hearing one man's opinion. So there would be no doubt, Wiśniewski provided some personal details: "Let me illustrate. Here I am in the shipyard. My block of flats is ringed by the militia, who tried to catch me in the car park. I go home when my family situation gets difficult, after all I have four children. People can be done in for anything. I know how careful you must be."

Another veteran activist, Andrzej Gwiazda, emphasized that the problem was systemic: "Prime Minister, until recently the press said all was well in our industry and our economy. The obligatory official line was that, despite some minor difficulties, everything was running smoothly. Now it transpires that great deficiencies were being deliberately concealed or glossed over. Don't you think, Prime Minister, that something similar may be happening in the administration of justice? Why should it be any different?"[12]

The first duality appears—the workers have some sort of agenda within their agenda. The twenty-one Gdańsk demands were so sweeping in their intent that most Polish intellectuals, including the most respected and militant of the leaders of the democratic opposition, concluded they went much too far. The first demand, particularly, was seen as excessive and unrealistic, an example of militancy carried to the point of "impossibilism." The workers were unmoved by this advice. Unbelievably, they acted as if there were issues even more vital than the twenty-one demands. The duality is starkly visible here: the status of the workers' "helpers"—and all the contentions about police repression that seemed to hover around that issue—somehow seemed more important than even the far-ranging first demand which called for free trade unions independent of the party. Why was the Strike Presidium so utterly preoccupied with the preconditions? What lay behind this "other" agenda?

To understand this first duality of the many that undergird the Polish movement, it is necessary to retrace briefly the initial and relatively unsensational series of events dating from August 14 that culminated in the strident opening moments of confrontation with the government commission on August 23.

On the morning of August 14, certain shipyard workers refuse to perform normal work. They gather together and then array themselves in the presence of the shipyard manager. They extemporaneously voice their principal grievances, hear familiar and predictable responses from him, and in due course declare an "occupation strike." They nominate a number of militants from within their own ranks to serve as a kind of *ad hoc* strike committee. When their representatives meet again with the shipyard manager, an immediate argument develops as to how many workers shall be privy to the discussion. The strikers make the surprising demand that everything be broadcast on the shipyard loudspeaker so that every worker will be informed as to what is happening. This unanticipated demand raises the tension between management and labor and precipitates certain maneuvers by the shipyard director. These efforts prove unavailing, and the loudspeakers are connected. Within forty-eight hours, a worker-generated network spreads word of the strike through the tri-cities and elsewhere along the Baltic coast. Not only do maritime enterprises in Gdynia go out on strike, but so does distant Szczecin near the East German border. At the other end of the northern coast, Elbląg near the Russian border joins in. Other, smaller enterprises soon follow. But the Lenin Shipyard is the fulcrum; delegates from striking enterprises come to the Gdańsk shipyard where an Interfactory Strike Committee is unveiled as the negotiating instrument for all working-class aspirations. It is in the name of the Interfactory Strike Committee that the twenty-one demands are announced. The strike takes on the appearance of a regional class action.[13]

At first glance, the escalating goals and organizational intricacies seem astonishing—almost as if some master hand lay behind the worker activism. Perhaps government propagandists are not being entirely disingenuous in asserting that sinister unseen "elements" are manipulating workers and tampering with "the honest postulates of the working class." Acting on this belief, security police round up Jacek Kuroń and other prominent opposition activists in Warsaw and keep them locked up for the duration of the strike.

Unfortunately for its cause, the party has the wrong target. Kuroń and his associates are astounded at the first of the twenty-one demands and are making plans to communicate their fears to the workers at the very time they are arrested. Similarly, as late as August 23, other Warsaw intellectuals, when making formal contact with the strike leadership in the Lenin Shipyard, have uppermost in their thoughts the intention of delivering the same cautionary advice.

But the simple fact is that the institution known as the Interfactory Strike Committee and its immediate objective of a free trade union independent of the party are worker initiatives that are well-developed products of the coastal movement itself. So is the occupation strike that gave birth to the interfactory committee and its bold first demand. All these organizational forms and strategic goals are, in fact, ten years old. All were energizing components of the 1970 worker rising on the Baltic coast. Indeed, lessons learned in that struggle—and even earlier ones—remain as controlling influences on the politics of August 1980. In this manner, the long-hidden history of what needs to be understood as the self-activity of the Polish working class makes its first appearance as the central causative element of the August confrontation in Gdańsk.

The demand for a free trade union independent of the party is but one of the more central strategic pieces of experience coastal workers have gained in their long struggle with the party-state. There are vital tactical nuances as well. The workers announce their readiness to negotiate—in the shipyard—with highest government authority. They are respectful, but firm. They decline to negotiate as individual factories; they decline to negotiate with local or provincial authority; they decline to negotiate in the national capital, Warsaw, or the regional capital, Gdańsk. They reiterate, politely, that they will negotiate as an Interfactory Strike Committee and do so in the shipyard. What they then proceed to avoid doing is as revealing as the actions in which they engage. The elaborate mechanisms of production in the shipyard are left untouched as workers concentrate upon their own presence in the workplace. In contrast to the past actions of other striking workers in other lands, they do not burn sugarcane fields as striking Jamaican sugarcane workers once did; they do not refuse to bank furnaces as striking English steelworkers once contemplated doing; instead, they post guards to see that harm does not befall any of the production facilities in the shipyard. The strikers also do not threaten the physical surroundings of those in power along the Baltic coast. They do not burn manor houses as peasants have historically done in agrarian risings; they do not wreck or burn the Polish United Workers' party headquarters as Polish workers, including Baltic workers, have done in the past; they do not march to the vicinity where power reposes. They stay in the shipyard and disseminate word of their actions to the government and to as much of the population as they can reach. Then, having done all the necessary things they themselves can do and having said all the necessary things they themselves can say, they apparently do and say no more. They wait.

The pressures that can be mounted against this sort of behavior are many. The party applies them, but the workers withstand them. Despite all coercive influences, the workers continue waiting day after day. They arrange for food, for shelter, for a system of security and, most important, for a system of internal communication that links to the negotiations all rank-and-file workers whose factories have affiliated with the Interfactory Strike Committee. They create, in short, all organizational forms necessary to maintain the social life and the collective strength of their voluntary community. They wait for nine days.

During this time, the shipyard appears a focus of what might be called a measure of cultural improvisation. The strikers sing, they write poems, they paint signs, they sit in ever-shifting groups and talk, and they listen through loudspeakers to their own strike committee chairman and to the voices of delegates newly arrived in the shipyard. From time to time, the license-plate numbers of secret-police and militia vehicles are announced over the loudspeaker for the benefit of the shipyard's outriders who fan out across the coastal region to intercept and provide escort for incoming delegates. None of this is said on the loudspeaker, of course; only license-plate numbers are read. Therefore, to view the shipyard in programmatic terms, the ritual of listening to loudspeakers and other modes of waiting are all that is visible. It seems a quiet, even a happy time. It is so reported in the world press. To the casual observer, all this could be interpreted as behavior that is quite passive. Beyond sitting around in a shipyard, workers do not seem to be doing anything. Perhaps the best that could be said of them is that they appear passively resolute.

Such a verdict, however, would constitute a gross misreading of what is actually happening. The shipyard appears relatively inactive only because the organizing energy of the strikers is now directed at points beyond the shipyard. It is aimed at workers in other enterprises on the coast who have yet to hear about the Interfactory Strike Committee. Thousands of leaflets are printed, and there is much coming and going of couriers. Hazards are encountered as leaflets are confiscated and couriers are arrested and beaten by those in authority.

Something approaching an all-out cultural war is being waged in unseen places throughout the coastal region of Poland. It involves underground printing, couriers, automobiles, highways, roadblocks, telephones, telephone blockades, and a form of physical terror in the interior of police stations where couriers are being beaten into unconsciousness. Since the details of this war are largely unknown, it becomes difficult for reporters and other observers to determine its course. There is, perhaps, one clue. The number of workplaces affiliated with the shipyard grows every day, from twenty on August 15 to 156 on August 18 to 370 on August 23. This occurs despite the fact that delegates literally have to infiltrate the shipyard rather than merely drive to it. Stories—repeatedly verified by incoming delegates—circulate that representatives of other striking factories across Poland have to escape the police in order to arrive in the shipyard to join the MKS. A number of delegates are stopped, and automobiles are confiscated by the police. Yet the MKS grows daily. Its growth is the verifying life's blood of the strike—and the movement the strike symbolizes. And the couriers, the entire worker-constructed apparatus, is the lifeblood of the MKS. It is precisely this appearance of "helpers" that the security service is trying to crush. Thus, the insistence by workers on their preconditions.[14]

Yet there is an even deeper dimension to the argument over courier beatings and political prisoners. The repression of the workers' helpers is merely a particularly brutal physical form of a much more pervasive underlying repression—the wholesale censorship of Polish life by the ruling party. Among other things, the movement of the Baltic workers is a mass mobilization against the censorship. The government's blackout of the specifics of the twenty-one Gdańsk demands is centrally relevant, and yet it is but one additional dimension of the underlying conflict about party control of all social information. The strike committee, acutely aware that the Baltic mobilization has become the most telling popular assault on the censorship in the entire postwar history of Poland, renews the confrontation with the deputy premier on this deeper element of the struggle.

GRUSZECKI: I read our local press. So far, I haven't seen a single item from which I could learn that an Interfactory Strike Committee exists. This name was nowhere to be found. If I haven't found it, then it doesn't exist for the readers.

A second point. People come to us from all over Poland, and are often deeply moved by what they see happening here. They frequently beg, with tears in their eyes, "give us some printed materials, we know nothing of all this at home." All that is happening here is kept secret: the local press says nothing and telephone links with other towns are cut. Workers come from very large enterprises. Delegates arrived recently from Świdnica and from Tarnowskie Góry, from factories with thousands of workers. Tourists come; others return from their holidays. They state unanimously that they knew nothing of what was going on in Gdańsk. They are astonished. How, then, can one speak of reliable information?

JAGIELSKI: I can answer you quite truthfully. I look at it like this: at 7:00 p.m. the radio news stated that a Jagielski commission is going for talks with the Interfactory Strike Committee. A communiqué will be broadcast tomorrow. Perhaps there wasn't one today—nor in the press—but there certainly will be, and television will probably report in tonight's late news that talks have begun. It may be in the press tomorrow.

GWIAZDA: Prime Minister, I second your proposal that we speak the truth. But the truth must be the whole truth. So let's speak the whole truth: what we are demanding, why we are on strike, what you and the Government Commission propose. We can report later what we have agreed.

There is realpolitik in this exchange, a central debate about the civil liberties of Polish workers. The issue is fundamentally alien to the government's agenda. Jagielski has only one purpose—to get the Baltic coast back to work. "But let us first start talks," he says, "After all, you say you want to talk with me, with us. So, let us proceed with the talks." The Strike Presidium, however, is engaged in driving home another point; talks cannot begin until the preconditions are met. Before the strikers will negotiate on the twenty-one demands, they are determined to protect their strategic position. Florian Wiśniewski ignores the minister's request and focuses instead on the censorship: "I consider a man of great standing and broad outlook capable of taking great steps. This means releasing a communiqué in the usual fashion, through the national press and television. Making our demands known in this way will restore our reputation after all the slanders that were heaped upon us."

But the strike leader also tries to acquaint the government side with the balance of forces present. Wiśniewski is not exactly making a request: "We won't tolerate prevarications. Truth cannot be concealed. The work force will make sure of that. The great barrier here is censorship. It is enough to call something political and it won't be published. But these are not political matters." Wałęsa endeavors to capitalize on this clear summary by forcing Jagielski into a direct exchange on the issue: "I have a question for the Prefect. We agreed at the outset that precondition for our starting talks would be restoration of telephone links with other cities. But it turns out that the telephones are still not working. There's something wrong here. May we know what the blockade is for?" But the minister cannot acknowledge such realities: "I can't answer now. I must look into it. I can't answer now." "But it was a precondition," Wałęsa insists.[15]

The accumulated impact of this worker pressure—a simple matter of logic openly presented—begins to have at least a marginal effect. Jagielski and Kołodziejski, the two officials who represent, respectively, the central power in Warsaw and the provincial power in Gdańsk, take different routes in replying to Wałęsa. Protesting that he learned about workers' objections to the telephone blockade only the day before, Jagielski engaged in the time-honored party habit of dissembling: "I made enquiries and it was established, technically, that telecommunication links between Gdańsk and Szczecin are open." But Kołodziejski is a bit more to the point: "My answer to the question is that this is probably a central decision. It is not for a Prefect to make. I cannot answer your question."

The strikers press on: "Let me read out the agreed conditions," says one veteran worker activist, Lech Sobieszek. "One of the first conditions for starting the talks is restoration of all telephone links. In my view, this has not been fulfilled. I propose we suspend our meeting. This is my opinion." Wałęsa intercepts this extemporaneous ultimatum by incorporating the point into the continuing discussion: "There should simply be a decision to reconnect."

The harried minister makes one more attempt: "I suggest you give me an opportunity to clarify this matter. I came to the meeting because I wanted to talk, to explain our point of view. So let's clear it up. Perhaps what cannot be settled today will be done tomorrow. After all, I am not in a position to clarify everything straight away . . ." The accumulating evasions drive Wałęsa almost to distraction and engenders within him a private war, "No, No, No," he blurts out. Then, regaining control, he says: "However, despite everything, I propose we go on. So that we will know your attitude to the remaining points." But Wałęsa then returns to the preconditions: "But you will have to speed up the ending of the blockade. It surely can't be so difficult. Of course it isn't."[16]

Neither the language of the party nor the language of the workers would be fully coherent without the other. Taken by itself in all its evasion and mediocrity, the ritualistic revealed wisdom of the party materializes as something that not only fails to persuade, it is patently insulting. Above all, it does not bear on the reality that appears in the shipyard; it does not make sense. Its most assaulting feature is that it diminishes anyone who hears it. As it surfaces in the Lenin Shipyard, this language succeeds in acquiring a certain vitality precisely because it is laced with counterpassages that come storming into the dialogue radiant with indignation. The words of the delegates on the strike committee serve as a kind of energizing background music to what otherwise would be an endless monotone of deception.

But the reverse is also true. In the opening negotiation, the language of the workers on the Strike Presidium, if isolated and set down by itself in the exact rhythms of the transcript, would seem too unrelenting for most readers to assimilate readily. The prose of the party is necessary to clarify the rules of the game and bring the reader up to speed by generating a level of disbelief and indignation sufficient to stay apace of the principal players. The limp dissimulation of the party acquires vitality because it stokes the outrage of the workers and abruptly calls forth furious ripostes. The result is a curious kind of harmony.

But the dynamics ultimately prove dislocating to the workers. The activists who comprise the strike committee inevitably go too far, allow their preoccupation with immediate crimes to divert them from the central crime. They must be invited back to the main topic by their own chairman. They know they should take his advice, but they proceed as if they are saying (one must understand this to understand the Polish movement), "the latest deception has not been conclusively pinned down yet." And so they are not quite finished. Their chairman has to remind them again of the main issue, but even he soon gets diverted by a new fabrication and adds his own voice to the overpowering litany about the preconditions.

There is a necessary human rhythm here—the pent-up needs of voiceless people when they are at last able to speak and a Chief Censor who is forced to listen. It is a fine moment in history, one that does not happen enough in any society or in any unbalanced human relationship. A bit of excess seems always to be visible the first time; its presence verifies the humiliation and tragedy of the past and signals that some basic realignment is in the offing, or is possible, or is at least passionately longed for.

At this juncture in the opening hour of negotiations, there occurred one of those absurd flights of pure fiction through which governments periodically denigrate themselves. All governments lie when it is considered necessary, of course, but the form and extent of deceit often illuminate the kind of functioning social compact that actually connects the rulers and the ruled. It might be said that the amount of truth imbedded in an official lie determines in some tenuous way the social presumptions of those in authority and their adherence, however insecurely, to professed rules of candor. The American governmental deceit, known as "plausible deniability," seems to rest upon—or float around—this tenet. But the Polish regime, trapped in uneasy fraternal friendship with the Soviet regime, has long since ceased to be plausibly legitimate, and many of its clumsier members seem to have abandoned even the need to *appear* plausible. Thus, in the discussion about ending the telephone blockade of the Baltic coast, a member of the government commission, Zbigniew Zieliński, suddenly transfixed everyone in the room by offering the following explanation of the telephone blockade:

> A hurricane passed through Warsaw last night, destroying buildings in large areas of the city. I was in Warsaw at the time, to be exact just after the hurricane. You can see whole streets—such as the avenue from the airport—where huge trees, huge limbs, beautiful lines are completely demolished along half the route. The central telephone exchange was completely demolished. So I don't think telephone links with Warsaw will be restored today—even though I haven't been in Warsaw today—and I don't know how advanced the repair work is, or whether it can be finished tomorrow . . . [protests from the hall] Excuse me, but there are many serious matters among the points you raised, deserving discussion. It is worthwhile for the several hundred sitting in the hall to listen to them and think them over quietly. Time should not be wasted on inessential matters.[17]

Zieliński's attempt to summon a hurricane as a cover for government repression was the kind of breathtaking public lie that, aside from being absurd, was subtly demoralizing insofar as it demonstrated once again the party's mature contempt for the population. As Zieliński made his claim that telephonic reconnection took time "to sort out," everyone on the strike committee knew that one telephone in the shipyard had remained in constant working order every minute of the strike. The telephone was on the desk of Klemens Gniech, the party functionary who served as director of the shipyard. (It had been used earlier in the day by an associate of the Strike Presidium. See p. 37.) Was Zieliński unaware the workers knew this? It did not occur to him to consider what workers might know. In daily party life, one was not habituated to thinking about what the recipients of instructions might not know.

It was precisely this aspect of governance in People's Poland that three decades of popular assertion and government oppression had brought home to the population. The party had created within itself an institutional environment of deception and self-deception that acquired an orderly form by the way it served the day-to-day needs of the bureaucracy: one lied when one had a problem. The activists on the MKS had a deep understanding of the pervasiveness of this style of governance; its presence fortified and refortified their determination that all Poland should know what they had achieved on the coast. They felt certain that workers in every region of the nation would stop production and form interfactory strike committees when they learned the entire Baltic coast had done so. When that happened, they knew they would either have their own institution or the army would come. The risks were high. But they were the right risks.

Accordingly, members of the workers' presidium, led by a 26-year-old nurse, Alina Pieńkowska, responded bluntly to Zieliński: "I point out that telephone links with Warsaw were cut off last Friday, a week ago. Nothing was said then of any hurricane." Another worker, Zbigniew Lis, added: "I would like to ask the Deputy Premier why today's press and television made no mention of this hurricane. We have heard nothing about it. The other thing is, we had to wait nine days. Why must we now rush through all the proposals so urgently?" Finally, Andrzej Gwiazda adds some technical advice: "I would like to inform the Minister that in modern telephone exchanges there is no need to push a block into the socket of every subscriber. Disconnection is done simply by transmitting appropriate information to the register from the centre, or by removing it. It really just amounts to issuing an instruction."[18]

The staccato sentences resound across the room, like doors slamming on an old, tired conversation. The objective, of course, was not to end the negotiation that the workers have tried to initiate for nine days; rather, the aim was to have a new kind of exchange—a real one—and to have it in an environment in which the organizational achievement of the workers, the coordinated mobilization of the Baltic working class, would be protected from censorship and repression.

The repeated rebuffs began to have an effect on Poland's deputy premier. The central political fact that hovered over the first day of negotiations was that production had ceased not only in the tri-cities of Gdańsk, Gdynia, and Sopot but also at an increasing number of places along the northern coast. Jagielski's overriding mission was to get people back to work. The time-honored management techniques for achieving this having failed, the appearance of talks had become a practical necessity. Gradually, Jagielski tumbled to the fact that appearances were not enough. Part of the news blackout, perhaps all of it, would have to be lifted. Otherwise, the talks would collapse. So Jagielski decided to offer a concession, then unwittingly vitiated its impact through another evasion. It was an instructive moment, one that revealed the extent to which deception had become an organic part of party rule in Poland: "I have a constructive proposal. We can agree that we have a radio transmission of our next meeting, as suggested, to let local residents know about our talks and their progress. Yes, I am in favor of this. I am not myself a technician, but we must work out how it is to be done." In his bid for credibility, Jagielski then canceled his gain: "Let me explain another point. I have concentrated on what are called here, 'Demands of the work forces on strike, represented by the Interfactory Strike Committee.' This one was not included. Well, maybe it was an oversight. We all make mistakes. So I propose . . ."[19]

The patent demagogy of this remark (there had been no "oversight"; the issue of telephone communications had been raised as a precondition) almost brought the proceedings to a close. Loud interruptions from the hall, where delegates from hundreds of striking factories were listening to the proceedings, and from the presidium itself, temporarily robbed the deputy premier of the floor. He was overheard to say "Are our talks to be suspended on this one issue?"

At the end, Wałęsa once again thrust to the center the prerequisites for the resumption of negotiations: "We may talk if you finish blocking phone connections of Gdańsk with the rest of Poland and stop arresting and imprisoning. Only then can we talk seriously. We shall prompt you, we want to help and we will. We shall make things that will make the world wonder how is this possible."

Wałęsa had one final connection to drive home: a settlement would come "only through independent free trade unions, which can make us prompt you in a rational, logical, reasonable way. Then we shall bring this nation out of this mess. Thank you. I suggest we end the discussion here." He added a tactical nuance: "We shall wait, here, in the shipyard, for we have to finish this business in a true, human way. And not postponing it by going back to work, for then we might meet again and never again have the chance to meet and discuss everything."[20]

As Wałęsa well understood, this mode of ending the opening negotiations in no way resembled the government's classic scenario. From Jagielski's perspective, the meeting had literally gotten nowhere, since its purpose was to get everyone back to work. But for their part, the workers had succeeded in conveying a great deal of their understanding of political reality in Poland to the country's deputy premier. More than a difference in reference points existed, there was an awesome difference in lived experience. Were judicial proceedings, in fact, rigged? Wałęsa said he *knew* they were because he had been present when it had happened in court. He could have added, though he did not, that he had been present at rigged proceedings as a defendant as well as an observer. He had, in fact, been arrested scores of times since 1970. When Wiśniewski and Gwiazda pressed for explicit guarantees that people would not be harassed by security police for expressing their legal right to voice grievances, they were speaking out of their own experience. They had often been abused for precisely this crime and by the same administrative agency, the security police.

Underlying the very idea of "negotiations" over the twenty-one Gdańsk demands were basic questions about the legitimacy of the negotiating process itself. Far from providing a measure of assurance, past history demonstrated that ordinary Polish citizens had no defense against the administrative power of the party and state apparatus—other than the temporary protection now afforded by massed thousands of strikers. Those memories, and the power relationships they enshrined, gave the meeting the distinct tone it possessed.

Many such moments of confrontation, recoil, and riposte were to occur through the subsequent week of frenzied maneuvering that accompanied the birth of Solidarność. They were to be succeeded by even more dramatic developments over the ensuing sixteen months and the long trauma of martial law that followed. All bear on the analysis of Solidarność and, indeed, are contributory to it. Nevertheless, the historically crucial ingredient of "the Polish August" were already in place on that Saturday evening when Jagielski entered the Lenin Shipyard for the first time. Indeed, far more was in place than he could have imagined at that moment.

The twenty-one demands the workers had accumulated ranged over wide sectors of Polish life. In the aggregate, they constituted a telling commentary on existing economic conditions, social relations, and the public policies that had engendered both. The allocation of national resources, production relations in the workplaces of the country, overall health conditions, and the underlying state of the culture itself all fell within the scope of the workers' commentary. Yet some of the demands did not appear to lie beyond fairly immediate remedy. They called for adjustments that can perhaps best be described as administratively simple while at the same time conceptually wrenching for administrators habituated to the authoritarian customs of a one-party state. Other demands, however, could not easily be implemented without instrumental changes in the society as a whole. Accordingly, they constituted serious sticking points.

But over and above this grim certainty, one of the workers' postulates—the first one—was so transformative, both culturally and politically, that it seemed to many outsiders to be utterly fanciful. One's perspective here controlled one's descriptive response: the proposal was alternately romantic, utopian, or insane. It was courageous but foolhardy. It could be interpreted, generously, as the product of understandable despair; less generously, as the product of adolescent yearning. As an appalling blunt commentary on clearly visible realities, the demand undoubtedly made a certain abstract sense. But, sophisticates argued, in modern states people did not relate to one another in the abstract; in concrete terms, the postulate was transparently revolutionary. Finally, in a purely tactical sense, its most ominous quality was the fact that the call for independent trade unions was not discretely buried in the body of the document or, better still, in an appendix or, even better still, in a footnote outlining possible future objectives. That is to say, it was not placed where it might more easily be watered down or otherwise domesticated through adroit bargaining by powerful adversaries. On the contrary, it had been stripped of softening linguistic accoutrements and presented nakedly as the first demand.

The first twelve words of the Gdańsk declaration challenged the basic structural component of the Leninist state. The workers called for "acceptance of free trade unions independent of the Polish United Workers' party." The strike committee publicly emphasized its centrality: "The first of our demands is of crucial importance. Without independent trade unions all the other demands can be ruled out in the future, as has happened several times in the short history of the Polish People's Republic. The official trade unions have not only failed to defend our interests, they have been more hostile to justified strike action than party or state organs."[21]

By August 23, the whole world outside the Soviet bloc had heard of the

Polish workers' demands for free trade unions. To Westerners who had managed to remain modestly conversant with global politics in the twentieth century, the intentions of the coastal workers seemed to be clear, if surprisingly bold. The idea of cooperative associations of workers outside the control of the employer had won a measure of sanction in Western industrial societies in the twentieth century—after many decades of brutalizing struggle dating back to the early nineteenth century. After the dust had settled, capitalists of both conservative and liberal persuasions had with different degrees of enthusiasm eventually gotten used to the idea of trade unions. Solidarność appeared to be cut from this familiar Western pattern. Obscured from most Westerners—during August and thereafter—was the fact that Polish workers had set forth upon a road that, unless they were sidetracked by force, led inevitably to a mode of democratic organization that was considerably in advance of Western-style trade unions. This was not at all clear on August 23, 1980, and in essential ways it still has not become clear. In any event, at the time, Westerners generally looked upon the efforts of Polish workers to form their own trade union as a familiar development.

The view from within informed circles in Poland—both in the government and in the opposition—was fundamentally different. To all those persons who had spent their lives in a Leninist culture, under Leninist systems of governance, communication, and production relations, and who had observed the relationship of the police and judicial systems to the maintenance of such "Leninist norms," the declaration by the Gdańsk workers represented a leap of the imagination that was hard to describe. The idea of popularly based, self-governing institutions in a Leninist state was something that one could conceivably *think about*, privately, but it patently was not something that could be put forward publicly—not if one wished to live within the confines of the received culture of People's Poland. Yet, now the idea had appeared, nakedly, as the first postulate of the Gdańsk workers.

Furthermore, Strike Presidium members had not only placed the concept of self-governing trade unions first on the agenda, they had issued a separate declaration underlining the demand as the centerpiece of all the worker proposals. Beyond this, they had singled out the government's most direct counterstatement—the declaration by the party's chief trade unionist, Szydłak, that "the authorities do not intend to give up their power or share it with anyone else"—as a repeated object of denunciation. On the first evening of negotiations, Florian Wiśniewski blurted out his indignation at Szydłak's pronouncement ("God in heaven, what a scandal!"). The Strike Presidium itself, in its first public explanation of the demands, had emphasized the implacable hostility to "all justifiable strike actions" traditionally manifested by the official trade-union structure, which Szydłak headed. The message was clear, even if most outside observers, including the government, had trouble believing it; the old trade-union structure simply had to be obliterated.

In short, Gdańsk workers were not engaged in issuing a bold demand and then demonstrating a gracious willingness to settle for something less; on the contrary, they had seized every opportunity that presented itself to make certain the party hierarchy understood that the proposal could not be talked to death or otherwise circumvented either through Szydłak-style bluster or Jagielski-style circumlocution. Over and over again since the demands had first been formulated, Gdańsk workers endeavored by words, extemporaneous gestures, and collectively planned actions to convey their message to the party: we are contending in a new way; the sophistry of old formulas about "consultations" and "partnerships" will no longer do. The shipyard worker who instructed the prefect of Gdańsk to "look straight into the eyes of the workers" was in his own way conveying this sensibility as fully as Wiśniewski, Gwiazda, or Wałęsa were endeavoring to do in the formal negotiations. A new idea had somehow appeared on the agenda of public discussion in Poland—a profoundly democratic conception of self-governance through self-organization.

If the result was clear, however, the causes were not. How did all this come about? At first glance, the sequence of worker assertion appears quite straightforward. The occupation strike had begun on August 14, an Interfactory Strike Committee had appeared on the sixteenth, the twenty-one demands announced on the seventeenth and adopted by the Strike Presidium on the eighteenth; the centrality of the first demand had been emphasized publicly on the twentieth (though censored by the government) and then printed at the first opportunity in the initial Solidarność strike bulletin. The stages in this evolving sequence are not difficult to grasp. The real question concerns a development that presumably had to have occurred earlier in time than any of these events—namely, how and when was the constituency for the first demand created? Overnight—that is, spontaneously? Was it perhaps created over the course of the difficult summer of 1980 when the economic crisis deepened—that is, was it a product of "hard times?" Could this high level of worker consciousness have developed over a longer period, many months or even many years? The answer to this question begins the process of describing how Solidarność came into being.

But first, a counterquestion: why should it really matter? Is it, in fact, a real question? Is it not a reasonable proposition to suggest that people living under the tyranny of a brutal and singularly malfunctioning oligarchy simply got "fed up" and rebelled? This suggestion has at the very least the unarguable merit of being time-honored. Scholars have for many decades employed this reasoning and this language to explain moments of political fracture. There is, of course, a second time-honored suggestion, in Marxist reserve, so to speak. Each of the developing workers' assertions from August 14 through August 23 may be understood as a benchmark in the "rising consciousness" of the coastal workers. This alternative language of description has been available for historians of all persuasions ever since Marx originally demonstrated its utility in *The Eighteenth Brumaire of Louis Bonaparte*, first published in New York in 1852. So, this descriptive terminology bears the weight of tradition also.

The difficulty with both propositions is that they do not explain anything. Upon inspection, it is apparent that both are, in fact, tautologies that can be easily exposed by asking questions grounded in the terminology of each. Did the workers rebel because they got "fed up," or is the fact of the rebellion the essential proof that they were "fed up"? Did the workers go through an alteration of consciousness in order to progress from a factory strike committee to an interfactory committee, or does the reality of the progress provide the essential proof that their consciousness altered? Concretely, what does "fed up" mean in terms of predicting political action? As a term of analysis, is "rising consciousness" any more concrete than "fed up"?

The two propositions have other weaknesses. In Poland, as in the Soviet Union and Eastern Europe generally, the idea of self-governing trade unions seemed to be in August 1980 a new public idea—a very large, very serious idea with profound structural implications for society and the state. In the course of history, enough large and serious political ideas have appeared from time to time to generate a considerable body of historical evidence as to their fate. Let us broadly consider this evidence as it might bear on the task of evaluating the Polish democratic movement.

All serious new political ideas, when they first appear in any culture, are, by definition, very surprising. Unsurprising ideas are those with which everyone is familiar; that is, those ideas sanctioned by the received culture. Serious ideas that challenge that culture are, perforce, beyond sanctioned limits. They are, also by definition, culturally and therefore psychologically unsettling inasmuch as they contradict what one has been educated to consider "normal." The first Gdańsk demand is a decisive juncture in the development of worker consciousness. In August of 1980, Polish shipyard workers somehow gave themselves permission to assert publicly a basic aspiration that other people in Poland could not bring themselves to assert. How did the shipyard workers achieve such a level of self-possession? By what process? Was it—is it—something finite and tangible and therefore exportable to people who do not speak Polish? Would locating this process, if it exists, and discovering its sequential dynamics, if it possesses any, prove relevant to the study of history or of democratic aspiration in history? Would it help clarify how popular democratic forms come into existence in modern industrial societies—both institutional forms and the energizing cultural dynamics leading to the promise of such forms?

Still other questions immediately occur. In the Gdańsk shipyard, was this self-permission—this consciousness—a relatively private possession, one that was restricted to "high-consciousness" worker activists who sat on the presidium of the strike committee? Or was it a possession held by shipyard workers generally? Or—since several hundred thousand of them had affiliated with the MKS—coastal workers generally? Such questions seem to direct attention backward in time in quest of some process of experiental self-education that took place in the months or even years leading up to August 1980.

To focus the task a bit, it might also be useful to determine what other Poles who were not shipyard workers thought in August 1980 about the idea of creating independent and self-governing popular institutions in People's Poland. It is abundantly evident that the party hierarchy did not take to the idea with positive zeal. Jagielski's evasiveness on such matters as the censorship, political trials, and even the telephone blockade simply verified the persistence of an entrenched attitude of party governance. Though Jagielski was at home with the idea of party-organized trade unions, his affinity did not extend to independent ones or to self-governing ones. In party verbiage hoary from ancient use, he talked about the need for the official trade unions to have "greater influence" and encouraged presidium delegates to look upon themselves as "natural activists" who, as "partners" of the government within the existing trade-union structure, should be encouraged to "consult with the authorities." A certain rhetorical flourish served as embroidery: "As I personally understand it, I repeat, I am speaking quite sincerely, the underlying purpose of this proposal is that trade unions should be the real, authentic and effective representative of the interests of all employees . . . [and] in accordance with the contemporary requirements and demands of the work force. The practical first step in this direction will be the drafting of a new law on trade unions." Jagielski endeavored to filter a measure of jocularity into what were essentially a series of remarks without content. "The present one . . . I discover . . . is more than thirty years old. Working conditions have changed and so, evidently, have demands. A law must be drawn up which will meet the present requirements of working people." He also attempted a bit of flattery: "I notice on the strike committee with whom I am negotiating a great number of, so to speak, new people, with genuine authority and real talent: natural activists. Why shouldn't such people join factory councils or unions at some level? But it's up to you, to your opinion on this matter." He then returned to remarks without content: "So, I express my sincere conviction—I don't know whether we agree on this, but I express my own profound personal conviction—that trade unions should represent the interests of employees and defend their rights more effectively. Talks between trade unions and the authorities on these matters should take place on a regular basis . . . a dialogue between partners." His conclusion was something less than a high note: "There will be wide public discussion of this draft in which factory work forces will participate."[22]

It is possible to conclude from these remarks that Jagielski's democratic drives were not well developed. The deputy premier was, rather, a traditional politician. His task at Gdańsk was to defend a regime based on police terror by adopting the posture that no such thing existed. Within this rather stoutly confined reality, the deputy premier maneuvered as best he could. He lacked the authoritarian zeal of prominent Warsaw hard-liners, but he also lacked the democratizing leanings of the embattled and largely ineffective band of functionaries commonly described as "party liberals." He was, easily enough, a servant of the party mainstream: control, through evasion if possible, through force if absolutely necessary; but in all events, control. He was an appropriate representative of a Leninist party. He took back to Warsaw from the first negotiating session some ominous news for the Politburo. The Strike Presidium was laced with militants who seemed both articulate and patient. Worse, the MKS now bulged with delegates from something over four hundred factories from across the whole of northern Poland. The site of the talks was surrounded by a veritable army of workers interspersed with literally hundreds of reporters and photographers from dozens of countries. Worst of all, the MKS seemed absolutely unmovable on the demand for a union independent of the party. In all the years since the Communist accession to power, the party had never had to face such an organized social formation. The party leadership in Warsaw agonized in indecision.

Clearly, the Polish party did not know what to make of the coastal mobilization. Did the nation's intelligentsia? Intellectuals had not been a relevant factor in the tense period between the first surfacing of the strike on August 14 and the government's decision eight days later to experiment with the negotiating process. But late in the evening of August 22, two representatives of the Warsaw intelligentsia appeared in the shipyard bearing a formal statement of support from many of their colleagues in the capital city. The two visitors were Tadeusz Mazowiecki, an editor of the Catholic journal *Wież*, and Bronisław Geremek, a historian at Warsaw University. Their arrival in the shipyard was the culmination of two days of frantic activity in Warsaw, during which time Mazowiecki had rallied representatives of three groups of the organized intelligentsia for the purpose of generating some sort of positive intervention into the crisis on the coast.

The three groups and their appropriate Polish acronymns were the Warsaw Club of the Catholic Intelligentsia (KIK); the activists of the Society for Scientific Courses, better known as the boldly experimental "Flying University" (TKN); and a party-tolerated group of reformers known by the title of its most prominent report of policy recommendations, "Experience and the Future" (DiP). In the complex world of Polish politics, members of all three groups had at one time or another been harassed by the authorities, yet all had a certain fragile legitimacy, traceable to their intellectual, social, or political standing in the nation. They set about the business of writing a collective letter to both the government and strike leaders suggesting that the crisis be settled by negotiations. Since the government was studiously ignoring the Gdańsk MKS and trying to settle strikes with individual factories, this intervention by the intellectuals constituted an unwarranted interference with the leading political role of the party. Working literally one jump ahead of the security police, they hurriedly generated an agreed text with three-score signatories.[23] What they produced was formally designed as the "Appeal of the 64." It offered public support and solidarity at a time when the Baltic workers were standing alone against the state. As such, the declaration endangered every one of the signatories whom Mazowiecki gathered in his apartment on August 21.

It was not a random group. In addition to editing *Wież*, Mazowiecki had been a leading figure in the Clubs of the Catholic Intelligentsia, which had provided some intellectual foot troops for both the venturesome "Flying University" and the much more "respectable" DiP grouplet. Comprised of an unofficial faculty of independent academics and intellectuals, the "Flying University" taught uncensored versions of Polish history and discussed unsanctioned ideas in philosophy and the social sciences. It functioned at night on the run—"flying"—in a rotating series of apartments throughout the city. The more cautious DiP group, on the other hand, brought together an assortment of party reformers and independent intellectuals with a view to broadening the agenda of discussion and analysis within official circles. Mazowiecki thus served as a kind of Catholic bridge between sectors of the democratic opposition and the party's aspiring reform wing. Out of the meeting in Mazowliecki's apartment came a decision to send representatives to deliver the appeal directly to the Baltic coast. Mazowiecki and Geremek accepted this assignment and left for Gdańsk the following evening.[24]

In the shipyard they made contact with Lech Wałęsa. He was immediately engaged in an extended discussion over the content of the first demand. Both Mazowiecki and Geremek had brought with them from Warsaw "grave doubts" about the feasibility and political credibility of the proposal on self-governing trade unions. Such incredulity was inevitable. The thought that unions independent of the party could be realized in the Gdańsk manner—through a strike of production workers—was not something that Warsaw intellectuals had ever really anticipated. So remote was the world of the workers from the milieu of the intelligentsia that the topic of broad-scale industrial activism to create independent social structures was off the scale of discussion. In the new political context created in Poland by the coastal mobilization, this way of seeing became an absolutely central circumstance. For not only was Solidarność a trade union, it was a singularly adventurous conception of a trade union. Warsaw intellectuals were almost totally uninformed on the subject.[25]

This is a crucial point of analysis, but one that cannot intelligently be further explored until much of the story of the Baltic mobilization has been set in place. For the movement, it is sufficient to note that, as of August 23, 1980, when intellectuals appeared ready to move into coalition with workers, the situation housed organically divisive ingredients. The "grave doubts" of Mazowiecki and Geremek concerning self-governing trade unions were a product of their political experiences in Warsaw; the contrasting view of Wałęsa was a direct culmination of his experiences as as worker activist. Wałęsa's world was invisible from Warsaw, and this reality produced diametrically opposed reactions to events on the Baltic coast.

For Wałęsa, the scenes in the shipyard on August 23 constituted an enormously gratifying verification of years of effort by himself and other coastal activists. The victories were everywhere to be seen. The MKS presidium was composed of men and women who were steeped in experience in dealing with the party in the factory and the state in society. In the full MKS, the militants from each of 370 enterprises similarly represented a strong cross-section of working-class knowledge. As Wałęsa knew, delegates who were elected by workers to come to the Gdańsk shipyard to represent them on the MKS tended to be the very ones who had proved by their conduct over the years that they "knew the score" on the shop floor in their own enterprises. Each day of the strike, a delegate from an enterprise would travel back to the plant to provide a progress report on the strike. Sustained effort was made to keep all workers in the 370 affiliated enterprises informed in the certainty (it was movement wisdom that such was the case) that information was the essential key to continued solidarity.

Nowhere was information more fully disseminated, of course, than within the Lenin Shipyard itself. But, here again, the relative sophistication of the shipyard workers was not the product merely of the preceding nine days. They possessed knowledge and experience that they did not have when the coast first went up in flames ten years before. As a result of this knowledge, they were able to mobilize the patience that was missing in 1970. Only veteran organizers knew that "patience" was a quality that could also be described by two other words—"sustained militancy." As Wałęsa surveyed the shipyard on August 23, things looked steady. Even the older married ones with heavy family responsibilities had found strength in the group and had settled in for the long pull. They were coming up with a number of good ideas about food distribution and security precautions and they had a treasury of knowledge about militants in other factories who could be contacted. Their experience was a positive benefit at this stage. Everywhere workers had things under control; they had come a long way since 1970. One thing was absolutely clear: they would either get it all this time or they would get the tanks. Given the situation in Poland, this was the very best choice the nation had had since the end of the war. Anyone who looked with care at the scenes in the shipyard could now see how clear it had all become—at last.

But—early on—Mazowiecki and Geremek did not know how to read the scenes in the Lenin Shipyard. That first night, they could not surmise what the throng of delegates on the MKS meant in terms of work stoppages across the whole of northern Poland. They did not, in fact, see either workers or work stoppages.

What they saw on the Baltic coast was the state. Most immediately, they saw the guns of the state. They saw the horrifying prospect, yet once again, of the spectacle of guns and blood that had drenched the Baltic coast in agony in 1970. They knew the exact sequence: first, workers dying in the Lenin Shipyard in Gdańsk, then dying in the Paris Commune Shipyard in Gdynia, and finally workers dying in the Warski Shipyard in Szczecin. And they knew that the revolt this time had once again begun in Gdańsk, had spread to Gdynia, and then had reached Szczecin. Indeed, it had been this deadly parallel that had generated in the intellectuals the will to create the "Appeal of the 64" back in Warsaw and had imparted to the statement its controlling message to the authorities: do not crush, negotiate. From the very first nine-word sentence ("The present moment may be crucial for our country"), the intellectuals pled for restraint:

> Everything now rests upon which way out of the present crisis is chosen. We appeal to the political authorities and to the striking workers to choose the path of negotiation and compromise. Nobody can be allowed to resort to lawlessness or acts of violence nor to throw down any challenges. The tragedy of ten years ago must not be repeated: there must be no more bloodshed.[26]

Every word in this litany of anxiety reflected the humanistic heritage of the Polish intelligentsia. It also reflected a subtle awareness of the rhythms of governance and the patterns of paranoia that had so often led the regime down the path of violent repression. In its immediate social purpose, the "Appeal of the 64" expressed the kind of engaged wisdom that a heritage of tragedy can instill in thoughtful people.

But this was also the statement's weakness as a political act. Its authors had gone too far in speculations about state power; a social presence other than the state was visible in the Lenin Shipyard. Unable—at first—to perceive the social formation that had come into being on the Baltic coast, they could not divine the seminal substance in the constituency's goal; this dynamic led them inexorably to shrink from the idea of self-governing trade unions. On the practicality of the first Gdańsk demand, their reasoning took the following trajectory: the idea of free trade unions was a structural goal that could come into being only gradually; one could not simply pretend that the Leninist state was not there; therefore, one could scarcely endeavor to move directly to the immediately intended goal. Politics did not work that way. Power had to be taken into account.

It was this perspective that arrived in Gdańsk from Warsaw in the early hours of August 23 and though it would change with great speed (Mazowiecki and Geremek were to prove they could quickly adapt to new political realities), it was with this original perspective that Lech Wałęsa had to cope when he sat down with Tadeusz Mazowiecki and Bronisław Geremek.

There was little Wałęsa could do in one evening to explain convincingly that power was, in fact, being taken into account and that the first Gdańsk demand represented all that the workers had learned in their years of intimate contact with the party and the state. Such could not be explained because a shared context did not exist between the three men. Where could Wałęsa begin? The student demonstrations of 1968? Would it be helpful for him to outline the specific details of what he and Lenarciak and the others had learned from the way the party and the party's trade union had filtered images of the 1968 student protest to the workers in the Lenin Shipyard? Or what they had learned in subsequent experiments to counteract such tactics? What could he tell them about the first principles of state power that became starkly clear during the worker rising of 1970 and the truly decisive things learned in 1971? What could Wałęsa tell them about the politics of the shipyard monument in 1971 and 1972; about the severe repression in 1973; about how they subsequently experimented in countering such modes of state control in 1974, a year punctuated with another strike in Gdynia and another round of police-station beatings? None of these experiences were a part of the shared heritage of the Warsaw intelligentsia.

In the aggregate, the first demand, calling for free and independent trade unions, was a product of all that coastal activists had learned about organizing a popularly based democratic movement in Poland. This included not only the tactical lessons of the period from 1968 to 1975—touchstones of experience that were important in dealing with the working class and the state—but also the lessons of 1976–77 (unknown in Warsaw) and the particular ways coastal activists had built upon them in 1978–79. These experiences were instructive for working-class activists in dealing with their fellow workers, the state, and with society as a whole. How could he fit such complexity into a defense of the first demand? The task would require that he teach them all he knew about organizing. How could anybody from Warsaw be expected to know about the commuter train from Sopot and Gdynia or any of the trains, trams, and buses where the movement had grown and taken on life? And what about movement politics at shift changes with party functionaries watching closely: could Warsaw intellectuals be expected to be alert to the sundry uses of shipyard locker rooms?

There was also the matter of the movement's people, men like Szczepański and Hulsz, the ones who had been killed or dispatched to remote provinces and the "old ones" from the 1950s whom veterans like Mośiński and Lenarciak could remember. What could be said about unknown activists like Nowicki and the many others who had been fired so many times and harassed for so long that they could barely hang on? What about the ones with steel in their eyes whom the authorities would never intimidate, men like Bury, Sobieszek, Felski, or Gwiazda? Or the ones who listened and nodded and said nothing, but who had been right there on August 14—in the morning, early, when it all either happened or it did not? What about people like Wiśniewski, whom Wałęsa had recruited at Elektromontaż, or Lis, whom Gwiazda had recruited at Elmor. Would they understand what it meant to have people with that kind of experience on the Strike Presidium? How could anyone who was not from the coast realize how much proof their experiences provided for a union free of the party?

But, none of this touched the deepest level of all—the real mass base of the movement. What could Wałęsa tell Geremek and Mazowiecki about the workers' quarter named Stogi? It was ostensibly under the thumb of the party like every place else in Poland, but in reality, it was Wałęsa's terrain. He was by a wide margin the best-known worker there. Stogi provided foot troops for the movement. But the Warsaw intellectuals probably did not know where Stogi was and, in any case, could not know about workers learning how to distribute leaflets and tracts or being schooled in handling the inevitable police interrogations. Could Wałęsa tell them how one acquired experience in these crafts? Would they be impressed if he told them he was one of the better leaflet men and had paid for his experience in numerous police detentions? Or that he knew from the inside all the grouplets in the movement milieu—KPN, Young Poland, ROPCiO, and FTU—and therefore could reach beyond the working class for talent and recruits in ways not customary for working-class activists? Could they fully appreciate the activities Szczudłowski, Niezgoda, Borowski, or so many others, who had become accomplished couriers and were now bleeding in detention cells all over northern Poland?

Could Wałęsa explain to them the importance of the preconditions the Strike Presidium intended to put to Prefect Kołodziejski that very afternoon? Would they think the purpose would be simply to let the world know about the bloodletting of workers by the workers' state, or would they understand that the preconditions were designed to protect the organizing lifeline of the movement? Would they understand that couriers were a central communications link for a movement in a country where all phone lines were cut and highways blockaded? Could they grasp the range of the movement—that Szołoch, Felski and Szczudłowski provided places where the militants met; that Sobieszek brought to the Strike Presidium of 1980 his experience as a spokesman for the Strike Presidium of 1970, that Pieńkowska ran the food delivery operation in the shipyard like a movement veteran because she was precisely that; that the movement recruited physicians like Słominski, who had been twice arrested and twice beaten for performing courier duty? Could they guess what kind of intimate knowledge about everybody's talents and nontalents that service in the day-to-day life of the movement imparted to activists (the knowledge that Krzysztof Wyszkowski had a subtle grasp of the respective priorities of the movement and the party but that his transparent anticlericalism tended to alienate people the movement needed to recruit; that Borusewicz had intelligence and imagination and often a fine sense of timing but that he was more at home with ideas of conspiracy than with the demands of rank-and-file organizing; and that Gwiazda was articulate and aggressive and, despite his desperate drives, had found a way to live with his depressingly small talent for organizing at the grass roots)? Could they recognize that Anna Walentynowicz was a down-to-earth symbol of what the Baltic movement was all about—a 51-year-old crane operator with years of movement experience, so that she knew almost as much about the rhythms of production-line harassment as she did about the motions of giant cranes? Could people from Warsaw know why buses in Gdańsk came to a stop when Henryka Krzywonos of the transport workers gave the word?[27]

There was, in fact, no way Mazowiecki and Geremek could know these things, no matter how learned and trustworthy they might be. They simply could not know the richness of experience represented on the presidium—not only Krzywonos but also Gwiazda, Kobyliński, and Wiśniewski; not only Sobieszek but also Walentynowicz, Przybylski, and Lis; nor could they understand the presidium's influence with rank-and-file workers, represented by the growing army of delegates from all the supporting factories—not only the 370 strike committees affiliated with Gdańsk but also the Interfactory Strike Committees in Szczecin and Elbląg that had their own proliferating affiliates. It was all too much, too long in the making, too subtle in its assorted contexts, too alien to the world of Warsaw intellectuals. The first demand was not just the first demand; it was the absolute centerpiece of the struggle with the party. Everyone on the presidium knew that. In the early morning hours of August 23, Mazowiecki and Geremek seemed to Wałęsa to be Poles to the core, maybe even great Poles in the historical sense. He hoped so; they would all soon be put to the test by events in any case.

Such was the "evidence" that Wałęsa could have presented to the two visitors that first evening but did not, because the organizing experience had no shared context between them and because at long last a review of the past was no longer necessary.[28]

The idea at the heart of the Polish movement—and, indeed, all serious democratic movements—was that broad sectors of people without power could join together to have a genuine say about those things that affect them the most. It is an immensely appealing idea but an embattled one, because people do not generally believe its goal is possible beyond very narrow limits and, therefore, are afraid to run the risks necessary to achieve it. It is not simply "fear" or "apathy" that immobilizes people; it is knowledge they have been taught about how power works in their lives, in their jobs, in their neighborhoods, and, indeed, in most of their social relations. They desire change and yet do not believe they can get much of it.

To understand this equivocal tension and to respond to it is the first task of authentic democratic politics. There are, unfortunately, many ways to avoid confronting this very human circumstance and the far-reaching political implications it carries. One can condescend to persons whose conduct reveals they suffer from this circumstance, this tension; and they can thereupon be dismissed for their apathy or their inadequate consciousness or their "peasant mentality." Even more overtly political terminology can be employed to characterize their cautious actions as those of "moderates" or "reformers" who cannot summon the will to face the historical trap in which they are caught. One can elect to understand them as mystified victims of effectively applied elite propaganda or socialization. In appropriate historical situations, all such appellations may be reasonably accurate as pure description of a practical result; indeed, the more compelling the (largely unknown) supportive evidence, the more effectively such description will conceal the concrete historical causality that applies. All such terminology is grounded in a historical conceptions of classes and strata and, as such, achieves the rare descriptive feat of being simultaneously rigid and vague. All such formulaic approaches slide past the equivocal tension itself and, thus, in characterizing a result, inadvertantly pass over the essential particulars in the process of self-activity and movement building. At bottom, such terms of description—remote, abstract, and condescending as they are—conceal historical causation. History quite literally disappears beneath the illusion generated by their suggested meaning.

Aspiration conjoined with anxiety—it is the essential political tension of modern society, as unwanted coupling deeply rooted in subjective experience and one that surfaces in many guises every day in every human being and in every public political action. A relatively simple condition to describe, it is also a pervasive condition, and its endurance through the generations mocks the pretension of much that passes as high political theory in the late twentieth century. But to understand this equivocal tension and to respond to it is the starting place for democratic politics—a juncture that marks the point at which democratic organization must begin. This understanding and the ability to act upon it represents the essential preconditions offering the possibility of altering structural and psychological stasis through democratic activity.

For a number of reasons that informed his political past, a Baltic worker named Lech Wałęsa understood the general shape of this organizing task as thoroughly as any other activist on the coast. Concretely, he understood how and why the coast had been organized—once—in 1970. On the day of the strike, it had been shaky. Though the workers had measured up, communications between factories were sadly deficient, and things had gotten badly out of hand. They had made some crucial mistakes and had lost everything. But they had learned many things about organizing in December 1970; and more about the party in January; and still more about the state over the succeeding year. Edmund Bałuka was not the only one who had learned the great truth of 1970–71; "We now know how to go on strike. We don't know how to win a strike."[29]

Now they did. They had gone to school on their own failure in 1970–71. In ten years of effort, they had discovered what they still needed to know and had proceeded to learn it. "Monument politics" had proved a very instructive schoolroom throughout the 1970s—as had Stogi, the workers' council of 1971–72, and the years of leafleting and organizing in the shipyard and in the supply enterprises like ZREMB, Elmor, Elektromontaż and others in the city of Słupsk. As Wałęsa understood things, the veterans of 1970–71—people like Sobieszek, Walentynowicz, Lenarciak, and himself—and the ones who had come along since 1970—the Gwiazdas, Borusewicz and Felski or Bogdański, Kołodziej, and Pieńkowska—had collectively pulled it off in 1980. It had been shaky again, as it always was in the beginning, but after settling in, the workers had been very solid. Fear was down to a manageable level; they had come to believe in their own effort. They would remain solid if they were kept informed of everything important that happened; of that Wałęsa was certain. They had made but one mistake, a failure of anticipation, and the whole thing almost collapsed; but they improvised frantically, and in the end, they survived the error. So now they were over the hardest part. They were going to get the union or the tanks. But it had to be a union free of the party. It had taken ten years to learn how to get within range of this objective, but now they knew. And their friends, the two good and intelligent men from Warsaw, did not yet know. But they could be tactically helpful in any case. As long as they understood that the workers' Strike Presidium would retain full control of the negotiations, they could be useful in checkmating the platoon of government technocrats they would probably have to face.[30]

So, that first evening, Mazowiecki and Geremek heard all the reasons Wałęsa could mobilize as to why the first demand could not be watered down; they heard little about how anyone on the Baltic coast had come to learn such things. But they performed a politically essential act: they listened with respect. They came away from the meeting impressed with Wałęsa's assertiveness, his poise, his confidence, and his knowledge of both the work force in the Lenin Shipyard and the police and party authorities on the coast. Of his intelligence, there could be no doubt. But they also experienced a firsthand demonstration of Wałęsa's handling of the Polish language. A classically educated man he was not.

But if the two men from Warsaw were probing Wałęsa in an effort to understand the situation on the coast, the strike leader was testing the two unknown intellectuals who had appeared out of nowhere with their statement of support. They said they had come to help and, under questioning, that they would stay through to the very end. That was straight talk, the kind workers wanted to hear. They also seemed to have a certain poise in talking about the state. In order to cut the giant down to size, Wałęsa always referred to the party-state as *władza*—the power—so as to strip it of any moral or ideological claims. He habitually reminded his listeners how one part of *władza* contradicted another part, so Poles could think of it as a confused and human institution rather than an abstract police monolith.

Did either of the two visitors have a case of the *władza* disease? It was too early to be certain, but the indications were they did not. It was possible to read their statement in a way that created some doubt, however, The concluding summary of the "Appeal of the 64" contained a certain unresolved tension that blurred its meaning: "Only caution and imagination can today lead us to an understanding in the interest of our common fatherland." Prudence was demonstrably necessary—and certainly imagination. But both could be vitiated by an excess of fear masquerading as caution. In worker terms, the point could be made much more simply: people needed to know what they were doing at all times. The two men looked sincere, though the case for them was not a prima facia one. The workers, in any case, were not overrun with allies.[31]

The next morning, Wałęsa saw that his two guests were introduced to the ideas of other members of the presidium, an opportunity both Geremek and Mazowiecki exploited with some vigor as they had been unprepared for Wałęsa's firmness the night before. Conversations with Andrzej Gwiazda from Elmor, Henryka Krzywonos from the Gdańsk transport workers, Lech Sobieszek of Siarkopol and other presidium members merely confirmed, however, that all of them, not just Wałęsa, were implacably committed to the first demand. The intellectuals, in any case, intended to be a force on the workers' side, a mediating force to assist in brokering an agreement that would both advance the culture and end the crisis short of guns.

Wałęsa had brought a different set of presumptions. He had met Deputy Premier Jagielski for the first time in downtown Gdańsk only three hours before he had met Geremek and Mazowiecki at the shipyard. There were going to be negotiations at last, and Wałęsa understood what that meant. He had been on the Strike Presidium in the 1970–71 crisis, had talked to and been flim-flammed by First Secretary Gierek and had engaged in "negotiations" with the party, the police, and the state ever since. He had been arrested scores of times in the course of these relationships. Until the presidium delegates left their first planning meeting in Gdańsk with Jagielski late in the evening on the twenty-second, there had always been the risk that the government would elect to settle the strike by resorting to the police truncheon. Now that they were going to sit down at the negotiating table, the government would try to drown them in facts, ostensibly drawn from a mountain of statistics from half a dozen government bureaus. Wałęsa had watched that process in 1970–71. The workers could use some experts of their own—if they could find compatible ones—to help sweep the government's facts aside and clear the way for serious conversation. The two men from Warsaw seemed compatible. Geremek appeared quiet, thoughtful, and closely attentive. As for the equally attentive Mazowiecki, he edited a Catholic journal, so the Church would like him. That should help. It was far too early, in any case, to tell the more subtle strengths and weaknesses of either man. They were present, and that certainly counted for much.[32]

The problem they represented—one that Wałęsa knew, and they did not—was the simple fact that the prospect of academics associated with the presidium would not be an easy proposition for the strike leadership to accept. As most workers saw matters, intellectuals could not be relied upon in a showdown. Under the guns of the Soviet army, communism had been institutionalized in Poland by intellectuals. They had profited as a class under it much more than workers had, and it had taken intellectuals much longer to see through the party and go over into opposition. Most of the time, it was hard for workers to tell whether intellectuals were in opposition or not, since they equivocated so much. As a ship welder put it, "They tend to blink at decisive moments and cover their tracks with a lot of words." The Gdańsk negotiation was going to be the most decisive moment of their lives as workers, a struggle against very long odds. Thanks to their own efforts over many years, the odds were better than they had ever been before; they had 370 enterprises out on strike as of August 23, and the government's maneuvering room had been sharply narrowed. But there was no room for error or for "blinking at decisive moments." Beyond this, workers felt college professors, in general, had a style of watchful reserve and a habit of being cordial in a polysyllabic way. Such customs made workers uneasy and sent an unintended signal: professors could not be trusted. There was a final intuition among workers, widely held and one that would have appalled the progressive intelligentsia had they sensed it. The tendency of intellectuals as one worker put it, "to talk around things and make them sound more complicated than they were reminded everyone of high party officials."[33]

Under the circumstances, it is perhaps not surprising that Wałęsa's proposal for the creation of a panel of advisors did not receive instant approval from the activists who comprised the presidium. The feeling among many was that they had not labored for so many years and come such a long distance merely to turn things over at the critical moment to a group of Warsaw professors. No one was suggesting that they do so, but they intended to exact every possible guarantee that such was, in fact, not the case. The reasoning behind this caution or this intransigence was richly textured. The experience of workers in dealing with state power over many years had vividly demonstrated the subtle variety of ways in which the party apparatus could blur meanings and alter forms in the course of domesticating any threat to its absolute authority. To anyone who could think clearly, it was obvious that the sole potential countervailing power to the party-state was the ability of the nation's workers to bring the economy to a halt through a massive strike that could open up a renegotiation of the country's social contract. The daunting prerequisite of a large-scale work stoppage had now been accomplished. The coast was shut down from Szczecin to Elbląg.

But to capitalize on this rare achievement required more than intelligence armed with formal mastery of language or specific academic disciplines; it required the experience of activists. Workers had seen many "new" forms of worker control sabotaged by the regime: after the war, after the Polish October of 1956, and after the great worker rising of 1970–71. They understood the structural forms of this struggle on the shop floors of Poland. Above all, they knew that government negotiators like Jagielski were very skilled in fashioning a verbal reform that seemed reasonable to reasonable people but which, after due manipulation, turned out to be one more means to preserve inherited power relationships. The worker activists knew this—because it had happened to them more than once. To them, therefore, intellectuals seemed "reasonable" in precisely the way the workers themselves had been reasonable earlier in their careers, and that is what worried the men and women on the Strike Presidium. The task at hand in the upcoming negotiations was to create an institutional form of worker organization that possessed sufficient clarity, strength, and independence to make as difficult as possible all subsequent government attempts to destroy it. Without such an institutional form to provide protection for the working population, all other reforms in the Gdańsk proposals, even if accepted by the government, would be helpless prey to future manipulation.[34]

Wałęsa, a Baltic activist for twelve years, fully appreciated the concerns of his fellow activists on the Strike Presidium. He was able to assure the committee that only advisory roles were projected for the intellectuals and that negotiating decisions would be made by the Strike Presidium itself. Once it was verified—clearly verified—that everyone including the Warsaw contingent understood these ground rules, the presidium moved, unanimously in the end, to accept Wałęsa's proposal for an advisory panel of intellectuals.

The shipyard workers themselves and the delegates from 370 affiliated enterprises also posed a possible obstacle to the intrusion of intellectuals into the negotiations, even simply as advisors. Wałęsa presented the issue to the shipyard rank and file over the loudspeaker: "In connection with—so that we shall all benefit—we must call into being a group of advisors. Well we must be, in sum total, sufficiently clever and good, What? So we have done this. Were we right?" This extracted a round of applause which, given the somewhat cryptic explanation, essentially expressed the delegates' confidence in their own leadership. Mazowiecki, having clearly received the message sent to him by the presidium, did what he could to help by assuring the workers: "Our role is purely advisory. All decisions will remain in the hands of your presidium." This, too, got a big round of applause.[35]

While Mazowiecki performed this chore, Geremek was directly by members of the presidium to the only available telephone that was connected to Warsaw, located in the office of Klemens Gniech, the much subdued director of the Lenin Shipyard. By long distance, Geremek recruited his team of advisors: economists Tadeus Kowalik and Waldemar Kuczyński; philosopher Bogdan Cywiński; sociologist Jadwiga Staniszkis; and Andrzej Wielowieski, leading spokesman of the Warsaw Club of the Catholic Intelligentsia. Though distressingly short of what the workers most wanted—legal expertise—it was a talented group. When they arrived in Gdańsk the following day, however, the issue of the first demand flared anew. Everyone in the group was centrally troubled by the first demand. Mazowiecki and Geremek endeavored to assure them that the workers were adamant on the issue, and Mazowiecki, perhaps sensing a presumptuous attitude lurking behind the repeated questioning on the issue, reminded them all that their advisory status did not permit them to alter the content of the strikers' demands.[36]

Geremek suggested the newcomers form their own opinions after talking to the workers, something he had already taken the precaution of doing on his own. Even though all of the academics acted on Geremek's suggestion and discovered, as one of them reported, "that the workers' resolve was too great to admit of any concession," some of the group felt they had an obligation as advisors to develop contingency plans that placed in reserve some milder versions of the proposal on free and independent unions. They discussed the matter among themselves and then, convinced, arranged a meeting with the Strike Presidium. The academics presented a short paper as a way of opening discussion on the question of "whether the strike committee would be satisfied with a radical reconstruction of the old trade union." They called their plan Variant "B." To their surprise, the strategy was categorically rejected by the Strike Presidium. The economist Kowalik, somewhat chastened, reported that "all who spoke opposed it." Kowalik later conceded that "during the course of a whole week's negotiations, I did not meet a single striker or delegate who was willing to compromise on this issue."[37]

So instructed (though some of them were still not convinced), the MKS intellectuals sat down with their academic counterparts on the government's side. The first meeting of the "working group," as they elected to call themselves, took place on August 26 in a "half-relaxed atmosphere" that contrasted sharply with the engaged tension surrounding the main negotiations between the Strike Presidium and Jagielski. For one of the participants—Staniszkis—the professorial camaraderie in the working group reflected, as she later wrote, their long association in connecting social circles back in Warsaw. Despite this suggestion of conviviality, the issue of self-governing trade unions dominated all discussion, as it had in the primary negotiation. Citing the relevant legal statutes, the government team adopted a very technical position in support of the proposition that a new union could only be registered under the existing centralized trade-union structure. The MKS advisors tried for a time to engage on this technical level—a most doubtful and even dangerous procedure. However, the MKS delegation also included three worker members of the Strike Presidium, led by a longtime coastal activist named Andrzej Gwiazda. He was joined by his two worker colleagues in a united front against the shared constraints embedded in the acceptance, by both teams of academics, of an artificially "legal" discussion. In an effort to redirect the conversation toward matters of substance, they pointed out that a way needed to be found to create independent unions, regardless of the statutes. Create new statutes and form the union, they said, or the MKS will simply "exist and strike." The government side, somewhat taken aback, promised to investigate this possibility. There matters hung—inconclusively but starkly.[38]

Nevertheless, the very structure of the meeting seemed to create an interesting precedent. With both the government and the MKS now buttressed by a team of academic advisors, an aura of balance had appeared which imparted a kind of sanction to the session and suggested a measure of equality within it. But after the meeting was over, Deputy Premier Jagielski raised a question about the credentials of the MKS advisors, some of whom he characterized as having "connections" to the democratic opposition and its driving force in Warsaw, the activist group known as KOR. Jagielski explained helpfully that their presence in the negotiating hall placed him in an awkward position given the government's known attitude toward KOR as an "antisocialist" element in Polish society. In response, Mazowiecki rather dutifully suggested that only two intellectuals—both suitably unconnected to KOR—be present to advise the workers. He named himself and Kowalik and suggested the others could withdraw to an adjoining room. When Wałęsa heard about the proposed arrangements, he recognized Jagielski's move for what it was, a transparent ploy to place the advisors in a defensive and therefore subordinate position. Such a psychological result, if achieved, might materially reduce the usefulness of the advisors to the MKS. Wałęsa categorically rejected the government move and insisted that all advisors be admitted to the negotiating session. Jagielski, of course, had no option but to accept this fiat if he wanted the talks to proceed. The incident proved rather instructive in other quarters as well. Though doubtless chagrined by their own conduct, the academics were pleased to discover the immediate impact of Wałęsa's intervention—the restoration of their status. Kowalik reported that "the atmosphere improved thereafter."[39]

Thus, in their first few days in Gdańsk, the Warsaw intellectuals attempted to soften the first demand and were firmly rebuffed by the entire Strike Presidium on the central issue of self-governing trade unions. They had to have their resolve stiffened a second time in the "working group," again by members of the Strike Presidium led by Gwiazda. With a kind of bemused sense of discovery that was unfortunately rather quickly forgotten by some of them, the intellectuals came to accept their secondary relationship to the workers' movement. At the same time, the deference of most of them to the regime, a habit unwittingly acquired as part of their socialization near the seats of power in Warsaw, came abruptly to the surface in the course of the colloquy with Jagielski over their credentials. The intellectuals were thereupon provided with a demonstration of forthright conduct by Wałęsa. They were able to observe at close range the effectiveness of this mode of procedure; they recovered their seats in the negotiating room. After these three separate stiffenings, the academics found the resulting freer environment much to their liking, grew in confidence within it, and became active and useful participants in the democratic movement. In this rather painless manner, a potentially injurious division in the ranks of the worker-intelligentsia coalition was overcome at the very moment of its formation.

It would be a mistake, however, to interpret the strategic differences between the MKS and its advisors essentially as a function of divergencies in formal ideology. By any reasonable standard that might generally be applied to Polish society both within the intelligentsia and within the working class, the academics recruited by Geremek were not "moderates" or "conservatives." Four of the group—Geremek, Cywiński, Kuczyński, and Kowalik—had been active participants in the "Flying University." This activity publicly exposed them to police harassment. Wielowieski, as secretary of the Warsaw Club of the Catholic Intelligentsia, may perhaps accurately be classified as a "dignified and judicious activist." The weakest link was Staniszkis, a Warsaw University sociologist who proved to be more at home with lofty abstractions than with the naked realpolitik of the Gdańsk confrontation. Her lack of political poise within a working-class milieu was decisively reinforced at a personal level by her enormous psychological distance from Polish workers whose politics she consistently misread. Haughty and imperial, Staniszkis brought few democratic tendencies to the advisory group. She was, however, to demonstrate a strange capacity for mischief making. Though she was not to remain among the group for long, her impact on the proceedings would be considerable indeed.[40]

The more one learned about the dynamics of working-class assertion on the Baltic coast, the more complex and even puzzling it all seemed to become. An apparent wellspring of mature confidence seemed to lurk somewhere in a highly usable form that the strike leaders could draw upon whenever the need for it arose. At first glance, there seemed a curious disjunction between the bold sweep of the workers' agenda and the patience and resolve they applied to the crisis that agenda provoked. They not only unveiled an elaborate set of objectives, they seemed to feel no compelling need for strategic advice as to how to proceed toward its attainment. To anyone who cared to look, a number of pieces to the puzzle seemed to be missing. From whence had all this purposefulness come?

If Jagielski and the ruling-party apparatus on the one hand and the Warsaw intelligentsia on the other had both experienced difficulty trying to size up the precise nature of the Baltic movement, the world press did not have much better success. The remarkable range of events which kept unfolding day after day in the Lenin Shipyard did not lend itself to easy characterization. For the official press in Poland, advice replaced reporting. In a remark that could have been taken from most American newspapers at the time of a local strike, the party paper in Gdańsk was soberly avuncular in advocating a back-to-work movement: "In the interest of every one of us, in the interest of society, in the interest of Poland, we should work when it is working time, and discuss at a time and a place appropriate for talking." For journalists from outside Poland, conventional political terminology proved inadequate as an explanatory aid. News dispatches grew thick with drama while remaining thin on clarity. Western papers reported that the strike movement, "unprecedented in the Eastern bloc," seemed to have materialized out of a "uniquely Polish kind of populism." It could also be seen as "revolutionary." But the more precise the attempt at ideological description, the more elusive the Lenin Shipyard became. The movement's leaders, for the most part, "seem to be socialists," but on the other hand, many were "deeply religious." Western reporters were careful to note that "Catholic symbols are much in evidence" (a fact that especially puzzled French and Italian trade unionists). Gdańsk Bishop Lech Kaczmarek's pious advice to the workers to "act in a wise and prudent manner" was passed on to the world. The strikers seemed to be trying to do just that, though the desk-bound wire services in Warsaw added the "pointed" assertion of a high party spokesman that "antisocialist elements and extremist groups" were active in the strike.

Somewhat more to the point, the *New York Times* quoted a shipyard worker in an effort to make one thing clear, the strike "is about a free labor union. That is the idea; the worker can go and ask for his rights and do it while he is protected against (repression)." The morning after the first negotiating session, thousands of citizens of Gdańsk gathered next to the shipyard gates to join the strikers in Holy Mass. The event got a considerable amount of ink in the world press. Symbols were indeed in abundance. Flowers were heaped across Gate No. 2 of the shipyard, as were pictures of the Polish Pope and the red and white flag of Poland. Political statements also proliferated. *Solidarność Strike Bulletin* no. 10 carried a deeply felt and highly ironic reflection of the corrupted socialist aspiration: "Workers of the World . . . my sincere apologies. [Signed] Karl Marx." A huge banner in the shipyard proclaimed, "Proletarians Of All Factories, Unite." Did such slogans signify two political ideas contending against one another or one idea presented in a new way or one idea split in half? Clearly, Solidarność was socialist, Catholic, patriotic, working-class, democratic, and insurgent. Clearly. What did it all mean? Radio Free Europe took a stab at a background explanation when its research department came up with the information that the first free trade-union "cell" had been organized in the Polish city of Radom in 1977 by an activist group of Warsaw intellectuals. They called their group KOR, an acronym for "Committee for the Defense of Workers."[41]

The most popular journalistic solution was to focus on personalities, particularly since such a vivid one had surfaced at the head of the presidium. Reporters paid close attention when the chairman of the MKS talked at length over the loudspeaker to the MKS delegates or to the rank and file of strikers in the shipyard. He offered short summaries of events and quick analyses of possible options; he took questions and spun off answers, and, afterwards, he moved casually through the huge throng, "listening, nodding, explaining, laughing." But in ways that extended beyond Wałęsa, the sights and sounds of the Lenin Shipyard provided other raw materials for vivid journalism: the elegantly attired government commission juxtaposed to the presidium members in their work clothes; the Greek chorus in the wings, growing larger by the day as additional delegates joined the MKS from newly affiliated factories; the loudspeaker that brought each subtle evasion of the government negotiators to the attention of every MKS delegate and shipyard striker; the second Greek chorus in the shipyard, the massed thousands armed with tape recorders, banners, and their impressive patience. Taken in all its parts, the Lenin Shipyard was as irresistible as theater.

But on August 23, the "spontaneous" chant which had seemed so self-explanatory at the outset began to take on some new meaning. The shipyard workers shouted it out, "Leszek, Leszek," two syllables to the word so that it had a certain rolling power: "Le-szek, Le-szek." Again and again, the chant thundered through the Baltic air. For the correspondents of *Le Figaro* in Paris and United Press International, such things helped establish that Lech Wałęsa was a "born leader" who had "charisma." The fact seemed to make the Baltic strike comprehensible.[42]

For those observers who could somehow seperate themselves from the spectacle itself, however, it became possible after awhile to realize that the stentorian chant was not really about Lech Wałęsa at all. It stood, rather, as a statement by the workers on behalf of themselves. When the government commission first came into the shipyard on August 23 and moved into range of Wałęsa and the presidium, thousands of voices shouting "Le-szek, Le-szek" reminded the authorities that the contest did not involve an all-powerful government against one little man with a moustache, but a government forced to confront self-organized workers. Emanating from those who for so many years had not been visible to anyone in People's Poland, the chant was a reminder to the party that "we are here now." The chant had become another and more effective way of telling government functionaries to "look straight into the eyes of the workers."

In this manner, what appeared to be pageantry concealed the operative political statement of the Baltic movement: the emergence of an intense and fragile new form laboring to find a home in People's Poland, a nascent democratic institution in passage from idea to social reality. Grounded in an occupation strike, the Interfactory Strike Committee was contending with the state to become the founding committee of a new democratic space in Poland—an institutional space created by the working class for occupancy by the people of the nation. As matters stood, no one outside the ruling apparatus had a dignified presence in Poland. The entire population was in the same powerless position as the most voiceless shipyard worker.

It is appropriate to grasp this moment in the Polish August. At the bottom of society, the first democratic utterance to sound in the presence of power was the first name of a worker spoken by another worker at the top of his lungs. If the space could be secured, if he could speak in the presence of power next week and next month, the whole nation might recover its voice again. This was the realpolitik of August 23 on the Baltic coast.

Through the prism of this political development in the Lenin Shipyard, it became possible to perceive two emerging social facts in People's Poland. The first could be simply stated: after thirty-five years of hegemonic control over the population, the Polish United Workers' party had thoroughly succeeded in isolating itself from the nation. The second was more complex. Poland's working class and sectors of its intelligentsia, each in its own way, had intermittently challenged the range of one-party rule since the end of World War II. Now they were trying to come together in support of a large and growing working-class movement. Nevertheless, despite many commonalities as victims—both had been suppressed and intimidated for long periods—the two groups had lived through vividly different experiences in their relationship with the party and state apparatus. Though they might agree in general terms that some kind of wholesale reorganization of society was needed, they possessed wildly divergent understandings of what was achievable and starkly different degrees of sophistication as to what had to be done to achieve it. Intellectuals had learned, in more grievous detail than most of them cared to remember, the suffocating reach of the censorship; workers had been exposed to the same instruction—and to many brutal and subtle additional lessons as well.

This experience of the Polish working class constituted the foundation of the second social fact to emerge in the Lenin Shipyard. The workers' road to Gdańsk had been a very long one that included journeys down difficult byways in every part of the country. Workingmen and women had on occasion run into police and army roadblocks that caused appalling bloodshed and briefly caught the attention of the party and intellectual elite. But most of the journey had been made in short segments and in a manner that was so unobtrusive and remote from the world of high culture that anything resembling progress was invisible to the party, the intelligentsia, or the outside world. This unseen but persistent struggle took place at the point of production in Poland's factories, mines, and shipyards. Slowly, incrementally, piece by piece, the body of knowledge assembled through this effort acquired transforming potential.

It was in the workplaces of Poland that the need for Solidarność was discovered, and there Solidarność grew. The knowledge was intricate, subtle, and bought with the individual striving of many workers, some of whom were no longer in Poland or even, in many cases, alive. It required over thirty-five years for Polish workers to amass the tactical and strategic experience necessary to challenge the party-state effectively.

The irony that hovered over the Gdańsk negotiations was that no one in Poland knew they had acquired it but the workers themselves.

2 ▪ The Workers Encounter a "Communications" Problem

> The strikes which we observe in different places are the signs of anarchy and are not in keeping with socialist liberties.
>
> Władysław Gomułka, first secretary of the Polish United Workers' Party, 1958

THE MOST BASIC social information on Polish society has been left ungathered, gathered and suppressed, or gathered and distorted for purposes of public disinformation. The principal repositors of information on the working population of Poland are in the Interior Ministry in Warsaw and in the provincial archival holdings of the security services. For thirty-five years, the police bureaucracy of the party has been focused so obsessively on controlling the working class that the identity of shop-floor activists has been a central research objective. Police files bulge. As for the other ministries of the government, even the most elementary statistical data have also been treated as classified material, apparently out of fear that they might somehow be made to serve an antiparty purpose. As a result, the kinds of written evidence on social activity that Western journalists and scholars routinely research is simply not readily available in Poland—either for Westerners or for the Poles themselves.

Since scholars, like journalists, are yoked to their sources and, perforce, cannot write about what they do not know, the ephemerality of the pre-Solidarność history of the Polish working class has encouraged outsiders to look elsewhere for an explanation of the Polish August. Polish intellectuals, of course, have been willing to provide all they know about their own contributions. The result of this circumstance has been an effective clouding over of the central role of workers in the creation of Solidarność. Given the size of the literature produced on the Polish movement, this condition is now a worldwide one.

The only approved window into the Polish working class is government-controlled polling data. A few shaky insights, very broadly defined, can be gained from this information, but on the whole, it is almost worthless for all the years before 1980. The value of citizen answers is contingent upon the context surrounding the interview process itself. This context is further framed by obviously relevant questions not asked, by the suggestive form in which the questions are frequently shaped, and, not least, by the power relationships that surround all contacts between citizens and formal interviewers in an authoritarian state. Matters of salience aside, Polish polling data consist principally of attitudinal surveys that are quite remote from the realities of power that inform organizational work within an industrial society. The students of the Polish movement will have to look elsewhere.[1]

The regime itself is a better starting point. Hanging over the postwar Polish working class has been a grim irony that serves quite well as a conceptual framework through which the complicated route to Solidarność can be reconstructed. This irony has at its base a Polish social reality rarely officially conceded in the West; in twentieth-century cultures that are nominally socialist, the working population possesses a level of public prestige that is a bit beyond anything present in capitalist societies. The difference exists in a number of concrete ways that extend beyond public speech making or the ritual invocation of sanctioned doctrine. In the societies of Eastern Europe, it is a centerpiece of popular wisdom that the welfare of the nation is dependent on economic productivity and that progress is dependent on the collective effort of the working classes. The party seeks out workers as party members and actively recruits from shop floors as many unskilled manual workers as possible. Such customs help to reinforce the general claim that the working class is an absolutely essential component of political legitimacy. Consequently, workers in socialist societies are not ideologically or culturally stigmatized in quite the way they are in the West. This is not meant to suggest that feelings of inferiority and superiority have somehow been obliterated in Poland or that some classless plateau has been reached. On the contrary, Leninism is grounded in the conviction that the working class cannot be a self-generating agent of social progress: workers can play a significant role in society only to the extent they play a role in the Leninist party. But because the societies of Eastern Europe are not as visibly stratified as societies in the West, a certain viscerally registered cutting edge is missing from prevailing folkways of condescension—and the morale of the working class is the better for it.[2]

Under such conditions of nominal public affirmation, the circumstance of finding oneself both poor and powerless is therefore especially galling—and especially ironic. If continued long enough, the effect of this degradation can destroy even those psychological benefits which the advent of socialism usually temporarily affords working populations. Such has long since happened in the Soviet Union and has become strikingly obvious through the events of 1989 in Hungary, East Germany, Czechoslovakia, Bulgaria, and Romania as well.

In Poland, the struggle of workers for dignity has been long and complicated. In its early stages after the war, the effort consisted essentially of testing the extent to which party control of the workplace was considered an essential ingredient of the new socialist life. The answer obtained—after considerable bloodshed—was stark: the party was to be the unquestioned authority at the point of production. The next phase of working-class activity, lasting over a generation, involved experiments in ways of breaking free. Workers in factories all over Poland endeavored to create some space between themselves and the long police-anchored reach of the party apparatus. All of these efforts ended in failure, but each of them left a residue of knowledge that confirmed at one level precisely what tactics had no chance of working and at a second level suggested those that could be useful. The decisive learning period came in a protracted struggle on the Baltic coast between 1970 and 1972 that experientially established the essential principles that had to be followed. Everything after that consisted of polishing and improving organizing techniques that were the necessary corollaries of these principles. At its climax of 1980, the political challenge involved very human equations—whether experienced worker activists could sustain both their poise and their organizational coordination for the duration of the national crisis that an engaged working-class presence had created. The entire thirty-five-year struggle mobilized all the ingredients that are generally understood to comprise "political consciousness," though perhaps not quite in the cosmic ways political theorists have traditionally employed the phase. In any event, to trace this struggle is to locate the dynamics that produced Solidarność.

The initial postwar environment in Poland was one of enormous chaos, physical destruction, and political tension. For no nation on earth was World War II the catastrophe that it was in Poland. For the nation's Jewish population, it was, as is now understood, a literal holocaust. Over 90 percent of prewar Polish Jewry perished in Hitler's extermination camps. After 250,000 survivors migrated to Israel, only 60,000 Jews of a prewar population of 3,400,000 remained in their Polish homeland. Three million other Poles also died in the war as German and Russian armies each thrashed through the nation twice during six years of conflict. No less than 22 percent of all Poles died in the war, a per capita loss far greater than that suffered by any nation. The physical destruction was also appalling. Before retreating in 1944, the Germans systematically leveled Warsaw, meticulously dynamiting block after block following the suppression of the 1944 Warsaw uprising. The city that was the nation's maritime lifeline—which the Germans called Danzig and the Poles called Gdańsk—was also destroyed during hostilities. Meanwhile, the countryside was denuded of livestock, fencing, and machinery. Similarly, over 80 percent of the nation's transportation and 60 percent of its schools and scientific institutions were destroyed. On top of everything, a massive geographical wrenching took place. The Soviet Union annexed seventy thousand square miles of Poland—the eastern one-third of the nation—for which the victorious powers offered partial compensation in the form of forty thousand square miles of prewar eastern Germany. The German lands had been subjected to a systematic double devastation, first by the scorched-earth policy of the retreating Nazis and then by Red Army cadres which dismantled every salvageable piece of machinery and shipped it back to the Soviet Union. A gigantic westward migration accompanied the end of the war as millions of people were forced out of their ancestral lands in eastern Poland, many of them relocating in the newly annexed provinces in the extreme western end of the reconstituted country.[3]

The first economic challenge was in these new provinces. Great shipyards were rebuilt on the coast all the way from Gdańsk to the border of East Germany where German "Stettin" became Polish "Szczecin." Painstakingly, debris was cleared, and the railroads made to function. Gradually, Poland's cities began to reemerge from the wreckage. As rapidly as possible, the vastly overpopulated countryside, bulging with tenants and smallholders farming tiny plots, was systematically cleared somewhat as peasants were directed to the new factories. Through herculean efforts, Poland struggled to its feet.[4]

Under the protective guns of the Red Army, a tiny Communist party provided the nucleus of the postwar ruling elite that supervised this enormous effort. Perhaps the most telling proof of its social failure over the next thirty-five years is the fact that the position the party filled after martial law in 1981 was almost precisely the position it occupied in 1945: a band of functionaries, despised by the population, maintained in power solely through military force, politically trapped between its dependence on the Soviet Union and its demoralizing illegitimacy at home.

The prewar party had suffered a convincing defeat at the hands of Polish Socialists for the loyalty of the nation's relatively small industrial working class. In the vacuum that existed immediately after the war, Polish workers simply took over factories and established their own shop councils. Following the first elections to these autonomous bodies, communists possessed only a fragile foothold in most shops and were badly overshadowed by the embedded socialist presence among Polish workers. Even with the state apparatus behind them and the entire population abundantly aware of the new postwar "geopolitical realities," the communists generally failed to gain control of local councils. In April 1945, communist-sponsored trade unions, recruiting in a war-shocked society dominated by Red Army troops, nevertheless, could gather in only 116,000 members. Independent unions, on the other hand, defiantly signed up 355,000 adherents. In all kinds of trades Communists on the slates in shop council elections went down to repeated defeats.[5]

What the party did have, however, was the backing of Soviet military power. The latter was used to ensure the immediate organization of party-backed security-police units. In turn, police terror throughout the summer and fall of 1945 was systematically employed against the independent industrial unions. Individuals were persuaded by police truncheon that public support of nonparty unions was an impolitic habit in People's Poland. As attendance at union meetings began to decline, intimidation was directed at the union structures themselves. By the time the first postwar snow began to fall in Poland, the independent unions could count only 159,000 members. In the new year, the remaining institutional structures were obliterated. Surprisingly, however, the ruling party still proved unable to consolidate its grip on the working class. Though independent candidates in shop council elections no longer had what could be recognized as organized support, they won office anyway. Socialist trade unions were gone, but a longing for a home of their own continued to course through the rank and file. In structural terms, the Polish United Workers' party remained an institution with a base in the security police. Political direction came from a thin administrative layer of old prewar Bolsheviks who had managed to survive Stalin's purges. The majority of the population, meanwhile, remained in sullen disassociation.[6]

Worker-run councils and party-run police forces obviously constituted a volatile mixture. Through the second half of 1945, throughout 1946, and into 1947, Poland's industrial regions experienced intermittent turmoil. The party improvised a variety of ruses and intimidations in an effort to take over factory councils. When the shop-floor militants were arrested, workers deprived of any negotiating rights resorted to the only weapon that remained: Poland was skewered by strikes from late 1945 to 1947. When the party forcibly took control of factory councils, the workers simply went into the streets. Intermittent conflicts between workers and police erupted in the mining regions of Silesia, in the textile center at Łódź, and throughout the maritime provinces. Workers poured out of shipyards in Gdańsk, Gdynia, and Szczecin, arrests mounted into the thousands, and some large but uncountable number of workers whom the police took to be "leaders" were packed off to prison. In shop elections, the stoutest resistance to communists tended to materialize among workers in heavy industries like mining and shipbuilding. But in class terms, there was a certain amount of pride and preserved integrity—if never victory—in a wide variety of blue- and white-collar occupations. The Railroad Federation, centered in Warsaw, earned a special measure of party enmity by rejecting *every* communist candidate for federation office, while honors for the most enduring resistance went to the Teachers' Federation, which held out until May 1948. The latter union, ironically enough, had been a special target of abuse by the prewar government, a coalition of landowners and colonels that had viewed the prewar teachers' union as a hotbed of communism. In any case, the single most important structural reality from 1945 to 1947 was the fact that a genuine movement of the country's working class had materialized by 1946. In the autumn of that year, a renewed wave of strikes engulfed the familiar centers of industrial Poland—Silesia, Poznań, Łódź, Gdańsk, Gdynia, Szczecin, and Warsaw. These mobilizations constituted the last real hope for any form of democratic future for Poland.[7]

Another way of seeing the immediate postwar politics of the country is to focus on the desperate maneuvering of the noncommunist political leaders of the prewar Peasant and Socialist parties and the factional tensions within them as they were slowly strangled by the nation's Leninist police bureaucracy. Most histories of postwar Poland published in the West have done this. Many well-known prewar representatives of various classes and interests did, indeed, put up a courageous struggle, and some of them paid with their lives. What needs to be stressed, rather, is the fact that once the struggle in the working class had been lost, the systematic arrest of noncommunist figures and the obliteration of their political institutions was a foregone conclusion.

The two scholars who have produced a cogent, book-length study of these centrally germane relationships, Jean Marlara and Lucienne Rey, suggest that socialist leaders, many of them deeply preoccupied with party matters, sometimes acted as if the defense of their social base in the working class was not the foremost priority of the struggle. Thanks to the presence of the Red Army, defeat at the hands of party forces was imminent whatever maneuvers were taken. But this tendency of members of the Polish intelligentsia to pass quickly over workers without seeing them was to settle in as a constant feature of the national culture down to—and beyond—the Polish August of 1980. Indeed, it was to resurface in 1989.[8]

Following the suppression of the demonstrations and strikes, the destruction of the independent shop councils proceeded with a relentless rhythm. By government dictate, the councils were forced in 1947 to become "rank-and-file bodies" under the party-controlled trade union; within another year, the now powerless councils were relegated to the chore of assisting in the government's "incentive" schemes for increasing productivity; public proclamation of their final degradation came in 1949 when the party's trade-union congress admonished Poland's working class as to the "dangers inherent in the idea of workers' management." Old socialist unionists who resisted this process at any point were fired, dispersed across the country, or imprisoned as the party razed trade-union traditions and domesticated the shop-floor militants who lived by them.[9] So strong was worker belief in shop-floor democracy that no less than 80 percent of elected local union heads had to be cashiered by the party in its quest for absolute hegemony. All were replaced by docile party functionaries.[10]

It would be wrong, however, to infer that Stalinism, especially in its earliest stages, united the entire society in engaged opposition. The Bierut regime was by no means without some measure of support, freely given. Indeed, among the most concerned people in Poland, both within the working class and within the intelligentsia, a significant number threw themselves headlong into what they were convinced was "the revolution." Many more, equally progressive and intelligent, held back. In any case, after lofty aspirations had become throttled and only raw terror remained, men and women who had spent the early 1950s in Poland as dedicated Stalinists withdrew into passivity or went over into the opposition and joined their comrades who had preceded them.[11]

There were, of course, still other responses. In the grip of the first wave of the national shock that greeted the police terror in the early 1950s, many workers in Poland simply lost their poise and groveled, earning minor party rank in the state trade unions as a reward. Many intellectuals groveled, too, and acquired a measure of cultural status. Whether as members of the intelligentsia or the working class, however, most Poles endeavored to maintain their dignity by recoiling into brooding silence. For intellectuals, activist politics came to consist of a constant search for ever more ingenious ways to circumvent the censorship or some of the censorship or some of the worst features of the censorship. For workers, activism consisted of inventing ever more creative ways to bring some dignity, fairness, and rationality to their lives at work. Their struggle, too, was against the censorship, the institutional censorship of the party-controlled trade unions. The options in both milieus were extraordinarily narrow: either pedestrian performance in tight-lipped silence or bold activism that was wholly defensive and conducted almost completely within terms fashioned by the state. In either case, the impact upon ongoing political reality was inconsequential. Beyond such limits, one had the option of going to prison.

The party leader who presided over the initial domestication of the Polish working class was a spartan and incorruptible Bolshevik named Władysław Gomułka. He was one of the many East European communists who did not understand that the policy options in the immediate postwar world were fated to be quite narrow and that the only sanctioned communist way was the Soviet way. In common with the members of the old prewar Polish Socialist party, Gomułka envisioned a "Polish path" to socialism, though he regarded most old socialist leaders as hopelessly fixated on bourgeois concepts of democracy. In presiding over the party's postwar consolidation of political power, Gomułka participated in a considerable amount of coercion and intimidation. This included the systematic extermination of the prewar Peasant and Socialist parties, highlighted by a massively rigged election in 1947 and culminated by the steady removal of all socialists and agrarian radicals in the government who persisted in the habit of occasionally disagreeing with his policies.[12]

The party faction under Gomułka that favored the "Polish path" was useful immediately after the war when the great majority of the citizenry as a whole had overtly noncommunist political allegiances, but any such popular inclinations as Gomułka might have possessed became superfluous after one-party rule was fully consolidated late in 1948. Shortly thereafter, the new "Polish United Workers' party" fell under the control of its Moscow-trained faction, and both Gomułka and the "Polish path to socialism" were on the way out.[13]

With their own wounds still fresh, Polish workers did not grieve when Gomułka departed. The prevailing belief was that his successor, Bolesław Bierut, could not be worse. They were wrong. With the workers brought to heel, Bierut's party embarked upon an adventurous six-year plan for the period from 1950 to 1956. It called for nationwide industrialization on a crash timetable.

The capital for such a project necessarily had to be squeezed out of the work force itself. Draconian measures were introduced. A program of ruthlessly planned underconsumption for the population was augmented by militarized work procedures. Working-class acquiescence was ensured through police terror as a fully empowered brand of Stanlinism descended upon Poland. In the course of this policy, the party and security police acquired a stranglehold on social life in Poland. Even the death of Stalin in 1953 did not seem to slow the pace of authoritarian momentum that had been achieved. The years of Bolesław Bierut were, quite simply, a time of national agony in which the only reasonable political hope for most of the population seemed to rest in the possibility that even the enforcers some day would lose their way.[14]

Oddly, something approaching this actually happened. In the natural course of events, the demoralization of society came to extend to the lower and middle echelons of the party. Increasingly, party functionaries deflected demands for ever greater production by simply lying to their superiors, who lied to theirs. Fissures opened between what was produced and what was reported as produced. Framers of a number of plans tried to incorporate the use of supplies that did not exist and sought to capitalize on the anticipated output from newly finished factories that were not finished. Planning—rationality itself—is difficult in politics under most conditions; it is especially elusive under conditions that amounted to a nationwide production speed-up conducted among unenthusiastic producers. Though an enormous amount of physical effort was, in fact, mobilized and much accomplished, Bierut's six-year plan became unbalanced in application and slipshod in administration at a very early stage. In social terms, it worsened thereafter to a norm that might be described as routinely chaotic. Even the most blockheaded party functionary eventually had to concede that many plans were failing not because of lack of proper socialist zeal at the level of production but rather because there was simply no humanly possible way for them to be fulfilled. Executive supervision lost its thrust, administrative coordination lost direction, and in due course, even the police became uncertain. Demoralization, having infected the intelligentsia and working class, moved inexorably through the party until it began to touch portions of the police apparatus, the militia, and the armed forces.[15]

In this manner, Stalinism slowly wore itself out in People's Poland. It is impossible to "date" this process. When people have been rendered mute out of fear, they remain silent even after the immediate cause of fear has begun to dissipate—out of habit if nothing else. In retrospect, it is now possible to conjecture that perhaps Stalin's death did have a certain invisible effect at the very top of society as the party leadership groped for a way to preside over Stalinism without Stalin.

Throughout the dark days, Polish workers waged a kind of rearguard action, wholly defensive in nature, that was designed to preserve at least a minimum of dignity and rationality in their involvement in daily production. Stoppages, negotiations, arguments, and grudging adjustments by both sides punctuated what party newspapers became fond of calling "the normal rhythms of work." The simple reality was that continuous contention on the shop floor more nearly characterized the normal rhythm of work.

It is helpful to pause at this point to spell out the socially flammable materials of modern industrial production. The managerial task of keeping an industrial work force in a state of persistent cooperation requires both a measure of skill and a measure of resources. All of the obvious ingredients routinely apply—safety conditions, pay schedules, work routines that seem to make sense. In their unfamiliarity with industrial production, nonworkers can easily overestimate the primacy of wages to the exclusion of all the other factors that bear on the dignity of workers. This is a familiar error. On the production line, people can be driven into a frenzy by systems that require them to expend more physical effort than would be required if production were organized differently. Anything that unnecessarily creates additional work or nullifies the normal effect of normal work can, if repeated often enough, drive production workers into shop-floor insurgency. Worker morale can be undermined by a number of causes—a sudden lack of parts that undermines work already performed, repeated failures of the same piece of equipment, recurrences of official disdain for worker safety. A supervisory folkway that always carries the potential for triggering a rebellion is the practice of ordering workers to put in long hours of overtime for no persuasive reason. For Western workers, shop stewards, local and international union officials, plant supervisors, and industrial-relations personnel, such matters of contention are the very stuff of daily industrial life. Over and above matters of wages, they comprise the issues that remain in constant contention and adjustment. In general, the process institutionalizes the permanent debate over the human cost of industrial production.

The engine driving the debate is the central element of coercion inherent in the work process itself—as that coercion is given particular definition at the point of production in a given factory. Whether conducted well or poorly, with surface amity or visible bitterness, the negotiating process is brokered through ever-changing perceptions that exist in the minds of the contenders concerning fairness, intensity of resolve, and relative power. Though significant differences have always persisted in the equities distributed to management and labor in various countries, the argument itself has never been a balanced one in any economic system or in any nation. But in an unbalanced form, it has defined much of labor-management relations in advanced industrial societies.[16]

The struggle of workers in Poland, as in any country, contained both a tactical and strategic dimension. The initial postwar effort possessed strategic meaning in that it settled the question of who—worker council or the party—controlled the shop. Between 1945 and 1948, the party won the battle, not on the shop floor but in the state. It then was able to enforce its own concept of production relations by imposing its own administrative hierarchy supported by coordinated police and judicial authority.

But controlling an industrial enterprise is not quite the same thing as running it on a daily basis. In the early postwar years, the party structure itself was too fragile and state planning was too extemporaneous to rule out a certain flexibility in the relationship of party supervisors to workers. What in the United States would be called "wildcat" stoppages abounded. Lacking an institutional framework of their own, workers generated job actions that tended to be confined within the circle of personal friendship. A section stopped work, and an argument began. The growing size of job actions in the mid-1950s verified the emergence of organizational work linking sections, divisions, and whole plants. The events in Poznań in 1956 signaled elaborate working-class organization reaching fifteen thousand workers on a plant-wide level. The nature of these early struggles can be summarized as ranging from a sudden explosion of exasperation by a single worker to the first signs of organized actions by larger groups. Long-smoldering grievances suddenly stimulated by some new absurdity in work and pay schedules or production quotas could activate anything from a respectful petition to a plant-wide mutiny. Still another form of insurgency was defensive, individualistic, usually futile—and massively applied. Frustrated workers changed jobs. Despite proliferating rules, job charters, and restrictions of all kinds, Poles developed ingenious ways to try something else. At one point, the party discovered to its dismay that 800,000 changes in employment had been officially registered. The bureaucratic response was a flood of new regulations designed to produce "socialist work discipline." The worker reaction was a vast increase in absenteeism.[17]

It is therefore appropriate to invoke an old descriptive term and characterize the Bierut years as a period of escalating class struggle. The first wave of anti-worker laws peaked in 1950, a year that saw a bitter confrontation in the

Silesian coal fields. The government—reeling but also determined—made significant wage concessions to the miners but tied them to a new series of production norms. The following year, an even bigger walkout shut down the Dąbrowa mining district and brought worried government and union heads running to the scene. Spokesmen for the miners were promptly arrested and sent off to prison, and the situation was brought under control only after three days of repression and reactive turmoil. Mining, textiles, shipbuilding—not a single major Polish industry escaped bitter contention at the point of production as Bierut's six-year plan moved through ever-tightening constraints of centralized control. Working conditions became almost intolerable. The Transport Workers' Federation, meeting in national congress in the early 1950s, passed fifty-three resolutions protesting working conditions. The grievances were spelled out with such stark specificity that they drew a harsh denunciation from Wiktor Kłosiewicz, the party's man in charge of the centralized trade-union structure.[18]

The party created more rules, the workers generated more grievances, absenteeism became a way of life, and prewar agricultural Poland slowly transformed itself into an industrial society. In the course of this combat, the somewhat extemporaneous mode of industrial supervision that materialized immediately following the war rapidly hardened to tight party control that was vertically integrated into the central bureaucracy in Warsaw. Fairly soon, workers learned that the judgments of local party people did not really count on most issues. Party spokesmen in the plant were often well informed—that is, they could recognize a production malfunction when they saw one as well as any worker—but they were powerless to respond. When workers complained, functionaries nodded agreement, shrugged their shoulders, and pointed in the direction of higher party authority. In this way, workplace actions in the early 1950s taught Polish workers that shop grievances were a political matter. If they were to do anything about their situation, they would have to find a way to get a hearing at some level of the party higher than their own enterprise director.

But this was not an insight that came to Polish workers as a class or one that extended throughout the country. On the contrary, the process of education proceeded quite slowly because each group of workers in each region of Poland had to learn through personal experience. This unwanted circumstance was a function of the peculiarities of social relations in a Leninist state. Party control of the official trade-union structure had the effect of blocking the only internal means of communication workers could normally possess as a class. Brooded over long enough, this problem contained the seeds of restructuring—a free trade union independent of party control—that would constitute the key conceptual breakthrough represented by the formation of Solidarność.[19]

It was in this context that Polish workers began to grasp the full dimensions of their initial postwar defeat. In the West, workers have traditionally learned tactical and strategic lessons about corporate power not only through their own efforts but through self-activity of other workers. The Polish state's stranglehold on the trade-union structure not only converted it into a transmission belt through which the regime could "instruct" its working class—a fact that was ominous enough in itself—but the one-way communication system destroyed lateral lines to other factories that workers have traditionally gained by forming their own institutions.

The official trade union was, workers discovered, one vast cloak of censorship. It would be very difficult for worker activists in Bydgoszcz, for example, to learn from the experiences of fellow activists in Kraków. Polish workers simply had no convenient way to communicate. This was the practical meaning of the concept of "Leninist norms," the phrase used to sanctify the leading role of the party at the workplace and everywhere else in society.[20]

The effect of these informational and organizational constraints was to create a kind of regional imperative in working-class Poland. The only experiential knowledge that Polish workers could pass down to the incoming generation on the shop floor was knowledge gained in the immediate vicinity; that is, knowledge that could be conveyed by word of mouth. For a working-class culture of self-assertion to develop, it would almost certainly have to materialize within the geographical constraints of a confined region. The circle of knowledge would be unlikely to extend much beyond the distance that workers could travel in the course of normal daily living.

Three comparative high points in worker mobilization illuminate the underlying dynamics that finally set Polish workers on the road to Solidarność. The first of these moments occurred at the end of the Bierut era in 1956 when workers at Poznań rose *en masse*. The Poznań workers paid an enormous price in the course of learning that their own strategic analysis was inadequate as a servant of working-class aspiration. Nevertheless, they created such massive material havoc in downtown Poznań and suffered so greatly in consequence that they engaged in extended postmortems. These conversations in Poznań proceeded throughout the summer and fall of 1956 and culminated in an attempt to revive the early postwar worker councils. The timing was fortuitous, for Bierut died in 1956, and a divided party loosened the social reins as preoccupied factions jockeyed for power. Much happened in Poland in 1956.

The second event occurred in December 1970. This time workers on the Baltic coast went on a violent rampage. In common with their predecessors in Poznań, the Baltic workers also paid a fearful price, and they, too, engaged in intensive postmortems. This, however, produced a different series of strategic conceptions from those formulated at Poznań.

The third signal from the Polish working class materialized in 1976 at Ursus, a suburb of Warsaw, and at Radom, a town forty miles away from the capital. This, too, was a violent eruption that was promptly and brutally suppressed. Systematic working-class reflections did not occur in Radom or Ursus, and therefore no strategic visions emerged.

The striking commonality in the three eruptions of 1956, 1970, and 1976 was that workers in each of them marched to the center of a city and burned or wrecked party buildings before engaging in violent confrontations with the police. Surveying this carnage after the fact, it is not immediately apparent that anything that might be called a progression of working-class awareness developed over the twenty-year period. To all appearances, workers in all three cases seemed to have acted before they thought—and nowhere more exclusively than in the final outburst at Radom where no structures of political assertion emerged beyond simple violence. If anything, a regression in consciousness seemed to be at work. If events at Poznań in 1956 or those along the Baltic coast in 1970–71 left a residue of increased shop-floor knowledge, workers at Radom do not appear to have been aware of it in 1976. The isolation of the Polish working class and its inability to rise above a guerilla warrior mentality seem fairly rooted at the core of these events.

Such, however, was not remotely the case. The differences in the three events were profound. Indeed, they go a long way to illuminate why Solidarność sprouted forth from the coast and not from Poznań, Silesia, Radom, Ursus, or from anywhere else in Poland. "Consciousness" lurks in the details surrounding this self-activity of Polish workers.

The "Poznań events" as they came to be called erupted in a city that is the site of one of Poland's most important machine and engineering works. Poznań's industrial and commercial prestige approached that of the Silesian coal fields, the shipbuilding complex on the Baltic, or the textile center at Łódź. In 1946, Poznań's leading plant, Hipolit Cegielski, was known across Europe. Renamed ZISPO in 1950 (the acronym conceals a tribute to Stalin) the huge works, employing fifteen thousand workers, manufactured a variety of machine tools, ship engines, locomotives, railway freight cars, and other metal products that were essential to Poland's entire postwar industrializing effort.[21]

The workers' grievances at Cegielski grew out of the systematic degradation of the Polish working class that accompanied the Bierut six-year plan. Statistically, the frantic industrializing effort that was the heart of the plan appeared "fulfilled." The party press triumphantly announced record production figures in almost every basic industry. But the realities of Polish life that were visible on the streets, in working conditions, in factories, in food on the table, and in the interior of houses mocked all these claims. The plan had proceeded especially poorly in food and housing. Agricultural output had been consistently "below plan," and supplies of a number of food items were scarce. Young couples contemplating marriage had to watch helplessly as the average number of years on the waiting list for apartments climbed to eight. It developed that the housing plan had been strikingly modest compared with the party's obsessive commitment to heavy industry; in any case, even the truncated plan for housing construction had been only 30 percent fulfilled. Housing production was one of the ways state planners convinced Poles that the party held the population in low esteem.[22]

Bloodless statistics, however grim, simply cannot convey the manner in which the party's ruthless industrialization degraded people and foreclosed the most elementary hopes. A foreign journalist recorded the dream of a 26-year-old Poznań woman whose simple longing gave the national housing crisis a human dimension that no statistical table could possibly impart. The young woman had for seven years been forced to share a small room with three other women not of her own choosing. "All I want," she said, "is a room of my own. I want that more than anything in the world. It seems to me that if I had my own room, no matter how tiny, I would have my own life. I don't need more money—what for? To buy clothes? But I have no boy friend, so I don't care about clothes. How could I have a boy friend? I can't possibly invite anyone to visit me."[23]

By 1956, living standards had fallen precipitously from their 1939 level. While the official media emphasized total production figures in steel, coal, and locomotives, the real value of wages, pegged at 100 for 1939, stood at 65 in 1951, 64 in 1953, and 67 in 1955. The lowest 30 percent of Polish wage earners lived on the absolute margin of survival. Almost every hard-pressed worker family in Poznań contained one or more aged pensioners as close relatives who had to plan very carefully to avoid hunger. The government did its best to conceal the evidence through the practice of annually increasing wages while at the same time also increasing prices at a higher rate.[24]

For workers, added humiliations materialized out of the irrationality of Polish industrial production. Monthly production quotas set in Warsaw were met on Poland's production lines by frenzied end-of-the-month labor that left men, women, and machines in need of rest and repair. Not much happened in the first ten days of the ensuing month. The slowdown meant that factories producing parts for production lines in other factories set in motion a chain reaction of routine shortages that pervaded the entire system. The phenomenon ensured that full-scale production rarely began before the fifteenth and sometimes even the twentieth of each month, insuring yet another round of end-of-the-month frenzy. The custom acquired a name—"storming." It was a folkway of the entire Eastern bloc, nowhere more intractable as a system than in the Soviet Union itself. In Poland, production averages were ludicrous; first ten days, 7 percent of monthly output; second ten days, 22 percent; final ten days, 71 percent. The practice was wasteful of human and material resources and yielded additional waste in the form of shoddy products that quickly wore out. Since waste was literally structured into labor itself, the incoherence of storming seemed to undermine prospects for a better future.[25]

These dynamics were particularly galling at an advanced engineering works such as Cegielski where workers possessed understandable pride in their skills and where they had, in common with their counterparts throughout the nation, made enormous postwar sacrifices in responding to the party's call to rebuild Poland. After years of effort, the nation seemed exhausted; storming in the factories that supplied Cegielski translated into routine beginning-of-the-month shortages that necessarily made storming the way of life there, too.

The locomotive factory, for example, routinely received steel for tie beams and links a fortnight before the scheduled completion date for a new locomotive. Unfortunately for the plan, machining of these parts required seventy-five days. Boiler sheets from the Batory Steel Mill normally arrived forty-five days behind schedule. Though Cegielski was scheduled to produce fifteen locomotives in July 1956, the factory did not receive from its Pomet supplier enough casts to make even two locomotives. With part of their income structured on a new and highly unpopular piece-work basis, workers whose sections did not fulfill quotas received pay cuts. In 1956, employees of the locomotive section were among the many Cegielski workers who were bristling with frustration. But grievances went far beyond the matter of bread on the table. Safety conditions revealed the regime's contempt for workers. Poor ventilation turned chemical warehouses into efficient producers of occupational diseases. No standard norms for "safe," "bearable," or "dangerous" gas pollutants existed in Poland, and whatever standards were applied tended to be disregarded in periods of high production—that is, during storming. Warm clothing and appropriate footwear were lacking for those forced to work outdoors in winter. The sickness and accident rates were high at Cegielski and in 1956 they were getting higher.[26]

Such is the view from afar—Cegielski as seen from the outside. It is relevant to recall at this point the duality about Solidarność that was at work in Gdańsk on August 23, 1980. The distinction emphasized was the sharp contrast between the visible happening—the almost pastoral scene of peaceful, song-singing, poetry-reading shipyard workers intermittently engaged in observing the formal negotiations of their Strike Presidium—and the invisible struggle waged against the security police by unknown numbers of leaflet printers, couriers, and traffic coordinators. It was a war that by August 23 had spread across whole regions of northern and central Poland. Detailed knowledge of such hidden events transforms the meaning of its public counterpart; gathered from the interior of the workers' movement, the invisible evidence explains why strike leaders were willing to risk bringing the talks to an end over the issue of "preconditions" even before the substance of their demands were formally considered. Properly understood, the workers' preconditions that their "helpers" be freed both from prison and from further repression constituted more than half the battle.[27]

The duality that surrounds the Poznań affair is just as stark and just as controlling. Knowledge of obscure dynamics at Cegielski transforms the meaning of the "Poznań events" that have generally become a part of the historical literature on postwar Poland. It is useful to summarize this conventional understanding of Poznań briefly and then—with what might be termed the "received history" in place—to probe into the society of Cegielski workers in pursuit of the deeper social dynamics that generated the crisis. The second body of evidence fundamentally alters, as it does for Gdańsk, the existing perception of the underlying politics of social life in Poland.

The received historical tradition about Poznań takes the following trajectory. Poznań workers had a number of legitimate complaints about the state of the economy, grievances that were compounded by the debilitating effects of newly revised tax, wage, and production schedules that had the practical effects of reducing worker income. In June 1956, after attempts to resolve their grievances at the local level had been rebuffed by party officials in Poznań, a delegation from Cegielski went to Warsaw where they were rebuffed again. The following day, June 28, 1956, angry Cegielski workers spontaneously gathered in front of the plant and marched on the city center where they performed the acts that constitute the "Poznań events." They assaulted party buildings, ravaged the local police station, released hundreds of prisoners, made a bonfire of their dossiers, besieged the stone fortress that housed the security service, wrecked the radio-jamming towers that blocked out broadcasts of the BBC and Radio Free Europe, and, finally, built barricades and waged a pitched battle with security forces for two full days and part of a third. Casualty figures, much disputed, were many hundreds dead or injured, and many hundreds subsequently imprisoned. In the months that followed, a period characterized by the political thaw that accompanied de-Stalinization and the return to power of Władysław Gomułka, almost all of the imprisoned workers were acquitted or pardoned. It soon became clear that the "Poznań events" had been far overshadowed by the national confrontation between the leaders of the Polish and Russian parties. The successful resolution of this conflict, on Polish terms, ushered in a veritable revolution known as "the Polish October," which achieved significant improvement in the lives of all Poles and transformed for a time the cultural environment of the nation. In this historical reading, high-level political events in Warsaw in October consign the June affair in Poznań to a minor role—through, admittedly, the spectacular nature of the Poznań eruption has occasionally received a measure of superficial narrative attention.

Overall, the Poznań workers came through as aggrieved rioters engaged in an outburst of indiscriminate anger. As objects of historical analysis, their moment on the stage of Polish history can be four days—one at a conference table in Warsaw, and three more in the streets of Poznań. That time period can be broadly treated in the West—up to twenty or so pages in a book on modern Poland, a half dozen or so in a comparative history of dissent in the Eastern bloc—or conversely, it can be reduced to a single, sanitized line in approved state textbooks in Poland. At whatever length it is treated, the Poznań events can be understood as a dramatic signpost of discontent, proof that Polish workers can get irrational over matters affecting their pay envelope but proof also of their preoccupation with "stomach issues" and their willingness to travel in violent fashion on roads that lead nowhere. Like certain insurrectionary Poles in other centuries, they appear as ardent proponents of lost causes. In terms of relative consciousness, the Polish workers who become visible in this portrait may be judged to be angry, even justifiably angry, but they also emerge—to put the matter gently—as strategically unimaginative and tactically prone to collective rage. The affair at Poznań was courageous but doomed—an unthinking festival of the unlearned.[28]

Workers in this portrait are seen as they might appear from a high window overlooking the square in the middle of Poznań. Though faceless, they busily torch the party buildings, earnestly smash the radio-jamming station, and generally run about like so many agitated ants. As was to be the case for the Lenin Shipyard in 1980, the picture that emerges for Poznań in 1956 is a view positioned securely outside the social formation it purports to illustrate. It is demonstrably a dramatic scene, but all the causal dynamics are somehow beyond the horizon. Historical causation has disappeared. A mob is discernible but not people.

Some questions are clearly in order. What did the workers of Cegielski think they were doing? What kind of people were they, and what could possibly have fueled such breathtaking hostility? What had life been like at the engineering works? And if the general situation had been sufficiently grievous to have generated such massive frustration, what, if anything, had the workers tried to do to improve things before losing their minds and taking to the streets? To ask questions is to steady the lens and offer the prospect of zooming in for a detailed view of this stunning moment.

For a U.S. audience, it is a fairly simple matter to create a verbal description of daily life on the Cegielski production lines that resonates within the U.S. experience. In general, social relations between low-level party foremen and rank-and-file workers at Cegielski can be likened to those in union-threatened plants in the United States in the late nineteenth and early twentieth centuries down to the time the Congress of Industrial Organizations (CIO) made its breakthrough in the mass-production industries in the 1930s. That is to say, managerial supervision at Cegielski was generally very strict and intermittently laced with a kind of institutional paranoia.[29] As was true on shop floors throughout Poland, section supervisors at Cegielski were especially sensitive about workers talking to one another. People whispering together on the job meant the sharing of grievances; if conversations were allowed to proliferate, grievances turned into trouble in the form of some sort of job action over work rules and production norms.

In 1956, party foremen at Cegielski had every reason to be on guard; under their gaze, production workers were *en route* to building a plant-wide structure of workers unprecedented in the history of postwar Poland. The results were to reverberate through national politics in 1956—not merely on the occasion of the "Poznań events" in June, but throughout the rest of that highly dramatic year.

The initial organizing conversations at Cegielski were conducted in places beyond the gaze of foremen—in trams and buses to and from work, in remote sections of the plant, at lunch breaks, and in the grossly inadequate cold-water locker rooms (which in themselves constituted one of the continuing grievances at Cegielski). As individual workers made judgments about the steadfastness and trustworthiness of one another, meetings among them could be conducted at night in worker apartments. Such a stage of organizing was, needless to say, a risky business in a police state, but it was necessary to the prolonged discussions essential to strategic planning. Because of their actions, these particular participants can be given a descriptive identity: we may call them "activists" or "militants." The pivotal tactical challenge facing them—the recruiting of rank-and-file workers—was necessarily a democratic activity that had to go on in the plant.[30]

Organization at Cegielski-Poznań was centered in the largest division of the engineering works—the railway freight-car section known as W-3. Unknown to anyone—it was the kind of industrial truth that became obvious only when it became visible—the railway freight-car section contained a unique conjunction of rank-and-file production talent, rank-and-file grievances, and rank-and-file leadership. The three thousand workers of W-3 contained a higher proportion of skilled craftsmen than any other division at Cegielski. Under the work rules and pay schedules revised in the late stages of Bierut's stepped-up six-year plan, these workers were, in fact—and certainly felt themselves to be—the most obviously discriminated against of the entire work force. Tellingly, the new piecework rules constituted a savage component of an evolving crisis in job safety. A report prepared after the "Poznań events" conceded that "highly qualified workers were exposed to more serious accidents because their work quotas were determined more precisely and, therefore, more rigorously than the work of ordinary laborers. In order to keep up the pace, workers have often removed all kinds of safety shields on machines' casings which hindered the performance of [speeded-up] technological operations." Despite working at a "storming" pace with unsafe equipment, Cegielski machinists in W-3 were so victimized by supply shortages and other vagaries of storming that real wages throughout the first half of 1956 dropped below those of 1955. The new wage structure, coupled with the persistence of the production methods, seemed to open up the possibility of a permanent reduction of a very modest standard of living.[31]

Many in W-3 literally risked their eyesight and their limbs in an effort to produce enough in the last part of the month to maintain their income. This way of living gradually became a routine risk, beyond the vision of other sectors of Polish society. For the expert machinists of W-3, the steady upward tilt in the charts of job accidents levered daily industrial life to a level of unprecedented strain. As many workers were hurt on the job in the first half of 1956 as in all of 1955. The statistic was a measure of worker desperation as much as one of job safety. Cegielski workers knew they were not building socialism: they were participating in a life of high-risk peonage.

The deepest problem at Cegielski, however, went beyond these hazards. It was not simply a matter of job safety or wages or housing or food. The biggest problem was that no one would listen. As a woman worker put it: "We would send out our ideas, we'd rush, speak, write and what'd become of it? What would those on top need a difference of opinion for?" In Poland, working without safety shields in a desperate effort to hold a modest income intact was an act that simply did not count because the views of workers did not count. As a section leader put it, "the game had to be broken off."[32]

It is helpful at this point to specify several underlying modern presumptions that have the effect of obscuring the dynamics that actually were at work at Cegielski. First, while specifying the grievances of Poznań workers, it is also necessary to emphasize that the historical evidence is overwhelming that grievances do not cause democratic movements. In comparable or lesser degrees, workplace grievances had been endemic across Poland in 1955, but mass organizing of the work force did not occur. Grievances do no translate into movements; they merely make them possible. Movements happen when they are organized. They happen in no other way. It is upon this activity, then, that it is necessary to focus, for it provides the early milestones on the road to Solidarność.

In hierarchical societies, self-activity is not something that outsiders, including scholars, can "date," though the effort is routinely attempted. To most participants in W-3, "self-activity" began when theirs did—when they first saw the corruption of a section foreman in, say, 1951; when a particularly grievous accident occurred in, say, 1953; when a singularly offensive injustice befell them personally in, say, 1955; or at that decisive psychological moment (it could have come as early as the end of the war in 1945) when they internalized a settled set of antiparty beliefs. All such dates can "fix" the moment of incipient insurgency. Every worker at Cegielski has his own date for the origins of the "Poznań events."[33]

But in practical terms relating to the process of democratic movement building, it was in March 1956 that self-activity at Cegielski may be said to have reached the stage where it became visible and thus historically recoverable after the fact. The first worker to "know" this was Edmund Taszer. Taszer was chairman of the party plant committee in W-3; in effect, the chief watchdog over the platoons of section-level watchdogs the party had strewn through the process of producing railroad freight cars. He was a conscientious conduit of shop grievances to higher authority. Quietly but resolutely, he passed on worker complaints about a score of issues vital to the self-respect if not the sanity of production workers.

The response of the party—rhetorical evasion usually followed by an absence of correctives—gradually politicized shop-floor activists. The most outspoken was a carpenter named Stanisław Matyja. He provided Taszer with so many grievances that Taszer had to become a compiler as well as a conduit. A thoughtful man and a careful functionary, Taszer pressed his positions upon the plant's top management in a calculatedly judicious manner—as Matyja pressed him (with somewhat more passionate specificity).

The rhythms of movements at the moment of birth are discernible at Cegielski-Poznań. When the same machine would cause the same accident for the same reason and for the third time, an increasing number of Cegielski workers, in the words of one of them, would "fall into thought." "They'd calculate in their own ways" about how to circumvent the shop foremen. In due course, "they'd gather in small groups." They would "discuss during lunch breaks" or "in railway cars" beyond the hearing of the party. And eventually, they would talk to Taszer, and he would talk to the party leadership, and nothing would change.[34]

Out of this self-activity came new worker postulates—quite reasonable amendments to existing work arrangements—and reasonable calls for revaluations of piecework and production norms. But shop section foremen were thrown into a frenzy by these coordinated suggestions because, quite simply, they had no power to formulate answers. An organic contradiction within the structure of "democratic centralism" became highly visible. The regime had grounded itself in a revolutionary history in which worker militants and social activists were portrayed as architects of the future; this official ideology did not make it operationally simple for party functionaries to oppose workers on theoretical grounds. It was necessary for the party to sell a distinction between acceptable activism, described as "healthy," and unacceptable assertion, characterized as "antisocialist." In real terms, healthy activities were those generated and controlled by the party; all other sources of activity and all other activities were subversive. Such a reasoning process not only controlled economic life by generating a production "plan," it locked party spokesmen into its context. There was no way that party functionaries—even should they have the desire—could at one and the same time be candid with Cegielski workers and loyal to higher party authorities who had produced the plan. The postulates of the workers could be evaded but not recognized as existing. Theory aside, the operational dynamics of democratic centralism translated into the need for party functionaries to polish the art of evasion as a central ingredient of managerial skill and of career advancement. Party officials in the Cegielski plant, therefore, behaved in a routinely equivocal manner. This tactic functioned in a limp-along fashion as long as workers thought of themselves as individuals and accepted the party trade union as the collective.[35]

But this was no longer the case at Cegielski. Indeed, it is timely as this juncture to describe the emergence of a certain democratic form in Poznań. By the spring of 1956, the immediate area of conversation around Stanisław Matyja, occupied as it was by self-selected worker militants from most of the shop sections of W-3, had become a kind of free democratic space in People's Poland. Herein, the norms of control were being fundamentally violated. People were talking to each other about what they really believed; indeed, they were talking about those things that mattered to them most. This space was not a gift; it had to be created by people who fought to create it. It was, in short, the product of self-activity. And if it could grow and somehow find a way to protect itself, it constituted a series of escalating possibilities that had transforming potential.

How should this "space" be characterized? As of May 1956, it can be called, somewhat grandly, the "popular movement"; it is, in any case, the only self-organized popular movement functioning in Poland.[36] Theoretically, if it can recruit well enough and survive, it can be a launching pad for a new more democratically structured society. Should it be able to maintain internal democratic forms, it is possible to conceptualize this "space" as an incubator of new modes of democratic social relations in Polish culture. Transparently, the people who occupied the space around Stanisław Matyja would have to know or quickly learn through their own efforts a great number of things for any of these potentialities to come to life. In any case, in theoretical terms, the potential may be seen to exist because the self-organized democratic space created by workers had not disfigured itself or collapsed internally or been repressed out of existence.

It was, nevertheless, an extremely fragile space at the beginning of June. As a self-generated social formation, it had no institutional presence. It did not even have a name. It was just a group of Cegielski workers (by that time representing the hopes of almost all of them) who had found a way to talk to each other seriously and had begun reaching agreement on what they talked about. These discussions, it turned out, were largely defensive, a restatement about the relationship to worker dignity or unresolved issues about production norms, wages, taxes, and job safety. In short, the original propositions put forward at

Cegielski can be judged to be cautious because the space in which they had developed was tentative, because it was not protected social space. Proposals for social change developed in that space were warily advanced because they were untested by experience and, therefore, were relatively unreflected upon. But because of their own subsequent efforts at self-organization, Poznań's workers gave their ideas a tangible social form, when previously such thoughts had been only an unspecified longing locked silently within the mind of each intimidated worker. Silence was the norm at the beginning of 1956; the new social form was the product of worker self-activity in the spring and summer.

The first recipients to feel the impact of the new form were, inevitably, the lower echelons of the party at the point of production on the shop floor. The workers' cause made sense to anyone who understood production. Because the trade-union foremen had no authority, they soon found themselves joining the workers in putting pressure on successive layers of the apparatus to "do something." Throughout May and June, worker representatives met with layer after layer of trade-union and party functionaries in the sections, in the plant, in the city of Poznań, and finally at a climactic moment with the party hierarchy in Warsaw. At appropriate stages of this process, more meetings of workers were called to receive reports of what had happened at each negotiating session and to debate what to do next.[37]

Such activity comprises the very essence of consciousness-formation and clarifies how the Poznań organizing effort fortified the self-respect of individual workers. It also clarifies how that development, in turn, generated the first fragile beginnings of collective self-confidence within the work force as a group and, as such, altered the sense workers had of their relationship with the party-state. Because their ideas had now found a place where they could be expressed and verified by peers, these workers no longer were forced to respond with routine acquiescence to routine party evasion. In Poznań, working-class "consciousness" was not a word that described some mystical point in an abstract Marxist paradigm, nor an empty category extruded from a static theory of bourgeois sociology; it was a developing social fact of seminal relevance, but one that was not comprehensible because the structural components of the "fact" were not visible to outside observers and thus, were not understood. The social fact that *was* seen in Poznań was a static one, described in static terminology, rather then a dynamic process understood as a sequential building block of a democratic social movement. The "fact" had to do with the specific organizing achievement of Poznań workers that was obscured by three distinct historical ingredients: (1) the sheer physical violence and drama of the "Poznań events" seemed to indicate that a "spontaneous" mass "uprising" had occurred rather than months of purposeful and highly relevant worker organizing; (2) the social dynamics generated within the ruling party in Poland by the events in Poznań took on such strategic importance, culminating in the "Polish October," that subsequent interpretive interest, worldwide, has focused almost exclusively on high politics in Warsaw in the autumn to the exclusion of an analysis of worker activism that originally set these dynamics in motion; and (3) there are prevailing intellectual assumptions—again worldwide—as to how social movements "happen'" that (as stated in the Introduction) constitute "views from afar." The relevance of this third factor is that it undergirds the kind of misreading of "consciousness" represented by, and organic to, the first two ingredients. Stated simply, the relevant social component of movement building represented by the Cegielski mobilization in Poznań was that, as a function of the actions of workers there, people had generated a place in which they could think, communicate, and—consequently—act.[38]

It was inevitable that Cegielski workers, in finding a way to push their grievances to higher levels of the party apparatus around them, would exhaust the evasions routinely available to local party officials. The day would come when provincial functionaries, publicly immobilized before a worker mass meeting, would in desperation call on their Warsaw superiors to step into the situation as the only way to prevent a plant-wide strike.

This moment came in June 1956 when the workers—after a tumultuous mass confrontation—received word to send representatives to Warsaw. Heartened by this accomplishment, the men and women of Cegielski took considerable care in electing delegates who would properly represent the far-flung shop sections of the engineering complex. Proper care in this sense seemed to embrace two criteria—a predictable fairness in representing different crafts and shop sections, and a special acknowledgment of the primacy of organizational experience and the clarity it generated. W-3, led by Stanisław Matyja, received two more delegates than any other section because it was the largest division at Cegielski and because of its role in developing the plant-wide communications network that was the heart of the entire worker effort.[39]

The elected delegation was augmented by the ritual inclusion of the party's troika—the plant manager, the head of the party in the plant, and the head of the official trade union. The threesome understood (whatever their superiors in Warsaw might have thought) that they would not—this time—be surrounded by silent delegates who would let the party leadership speak for a voiceless work force. In real terms, the delegation effectively remained a body of worker activists.

The trip to Warsaw produced what appeared to be unexpectedly clearcut results. There was, admittedly, an initial effort to intimidate. Interspersed among the workers' negotiating committee were "some strangers who never spoke." It was obvious to all that the pronounced bulges under their jackets concealed—or in this case revealed—pistols. After waiting to see if the workers were suitably subdued and determining from their statements that they were not, the minister of machine industry, Roman Fidelski, blandly accepted all the worker postulates. The high party official thereupon proposed to the delegates that they go home and return to work. The delegates, far more poised in June than they had been in March, insisted that Fidelski return to Cegielski with them to make the agreement clear to the work force in a mass meeting. This initiative from below produced what Matyja later described as "consternation" among party officials in Warsaw. Fidelski's task was at all costs to avoid a strike, because such an event simply translated into lost production never to be recovered. The party threatened and intimidated: "If you strike you will get tanks."

But the workers held their ground—a collective stance made possible by months of psychologically enhancing self-organization. Their enhanced sense of prerogative was, of course, both hard won and still tentative, but it armed them with enough sense of self that, in opposing Fidelski's tactics, it left the minister with only two options. He could nod to the security policemen in the room and have the delegation arrested or he could go to Poznań. After some consultation of the Central Committee of the party and some final threats, he agreed to journey to Poznań and appear before the work force. The Cegielski delegates were deeply satisfied with their work. Victory appeared to be at hand.[40]

On the train home that night, the delegation gathered in a compartment and talked through the next day's agenda. The basic decision was that "everything was to be clearly explained to the workers." The next morning, June 27, 1956, the people of Cegielski heard the good news in mass meetings. In the afternoon, Fidelski duly appeared. But things had changed. Like all party negotiators before him in the months of contention in Poznań, Fidelski suddenly became evasive on all relevant points. "I simply lost breath," said Matyja. "He was saying things quite different from what he had promised to say." An "indescribable uproar" promptly broke out. Members of the plant delegation to Warsaw angrily replied to Fidelski, and workers added their views from the floor. In the resulting din, Fidelski held his ground, strongly implored the workers to return to work the next day, offered some veiled threats, and, in due course, retired from the fray. In leaving, he assured anxious local officials and dismayed worker spokesmen that "everything would be all right."[41]

He was wrong. When Cegielski workers reported in at 6 a.m. on June 28, no machines were started up. Workers gathered in meetings throughout the plant. Many were aware that an international trade exposition was in progress at the time in Poznań. Thousands of influential participants in the world of commerce and politics, including foreign journalists, were present in the city. Here, then, was a rare chance for Poznań's workers to break through the censorship and let the world know about the true situation in Eastern Europe. The way to exploit this opportunity was through a politics conducted, as one participant put it, "out of doors" in public "where it could be seen." Led by the workers in section W-3, they began lining up in the factory streets, paraded by W-4, and then, massively augmented, marched to W-2, which was located near the main gate. The plant siren was liberated and turned on, a previously agreed upon signal for the start of a walkout.[42]

The party learned that morning that self-organization in Poznań in the spring of 1956 had extended beyond Cegielski workers. Militants in W-3 had made contact with known activists in many of the other leading industrial plants in the city. Thousands of Cegielski workers marched from Wildecki Market at the same time that the work force at ZNTK was assembled in a mass meeting. The leaders of the Cegielski march halted and quickly entered the gate to confer with the worker spokesmen of ZNTK. In short order, the march resumed with Cegielski's thousands, augmented by ZNTK's workers, moving toward the car barns of the streetcar workers. The party director there interposed himself and tried to speak. According to an eyewitness, he was pushed into a gutter and doused with "a barrel of old machine oil." Cegielski, ZNTK, and transport workers turned toward Kraszewski Street to meet there the women workers of the Komuna Paryska garment plant. The Komuna party committee had locked all the doors, and textile workers were yelling from the windows: "We want bread . . . freedom . . . religion." The gate was forced open, and the plant committee members bodily moved aside. The textile workers poured out, and the marchers moved on. The procession transformed the city as it moved through it. Onlookers waved, shouted encouragement, and in many cases decided to join in. Finally, what had become a civic mass reached one side of Wolność Square where, worker activists learned, their prior organizing had again borne fruit. Waiting patiently on the opposite side of the square stood slaughterhouse workers and men and women from other enterprises whom they had recruited. The marchers now constituted an army—probably 75,000—as they turned on Ratayczak Street and moved toward the center of the city. Into view at last came the party headquarters, popularly known as "the Castle." Suddenly, militia vehicles loomed in front of the marchers. Hurried conversations ensued between marches and militiamen. The latter got out of their cars and disappeared into the crowd. The word went through the throng, "The Militia is with us. Militia with the people." A precaution, however, was taken; some young workers deflated all tires on the militia vehicles.[43]

It became evident that Poznań's high schools and the university were not functioning, for a great many students appeared in the square, standing among the workers in front of party headquarters. Poznań at that time was a city of 380,000; estimates of the crowd at the great square in the heart of the city ranged above 100,000. Foreign visitors in nearby hotels received an instant insight into the regime's unpopularity. But Poles were even more impressed. In a society in which the most rudimentary needs could not be talked about, the assertive signs of the assembled throng were deeply satisfying. "Bread and Freedom." "We Ask the Party Where Has the Food Gone?" One sign acknowledged that Poland's Catholic Primate, Stefan Cardinal Wyszyński, had been under house arrest for years: "Free Poland's Primate." A central political truth visibly surfaced: for workers to express themselves on issues of economic survival or on job safety and housing, the censorship itself had to be challenged. Every issue was a political issue that touched upon the most fundamental questions of civil liberty. "Free Poland's Primate" was transparently an assertion of religious loyalty, but it was also inevitably a much more general cultural statement about party control and thus a political postulate on a fundamental level.[44]

The crowd called for the appearance of Fidelski, or anybody, to provide an explanation of the party's betrayal. A half hour passed, then another and another. A white flag finally appeared from a window signifying, perhaps, that party spokesmen were in short supply. If Fidelski, representing the central power, could not pacify them, no underling wished to volunteer for the assignment. Some youths then entered the building and soon reappeared at windows above the crowd. "Look how they are living here," they shouted, displaying fine hams and vodka of a quality workers rarely saw. In the lightning-like manner in which rumors capture crowds that have no clear purpose, it became more or less instant wisdom that the fine food and tableware had been set up for the visiting ministerial delegation from Warsaw. In a climate of thwarted hopes and deepening suspicion of the party, a second rumor circulated that the workers' delegation had been arrested. (They soon were—after these "events.") The thought at least had the merit of giving expression to society's long muted resentment of the party's police. Within an instant, the "Poznań events" began.[45]

Worker anger at law enforcement in Poznań had two targets—the provincial police headquarters housing the local prison, and the security police, the hated "U.B." To shouts of "free the prisoners," the more excitable individuals in the crowd began an attack on both. The provincial police headquarters was simply overwhelmed. Every prisoner was released, and police records were thrown out the windows, file upon file. The racks of dossiers were not only of prisoners but of sizable sectors of the population. A bonfire of dossiers was started to the cheers of "Freedom from police spies" and "Freedom from the Party." But the security-police building turned out to be a fortress that could obviously not be stormed. The crowd stood and shouted taunts. Stones were thrown. The police replied with fire hoses from upstairs windows. Some people in the crowd knew the location of the main water supply and promptly cut it off.[46]

That is about as far as an orderly sequence of the Poznań events can be reconstructed. Firing broke out, news of it spread, people with guns came, and something approaching a fixed battle began. A thirteen-year-old boy suddenly fell over, hit fatally by a police bullet. A Polish flag was dipped in his blood, and things became considerably wilder after that. Yet if a mass civic eruption can ever be said to possess any kind of order, there was a logic of sorts to the remarkable series of events that ensued. Crowd actions were self-evidently grounded in ideological moorings. "Down with the U.B." soon became "Out with the Russians." "Freedom from the Party" quickly turned into a mass attack on the eleven-million-dollar radio-jamming station that the regime had erected in Poznań to combat Radio Free Europe and the BBC. The jamming tower, rising from the top of a four-story building, was protected by one maintenance man. He was quickly persuaded to be cooperative. Electronic equipment was ripped off the tower and tossed into the street below. An enlarged sense of the possible seized members of the crowd. Not only was the assault on the police intensified, but the party headquarters was invaded. Streetcars were overturned at downtown intersections to serve as barricades. Some workers left and returned with huge trucks loaded with cement. Like the streetcars, the trucks were also overturned and converted into street barricades. About noon, the police, badly outnumbered and pressed into side streets, were augmented by army troops. The latter proved unreliable, however. A tank crew turned its vehicle over to the crowd and left. It was probably fortunate that no one knew how to operate its guns. Sharp distinctions appeared in the way the crowd responded to different sectors of power. While no soldier was treated as an opponent until his actions so demonstrated, any member of the security police was regarded as an enemy on sight. One security agent, rumored to have shot a woman, was captured and terribly beaten. An ambulance arrived, but the crowd refused to let anyone through until his attackers assured themselves he had bled to death. It turned out to be a case of mistaken identity.[47]

Throughout the afternoon, the population slowly took possession of the entire center of the city. In scores of isolated incidents, individual Poles made political statements in their own way. A "For rent" sign was hung over party headquarters. Another group of workers tore down the "Stalin Works" sign over the Cegielski plant and smashed all the statues of the Soviet dictator they could find. Still another group went to the international exposition and ostentatiously tore down every Soviet flag in sight, climbing rows of flagpoles to do so. To recapture Poznań, the government finally had to import special security units equipped with heavy weapons and tanks. The revolt, which began early on June 28, staggered on until the last snipers were rooted out on June 30. As in all cases in Poland's postwar history, casualty figures were much disputed, but a total of six hundred to seven hundred injured and dead is probably a conservative estimate. Aside from the awkward fact that the Poznań uprising was seen by hundreds of foreigners, including a considerable number of Westerners, it was simply too big an event to be hushed up. Though organized by Cegielski workers and augmented by thousands recruited from other factories, the eruption was a multi-class expression that ranged across all age groups and included women as well as men.

In the shocked aftermath of Poznań, the party hierarchy elected to interpret the worker rising in classic Stalinist terms. Prime Minister Józef Cyrankiewicz said: "Imperalistic centers and the reactionary underground hostile to Poland are directly responsible for the incidents . . . Every provocateur or maniac who dares raise his hand against the people's rule may be sure that in the interest of the working class . . . the authorities will chop off his hand." *Trybuna Ludu*, the official organ of the party, echoed this perspective, terming the Poznań rising a "provocation" that was "organized by the enemies of our fatherland at a time when the Party and the Government are greatly concerned with eliminating the shortcomings in the life of workers and making our country more democratic." There seemed no limit to the piety the party press could summon at moments of popular politics. "The people's enemies, foreign agents, chose that moment [of the international fair] to discredit the working class and the Party whose main effort is the improvement of the standard of living of each of us."[48]

But the political tides were running against this time-honored way of seeing, both within Poland and elsewhere in Eastern Europe. In his own time, Stalin had been able to impose his monomania on the entire bloc. Now, Stalin was dead and Poland's equivalent, Bolesław Bierut, had died in February 1956 in Moscow. The latter event came barely two weeks after Nikita Khrushchev's famous exposé of Stalin's crimes at the Twentieth Party Congress. Khrushchev's speech not only inferentially discredited the whole system of rule through terror that had come to be institutionalized in the bloc as Stalinism, it also vastly complicated the political future of every (necessarily Stalinist) ruling-party clique in Eastern Europe. The entire year of 1956 was one of uncertainty at the top in Poland, a circumstance that encouraged a divided Politburo to relax the censorship slightly and pay closer attention to working-class discontent in Poznań and throughout the country. The intelligentsia, numbed by years of doltish Stalinism, also began stirring, a folkway particularly visible among young, bright, and sanctioned members of the party. Their "revisionist" efforts were augmented by signs of restlessness in the literary world, visible since 1955, and also extended to the technical intelligentsia. Economists, particularly, had long been embittered by the systematic statistical deceit employed by the party to conceal the extent of mass privation enforced under Bierut's six-year plan.[49]

But the most tangible change came at the top. Hard-liners who had in 1948–49 utilized Stalinist dogma to topple Gomułka and usher in the dark years of Bierut now stood virtually naked in the presence of the Khrushchev line. Acutely aware of their vulnerability, they resisted calls for a party plenum and clung to power. But the unanticipated Poznań events constituted a mortal blow. Orthodox hard-liners pretended it was not such a blow at first; the prime minister's scathing denunciation of Poznań workers can best be understood as a blunt attempt to maintain business as usual. But many Warsaw insiders knew that trouble at Cegielski had been brewing for months and that all talk of "foreign agents" in the Poznań working class was just handy nonsense to conceal the blockheaded style of governance of the Stalinists.

In the immediate aftermath of Poznań, even as some three hundred workers were being packed off to prison, the pressure for a plenum became irresistible. The long-delayed affair, held in July, verified that a new balance of forces existed at the top of the hierarchy. It was in the direct interest of an emerging anti-Stalinist faction that both the popular suffering under Bierut and the discontent that sprang from it should be quickly acknowledged. Such public revelations, promptly unveiled, had the effect of locking the old party leadership in an even narrower defensive position. In August, the word went out that the imprisoned Poznań workers were to be permitted to defend themselves in open court. Furthermore, they were to have good lawyers, who were suddenly given a relatively free hand. Meanwhile, Edward Ochab, the party's first secretary, quietly sought out a man who had for years been under nominal house arrest—Władysław Gomułka himself. Some kind of changing of the guard at Poland's highest levels seemed increasingly probable. That it might involve Gomułka merely added a fine edge of irony to the politics of 1956. When Stalin had read Tito out of the Eastern bloc for his "deviations" in 1948, only Poland's Gomułka had not dutifully joined in the chorus of anti-Tito propaganda that coursed through the satellite nations. Gomułka promptly paid the price—he lost power. Now, for reasons having to do with Soviet domestic policy, Khrushchev had begun reversing course and had personally flown to Belgrade to dramatize the new policy of tolerance toward Tito. It was a move that implicitly acknowledged the possibility of different roads to socialism. Each satellite apparently was to be permitted a measure of latitude in administering domestic policy, within what limits no one could say. Tough, incorruptible old Gomułka suddenly seemed highly useful to the Polish national cause. On August 5, he was readmitted to the party. The Stalinists who had denounced Poznań's workers were suddenly reeling. When the Poznań trials opened in September, it seemed possible that the Bierut era might be more on trial than the worker defendants.[50]

In a society as forcibly constrained and therefore as latently explosive as Poland, the possible routes to de-Stalinization were by no means clear. Most options appeared dangerous. The Poznań events, alarming in themselves, indicated a concealed time bomb lurking in other parts of Poland. Two possible courses seemed open. The party could placate, but it also needed to control. In the immediate aftermath of Poznań, hard-liners hung on desperately to power. To them, it seemed imperative that Cegielski-style organizing not be duplicated elsewhere for the simple and obvious reason that the Poznań precedent had frightening implications. But denunciations of workers as antisocialist agitators merely heightened the problem, because self-organization at Cegielski had proceeded with what might be called democratic non-discrimination. Among the rank and file of the shop floor, party and nonparty people spoke at meetings, lobbied together in Warsaw, and marched shoulder to shoulder to downtown Poznań—facts that simply verified the broad-gauged alienation rampant within the Polish working class. When Prime Minister Cyrankiewicz threatened to "chop off the hands" of the instigators of Poznań, the party's logic simply obliterated party members within Poznań's working class. At best, they had become nonpersons whose actions could not be seen; at worst, they had become foreign agents. In either case, the cynicism embedded in the hard-line approach was thus exposed within the party itself with a clarity and intensity that had happened only once before—at the time of the initial consolidation of power from 1945 to 1948. Added to Fidelski's bald deceptions as well as the mass arrests of workers, Cyrankiewicz's bluster destroyed the last vestige of rank-and-file allegiance in Poznań. The party simply melted away. Among Cegielski's fifteen thousand workers, scarcely a handful of believing communists remained.[51]

But the party's new tactic of trumpeting a worker-party coalition also carried its own dangers in the aftermath of the June upheaval. At the July party plenum that the Poznań events had precipitated, the anti-Stalinist faction, evolving into a pro-Gomułka faction, hurriedly reorganized the new five-year plan, allocating less for capital investment and more for consumer goods. But out of personal ambition, the conciliators went beyond these measures and started playing with fire. They sent signals through the Warsaw party secretary to the factories in the capital that worker councils were looked upon with favor. An activated working class could be counted on to be an active anti-Stalinist force. What other active things it might do was an open question. Loose dynamite was lying everywhere in Poland's working class.[52]

This maneuver by the party introduced some complicated new dynamics into working-class politics. To the Warsaw party hierarchy, the most visible enterprise in all of Poland was the industrial jewel of the six-year plan, the (old-fashioned) "ultra modern" Żerań automobile plant in Warsaw itself. Employing thirty thousand auto workers, Żerań was one of the superplants designed to propel Poland to a world-competitive level in autos, as Nowa Huta was supposed to do in steel, Oświęcim in chemicals, and Zambrów in textiles. Odd planning oversights, interference by the Soviets, and simple management blunders had marred all of these undertakings. The bitter truth, which the party attempted to conceal from the nation and in some ways from itself, was that because of planning errors, the Żerań factory in 1957 required 230 workdays to produce one Warszawa automobile (a Soviet-modified prewar model). It required 75 workdays to produce a smaller but more modern Skoda. But only 23 workdays were needed to make one Volkswagen at a state-owned West German enterprise. One billion złotys of the nation's wealth was invested in the Żerań facility. The state, however, controlled all social information; blandly, Żerań was habitually pushed forward as a symbol of new socialist Poland.[53]

Thus—improbable as it seems—the first worker council in the Warsaw region was organized in September at Żerań and organized by the plant party committee! How risky this business was soon became clear. Open meetings of workers swiftly became loudly democratic. It turned out that Żerań workers, who, after all, lived under the same Leninist norms as Cegielski workers, had an imposing array of grievances. The party committee was overwhelmed. Their credibility in doubt, party members called on high trade-union officials to come in and calm things down. It was time for a bit of time-honored "worker consultation." Actually, this practice was more than a simple fire-fighting tactic; it was a richly brocaded tradition. As "consultation with workers" functioned in Poland, trade-union administrators responded to worker disaffection by explaining that the official union had, indeed, made some "mistakes." Local union representatives had "drifted away from the masses." The party announced that it appreciated this fact and understood the protests of workers as needed signs that the party trade union "had to get back to basics." The folkway of consultation contained a traditional conclusion: the signs of worker disaffection could only be regarded as "healthy" if the workers avoided falling prey to "anarchistic and antisocialist elements" that were trying to exploit the "just grievances and postulates of the working-class." Antiparty manipulators possessed hidden political motives rather than a genuine concern for the welfare of the working class. The actions of such elements stood in the way of reform and inhibited the party's "return to Leninist norms." Such was the rhetoric of consultation.

But in Poland in the autumn of 1956, these party clichés simply enraged people. At the Żerań factory, angry workers forced trade-union leaders from the party center into a bitter public confrontation that lasted seven hours. "The unions are a cancer on the body of the working class," the officials were told. An old socialist activist from the prewar unions confronted Wiktor Kłosiewicz, the national head of the trade-union apparatus, with these words: "Nobody ever asked us if we want to belong to the union. The dues are deducted from our pay and no one asks us. And what do you do with our money? It goes to provide the comfort of our bosses . . . Well this is not socialism and not trade unionism the way we old workers see it." When the subsequent discussion centered on a comparison of Kłosiewicz's salary with Żerań workers' incomes, it became abundantly clear that the old formulas of "consultation" had lost much of their utility. This was evident not only at Żerań but also in Silesia, on the Baltic coast, and in a dozen other centers of industrial concentration, including, of course, a now notorious engineering works in the city of Poznań.[54]

October 1956; it was the month the Poznań trials came to an end with the world press reporting a flood of acquittals and suspended sentences; it was a month of unveilings of factory commissions, shop councils, and new self-management formulations in some of the largest factories in Poland; and it was the month of the return to power of Władysław Gomułka. It was, in short, a moment of nationwide hope and expectation that the time had finally come for the "Polish path to socialism." In the calendar-conscious lexicon of the nation's history, it was an epic juncture in postwar history—"the Polish October." The momentum of October 1956 was to shatter the ingrained habits of Stalinism and bring much of the Eastern bloc to political crisis.

As the chief executive of a state corporation saddled with a more or less permanent labor problem, Władysław Gomułka influenced in decisive ways the direction the Polish working class was forced to take along the road of democratic experimentation that led to Solidarność.

The radical milieu that nourished Poland's prewar socialists and communists, including men like Gomułka, was not one that could live with the time-honored bourgeois assessment that all of Poland's troubles could be explained simply in terms of geography. The Polish émigré writer V. S. Karol characterized the world view of the Polish Left at the end of the war: "The basic premises of the situation in Poland had not changed; without a socialist revolution she was condemned to vegetate in mediocrity, and sooner or later to disappear from the map of Europe. The dilemma of the country was that in order to live, crushed between Germany and Russia, she had to be powerful and internally healthy. Without social justice, industrialization and an equitable distribution of the national income, her independence was a dream. Geography would not allow Poland the luxury of a new regime of colonels."[55]

This seemingly straightforward assessment was undermined by the impact of Leninism on the socialist dream; the lesson of the postwar years in Poland corroborated the earlier experiences in the Soviet Union: one-party rule produced a regime of colonels. While this historical imperative quickly became clear to a socialist journalist like Karol, it was not clear to a Bolshevik colonel like Gomułka. The tragedy for Poland of Gomułka's politics was that though his "Polish path to socialism" offered at least a chance of freeing the nation from the suffocating embrace of the Soviets, it could not, given Gomułka's own preferences, dislodge the subtle and equally suffocating embrace of Leninism. Gomułka's encounter with the Polish working class merely added one more convincing verse to this continuing litany.

The incessant irony in Gomułka's relationship with Polish workers was vividly illustrated by the decisive impact of the relationship upon his personal career as a communist politician. He originally came to power in 1945 without the support of the Moscow-trained party faction because his reputation as a Polish patriot gave him popular credentials no other prominent party leader possessed. It was Gomułka who in the name of the party presided over the initial intimidation of the nation's working class. By the time that task was completed in 1949, Polish workers hated him. But for that very reason, as well as his "Titoism" (which made him vulnerable to Stalinist attack), his usefulness to the party had been exhausted, and he became dispensable. Yet, it was the working class that unwillingly played a tactically critical role in his return to power in 1956, and it was to be the working class that would bring his career to an end in 1970.

Given the intricacy of these relationships, it is necessary to review certain features of the intraparty struggle in 1956 and the role of Polish workers within that struggle in order to account for the enormous emotional meaning of "the Polish October" to the entire nation. The popular position of party hard-liners had been fundamentally undercut, of course, by the simple ruthlessness of Bierut's first six-year plan. But their rule through police terror was somewhat weakened by Stalin's death in 1953 and much further weakened by Khrushchev's detailed revelations of the crimes of Stalinism in February 1956. With all this, it was the Poznań events in June 1956 that fatally eroded the strength of Politburo hard-liners and permitted a new balance of forces to emerge in July. The climax came in October when anti-Stalinist forces in the party made their move to install Gomułka at the head of the party as a symbol of the party's hunger for greater independence from Moscow.

Khrushchev brought to bear upon the Polish party all the pressure he could in the autumn of 1956; when this seemed to be unavailing, the entire Russian Politburo descended on Warsaw in mid-October to prevent Gomułka's ascendancy and to preserve firm Russian dominance over the Polish party. The contest of wills that ensued was tense, enormously complex, and ultimately exemplified by military demonstrations of force. Soviet troops and Soviet-officered Polish troops moved on Warsaw and other cities, while Polish-officered troops and security formations were ordered into positions to intercept them. At the climax, the Polish party played its last card and dispatched couriers to Warsaw factories with full particulars on the struggle, including the names of the reshuffled Central Committee that Gomułka sought and the Soviets opposed. Prominent among those *not* included was the Russian officer who headed the Polish army, Marshall Konstanty Rokossovski, whose presence on the Central Committee symbolized for the entire nation its abject subservience to Moscow. In angry moments of confrontation, the Soviets bluffed and threatened, and the Poles parried and remained firm, Khrushchev and Gomułka excelling in their respective roles. Khrushchev's demand that Rokossovski be reinstated along with other Soviet favorites was turned aside with the counterannouncement that the Polish list of the new Gomułka-led Politburo had been made public and had been circulated among the Warsaw working class. In this way, Gomułka moved to define the struggle as a test of Polish patriotism in behalf of a genuine measure of national independence. Such was a cause in which the entire society could participate with heartfelt enthusiasm. Workers remained in the factories throughout the tense weekend as a wave of national longing swept through the Polish capital. And then the announcement came that the Russians had gone home and Gomułka had come to power. The victory of the Polish party was in a very real sense a victory for the nation and explained the immense outpouring of popular ecstasy that accompanied Gomułka's accession to office on October 22. These emotions imparted to the Polish October its fervent power. In such a climate, the workers' drive for autonomy in the factories simply could not be deflected.[56]

It was in this setting that Władysław Gomułka addressed the party's Central Committee and a nationwide radio audience in his first public declaration as first secretary of the party. His words seemed to put an official stamp upon the Polish October by signaling the beginning of a radical new relationship between the government and the population. "The cause of the Poznań tragedy and of the profound dissatisfaction of the entire working class," Gomułka said, "are to be found in ourselves, in the leadership of the party, in the government." He noted with approval that following Khrushchev's denunciation of Stalin at the Twentieth Party Congress eight months before, the party's "silent, enslaved minds began to shake off the poison of mendacity, falsehood and hypocrisy." He went much further:

> The Poznań workers did not protest against People's Poland or against socialism when they went out into the streets of the city. They protested against the evil which was widespread in our social system and which was painfully felt by them . . . The six-year plan has not brought the rise in the standard of living promised by its promoters. To pretend that real wages had increased 27 percent only irritated the people further . . . The working class recently gave a painful lesson to the Party, the leadership and the government . . . They shouted in a powerful voice: "Enough! This cannot go on any longer! Turn back from this false road." . . . I am convinced that the workers would not have resorted to a strike and that blood would not have flowed on the Black Thursday of Poznań if the Party—that is to say, its leaders—had spoken the truth . . .
>
> It is necessary to tell the working class the truth about the past and the present. There is no escaping from truth. The leadership of the Party was frightened of it. The loss of the credit of the working class means the loss of the moral basis of power. It is possible to govern the country even in such conditions. But then this will be bad government, for it must be based on bureaucracy, on infringing the rule of law, on violence . . ."

Gomułka went on to define three specific relationships that a successful party needed to achieve—the first with its own working class, the second with the Soviet Union, and the third with the Catholic Church.

> These relations should be shaped on the principle of working-class solidarity, should be based on mutual confidence and equality of rights; on granting assistance to each other; on mutual criticism, if such should be necessary . . . Abolition of the exploitation of man by man is the essence of socialism . . . Within the framework of such relations each country should have full independence, and the right of each nation to a sovereign government in an independent country should be fully and mutually respected. This is how it should be and . . . this is how it is beginning to be.

Gomułka signaled that he had already decided to fashion some kind of accord with the Polish episcopate and to free its head, Stefan Cardinal Wyszyński, from his long house arrest. Catholics, Gomułka told his fellow party members, could be enlisted in the national cause: "It is a poor idea to maintain that only Communists can build socialism, only people holding materialistic social views." Two nights later, Gomułka reiterated the same themes in an effective speech to an ecstatic Warsaw throng of half a million people. He even augmented his previous praise for the working class by commending "its noble behavior during its difficult days." But he ended his appeal for an immediate return to work with a revealing choice of words: "Enough demonstrations, enough meetings."[57]

This was the heart of the matter. The workers were talking too much to each other. With a push from the fallout over Poznań, the end of Stalinism had essentially been achieved within the framework of the party hierarchy itself. Gomułka's return from the wilderness symbolized this party achievement. But other social forces had been unleashed. Indeed, the deepest strategic meaning of the Polish October lay, in fact, in the far-flung self-organization that the Polish working class had embarked upon. And it was this specific dimension of Polish events that could not be invested with official praise by Gomułka or any faction of the party he headed. Popular expression among Polish workers had come into existence in spite of the party and in opposition to it. As a cultural ingredient in Polish society, it was still quite fragile and not at all grounded in experience over time. Its driving impulse could possibly be manipulated, and there were a few thoughtful activists in working-class ranks who felt it already had been to some extent—in the course of the Byzantine maneuverings that surrounded the party's confrontation with the Soviet leadership. But if so, that utilization of the working class by the Gomułka faction had been in an acceptable cause, and it did not prejudge the outcome of the task that now loomed before social activists in the working class.

One thing was abundantly clear at the moment of Gomułka's triumph in October: the working class was in motion. The momentum toward self-managing shop councils was both startling in its intensity and ominous in its implications. The assertive stance of Poznań workers was in danger of becoming a national folkway. As for the mood at the source, almost all the Cegielski militants had been discharged from prison amidst the early flowering of what was later to be called the "Spirit of October." They had become well instructed by events so that they possessed a sober new understanding of the precise relationship between the working class and the party. Since the points of contention were quite concrete, the lessons learned were quite specific. An essential dynamic was clear: self-organization inevitably generated self-education about political realities in Poland.

The level of "consciousness" thus obtained could be specified with a certain precision. Cegielski workers had begun in early March with the knowledge that low-level supervisors in the official trade unions at the plant could not be trusted. This knowledge was an inheritance from the past, confirmed by the totality of union-worker relationships that had transpired at the point of production during the nation's first six-year plan. As one worker explained, "There were several supervisors who would . . . suck up to the manager and cast blame on us, and then tell us that it was the manager who was responsible for the trouble. But as to their political views, oh, they were just crystalline. That world outlook was just strong elbows." Between March and June, as the base of worker involvement broadened through Cegielski's sprawling shop sections, the increasing pressure had the effect of discrediting successive layers of trade-union bureaucrats who were forced into the role of providing evasive answers. The trade-union apparatus, as one worker effectively put it, "was like a theater, like a mock democracy . . . It's just a satire, those seemingly democratic elections of representatives who are supposedly to take care of the workers' interests." And yet it was not merely that the union bureaucracy had become generally discredited; rather, specific levels of the bureaucracy had become specifically discredited on the concrete issues raised. The same fate gradually befell successive layers of the party, from functionaries in Poznań to ministers in Warsaw.[58]

But this was as far as the activists were able to take themselves in the initial mobilization during the spring months. Knowledge of power relationships at Cegielski did not extend beyond the lessons learned in formulating the specific tactics that propelled them into the office of Roman Fidelski in Warsaw on June 26, 1956. Until that moment, their contest against state power had been a distant, almost anonymous confrontation rather than a face-to-face experience. Fidelski's ministry was thus new terrain, a higher plateau of experiential struggle. Understandably—perhaps it is fair to say predictably—they had not thought through to the crisis that would immediately occur should Fidelski turn their proposals aside. Should such an impasse develop, the Cegielski leadership of 1956 would not know what to do. They faced an organic problem: How to acquire a voice, a presence, in People's Poland? They faced a mechanical problem: How to get the word out. How could they break through the censorship and let the world know?

In venturing into the Warsaw offices representing the highest levels of the Central Committee, Poznań workers were encountering a new experiential reality. Perhaps it is clearer to put the matter in negative terms. They had gradually equipped themselves to see through the evasions of party functionaries in the Cegielski plant and in the city of Poznań. Once party evasion was uncovered at one level, worker activists had simply kept the organizing pressure on until a higher level of the party had to be brought in to "pacify" them. The workers had persevered in this process until they had created a crisis that propelled everyone—provincial party leaders and provincial workers—into the office of Roman Fidelski in Warsaw. If Fidelski merely continued the process of deception, the workers, without prior analysis and planning, would not know what to do; there was no "higher level" of the party to which they could appeal. Their problem at that juncture, therefore, was both tactical and strategic: tactical in the sense that they had to find a way to "get the word out," that a new round of protest had to be attempted; and strategic in the deeper sense that they now had to find their own voice—one beyond the confines of the party. In short, confined by the party, they had to find some way to break out, to break through the censorship and let the world know of their plight. How to do this was the problem.

Matyja and others had talked about the possibility of seizing the occasion of an international trade exposition in Poznań to conduct some out-of-doors politics that Western visitors and everyone else could see. It was less a strategic plan than a kind of musing. They might turn on the Cegielski siren so the whole city would know that "Cegielski is out." It made a certain kind of sense, and it was uplifting just to talk about it, but what could an outdoor march actually achieve? Though they talked about it with activists from other factories in the city, no one could be certain, and no decisions were made. It was just an idea that hung in the air. Thus, when Roman Fidelski betrayed them all on June 27, the worker activists had no strategic plans. So they seized the idea. The siren blew, Cegielski marched, and the "Poznań events" became the chaotic result.[59]

In the evolution of movement consciousness, worker postmortems on June 28 can be understood as launching the second round of the struggle. The "Poznań events" were highly instructive in detailing what would not work. Not only were street rampages revealed as counterproductive, but a much subtler organizational imperative came into view for the first time: large numbers of people with little experience in self-organization had been recruited to the movement on a pivotal day of assertion when they would be asked to assert collectively.

If the richness and complexity of Solidarność in 1980 is to be understood, it is absolutely essential that the complexity of Poznań in 1956 be understood as rooted in the same dynamics. A misreading of Poznań inevitably portends a misreading of Solidarność. The social ingredient that united the people of Poznań on June 28 was their emotional identification with Cegielski workers. When the word first flashed that "Cegielski is out," the response of most people in Poznań's working class was one of instant attentiveness and shared hope. But that was about all it was—a heartfelt but vague impulse. Most people in Poznań knew little or nothing about problems of job safety or production irrationalities or the weighted work and wage norms that disfigured life in Cegielski's shops. But they felt they "generally" knew because they had the same general problems. Were work areas at Cegielski overcrowded? So were living conditions throughout the city. The woman whose deepest hope was for "a room of my own" was in the march on June 28. There were many such Polish workers, veterans of the organized madness of storming among them, who knew that morning that Cegielski was marching for them.

How did they respond? Many of Poznań's citizens did not join the march because of their fear of what a Baltic electrician would later call *władza*—the power. Yet it would be incorrect to say that they suffered from "apathy." They waved to the marchers and shouted encouragement. To the extent they were able, they participated in the shared hope. Cegielski was marching for them, too. To characterize the 100,000 or so who did place their bodies publicly in view of *władza*, other distinctions need to be specified. A number of shop-floor activists from throughout the city had been in touch with the Cegielski leadership and had spread the words of the movement to their own co-workers. The work force at ZNTK, for example, had mobilized themselves in a mass meeting on the fateful morning. Internal communication at ZNTK was rather easily activated. "Do we join Cegielski or not?" went the shouted question. The answer, unsurprisingly, was yes. Similarly, the transport workers at the car barn knew about the siren and were ready when it sounded. The slaughterhouse workers were waiting at an agreed upon crossroads when Cegielski's thousands appeared. Clearly, many activists had performed many organizing tasks in Poznań's factories.

It is necessary to be as precise as possible about what is and what is not present in these social facts. Taking the ZNTK factory as an example, the shouted question "Do we join Cegielski or not?" is scarcely the opening and closing of an engaged and thoughtful dialogue. Though almost all the ZNTK workers responded affirmatively to the question, the political statement they thought they were making as they marched down Rokossowski Street depended upon the range and depth of prior discussions in their own shop sections. Quite obviously, the general level of such prior discussions was as varied as there were shop sections at ZNTK or factories in Poznań; and quite obviously as well, the response of individuals to these discussions, at whatever level of intensity they had been conducted, varied greatly. Everyone knew or thought or hoped that they were marching—as the textile workers put it—for "bread" or "freedom" or "religion" or for what can be summarized as a measure of dignity.

Such a description, however, is quite general, and therein lies its interpretive weakness and ultimately its concrete political inadequacy. As political agendas, what did shouts for "bread" or "freedom" mean? What did the thousands of students who suddenly materialized in the downtown square know about shop-floor grievances and the precise stages of negotiations between Cegielski militants and party functionaries? Obviously, not very much. In contrast, the worker activists who had already achieved in their own plants a level of internal organization sufficient to initiate mass meetings in coordination with Cegielski's timetable had traveled a measurable distance down a path of democratic self-activity. It may be said that their consciousness of the issues at stake was relatively high. But among activists, differentiations in experience persisted throughout the city and even larger differentiations persisted among their tens of thousands of co-workers.

To penetrate beneath this very general level of interpretation, it can be said that the most highly politicized worker activists in Poznań on the morning of June 28 were the militants around Stanisław Matyja who had created, and for months had lived in, a certain democratic space—a self-created arena of discussion and striving—at Cegielski. In concrete terms, they knew more about the party and about the specific dynamics of negotiations between workers and the party than anyone else in the city. And among the city's thousands of workers, it was once again Cegielski's fifteen thousand who were most concretely informed, most experienced in collective assertion, and therefore most experientially politicized. The tens of thousands of other workers joining in the march had done less and therefore knew less.

In short, a central problem of democratic movements was made visible by the Poznań events. Cegielski workers stumbled into a crucial error on June 28, one that uncovered the extreme fragility of all that they had previously achieved. What the "Poznań events" revealed was the strategic vulnerability of a relatively high-consciousness group of people (self-organized Cegielski workers) when they voluntarily placed their politics into the hands of a much larger unorganized mass of people. The political price was high. The painfully constructed democratic space that had been created at Cegielski was made hostage to the unpredictable actions of friendly, unknown people. The internal connecting links essential to democratic movements did not exist between the Cegielski work force and all the other people in the march. Thus, an unanticipated gap in democratic organizing practices that guided Matyja's militants abruptly came into view. The activists had thought some things through, but not enough things. For them, internal communication was an organizing practice at Cegielski; that is to say, a practice restricted to Cegielski. Militants and rank and file alike had jointly formulated their position and carried it through successive layers of Poland's industrial hierarchy all the way to Warsaw. They had learned new things in this process, including the fact that "sustained militancy" is merely another way of saying "democratic patience." They had proved they had acquired it. Now suddenly, on the morning of June 28, after four months of sustained self-organization, Cegielski workers turned over control of their own political destiny to people they did not know, who had not shared their organizing experiences, and who had not internalized the knowledge gained therefrom. In an instant, the Cegielski movement lost direction over its own politics. When the Poznań crowd got restless and the rumor spread that the workers' negotiating committee had been arrested, Matyja, lost in the throng, tried to identify himself to those around him as a delegate who had obviously not been arrested. It was a meaningless gesture in the midst of 100,000 people.

The subsequent mass rioting was a heartfelt but essentially authoritarian statement by people whose experiences were less seasoned than those of the Cegielski workers and whose identities were concealed in an enormous crowd—an energized mass that felt it was acting politically. While such terminology offers every appearance of being a workable political definition of a "mob," such a conclusion would be inexact because the recruits intuitively associated themselves with the Cegielski movement and felt they were acting in defense of its leading representatives. But the connection, while emotionally strong, was institutionally almost nonexistent: internal communication links that are essential to democratic assertion were distressingly absent. As a direct consequence of this flaw in the structure of movement building at Cegielski, the democratic polity at the core of the throng had been rendered mute and therefore immobile.

Matyja's helpless stance symbolized this sudden marginality. In consequence, rather than giving expression to all that Cegielski workers had learned by their own efforts over four intense months of shop-floor activity, the Poznań upheaval was a destructive assertion by a huge mass of associated strangers who were well-meaning but comparatively innocent in political terms. What June 28 revealed was both simple and devastating: that which Cegielski workers had *not* learned about building a democratic movement contributed to unanticipated actions by others that undercut all that they *had* learned.

So much for the "Poznań events." The stunning moment is perhaps best understood as the unplanned culmination of four months of sustained self-organization. This time period may be designated as the "workers' movement unpoliticized." It represented the first step toward Solidarność.

The next stage began immediately. It constituted a reprise on the 1945–47 struggle for control of the shop floors of industrial Poland. This second phase of insurgency lasted from July through November 1956 and suffered through a long afterlife that extended into the spring of 1958. Ironically, it began on the defensive and in what presumably was a leadership vacuum. In the aftermath of the uprising, government security forces simply arrested the entire Cegielski negotiating committee and introduced them to a systematic routine of police-station beatings. While Matyja and the rest of the worker delegates bled in detention cells throughout the city, Poland's prime minister led a chorus of denunciations in the mass media. But as this repression began to soften under the impact of a steadily escalating intraparty struggle for power, Cegielski workers began to recover their democratic organizing drives. The anger at Roman Fidelski and, indeed, at the entire party apparatus remained undiminished. But more important, the enormous social energy generated at Cegielski in March not only remained intact, but actually became more intense as a function of the Poznań events and their complicated and ongoing national aftermath. Self-evidently, the task at hand was to find for this energy a channel of expression that would produce more tangible and permanent results than those that could be expected to materialize from an extemporaneous riot.

At Cegielski, it rather quickly became clear what precisely had to happen for workers to create a more secure social space for themselves. It was a simple process of elimination. They knew from the past that the official trade unions did not represent a means to this end; indeed, the very *raison d'être* of the party unions was precisely to prevent independent workers' assertions from interfering with "the leading role of the party." Cegielski workers had learned how to get beyond the trade union only to fall victim to high-level party officials. The challenge now was to find a way around the party at this decisive level. Toward this end, the only dependable source of democratic possibility obviously was themselves. But after the Poznań events, self-organization could no longer be understood as a temporary phenomenon, a mode of "getting someone's attention." Rather, it had to be a permanent feature of the working-class presence in Polish society. To those in Poznań in 1956, this meant an organization of workers that was not submerged under the official trade unions and was otherwise free of party control. An independent union was needed. The way around manipulation by both the party and the official unions was through self-management. This was the concrete "consciousness" of July that was not present in March.

Here, indeed, was something new. The Cegielski agenda was no longer essentially defensive. An idea about self-activity had emerged from practicing it. How strongly was the idea held? The answer is a bit complicated because the thought had emerged in the face of two severe problems. The first was the inherited thinness of historical experience as to precisely what "self-management" actually meant at the point of production. There were not very many precedents—in Poland or in other countries—that provided a detailed guide as to how Cegielski workers should shape an independent structure of economic democracy for themselves. The second problem was unforeseeable. Whatever experiments Cegielski workers might attempt in post-Poznań Poland, they were fated to take place within rapidly shifting international currents that made the entire challenge considerably more difficult than it already was.

The first task was to visualize a concrete plan of self-management that would work; the issues involved were both easy to enumerate and at the same time maddeningly complex to define. The basic question could be simply put: To what extent was popular democracy possible in individual economic enterprises in an industrialized state? Since enterprises received supplies and raw materials from a number of factories and produced products that went to other factories and to domestic and foreign markets, all production units were necessarily integrated into a nationwide network of production and consumption as well as an international network of imports and exports. If workers in a factory were alienated because the state had appropriated the means of production, did this mean that each factory work force that assumed control of the shop could treat itself as the sole owner? If so, was the surplus a "profit" and was it theirs? An affirmative answer was, of course, impossible to justify. Such a solution not only would produce vast inequities in income between workers in different enterprises, but the unpredictable accounting practices involved could not be integrated into any coherent national economic policy. But if the answer was not affirmative, what then? If workers elected the factory manager, was that executive to be the powerless servant of the factory's workers, or was he to be a friendly local adversary representing some (popularly constructed?) national program? In any event, what was a "friendly adversary"? In summary, what did "workers' self-management" look like structurally either at the plant or the national level? At bottom, could anyone say with clarity what either an institution such as a workers council or a concept such as "economic democracy" actually encompassed until such prior questions were addressed in some sort of tangible way?

Such problems were sobering in their complexity, not only for Polish workers but for anyone else interested in the dilemmas of popular democracy in a technological era. Questions of this kind were more than sobering, in fact; they were dismaying. When viewed in such nakedly concrete terms, the recorded historical experience necessary to provide an evidential basis for detailed speculation appeared quite thin. Granted, a number of abstract speculations had been formulated, thanks to the efforts of such theorists as (to name a representative sample) Adam Smith, John Locke, Jean-Jacques Rousseau, Karl Marx, and Mikhail Bakunin. But to go beyond the less-than-ideal plateau to which these efforts had carried the human community, a rather imposing amount of concrete information seemed needed to provide a point of departure for future efforts. Democratic (or at least formally representational) political forms were one thing—there were a number of those around in Western countries—but a democratic economy grounded in democratic production relations was quite another matter. Except for a few speculations during a handful of revolutionary moments in history, very little experience existed. Theories, yes; experiences, not so much. While the word "democracy" was and is on everyone's lips, concrete achievements in the area of democratic production relations and economic distribution are far less reassuring than theorists have cared to admit. The problem existed not only in Poland and not only among socialists but also in the West and among capitalists. Indeed, the lack of historical experience in these areas is treated at some length in Chapter 7. Solidarność was to generate much helpful evidence in this respect, and particularly in the context of ideas and practices yet to be tested elsewhere—anywhere on the globe. In any case, the events of 1956 in Poland brought this shortfall into view with shocking clarity. "Democracy" was obviously much easier to talk about when one stayed on the level of general principles. In terms of structure, the idea of economic democracy got sticky when the subject got concrete—very quickly so.[60]

Cegielski workers had come to the point where they understood they needed an "independent union," but what that meant structurally was as unclear to them as it was to a watchful and worried party. After the Poznań events, in any case, the party was sufficiently rattled (and its anti-Stalinist faction sufficiently emboldened) to propose in July a vague suggestion about the need for worker councils. To the party's anti-Stalinist reformers—Leninists all—the great merit of the party's July plenum was that a vaguely worded call for worker councils offered the nicely balanced prospect of appealing to workers while at the same time providing a new way to contain them under party domination. A party-sponsored worker council would constitute a nice enclosure around Cegielski and would-be Cegielskis. Party leaders could tell themselves that both workers and intellectuals had become so habituated to thinking in terms of small-scale reforms of the various modes of censorship generated by the state that a bit of latitude for the population seemed indicated, especially in view of the turbulence in the working class. Individuals within the party's Central Committee essentially differed among themselves only about the degree of latitude that seemed prudent. On the other hand, from the workers' standpoint, the politics of 1956 provided some sort of opening for self-activity, but how much and in what form it could be exploited remained unclear.

But if this practical dilemma were not enough of a problem for Polish workers to chew on, either at Cegielski or anywhere else, the democratizing task facing Poles in 1956 was further complicated by unanticipated developments in the international political arena. Swiftly summarized, the Cegielski insurgency unleashed a whole series of reciprocal effects upon the party, upon the working class, and upon the overall political climate in Eastern Europe. First, the June uprising shook the party leadership, emboldened its self-described "Polish" or "progressive" wing, and precipitated July party decisions to permit worker councils and to legitimize a fair trial for the Poznań prisoners. These developments, while suitably embarrassing to party hard-liners, also unavoidably encouraged renewed working-class initiatives. These now came not only from Cegielski but also from other large enterprises scattered across Poland. Such reinforcing momentums, emanating sequentially from the working class and, in reaction, from the party, carried Gomułka to power in October. The rise of Gomułka and the immediate unfurling of the "October Revolution," in turn, dazed the Eastern bloc, raising hope in whole populations and raising fears in Stalinist regimes. Finally, events in Hungary promptly reverberated back into the Polish situation in ways that severely circumscribed options, altered the character of the struggle between the party and the working class, and thus threw the nation into crisis. Under the circumstances, it is necessary to fit dramatic events in Hungary into a Polish context.

No sooner had the "Polish October" burst upon the scene, no sooner had Gomułka made his historic speech promising an end to "the evil which was widespread in our social system," than hope quickened in other satellite populations. On October 23 in Budapest, a crowd of students made their way to the Polish embassy and to the statue of Józef Bem, a Polish general who had aided the struggle for Hungarian independence in the nineteenth century. Within hours, an astonishing number of people joined in; three hundred thousand were in the streets praising the Polish achievement and calling for Hungary to do the same. The pace of events was startling. The Russians had flown out of Warsaw on Saturday, October 20. Gomułka, confirmed in leadership, made his dramatic speech to the Central Committee on Sunday and to the Warsaw populace on Monday; the Hungarian rising began on Tuesday, October 23. Imre Nagy, the Hungarian Gomułka, came to power on the twenty-fourth and suddenly both Warsaw and Budapest were convulsed by that joyous social electricity that accompanies the promise of a momentous historical breakthrough. The red, white, and green flag of Hungary began to appear among Warsaw crowds, and eventually even the buses and trams began to fly the Hungarian national colors.

The wondrous possibility that the Soviets would be forced out of the entire Eastern bloc suddenly loomed as a tantalizing prospect. Polish crowds moved into the vicinity of Soviet army bases and demonstrated menacingly. Units of the Polish army, meanwhile, met and passed resolutions calling for democratization in the armed forces. Less than a week earlier, at the height of the confrontation between Khrushchev and Gomułka, Soviet warships had appeared in the bay of Gdańsk but had been denied permission by Polish authorities to enter the harbor. Now, massive street demonstrations occurred in Gdańsk with thousands cheering for the Russians to get out of Poland. High-ranking party members were dispatched to Gdańsk and (instinctively) to Poznań in a desperate effort to cool things down.[61]

But it turned out that Poznań workers had learned a few lessons from the June events. While rioting erupted in Gdańsk and also at Szczecin, worker self-activity in Poznań was both orderly and considerably more purposeful. Adventurous ideas about an independent union, ideas that at mid-year had been visualized by the plant's most articulate activists, were now on the lips of the vast majority of Cegielski's rank and file. Self-education through self-activity had come a very long way since March. While laying some future plans, the foundry workers took advantage of the "Spirit of October" to drop the name of Stalin from the title of their plant, which ceased being ZISPO and once again officially became H. Cegielski-Poznań. (In the popular culture of Poznań, no one had ever called the engineering works anything but "Cegielski" or "HCP" anyway.)

Back in Warsaw, Cardinal Wyszyński, released from years of house arrest, unobtrusively resumed his duties in the Polish episcopate. For dedicated Catholics, the event was an absolutely confirming moment in the Polish October. Suddenly, new air could be inhaled on one's own—in this case Catholic—terms. But the readmission of Poland's Primate to civil society was but one of many striking moments in an extraordinary month. So many things happened that had been sought for so long that simple reality took people's breath away. From the head of the ruling party came words that resonated with every Pole: "The Poznań workers did not protest against People's Poland or against socialism when they went out into the streets of the city. They protested against the evil which was widespread in our social system . . ." How such words lifted the hearts of peasants who had never been to Poznań, of intellectuals who had never known a shop-floor grievance, and of party members who had lived for years with the hypocrisy organic to their jobs and to their daily social relations with each other and with the rest of society. Virtually every Pole was moved by the speech and had reason to be. At a moment of national crisis, Gomułka had looked the Russians in the eye and had not blinked. Now the Russians were gone. In the Polish October, all things seemed possible. It was a moment that made one feel good to be alive and to be Polish. No subsequent "historical outcome" could ever deprive the Polish people of that moment. It was too real.

Meanwhile, in Budapest, Imre Nagy maneuvered desperately to try to contain the popular emotions that simultaneously had brought him to power and now threatened the possibility of a peaceful consolidation. As many would subsequently say, it was a moment of "world historical importance." Perhaps. But if the historical meaning of Solidarność is to be grasped, it might be prudent to slow the film down and describe this Hungarian historical moment in less emotional terminology.

Whatever the outside world thought, whatever the Soviets thought, there was little resemblance between the Polish and Hungarian movements. In Poland, many actions and reactions had occurred over eight months of complex political engagement. In contrast, things in Hungary happened with not much more forethought than usually preceded a street riot. The comparison is appropriate. The "Hungarian Revolution" contained many of the dynamics at work on June 28 in Poznań, and very few of those that either preceded or followed that moment. At Cegielski, the processes in motion from March through June had essentially been grounded in working-class self-organization and self-education; the processes that followed between July and October turned on a reciprocal and complex interaction between the working class and contending factions in the party leadership. The party, though giving the appearance of having been transformed by these dynamics, remained in power. It performed a number of concrete acts that carried the society in a new direction, and it also issued various promises about future acts in the same new direction. In sharp contrast, the Hungarian October was an undifferentiated cry for "freedom." Nagy was a dazed beneficiary of a moment of passionate expression by a energized mass of politically unassociated persons. It is imperative to grasp what the word "unassociated" meant in the context of October 24 in Budapest. Small groups—different small groups—in the crowd had been talking to one another, including sundry students as well as the nucleus of dissenting intellectuals called the Petöfi Circle. In September, the union of writers had displaced its presiding Stalinist leadership and on October 22, two days after the Soviets had lost their struggle to intimidate Gomułka and one day before the Hungarian Revolution had begun, large numbers of students resigned from the party's student organization. However, all these things had happened in separate milieus; that is, small groups of people had talked to one another about their specified political objectives. But the people in the streets of Budapest had literally not had a single minute of collective political dialogue with one another before October 23. As a collectivity, they had generated among themselves no shared experience of self-activity. The events of that day did not constitute the unveiling of a maturing or matured democratic polity, but rather the first public display of a suppressed longing. It was a deeply felt emotion, but it was one that was politically inchoate. In consequence, Nagy was necessarily cast in the role of *ad hoc* spiritual broker for the entire population. He found himself a "leader" who was forced by events to function through symbolic acts. His was a politics of flash cards, improvised to mollify a politically frustrated and passionately assertive mass of ethnically harmonious strangers. However long Hungarians, like others in Eastern Europe, had chafed under Soviet-imposed Leninist government, the precipitating impetus for public assertion in 1956 came not from themselves but from elsewhere. The Hungarian Revolution was an event grounded not so much in Hungarian self-activity as it was vicariously grounded in Polish self-activity. As such, it constituted an authentic expression of desire. There is a very real form of politics embedded in this circumstance, but it should never be confused with the much more difficult dynamics inherent in democratic self-organization. The Hungarian rising was precipitate because its accelerating goals had never been democratically reflected upon by multiple Hungarian constituencies before they became "goals." This is a very frequent historical circumstance; permanent democratic gains rarely materialize out of such circumstances.

To interpret 1956 in Hungary, as well as in Poland, is a necessary step toward illuminating the essentials of Solidarność in 1980 and is a necessary forerunner to a more precise way to view the complex dynamics animating the politics of the entire Eastern bloc in 1989. What the people in the streets of Budapest shared was a profound sense of Hungarian culture and nationhood. For years, these elements of national life had been systematically distorted and diminished by a police-driven bureaucratic state responsive to an alien military presence. In a move that reflected and also served to heighten nationalist desires, Nagy acquiesced in, or possibly encouraged, the hurried reconstitution of prewar noncommunist parties. "Representatives" of such long-departed entities promptly appeared and were admitted into the circle of national leadership. In this manner, Nagy became a symbol of an "independent nation" that had unfortunately not taken time to talk to itself about various routes to "independence."

The Soviets watched these headlong happenings with an interesting measure of patience and forbearance. On October 30, they even went so far as to reaffirm from Moscow the "complete equality" and "sovereignty" of the communist parties in the satellites. It was a promise (of undetermined substance) of Finlandization. As such, it offered to Hungary the fruits of the Polish party's courage in staring down the Russians in Warsaw on October 19. Nagy's first flash card seemed to have produced a real result. But two days later, Nagy responded to what he took to be the determining pressure around him. He pushed the matter of sovereignty to the limit by proclaiming Hungary's international "neutrality" and verified this stance by taking the nation out of the Warsaw Pact! This was a very daring flash-card politics, indeed.

Nagy's act was so stunning and so patently headstrong that it has not subsequently received appraisal beyond being summarily dismissed as foolish. But beyond cold-war statecraft, the criteria of democratic politics involved in this issue are too important to be treated summarily. It is at once apparent that events in Hungary from their very inception were deeply and almost obsessively driven by a mass preoccupation with the top of the nation's structure. And it was precisely this obsession to which Nagy gave voice. He seemed unable to pause in the presence of the obsession long enough to consider what was needed—including time—to create democratic structures at the base of Hungarian society. Among the unassociated mass of people who had engineered several sustained moments of street politics was some undetermined number who longed for a re-creation of prewar parliamentary politics; an undetermined number of others who conceivably might have begun thinking about taking the first halting steps toward creating workers' councils; and an undetermined number of others who had merely begun to weigh various options. The country was energized, certainly, but it was also at the very beginning of genuine political life. If any kind of authentically democratic result was to emerge, the nation desperately needed time to talk to itself.

Instead, from the apex of society, Nagy made his symbolic gestures. The act of abruptly bringing into a hastily contrived "national leadership" individual representatives of prewar parties was transparently cosmetic rather than structural. The parties themselves did not exist. The individuals Nagy summoned were tokens of remembered structures that possessed no contemporary social substance. To have created popular institutions, either inside or outside of parties, would have required hard and sustained democratic labor and debate. The name for this activity is "organizing." Indeed, popular dialogue was essential both to verify and to give life to the democratic quality of whatever structures that might be created. If, after the creation of such popularly based institutions and after argument within them and among them, the majority of Hungarians consciously decided to move toward "international neutrality," that would be one thing; Nagy's arbitrary assumption of that step was quite another. His move may be judged, therefore, not only as precipitate but also as organically undemocratic. To see it as "freedom-loving" or, worse, "courageous" or, even worse, as "militant" is to fail to honor the difficulty of the process itself. Nagy's democratic hungers were so great that he moved to satisfy them with authoritarian speed. His actions exposed his innocence about the prerequisities to the creation of democratic societies grounded in functioning democratic polities.

In calling attention to the structural vacuum that surrounded Nagy's frantic attempts at high politics, in no sense do I wish to pass over the political significance and emotional power of the Hungarian Revolution. As matters developed (historical events often have a contingent "unfairness" about them), there was no time for the population of the city to have acquired the self-generated democratic collectivity that Cegielski workers had fashioned over many months. While street politics can be and often is an energizing component of the democratic process through which people begin to wrestle with the many inherited forms of deference and resignation that they have consciously or unwittingly internalized, it is self-defeating to view such actions uncritically. It is still necessary to sift through the social evidence to discover what is innovative and democratic and what is traditional and hierarchical in the sundry modes of conduct generated by the members of the social formation itself. The Hungarian Revolution was a vivid insurgent moment in history; it was not an unalloyed democratic moment. And, in the end, some decisions made by Imre Nagy were not helpful.[62]

Whatever others thought of Nagy's extemporaneous lurches, it came through to the Russians as simply "too much." On November 4, Soviet tanks rolled into Budapest, the Hungarian "revolution" instantly became very bloody and, within a week, a crushed effort.

All of this brought the Polish nation down to earth with sobering speed. It now became much more widely understood just how dangerous public displays of anti-Russian exuberance were to Poland's immediate future. The entire society belatedly endeavored to recover its poise. With considerably less élan but with somewhat more mature determination, the Poles set about to consolidate all they could from what appeared only a few short days before to be the unlimited promise of the Polish October.

It took awhile before people could be certain, but when the evidence was all in, it became possible to measure the gains of October. The gains were real for the Church, for the intelligentsia, for the peasants, for the economic circumstances of the population, and for the pride of the nation. Religious instruction was reintroduced in the schools; the regime abandoned its control over Church appointments; all imprisoned priests were released; chaplains were once again authorized to enter hospitals and jails; a Catholic national weekly newspaper was authorized, and the ban on Catholic lay groups was lifted; a signal that the censorship was to be reduced was sent when the jamming stations were all shut down; restrictions were soon ended on importation of Western books, plays, and films, and an avalanche of all three art forms soon flooded the country; peasants were permitted to abandon their government-imposed cooperatives (very quickly the ten thousand in existence shrunk to fourteen hundred, and Poland once again possessed the most far-flung system of privately owned agriculture in the Eastern bloc); bureaucratic harassment of peasants was officially discouraged, and some new incentives structured into the system of agricultural production; the new five-year plan was revised so that it reflected increased allocations for consumer goods and decreased allocations for capital investment; a new emphasis was placed on housing construction to cope with the apartment shortage in the cities; Rokossovski was removed from the Politburo and from his post as commander in chief of the army and was sent home; all Russian "advisors" in the police and the army were also cleared out and sent home; Stalinogród became Katowice once again, and all the Stalin streets, Stalin squares, and Stalin parks reverted to the Polish names they had originally possessed. In short, the nation reclaimed its identity in the Polish October.[63]

And the worker councils were destroyed.

With strategic skill, Gomułka spent his first year in office patiently consolidating his base in the party and in society. He made what for him was an easy peace with the hard-line faction in the party, placated the Church, tossed an early bone or two to the intellectuals, a bit more than that to the peasants, and then, secure in his position, brought the working class to heel. It was not effortless, and it did not happen overnight, but that it happened as cleanly and as totally as it did indicated the limits of working-class self-organization in 1956. It needs to be said, however, that the workers faced, in Gomułka, the most formidable foe the party had ever generated within its own ranks. His capacity to govern was measured by a personal integrity, a keen sense of strategic priorities essential to Leninist modes of governance, a tested mediocrity in governance itself, and a pedestrian imagination. It was also fortified by a profound sense of self, a trait familiar to those who had known him intimately during his ascendancy from 1945 to 1948. For a man who had been in internal exile for eight years, Gomułka reacted with remarkable reserve to the party's overtures in the first week following the Poznań events. He did not oppose, of course, his own readmission to the party, but he resisted committing himself to any sort of personal alignment with existing leadership factions. He was a lonely and incorruptible old Bolshevik, energetic at fifty-one, and above all, self-confident.

Gomułka well understood that his essential adversary in Poland was the working class. With hindsight, it was evident in October that the so-called "Polish" faction of the Central Committee that engineered Gomułka's return had gone too far in cozying up to the working class. The July decision to sanction worker councils, while tactically useful in isolating the Stalinist Old Guard and taking some steam out of the Poznań insurgency, had been premature. Poland's workers were restless enough without adding a measure of party sanction to their proliferating ambitions. This soon became evident. Nowa Huta, Gdańsk, Silesia, Poznań again, Łódź, Warsaw—the roll call of centers of worker activism was sizable in August, longer in September, and still longer in October. Poland's largest enterprises were awash in newly forming shop councils. The party now owed so many due bills that the achievement of political stability would require both time and skillful maneuvering. In the short run, Gomułka had no option but to signal an official go-ahead on worker councils. The party promulgated and the Polish parliament, the docile Sejm, dutifully passed a new trade-union law in November. It formally dislodged the old trade unions as the sole representatives of workers and authorized self-management in factories. But Gomułka saw to it that a series of necessary precautions was built into the new trade-union law, which turned out to be both sweeping in its apparent scope and adroitly vague in a number of crucial areas. The ecstatic popular response in Warsaw to Gomułka's effective October speeches had served to meet the most immediate prerequisite—to give people hope and buy some time for himself. But the task of rebuilding a proper institutional wall around the emerging worker councils could not be done with speeches; nor could it be delayed if the pace of shop-floor activity in factories was to be contained before the entire working class of Poland had enrolled itself in self-managing institutions.[64]

"Independence" was the operative word for October—independence from the old censorship, independence of the peasants from the state cooperatives, independence for the Church, and (as the Stalin signs came down all over Poland) a measure of independence for the nation. Meanwhile, Polish workers on their own had provided rather ample evidence of their thirst for the same commodity. In the first step of the Gomułka counterrevolution, the state-owned trade unions announced a new "independence" for themselves. As orchestrated in the official media, the government trade unions, too, intended to partake of the Spirit of October by working closely with the emerging shop councils. Gomułka added some political heft to the new approach by formally calling upon the trade unions to "assist" the worker councils. Experienced shop-floor militants understood what all this cape waving meant: a showdown between the party and the working class was clearly very near.

By the end of October, the practical result for Cegielski workers of all the months of self-organization was clear. Militants had come to know as much about the state and about what workers required to contend with the state as any other group of workers in Poland. Their hard-won sophistication can perhaps be most visibly seen when arrayed alongside the awareness of other Polish workers whom many took to be in the vanguard of working-class assertion in 1956.

At the great automobile works of Żerań in Warsaw, the leading worker spokesman was an activist in the party plant committee named Lechosław Goździk. The emergence of Goździk, so different in its context from the evolution of Matyja, provides revealing insights into the dynamics of democratic movement building. Goździk's eventual tactical difficulty was that he was not a typical party functionary primarily loyal to his own career and to his relationship to the apparatus. Rather, he was a believing communist with a clear tendency to identify with workers as well as with the party.

In October 1956, amidst the patriotic emotions generated in the country by the Polish Politburo's confrontation with the Soviets over Gomułka's accession to party leadership, Goździk, along with Gomułka and his Central Committee supporters, was able to capitalize on this rare conjunction of the party with the nation. At the peak of the crisis with the Soviet Politburo, Żerań workers actually mounted a caravan of automobiles to drive to the suburbs of Warsaw, where Soviet troops were demonstrating as part of the Khrushchevian pressure being applied to the Polish party leadership. On that passionate weekend of hope and anxiety in Warsaw, the issue of shop-floor democracy at Żerań, while present, thus became deeply submerged beneath the emotions of Polish national dignity; one could be a good Pole and a good worker by taking orders from the Żerań party plant committee. Indeed, the entire nation rallied to Gomułka's effort to extricate the party from the suffocating reach of Moscow's heavy fraternal hand. The night after the Soviets grudgingly withdrew, excited crowds roamed the streets of Warsaw paying their respects to the Russian general who served as Polish minister of defense: "Rokossovski to Moscow" and, in joyous irreverence, "Rokossovski to Siberia." In the less volatile next few days, Goździk felt the twin pressures of the workers' drive for shop autonomy arrayed against the Żerań party committee's reluctance to give up its authority. While endeavoring to tilt a bit toward the workers, party member Goździk walked the chalk line between his two constituencies and endeavored to retain a measure of standing in both. But his role was fundamentally different from Matyja, whose dialogue with the party, like that of scores of other Cegielski activists, was largely confined to being beaten into unconsciousness in police stations.[65]

Matyja's constituency, grounded in itself, understood October politics differently than did Goździk's rank and file. Cegielski's working class was self-organized; Żerań's was party-organized. The Cegielski experience was one of constant struggle between activist and rank-and-file workers on the one hand and trade union and party functionaries on the other. It was an argument that went on at the plant every day and continued on trains as party and nonparty militants traveled together to meetings as members of the same worker delegations. To rank-and-file activists at Cegielski, whether party members or not, the state bureaucracy had been uncovered and revealed as a palpable and demystified adversary, both powerful and alien. In vivid contrast, thanks both to the short organizational life of workers' councils at Żerań and also to Goździk's equivocations, the Żerań experience unearthed far fewer contradictions in the party's life and structure and thus was a far less effective instrument of education for the plant's workers. Goździk was to confess ruefully years later that Żerań activists like himself had made the error of being "too full of faith in the party." Goździk eventually came to understand that the collective experience of Żerań workers had been "only days" compared to the eight months of self-education generated by Cegielski workers. The ultimate result was that Żerań workers had worked through fewer layers of their own confusion about the structure of the system of social control in which they were caught. This was the heart of the difference between the two groups of workers. By the end of the Polish October, Goździk was personally confused and tactically immobilized; in contrast, the Cegielski leadership, released from prison, had merged with an increasingly embattled but also increasingly sophisticated body of activists from dozens of Cegielski shop sections. As their actions indicated, unbalanced encounters with party police cleared the mind.[66]

Meanwhile, a political position similar to Goździk's materialized in intellectual circles. This was the "October Left," an intellectual current that existed more or less as an appendage to the party during the decisive months when the worker councils were struggling for life. The loosening of Stalinist reins upon the party apparatus permitted the appearance of this democratizing force. Emanating primarily from Warsaw University and Warsaw Polytechnic but also including a splinter of mid-level personnel within the party apparatus, the group centered itself around the journal *Po prostu,* itself a creature of the de-Stalinization climate. But though the October Left applauded working-class aspirations, it associated itself with a technocratic stratum that hoped above all to bring a measure of coherent planning to the national economy. As Jacek Kurón and Karol Modzelewski later understated it, the Left "did not set itself apart from the general anti-Stalinist front as a specifically proletarian movement." Instead of focusing its attention on the working class, the October Left concentrated upon the young intelligentsia. *Po prostu* initiated an independent student organization unconnected to the moribund party-controlled youth association.[67]

While communist reformers focused their energies upon the intelligentsia, the Church supported Gomułka. In the words of an experienced East European observer, Chris Harman, "from October onwards the Church used its resources to marshall support for Gomułka against those who wanted to carry the revolution further." Such a position found support in official circles in the West as well. In the words of one dismayed observer, Radio Free Europe "advised Poles to vote exactly as Gomułka outlined."[68]

When it became apparent that Gomułka opposed active Polish association with the Hungarian cause, Cegielski-Poznań moved to the brink of a strike in late October. But that moment constituted the peak of working-class self-activity during the Polish October. Such signs of autonomous working-class energy alarmed the newly installed Gomułka regime. In November, Poland's new leadership decided to launch a massive party intervention into the emerging world of worker councils. Amid much fanfare, the party announced a decision to convene a giant worker conference in Warsaw to which the nation's forty-nine largest enterprises were to send representatives. The Warsaw conclave would proceed to "perfect the new worker councils." The new plan, apparently well thought through by somebody, already possessed a catchy name—the "experimental program." The conference was to convene right away, in early November.[69]

The underlying meaning of the party-sponsored Warsaw conference becomes most clearly visible if the proceedings are viewed through the perspective of serious advocates of workplace democracy. The first feature of the conference to stand out was the remarkably high percentage of party members in most of the delegations. Further, they were party members who tended to talk docilely about what the party needed rather than forthrightly about what workers needed. It became evident that few plants in Poland had achieved the degree of rank-and-file self-consciousness that the long months of organizing had wrought at Cegielski. The official trade unions were still an influential factor in most enterprises—Żerań, Nowa Huta and Huta Warszawa among them. A second difficulty was the elaborate stage managing of the conference. In this and in subsequent regional and local organizing meetings across Poland in which the party engaged in "perfecting the experimental program," speeches by trade-union functionaries were the order of the day. Little time was allotted for sustained programmatic discussion. Indeed, the very fact that management of the agenda was so strict seemed to throw into question whether any worker autonomy had, in fact, been achieved with the passage of the new trade-union law. To serious worker advocates—that is, those not "too full of faith in the party"—the theater surrounding the experimental program was like watching an old movie from March or June, not a new film made in the "Spirit of October." A few staunch worker activists—the Matyjas of the November working class in Poland—tried to stem the tide of party manipulation, but it soon became clear they could not get issues framed favorably and that a majority of delegates (for a variety of reasons at this early stage of their politicization) could not keep up with the most experienced activists. It dawned on some thoughtful militants that discussions about how to structure the experimental program emanated from a party-controlled agenda and took place at party-called conferences. Such dynamics, of course, left precious little room for those still-believing worker communists, such as Lechosław Goździk, who regarded themselves as sincere party activists.[70]

The October Left found itself in precisely the same powerless position. The student organization inspired by *Po prostu* got itself sufficiently organized in November to issue a manifesto, promptly featured in the journal. The young reformers promised "to preserve the line of the Marxist Party of the working class in our actions, reserving for ourselves the right to interpret the line of the party and influence the decision on this line." While recognizing "the political role and importance of the leadership of the party" as the "directing force in our nation," the student activists also announced themselves as "opposed to our organization receiving orders from the party." This declaration of independence constituted the high point of the intelligentsia's self-activity during the Polish October. Three weeks later, the party leadership forced the students to merge their new organization into the official party youth organization.

Patently uncomfortable, but doing the best they could to remain hopeful, participants in the October Left watched helplessly as the Polish working class was inexorably shepherded toward intimate "cooperation" between the official unions and the nascent shop councils—not in the direction of worker autonomy.[71]

Whether they knew it or not, the world of the Goździks had become untenable in post-October Poland. They were sincere party members and sincere worker advocates, and they found themselves at party-called conferences of worker activists; yet they could not effectively shape events. It was all a bit puzzling.

The militants of Cegielski were not puzzled. They were beaten, and they knew it. Matyja objected to the party's control, an action that cemented the party's hostility to him and set him up as a marked man.[72]

Herein lay the irony and the tragedy of the Polish workers' movement as of 1956. The events of November provided the first evidence of the constricting impact of the regional imperative that historical circumstances imposed on Polish workers. The phrase should be capitalized: Regional Imperative. The lessons of Poznań were learned—by workers in Poznań. They were learned far less clearly by textile workers in Łódź, automobile workers at Żerań, tractor builders at Ursus, or maritime workers on the Baltic coast. A geographical barrier had been encountered. Lacking lateral lines of communication, Polish workers could learn from their own individual participation in their local job action, but they could not learn from each other. Where self-organization had transpired over many months—as at Cegielski—the lessons were many. Where worker councils had been initiated by high party officials, as at Żerań and other factories in the Warsaw metropolitan complex and elsewhere in Poland, the signals were infinitely more difficult for workers to sort out.

Viewing the new Poland from the perspective of Żerań, there could be no question that from Gomułka on down, almost everyone in the party was speaking in a new way—about "new roads" and "new paths" and Poland's new "independence." With no experience of their own as a guide, workers who had not engaged in prolonged efforts at self-organization could not possess sufficient criteria to judge post-October politics. It was not easy to discern where to take a forthright stand and where to go along with people in authority who, at least part of the time, seemed to be saying the right thing. A worker had to possess some experience of his own in self-activity in order to see the sharp distinctions between men like Stanisław Matyja of Cegielski and Lechosław Goździk of Żerań. Both seemed "sincere" and therefore trustworthy, yet the Goździks were silent at crucial junctures while the Matyjas of Poland spoke up for policies that did not prevail. When the euphoria of October passed, it was hard for newly activated worker delegates to know where the country was headed. The editors of *Po prostu*, activist students, and the October Left were in precisely the same state of confusion.

In point of fact, post-October politics was a bit puzzling for all Poles. Evidence of real change, after all, was everywhere to be seen: in the new public freedom for religious activities, in the flood of Western books and films that increasingly were available, in the proliferation of cultural and intellectual clubs, and in the new freedom for peasant agriculture. Though the wildest hopes of October had been forced into the background in the aftermath of the failed Hungarian Revolution, Gomułka's Poland was self-evidently a very different place from Stalinist Poland. Still, the general shape of the nation's future direction seemed hard to foresee.

Viewed from the standpoint of the party, however, the substance of 1956 was much easier to fathom. Inside the party, a strand of "revisionists" appeared in 1956, politically organized around *Po prostu*. Its leading figure was philosopher Leszek Kołakowski, who criticized the party's overcentralization and was himself severely rebuked publicly by Gomułka. Party revisionists and the independent intellectuals who jointly comprised the October Left thereafter possessed a considerably reduced silhouette and, in any case, proved organizationally easy for the party to isolate.[73]

In terms of mass constituencies, the party had to cope with three entities: the peasants, the workers, and the overlapping category of "the Catholics" with their organized leadership in the episcopate. The nonparty intelligentsia as a constituency was comparatively much smaller and, in any case, remained essentially unorganized. The intellectual mainstream proved fairly easy to placate through a moderate relaxation of the censorship. The Church and the peasantry had also been consigned to the sidelines through concessions and, should the need arise, could be further dealt with later.

The real task of the party was to re-domesticate the working class. Independent workers' councils were a distinct problem; but a formalized "conference of worker councils" was more than a problem, it was a long-term structural threat. It created lateral lines of communication between workers that opened up the possibility of an authentic working-class movement on a national scale. The state-owned trade unions represented an organizational blanket of silence that had to be held in place. The goal was effective censorship of workers. Achieving this necessity was the authentic centerpiece of party politics in 1956.

The ensuing six-month struggle between the party and the shop councils constituted a vital chapter in the postwar education of the Polish working class. The sustained effort to achieve democratic administration of factories was a great learning experience for Polish workers, one that was to carry them a bit further down the road to Solidarność.

Ironically enough, it was also a learning experience for the party as well, though scarcely in the way workers hoped. The party's resistance to the new labor law of November 1956 had been immediate. Indeed, local party functionaries were so intractable in a number of plants that workers had to engage in strikes, demonstrations, and forcible takeovers merely to establish the right to meet among themselves to plan a shop council. Some plant managers, upon being voted out, simply stayed on until they were bodily evicted. But muleheaded resistance was not the party's uniform tactic, nor its best one. The party adopted the practice of wearing out shop councils by watching passively as workers democratically went through many meetings to establish certain procedures; the party then delayed, undercut, or otherwise tried to alter and divert any initiatives from below. The object was to lower rank-and-file enthusiasm and thereby diminish both the democratic character and the operational thrust of shop-floor energy. Battles lost on one level by either the old trade unions or by the party apparatus in a plant or even a city could be refought at a regional or, if need be, a national level. The party had a vast structure and an elaborate media apparatus; the councils were isolated pockets of incipient democracy. It was not an even contest.[74] Fortunately from the party's standpoint, there was no coherent body of thought in the society capable of articulating a practical political case for an autonomous and internally democratic national association of the working class.[75]

It is important to assemble with some precision the applicable cultural elements that defined this impasse. The political challenge was to conceptualize precisely what structural forms existed in Polish society that could provide a possible future basis for a freer society, one liberated as much as possible from the suffocating yoke of undiluted one-party control. Actually, once the question was formulated, the answer was remarkably simple: the only entity in Polish society with the political and economic strength to temper the party-state in any structural way was the working class. It was the only social formation that could bring the entire economy to a halt and force onto the agenda of public discussion a review of the nation's social contract. But while the answer was obvious, nonparty circles in Poland did not formulate the question. Speculations in this area seem to have been beyond the range of what was perceived as permissible discussion—either in the intelligentsia or in the working class. This injunction applies with similar power to Western observers. Suffice it to say, here, that Polish society in 1956 was not uniquely opaque in terms of its capacity for self-analysis. What was significant about Gomułka's Poland in the 1956–57 period concerns the extent to which popular self-education became possible, was partially realized, and then ultimately crushed.

A seminal political truth was learned, however. In the diverse struggles through the spring of 1957, a collective understanding became plant-wide wisdom in widely separated enterprises throughout Poland: workers' councils, in themselves, were too fragile to stand up against the party. Workers who did not reach this conclusion as a practical result of the abortive shop-floor insurgencies of 1956–57 had the lesson driven home through the futility of various party-sponsored "experiments" in the early 1960s under the obscenely misnamed Council on Worker Self-Management. Whether defeated by the party's trade-union bureaucracy or the party's police, workers across Poland came to see the unserviceability of plant-sized worker institutions, such as shop councils. Something larger in scale was needed. The limiting element in this realization, however, was the fact that almost no one in the working class in any one region was in a position to realize how widespread the idea was throughout the whole of working-class Poland. Other events, happening years later, would create the conditions wherein the entire population could learn how disaffected from shop councils almost everyone in Poland's working class had become as a function of the Polish October and its bitter aftermath.

The immediate threat to the regime was, ironically enough, a product of the party's obsessive preoccupation with maximizing large-unit production as a perceived cornerstone in "building socialism." Factories in Poland were huge. A large part of Poland's industry was located in giant enterprises employing five to forty thousand workers. The very names of these sprawling installations rang with honor and prestige—the Wujek mine near Katowice, the Ursus tractor factory and the Żerań automobile plant in metropolitan Warsaw, Cegielski-Poznań, Nowa Huta in steel, Zambrów in textiles, Oświęcim in chemicals, the

Warski Shipyard at Szczecin and the Lenin Shipyard at Gdańsk, to name only a representative sample of the most famous. It was vital that these giant enterprises be inoculated against the Cegielski disease and that the work force at Cegielski itself be hemmed in and tranquilized. But in the highly energized climate of 1956–57, such an objective could scarcely be achieved with a simple wave of a Leninist hand. No matter how many "experimental programs" were trumpeted or how many "worker conferences" were stage-managed, the most significant social reality of 1956 was the frightening amount of democratic space that had been created on shop floors across the entire nation.

Party functionaries discovered, unhappily, that not only an immense amount of energy but also a number of new ideas and contentions began to percolate upward from the bottom, through the trade-union screen and out into society. As quickly as possible, shutters were hammered into place. The national director of the trade unions issued instructions to local administrators to "assist" in adjudicating any disputes between shop councils and the sundry agencies of the trade unions or the party. An entire new layer of "arbitration" was created that established judicial machinery within which representatives of workers' councils were uniformly in the minority. It soon became routine for the entire trade-union apparatus in a factory to intrude into shop-council functions, particularly elections of delegates. Trade-union newspapers disseminated earnest articles emphasizing the need for workers to avoid electing "undesirable elements" to shop councils. The official unions were zealous in their assistance to the shop councils.[76]

However, the party's trade-union apparatus occasionally moved too rapidly or too heavy-handedly and brought on a large-scale strike as a result. The regime soon learned to apply pressure on a plant by plant basis, at different times, and in widely separated parts of Poland. The tactic proved an effective means of coping with any stray lateral communication among workers until such communication could be permanently severed.

Thus, in the natural course of events in late 1956 and in the early months of 1957, the first Poles to see through the mystique of the Polish October were members of the working class. Among the very first to so perceive was the work force at Cegielski-Poznań. Within the first month of the new Gomułka era, Cegielski workers understood the "experimental program" for what it was. Fairly rapidly, workers elsewhere felt the constraints and drew the proper conclusion: the democratic movement on the shop floors of Poland was in deep trouble.

This process worked itself out in the six months following the Polish October. The country's working class did not surrender quietly. The spring of 1957 was a time of unparalleled working-class insurgency across the length and breadth of Poland as workers fought local battles for shop-floor democracy. The regime rushed in with material concessions—pay raises, promises of accelerated housing construction, and assurances related both to job safety and to reform of the "storming" system of production. The regime was willing to make almost any kind of verbal concession as long as it did not concern structural change in the subordinate relationship of the working class to the state-owned trade unions. Conciliation alternated with repression. Persistent worker activists suddenly found themselves detained for prolonged questioning about their "anti-socialist" motives. If detentions did not prove a sufficient educational device, beatings were administered, sometimes in police stations, sometimes by "unknown assailants" on the streets. Only the most highly politicized worker activists survived these lessons. Many such activists were simply transferred to distant—and usually demeaning—jobs at the opposite end of Poland. Still another favored tactic involved what might be called "engineered corruption." Some of the most prominent (and therefore more dangerous) worker activists were praised by the party while at the same time being gently admonished to "calm down a bit." They were granted new apartments and access to party stores, party cars, and party vodka. An inebriated Goździk on two occasions wrecked brand-new Żerań-produced automobiles. Conveniently discredited, he was confidently eased out of the Żerań party committee in 1957. At Cegielski, the hard-earned sophistication of the workers prolonged the struggle; it was not until 1958 that the party felt itself to be so thoroughly in control that it could safely do what it had longed to do for almost two years: fire Stanisław Matyja and the rest of the Cegielski militants. Matyja and forty others were summarily dismissed in the spring of 1958.[77]

Meanwhile, Roman Fidelski, the minister whose deception had engendered the Poznań riot, wrote an authoritative piece in the party's theoretical monthly, *Nowe Drogi (New Roads)* that specified with statistical precision the progressive impact on production of the experimental program. If not all readers in the Warsaw intelligentsia were sufficiently informed to appreciate the irony of the source, workers at Cegielski-Poznań were well equipped to identify Fidelski. One Poznań worker recalled years later: "With friends like Fidelski as party progressives, who needed Stalinists as enemies?"[78]

Party members, of course, viewed matters differently. A writer for *Przegląd Kulturalny* was dismayed to discover the extent of worker disillusionment with Gomułka after only a few short months. The writer complained in print about the recalcitrant attitude of Polish workers. This attempt to blame the workers was not uniformly accepted in intellectual circles. Irritated by such self-serving party perspectives, a sociologist in the Warsaw intelligentsia rushed in with a behaviorist explanation of the workers' untoward conduct. Polish workers had a "frustration complex," he said. "No party and no government can with impunity make exaggerated promises without sowing demoralization." The party, in short, was also responsible. It may be seen that both sides of this argument took up positions that were tangential to the real issues at stake. There was nothing "exaggerated" about the kind of protected democratic space represented by autonomous worker councils. Though the milieu of Warsaw sociologists did not seem to encourage such intellectual generalizations, it was axiomatic that democratic activity began with the attempt to create democratic forms. It was remarkable the way Polish intellectuals argued about the meaning of the workers' reaction to their own defeats. Party journalists, seeing no defeat, felt the workers to be unappreciative of Gomułka. Independent intellectuals, uninterested in pandering to Gomułka but also seeing no worker defeat in terms they could understand, explained the workers' reaction as predictable, if unsophisticated, "frustration" at being taken in by exaggerated promises.[79]

It may be observed that party and nonparty intellectuals were closer in spirit than either realized. Critics of the party and critics of the workers shared an enormous distance from the realities of life in the working class, and this circumstance caused both to misread in different ways the meaning of working-class self-activity in 1956–57. Neither group of intellectuals seemed to grasp the fact that the nation's workers were waging a complicated but nevertheless desperate struggle to circumvent the censorship of the official unions. The Polish intelligentsia continued to believe that Polish workers were so preoccupied with "stomach issues" that they could not be mobilized against the censorship.

In fact, Polish workers had launched what may most aptly be described as an institutional escape attempt. Had this strategic objective been understood by the intelligentsia, the workers' reaction to their own defeat—disillusionment and bitterness—would have easily been understood as well. But such was not the case. Unsympathetic observers in the party saw only a brute stubbornness and recalcitrance in the workers, while sympathetic intellectuals could do no better than discover a kind of psychological malfunction of the unlearned. Conjoining the two divergent opinions was a cultural habit common to intellectuals: condescension toward workers. As a habit, it was always fashionable, but it was also, at certain historical junctures, crucially blinding. The example of Poznań was a harbinger of things to come in the era of Solidarność.

In any case, the government's campaign to turn "worker control" back into control of the workers culminated in a 1958 law that created a permanent Council on Worker Self-Management. The new institution was dominated by the party and the state-owned trade unions. Without much hope, workers tested the new form and quickly learned its stifling limits. Most fell into silence under effective institutional censorship. The 1958 law represented the formal consolidation of a successful party struggle that had achieved its decisive practical victory by the late spring of 1957.[80]

Once the bulk of the nation's workers were subjugated, the limited space granted the intelligentsia could be summarily closed off as Gomułka proceeded to tidy up the remaining loose ends of the Polish October. The few stray quasi-independent journals like *Po prostu* that had been permitted were suspended in the autumn of 1957. Gomułka himself appeared before the Writers' Union to confirm that the brief era of free thought was over. A celebratory orthodoxy once again became the official style of public dialogue. The intelligentsia thus joined the working class in unwilling servitude to the party bureaucracy. The Russians were gone, but Leninist norms remained in place in Poland.[81]

The gains of October had shrunk. The Church had a bit more room than previously, and so did the peasantry. The grip of the Kremlin was also loosened, at least in its most humiliating forms. As a nation, however, Poland remained a long way from the semi-independent status of Finland. The gains of October were real, but in the aggregate, they were disappointing. Slowly, the nation's intelligentsia was forced to confront the unhappy fact that Gomułka's Poland remained a constricted and unimaginative Leninist state.

For Poland's working class, however, there were some invisible gains—obscured because they were concealed beneath the appearance and the reality of defeat. The principal lesson was a negative inference that could be drawn. Working-class self-activity could generate worker leadership, but it also exposed that leadership and provided no protection for it. Unless some tactical solution to this problem was discovered, Polish workers could look forward to the unhappy prospect of losing their most informed and politicized advocates in the never-ending cycle of repression.

This reality can perhaps most effectively be illustrated through the fate of Stanisław Matyja. Because of the space opened up in Polish society by Solidarność in 1980, details about Matyja's activism finally found their way into print a quarter century after the fact. Interviewed by social scientists in Poznań in 1981, Matyja provided intimate evidence about the early months of organizing at Cegielski. But perhaps his most telling testimony was inadvertent and had to do with his own sense of the politics of 1956. What this testimony reveals about the process of consciousness formation is that—for Matyja personally—things came to a stop on June 28, 1956. Subsequently, he was able to act only on the basis of those understandings that he had internalized by that date.

Why was this so? In the immediate aftermath of the riot, Matyja was arrested, confined for sixteen days, interrogated for long hours, badly beaten, hospitalized for six weeks, and in convalescence for six more. Meanwhile, his comrades at Cegielski and the rest of Poland lived through the dramatic evolutions of the summer and early autumn that culminated in the Polish October. It was not until that month that he was reinstated at Cegielski. He did not return to his old position; he had been so badly injured by the police that he could no longer perform as a carpenter.

Aside from this, a second relevant element was the effect that being out of touch had on Matyja's understanding of subsequent events. The work force at W-3 tried to take care of him. Agitation at Cegielski played a role (Matyja thought it the decisive role) in his release from prison and in the care he received in the hospital. Cegielski's rank and file further showed their esteem by electing him the post-October chairman of the plant's newly formed "commission on workers' councils." He went to worker conferences in Warsaw, "talking all the time on the train" about "bandits" in the party, and lived through the heartbreaking process through which the Cegielski workers' council was redomesticated under the official trade-union structure. Though he provided few details of this ultimate defeat, he seems to have taken a strong stand at some juncture, for he remarked in passing that he was told that he was fired not only for his connection with the Poznań events in June but for his actions in October and November. But his political efforts after his release from the hospital were passed over quickly in his published interview; there was no printed evidence that he grasped the strategic significance of the organizational setbacks the workers encountered in the fall months. What might be called his "education on the structure" seems to have ended at the point it had reached on the day of his first arrest in June. He reported, with an edge of enthusiasm in 1981, that Cegielski activists had "formed all kinds of commissions" during 1956 and 1957 without the slightest hint that he understood how such a scattergun approach played into the hands of the party. He was a man of June, assertive, self-respecting, and worker-oriented, who was defeated by the politics of November. He remained a man of June the rest of his life. His taped narrative ends with a one-paragraph summation of the nature of that life after he was fired in March 1958. It is a moving statement, and one grounded in the realpolitik of People's Poland:

> Wherever I went, a plainclothesman followed. I looked for a job. My only chance was a private employer. In fact, I would get hired in all kinds of workshops, but I'd get fired on the first day of work as they did not want to have their business locked up because of me. It was a losing game. I was harassed in many ways. As a boy scout instructor in model plane building, I had a radio operator license, III category, required for construction of remote control aircraft. Even that was taken away from me, without explanation. I could hardly believe one could hate so much. I was harassed as late as 1973. Only now, after August 1980, did I become an honorary member of Solidarność at Cegielski. I found protection and respect there.[82]

Matyja's fate illustrated a great organizational truth about life in a one-party state: only a broad-based worker-centered institution could protect worker activists. That was one certain inference that could be drawn from the experiences of 1956–58. There was also a second inference. As long as the lack of lateral communications among workers ensured the persistence of the regional imperative in determining levels of politicization existing across Poland at any one moment in time, workers would be well advised to begin thinking in terms of regional organizing. Otherwise, high-consciousness plants would be in permanent peril of being co-opted through party-manipulated institutional association with less-organized work forces.

The lessons, then, were sometimes quite subtle, but they were there to be discovered and pondered. These lessons were virtually the only tangible residue workers possessed—as workers—from the Polish October. As Poles, of course, they got at least part of their country back. The Russians were now out of sight, if not deprived of powerful influence.

The early worker struggles of 1945–47 and 1956–58 had a complex impact on Poland's style of governance. The working class succeeded in so frightening the party leadership that repression was massively reinstituted. At the same time, the experience robbed the party of self-confidence. The prospect of serious internal dialogue receded once again as a centralized bureaucracy protected itself from criticism by depriving the rest of the population of social space.

It is necessary to emphasize, however, that not all of the blood-soaked lessons were immediately comprehensible. Indeed, there was one final dynamic lurking within the Polish working class that did not become visible in a tangible form in 1956 and therefore had no opportunity to become a "lesson" that could be internalized by broad sectors of the population. If the social contentions surrounding the Polish October verified the existence of a regional imperative at the heart of the organizing challenge, those events did not succeed in making visible some important cultural differences between various regions across Poland. Some geographical areas in every society have been historically shaped to be more culturally predisposed to encourage creative self-expression than surrounding regions. In Poland, that predisposition was most advanced in the maritime centers of the Baltic coast. This was the hidden truth concealed in the politics of 1956. It would flicker briefly in 1957. It would become broadly visible in 1970. It would bedevil the Polish regime in 1971 and resurface intermittently throughout the remainder of the decade. Finally, it would break into free democratic space in the summer of 1980.

After a sufficient time for reflection, the Baltic movement's links to Poznań would become less obscure than they appeared to be during the dramatic August of 1980. When that time came, it could be seen that some of the paving stones on the road to Gdańsk had been quarried by the men and women of H. Cegielski-Poznań. Stanisław Matyja and his fellow activists who had been arrested, beaten, and ultimately cashiered from Cegielski had earned—more deeply than anyone knew—the "protection and respect" they one day would receive from the organized workers of Solidarność.

3 ▪ The Movement Finds a Democratic Form

A great deal of knowledge about the revolution is not on paper but only within ourselves.

Thomas Jefferson, 1802

AFTER THE SOVIET army had provided Władysław Gomułka's unpopular party with the means to seize control of Poland's social life in the late 1940s, a new conception of geography emerged in the nation. It soon became effectively packaged as one-line Warsaw street wisdom: "Poland is bordered on three sides by Russian tanks and on the fourth by Russian cruisers." Such was the ever-present geopolitical reality of postwar Poland. Yet it did not constitute total closure. If three of the nation's borders were patrolled and the bordering nations were all yoked to the Soviet bloc, the fourth side was an open watercourse to the world that no fleet could effectively blockade in peacetime. In the coastal provinces and especially in the seaport cities of the Baltic, Poles were in constant contact with all sorts of people who lived beyond the reach of the Leninist state.

An effective monopoly of social information was simply not possible on the Baltic coast. In an unending stream, ships from all over the world came to Gdańsk, Gdynia, and Szczecin: ships with cargoes, ships needing repair, above all ships with crews. Over the decades, the dialogue that ensued between foreign seamen and Polish maritime workers inexorably affected the political culture of the coastal provinces. Many were from the Soviet Union, and this did not prove noticeably broadening. But many other seamen were from the West, and they brought a political outlook of extreme complexity. Almost all seemed to regard themselves as "socialists" of one kind or another, and in the early postwar years, a number were Stalinists. Whether in Marseilles, London, New York, or Gdańsk, political conversation in seamen's bars could be counted on to be varied and passionate.

A bit of history helps explain this folkway, for certain common threads are visible in the embattled pasts of the world's sundry working classes. While workers in heavy industry have generated for themselves a heritage that is more "militant" than that amassed by those who labor in offices or in light industry, participants in certain occupations may be seen in retrospect to have generated a particularly intense brand of assertion. Among them are miners, seamen, and longshoremen. In the case of miners, a certain militancy was built into the trade itself. They worked and lived in a separate world and were locked into a particularly stark mode of combat with mine owners. In America, the United Mine Workers put up their own strike funds as seed money to start the electrifying CIO organizing campaigns in America's mass-production industries in the 1930s. The mine workers' chieftain, John L. Lewis, hired communists to help organize the steel industry, rewarding them when they acted like militant unionists and firing them if they persisted in acting like communists. Similarly, from Silesia to China to South Africa, miners have played aggressive roles in the struggle for human rights.[1]

In Western Europe and North America, seamen and longshoremen have had an equally storied past. The great London dock strikes of 1889 and 1890 foreshadowed the militant unionism of the first half of the twentieth century and established a historical precedent for the emergence in the United States of such radical unions as Harry Bridges's Longshoremen and Joe Curran's Maritime Union. But a distinction should be noted here. In the case of miners, necessarily provincial by occupation, militancy in Western nations has been a distinctly national phenomenon, largely uninfluenced by organizing rhythms from other countries. Aggressive miners appeared in many nations as a product of the hazardous working conditions and occupational identification of each nation's miners, regardless of the specific nature of each country's political system.[2]

But seamen and longshoremen did not live in an isolated world; they themselves traveled or met workers who traveled from other lands. Nowhere in the world were lateral communications among workers more fully developed or more constantly reinforced than among coastal working classes. Innovations, ideas, and manifestos that worked in one nation quickly traveled by ship to the needy, wherever they might be. It is little wonder that in nation after nation, unions of longshoremen and seamen have been viewed by sanctioned authority as hotbeds of syndicalism, communism, and socialism. Even when unions grew staid and traditional, many a rank-and-file enthusiast remained to give full expression to the original militancy out of which the union had taken shape.[3]

It was foreign workers from such milieus who regularly disembarked at Gdańsk and Szczecin. Many took a deep interest in the Stalinist experiment in Poland following World War II. Their questions were matched by Polish workers' questions about the workers' struggle in the West. Gradually, in barely detectable ways, Baltic workers became a bit different from their comrades in the rest of Poland. The Polish October of 1956 on the coast reflected these distinctive regional influences. Coastal workers neither marched with the party, as so many workers in Warsaw enterprises did, nor acted with the tested anti-party militancy of the workers of Poznań. In common with their counterparts in much of the rest of Poland, coastal workers did not—in 1956—generate a sustained eight-month period of self-activity as was done in Poznań. They consequently had neither the self-generated experience nor the maturing judgment that Cegielski militants came to possess at the moment of Gomułka's "October Revolution." Rather, like Poznań workers in the early stages of their self-organization in June, the working class in Gdańsk took to the streets in October in an undisciplined expression of emotion that had been too long bottled up. "Out with the Russians" was a street song of release and splendor, but it scarcely substituted as a working-class program for liberation from the party-state. In Gdańsk, as elsewhere, there was still hope in October that Gomułka would "do the right thing" and renew once again the Polish path to socialism. This is merely to say that in their politics, coastal workers were unconsciously relying on others rather than upon themselves. Thus, it was not necessary for the party to sequester coastal workers inside some "experimental program" to contain their programmatic militancy, for they were not, in fact, programmatically militant. In strategic terms, workers on the Baltic coast did not require domestication, for they had not sustained a period of self-organization that was the prerequisite to "consciousness-raising." The Baltic coast in October of 1956 was not Poznań.[4]

But it was also not Warsaw. If coastal workers had not yet quite learned to look foremost to themselves, they certainly did not look to the party in the fashion of Goździk's comrades at the Żerań plant. Rather, in their quest for a proper model for shop councils, coastal workers looked to the sea. In the Gdańsk-Gdynia region, they hit upon the idea of forming workers' councils and affiliating them with the international union of maritime workers and longshoremen which grouped together predominantly left-wing unions from sixty nations. The solution seemed to offer a quick way to escape the party. A future strike on the coast could be expected, as always, to bring the party down upon them, but what would the party do when no Polish ships could be loaded or unloaded at any foreign ports where longshoremen belonged to the same union? It was a tantalizing prospect. But party officials were quite capable of projecting the same speculations. They set in motion a vast tide of machinery to defeat the idea of Polish workers participating in international unions. In this struggle, organizational cohesion was considerably more important than mere inventiveness. Coastal workers did not possess the necessary cohesion; prior worker self-activity on the cause was revealed to be quite fragile. No Poznań-style solidarity was visible because no Poznań-style organizational effort had been previously achieved over many months to generate the experience in collective action or to encourage in workers the collective self-confidence necessary to wage a sustained conflict with powerful adversaries. The Polish October had come as something out of the blue on the coast, as it had in so much of Poland. The party, in contrast, was well entrenched organizationally. Coastal workers were rather quickly persuaded to put aside wild ideas about associating with foreign workers and to accept the more traditional concept of functioning within the official Polish trade unions. In Gdańsk, Szczecin, and Gdynia, workers did not like this solution, but they had no ready alternative that they had discussed, labored for, and spread within their own ranks as a long-term working-class objective. In sum, the coast was not prepared for the opportunities, however slight they might have been, that appeared during the Polish October of 1956.[5]

Thus, the mere presence of a cultural opening to the world did not in itself provide Baltic workers with some sort of instant panacea. But in even more subtle ways, geography was a relevant factor in what might be called the de-provincialization of the coastal regions. Most visible were certain material conditions that coastal workers were forced to confront. Poles learned that workers from Sweden and Denmark or from the Netherlands and France dressed better and ate better than they did. The particular "detail" in comparative living standards that had the most impact on Poles varied according to individual hungers and needs. One Baltic worker was struck by the quality of off-duty clothes of Swedish seamen; another was transfixed by stories of dentistry in Denmark. Still another instructive comparison was derived from the custom of exchanging meals. The officers and men of ships that underwent repair in Baltic shipyards often threw a farewell banquet for the Polish workers they had come to know. Sometimes, Polish authorities arranged for a reciprocal meal, hosted by shipyard officials. In the 1960s and 1970s, Polish workers were thunderstruck by the elegance of both kinds of banquets. The visitors, necessarily drawing rations from their ships' stores, invariably displayed a wide variety of soups, main courses, and desserts—in the aggregate far superior to anything available to Polish workers. But the Polish banquet was instructive as well. The zealous efforts of party officials in the shipyard to present a good face to their visitors resulted in meals that—while not up to foreign standards—were so far superior to normal Polish fare as to reveal the sense of inferiority of Polish officialdom. "What got me," one worker said,

> was the lies of party officials. They acted like the meal was an everyday thing. But every Pole there knew better. All the party wanted was to impress the Dutch crew which had fed us so well the night before; they wanted it so much they didn't care about lying in front of us. They made me ashamed of being Polish. As soon as I discovered that, I got mad. I'm not ashamed of being Polish, I'm proud of being Polish. I discovered that what I was really ashamed of was the party's lies.[6]

The details of disillusionment: one worker remembered exchanging packages of cigarettes with a visiting seaman from Sweden and then later finding the Polish package, with one cigarette used, in the trash container. "It is not only that Polish cigarettes are no good," he said,

> but even the packages fall apart before you finish using them because Polish cigarette factories don't use enough glue. The plan you know—there wasn't enough glue in the plan, you see. I used to take empty foreign packages and put my cigarettes in them and use them over and over. Building socialism! We weren't building socialism; we were building shit.[7]

This education in what might be called comparative materialism, while somewhat richer in its instructional nuances on the coast, was one that every Pole eventually absorbed to some extent. The Polish army was undoubtedly the nation's most effective teacher, and it extended its lessons to every corner of the land. The subtleties the army conveyed were precisely opposite to the ones the party taught. In any province of Poland, the young of the nation had no option but to imbibe the official teachings of the state, with all its attendant propaganda about the rate of progress the country was achieving. Through sheer repetition, this steady informational pounding had a discernible impact, even though corroborating evidence was unavailable in one's personal life. One worker explained,

> I believed that Poland as a nation was improving. I could see the television films of Nowa Huta and the Żerań plant and new worker apartments. I thought everybody was moving ahead except in my little village. Our time was still to come, I thought. But when I served my time in the army, I learned that all Poland was like my village—not like television. We talked in the barracks at night. The stories you heard! Everywhere was the same. The only ones who lived well were the authorities! In the Army I learned that everything the party said was a lie.[8]

The material lessons, then, were national, not regional. But what imparted a special cutting edge to the stories heard around coastal shipyards was the difference in living standards for workers, East and West. It was one thing to suffer when everyone suffered—this was the message learned in the army. But it was quite another to learn that "everyone" was not, in fact, suffering. The message learned on the coast was that socialist workers suffered more than capitalist workers. Exposure to Western material achievements cracked the party censorship in a sensitive area and revealed with certainty that mass poverty did not have to be the unavoidable end product of Polish nationhood. Prolonged fraternal relationships with Western trade unionists also provided a fund of ideas about how workers might soften the suffocating impact of undiluted managerial authority. But if the same Western militants could—particularly after the Soviet invasion of Hungary in 1956—provide a searing left-wing critique of Stalinism, that was a capability that was scarcely needed, given how deeply felt the same folk wisdom had become in Poland itself.

What exposure to the West brought was not a specific formula for working-class action but a generalized conviction that action was possible—and through a number of alternative routes. The association of foreign and Polish workers confirmed the Polish conviction that the party had to be circumscribed by one method or another. When viewed at close range, the party simply was unable to win many committed defenders among maritime workers, whether they came from Poland, from the Netherlands, or from anywhere else. This was a helpful corroboration and imparted a certain long-term resolve to the workers' milieu on the coast.

This circumstance led—in 1957—to developments on the coast which set the Baltic region somewhat apart from the rest of the nation. As was the case in Poznań, Warsaw, and elsewhere in Poland in 1956, the struggle between the party and the working class on the Baltic coast was essentially a series of sharp local engagements that extended through the spring months of 1957 before slowly losing momentum in the face of overwhelming counterpressure. To the great advantage of the party, the regional imperative prevailed. But on the coast, worker insurgency did not meekly come to an end with the domestication of shop councils under the official trade unions.

The fishing fleet out of Gdynia went on strike in 1957 through the simple expedient of leaving the fishing grounds and returning to port. The party promptly moved to sack the strike committee and the ship captains responsible. But enough lateral lines of communication had been established—within the regional limits of the coast—to provide for a timely worker intervention on behalf of the beleaguered fishermen. In the words of a participant, "the Gdańsk and Commune (Gdynia) shipyards threatened to strike if a hair of these striking fishermen was touched. We didn't win much, but the decisive stance . . . ensured that there really was not a single victimization." One might add, a memory was established. For a moment, there had been an independent Seamen and Sea Fishermen's Union on the coast.[9]

Such organizational sophistication as was demonstrated in support of the fishermen constituted a type of job action on the coast that was not obtained at the time of the Poznań events in June 1956 or even at the time of the Polish October. The achievement of "not a single victimization" was a product of the lateral lines of communication developed among coastal workers *after* October in the course of the struggle with the trade unions over the shape of the new worker councils. This was a singular development. It indicated that coastal workers in 1957 had equaled the level that Poznań workers had achieved in 1956. Not remotely as collectively politicized as Cegielski workers in November 1956, shipyard workers utilized the ensuing months to develop the first fragile links between enterprises and between occupation groups. It all came a bit late—after the essential national battle for shop-council autonomy had been lost—but it nevertheless revealed organizational affinities that were not common elsewhere in Poland. Coastal workers did not tend to divide proudly—as Polish miners did—along craft lines. They were not exclusively "shipyard workers" or "longshoremen" or "fishermen." Rather, they saw themselves as workers in maritime enterprises who could cooperate with one another. Shipyard workers who could threaten a solidarity strike in support of striking fishermen were demonstrating—as early as 1957—a certain class orientation that could lead in its culminating extension to an interfactory strike committee. And the culminating expression of an interfactory strike committee was the almost mythological weapon of the working class: the general strike. No interfactory strike committee appeared on the Baltic coast in 1957, but the cooperative intervention of shipyard workers in support of seamen and fishermen constituted the idea itself in embryo.

With hindsight, it is fairly easy to see what Poland's working class needed in order to be properly armed for a sustained struggle with the party-state: the broadly creative vision that was haltingly visible on the coast in 1957 plus the sustained organizational self-activity that was visible in Poznań the previous year. The two in combination would be a potent force. Indeed, to combine these two provincial achievements into one would be to chart a path to Solidarność. One thing was clear in the aftermath of the events of 1956–57; workers in the maritime provinces of Poland were not afraid to propose adventurous alternatives to the one-party state. Their contact with workers from other nations and the instructive comparisons that such contact afforded produced a degree of demystification about "Leninist norms" that was unmatched anywhere else in Poland. In the absence of lateral lines of communication within the Polish working class, this development constituted a cultural fact about life along the Baltic coast that was unknown to the rest of the nation. But the task of building an independent structure upon this cultural foundation proved to be awesome. It required thirteen more years.

The reconfinement of the coastal unions under state-controlled structures left the same bitter taste in Gdańsk in 1957 as in the rest of working-class Poland. But in addition to the generalized oppression, each region of the country suffered in its own way and had its own local grievances. After Gomułka had consolidated his hold on the party and society, the central plan once again took on the same general configurations that had distinguished it in the era of Bolesław Bierut: resource allocation emphasized capital investments over social needs. On the coast, this translated into widespread investments in shipyard facilities, even as housing for shipyard workers continued in a state of crisis. With working-class families confined in cramped and dilapidated apartments, the priorities of the regime succeeded only in deepening the general resentment.

Given the government monopoly on social information, it is not clear what precise economic statistics Gomułka had to ignore in order to sustain the party's long-term policy of planned underconsumption throughout the 1960s. Such statistics as have come to light vary greatly, though they share a common thrust: Polish workers were very poorly dressed, particularly for winter; meat was an exception rather than the rule at Polish mealtimes; housing was in acutely short supply; and the great majority of Polish workers did not have hot running water, refrigerators, or telephones in their homes. The environment of Poland was one of shared poverty.[10]

In overall terms, it might seem that the situation of Polish workers over the quarter century from the end of World War II to the Baltic rising of 1970 had become marked by a rather profound continuity. The distinguishing features of this period were authoritarian production relations, a sharply confined public life, and general material deprivation. But the appearance of continuity concealed a profound alteration in the climate of popular aspiration. In consequence, a workers' movement very gradually began to take shape on the Baltic coast in the second administration of Władysław Gomułka.

The dynamics worked themselves out essentially in the following manner. Polish workers had waged a prolonged struggle right after the war in hopes of staving off a Stalinist future. When they lost, they had little choice but to submit grudgingly to the intense regimentation that accompanied Bierut's 1950–56 central plan. Throughout this first postwar decade, one thought was consoling: the physical evidence that materialized around every Pole provided positive verification that the nation was, in fact, industrializing. New enterprises were everywhere—in steel, chemicals, textiles, and in the Baltic shipyards. The giant steel complex at Nowa Huta genuinely appeared to embody transformative potential for the whole society; Żerań seemed a glittering jewel; and millions of peasants took their places on the proliferating production lines of People's Poland. All such tangible realities seemed to indicate that the government was right when it continually announced that "progress is being made." The contradictory evidence in housing, food, and clothing that disfigured every Pole's daily life could reasonably be explained away as the short-term price of a redeeming industrialization. Unfortunately for the regime, while this explanatory litany served for a time, it carried its own implicit corollary: the day would come when daily life itself would also begin to show a measure of "progress."

It was this latent expectation that fueled the emotional drama of the Polish October in 1956. In the popular mind, the Soviet presence in Poland was directly connected to the poverty and confinement of the nation. In removing the Soviets, Gomułka unwittingly assumed an obligation to deal with the second part of the equation. It was an obligation he explicitly acknowledged in his first public speeches upon assuming power. But when Gomułka cancelled the Polish October through the repressions of 1957 and resumed a national policy of planned underconsumption, the result in the popular mind was a deep resentment that proved enduring. After 1957, Poland's United Workers' party lost the freely given support of almost all its serious intellectuals and its most committed worker activists. In the early years of the Gomułka era, an increasing number of Poles, from left to right, came to the conclusion that they themselves would have to perform the acts that would turn the country in a new direction, because the party simply did not have the capacity to do so. Indeed, among the more adventuresome sectors of the working class and the intelligentsia, this thought had begun to surface even before the coming of Gomułka. It became visible culturally—in such genres as poetry—but it also could be found in ordinary conversations in the living rooms of the Warsaw intelligentsia and on lunch breaks at places like Cegielski-Poznań. After the collapse of the "Spirit of October," the understanding attracted an even broader sector of adherents.[11]

As with many other cultural phenomenon, this social "fact" cannot be dated with the kind of precision historians routinely prefer. Though the incipiently insurgent attitude constituted a crucial causal element in the emerging dynamics of Polish life, a moment of birth cannot be fixed. Moreover, it is important to recognize that the expression of this attitude did not at the outset take a public form. Rather, it was something individuals privately thought about and later mumbled to one another in the privacy of their own apartments.

We may characterize this moment of time as one of "private insurgency." It is a pivotal juncture, but in the literature of political theory, it has no name. As a component of social relations, it is one of the most uninvestigated areas of political science, a palpable lacuna rife with theoretical and practical import. The "moment" of private insurgency generates no historical records; it is, in a sense, invisible; it may therefore be understood to be pre-political. Nothing that might be characterized as a movement has yet emerged. It is, in fact, the last moment before movements become historical. It is that juncture when future activists are talking to each other and have not quite begun to ponder how to reach out and connect either with the larger society or with social groups other than their own.[12]

It is not an easy task to dramatize this lacuna in such a way as to make it visible. In societies the world over, abused people are forever discussing their grievances; some among them eventually proceed to create agendas, manifestos, open letters, platforms, and other plans of action designed to address such grievances. These earnest planners and manifesto writers may be pictured metaphorically: they sit around kitchen tables in East Berlin, New York, Managua, Warsaw, Buenos Aires, and Teheran—everywhere—and they analyze the established order. In such domestic settings, they think together, drink coffee or wine or some other national drink, argue, interpret power, and pull their chairs closer together for comfort in their isolation. It can be serious work, but historically it has usually not proved very rewarding. Aside from all the political problems that history verifies such people have had, their most maddening challenge has been to get out of the kitchen and connect with the larger society. Whether or not their discussions carried them to a political position that one might regard as "correct," their enduring practical task lay elsewhere—in the area of recruitment. However appropriate their plan of action might have appeared around a friendly table, the reality they discovered in the outside world was that few people cared or dared to listen.

It may be noted that unless something happens to alter this circumstance, the kitchen planners will never appear in history books. They will grow old around their table, dreaming of the revolution, and no one—save possibly file clerks in the security services—will take note of their existence. But should they somehow succeed in breaking out of their isolation, some of them, at least, will become historically recoverable. Journalists in the near term and scholars at a later date will probe into the seed time when the activists talked around kitchen tables. In this manner, the movement will become visible in its final pre-breakout phases. In the nature of journalistic and scholarly research, what transpired before then is likely to remain unexplored.

In the case of the Polish August of 1980, the relevant preceding phase is taken to be the few months or the few years immediately prior to 1980. Should one probe for this "relevant" period in Gdańsk and settle, let us say for a glimpse of incipient insurgent activity in 1979 or 1977, one can find many well-populated tables in worker apartments, and one can also identify with precision a number of people who sat at them: Krzysztof Wyszkowski, Bogdan Borusewicz, Andrzej Gwiazda, Lech Sobieszek, Florian Wiśniewski, Kazimierz Szołoch, Lech Wałęsa, Stefan Lewandowski, Anna Walentynowicz, and Bogdan Felski, among others. Should one probe a decade earlier, fewer tables can be discovered and some of the people around them will have changed: Wyszkowski, Sobieszek, Wałesa, Walentynowicz, and Szołoch are there, but also others named Mośiński, Nowicki, Jagielski, and Lenarciak. A decade further back, in 1959 or 1960, there are even fewer conversations that can retroactively be documented through the techniques of historical research with tape recorders. Lenarciak is there, a young man cautiously listening. Continuity cannot be expected because there are so many casualties in Eastern Europe. Many Matyjas and Goździks of the 1950s have vanished by the 1960s and 1970s.[13]

The central association that these men and women of coastal kitchens shared throughout these years was their potentially provocative social circumstance: they may be seen as incipient worker activists who are directly connected to their constituencies. As such, they constitute—at the instant they break out from the isolation of their kitchen tables and connect with their constituencies—a social movement.

At the center of this still unrealized social formation on the Baltic coast in the 1950 and 1960s was a compelling idea. It had been present after the war when the shop councils fought the party police, and it had resurfaced in 1956 at Cegielski-Poznań and in 1957 in the solidarity strike in support of the Gdynia fishing fleet. The idea, quite simply, was this: Poland needed a union of workers that was free of party influence. Historically, the idea came to have an inexorable rhythm in postwar Poland. It would appear in the working class, acquire some sort of institutional substance, take tangible public form, and then be crushed by the state. Though the idea would then disappear from public view (in 1947, in 1957) it would remain a part of conversations around kitchen tables. Over time, and particularly on the Baltic coast, the idea would once again edge back to the center of discussions, so that familiarity with it bred a renewed poise, so that workers could again talk about it without descending into immobilizing personal resignation. It was an idea that had the deepest kind of appeal, but it was also extremely difficult to carry beyond the kitchen because one's fellow workers had trouble believing the goal was socially possible. In truth, even the most committed activists intermittently had moments of doubt; it was an objective in constant peril of being abandoned as hopelessly fanciful. A union free of the party? After the national defeats of 1945–47 and 1956–57, after all the bitter local defeats across Poland in between those times, it seemed almost silly to believe in the idea.

What gave this impossible idea continued life and kept it bouncing around kitchen conversations for a generation on the Baltic coast was that its practical relevance was verified every time a worker was seriously hurt because of storming, every time an inappropriate worksheet or unworkable timetable appeared, every time work and pay rules were changed in response to the whims of a far-off bureaucrat, every time any such complaints were waved off by a party functionary habituated to helpless shrugs and ingrained deference to the central planners in Warsaw. The idea took on renewed meaning because of the simple incoherence of daily life on the job, and it was given added substance by the incompetence and corruption of the trade-union apparatus that hovered over production. The idea lived because in whatever manner discussions began around the kitchen tables of the Baltic coast, the inevitable end point, the only end point that made the slightest sense at all, was the idea of a union free of the party. It was really quite simple. The organizational challenge for activists also had a certain simplicity: how to take the idea out of the kitchen and into the workplaces of Poland. Whether along the Baltic coast or anywhere else in the world, it was a task whose difficulty could scarcely be underestimated. In the late 1960s, after years of talk, Sobieszek, Nowicki, Mośiński, Lenarciak and many other workers on the Baltic coast could attest to the reality of that unwanted fact. People had serious complaints, but seemingly nothing much could be done. So "politics" became the art of complaining to one's acquaintances.[14]

This, of course, was the problem. Coastal activists talked, but none of them did anything—that is to say, none of them attempted to organize. Rather, they went to work, raised their families, and occasionally in the evenings sat around kitchen tables. The days of a conscious postwar effort to rebuild Poland were in the distant past. The economy was stagnant, and the world of the Polish working class had become one of subsistence living. As in the era of the Poznań events, a newly married couple could look forward to a wait of seven years for an apartment. Poland was becalmed.

It seemed that Władysław Gomułka had been in office forever when the government suddenly announced on the brink of the Christmas season in 1970 a steep rise in the price of food and a wide variety of other consumer goods. It was an act that shattered the only remaining element of the national economy that had a shred of popular appeal—ten years of stability in food and meat prices. Suddenly, a food-price increase in the range of 12 to 36 percent was proclaimed by Warsaw radio. The act was stupidly timed as well as clumsy. A party leadership without credentials had picked the Christmas holidays to demonstrate yet one more time its disdain for the population. In utter exasperation, Polish workers exploded all over the country. In complexity, passion, and ultimately in violence, the risings in the maritime provinces stood far above all other protests around the nation.

The pace of events in Poland in December 1970 was both breathtaking and tragic. The government decision, announced on December 12, triggered a kind of despairing anger everywhere and threw local party officials on the defensive. In such a volatile climate, time-honored practices of party evasion simply had no chance of working. There was also a compelling practical reason: because Polish workers spent over half their incomes for food, the price increase meant a substantial reduction in an already marginal standard of living. Since by 1970 it was widely understood that plant-level party functionaries had no power, the dynamics of the situation seemed to propel workers into the streets to lay their grievances, Poznań-style, before metropolitan party headquarters. In light of the sequence of assertion on the coast during the Polish August of 1980, the evolution of collective action in 1970 established a pattern that was to endure: first, Gdánsk workers took to the streets on December 14, then Gdynia workers on the fifteenth, followed by Szczecin workers on the seventeenth. The form and the sequence of local government responses varied in each case, and the resulting collective reaction of workers varied also. But a grim commonality informed the results: buildings housing the party headquarters going up in flames, government troops firing into crowds and killing workers, mass rioting and prolonged street battles, frenzied regroupings of workers in their different shipyards, the appearance of masses of armored troops and tanks, many more deaths, and a culminating atmosphere of disbelief, hatred, and permanent alienation. Besides scores of deaths, hundreds more were wounded or badly injured in all three cities as a state of insurrection clouded the coast for days.

These cataclysmic happenings left an immensely complex residue. The legacy of 1970 extended to a complex series of intermittent negotiations, strikes, and organizational experiments that continued through December and into the spring and summer of 1971. Collectively, these experiences formed what can perhaps best be described as an institutional working-class memory, one that was decisively to shape events in 1980. But there was also another powerful social memory that extended beyond workers to include the whole population of the coast. This second aspect has been vividly captured and preserved by Lawrence Weschler, an American writer who journeyed to Poland in the era of Solidarność. In Warsaw, Weschler encountered a cassette tape recorded on the Baltic coast at the height of the 1970 turmoil—a tape carefully and angrily preserved for years by its owner. Weschler described what he heard:

> At first the impact of the sounds on the tape—the crackle of bullets, the wail of sirens, the chanting of crowds, the scatter of people—was once again dulled for me by Western familiarity with the aural vocabulary of violence. But as the tape continued—on and on, shooting, sirens, screams, more shooting, and then *more* shooting—the horror began to bleed through. It occurred to me that our Western newscasts always offer us bite-size morsels, little digestible snippets that disguise the true horror of conflict—that is, that it just seems to go on and on and you have no idea when it's going to stop. The man playing the tape for me seemed no less moved, even though, surely, this was the hundredth time he'd heard it. After a while the tape ran out in mid-wail.

Weschler contrasted the Polish memory of 1970 with the very different manner in which most Americans wave away the past—the past, say, of the Palmer raids, the bonus march, Selma, Birmingham, and Kent State—with quick nods that say, "That's gone now, that's history." In Poland, Weschler reported in 1981, "when people talk about the massacre in Gdańsk and say, '*That's history*,' the veins in their foreheads pulse and conversation stops dead cold. They *cannot* get over it. Poles, especially the residents of Gdańsk, talk about 1970 the way Jews talked about Auschwitz: it's almost a transcendent category." The irony here is that the other memory, the worker memory of worker organization, while considerably less powerful emotionally, was to prove even more "transcendent."[15]

On the Baltic coast in 1970, the workers internalized two experiences of their own—an effective democratic method of mobilization and an effective democratic method of assertion. The first was the occupation strike, and the second was the interfactory strike committee. Each was important, but in tandem they constituted a successful transcendence of the crippling working-class limitations first exposed at Poznań in 1956. The occupation strike was partly defensive, a way of assertion that protected organized workers from the violent policing power of the state: the mass of striking workers, remaining together in a factory, protected their own strike committee from arrest. The interfactory strike committee was offensive—a mechanism of internal communication that offered the prospect of bringing every rank-and-file worker operationally "on stage" in ultimate dialogue with government power; the party was forced into dialogue not with an organized factory but with an organized region. The economic cost and, therefore, the political power of the work stoppage was multiplied exponentially. Moreover, through this mechanism of internal organization, the interfactory strike committee could speak with a single, strategically compelling voice. The problem in lateral communications that had wrecked the Cegielski movement in Poznań had been overcome.

The occupation strike has a storied reputation in the American working class, among whose organizers it has long been known as the "Polish strike" and more widely as the sit-down strike. Its American popularizer was a militant trade unionist named Wyndham Mortimer, who in 1934–35 successfully organized the work force at Cleveland's White Truck Company, which contained a substantial number of Polish-Americans. In 1936, Mortimer was the organizing strategist behind the "great sit-down" at the Fisher Body plant in Flint, Michigan, that finally brought General Motors to the bargaining table and launched the United Automobile Workers of America. In turn, the Flint sit-down promptly became the organizing model for the CIO in other mass-production industries in America; it was the tool successfully employed to combat police and company thugs in the tense and often bloody recognition strikes in the steel, electrical, and rubber industries in the late 1930s. The CIO's debt to the sit-down strategy was sizable indeed—and it was a debt to the Polish working class. The occupation strike had gained great fame in the early 1930s in Poland where socialist and communist trade unions battled the nation's right-wing prewar government. By 1936, it had become a finely honed and widely used instrument of working-class assertion, a tactic so tested and so heralded that it made the long ethnic journey across "Polonia" to the CIO in America.[16]

How did Baltic shipyard workers recover in 1970 this vintage organizing tool first popularized internationally by their Polish grandfathers? The most informed student of the 1970 events on the Baltic coast is the American political scientist Roman Laba. He has brought forth evidence that the recovery by the Polish working class of the occupation strike can most reasonably be attributed to the unwitting efforts of the Polish United Workers' party in 1970 forcibly to keep shipyard workers confined to their work areas during the crisis: "With dozens of tanks, armored transports and automatic weapons trained at the shipyard gates, this was a powerful incentive indeed to have an occupation strike."[17] Thus do authoritarian regimes instruct their victims in the nuances of countermeasures. The irony here is compelling. Irrespective of the precipitating cause, however, the enduring result for Polish workers was experiential. After 1970, shipyard workers on the Baltic coast had relearned what an occupation strike was all about and how to conduct one under the escalating pressure of the state's police power. A tactical cornerstone of the Polish August of 1980 was in place.

Similarly, the interfactory strike committee constituted a new plateau of self-generated democratic forms. It, too, grew out of the logic of the engaged experiences between the party and the working class on the Baltic coast in 1970. To grasp the significance of this form of popular assertion, it is necessary to keep firmly in mind certain sequential imperatives that historical experience has clarified as functioning components of democratic movements. First, it is necessary to put aside the sanctioned assumption that movements of effective democratic assertion can be explained away as "spontaneous" happenings that suddenly materialize in heretofore tranquil places. As has repeatedly happened in other societies (and as specifically happened in Poland at the time of the Poznań rising on June 28, 1956), street riots can be seen as unplanned and therefore "spontaneous." But sustained and effective democratic assertion cannot be unplanned; it has to be premeditated, a result of prior conversation among the asserters. In the same way, large-scale strike actions cannot be and are not spontaneous. Transparently, the idea of a strike has to begin somewhere in a small group that subsequently endeavors to spread the idea either prior to or during the strike itself.[18]

For a popular mobilization to grow large in scale and at the same time maintain democratic coherence, strong mechanisms of internal communications must be developed. Such communications links are essential to permit the accumulated experience of the group to be transmitted from those who have thought about that experience a great deal to those who have considered it less intensively. If such thick webs of communication are not present, democratic self-activity by one group of people will be in constant peril of being undermined by less-informed self-activity of those they recruit. This organic relationship may be taken as a fundamental axiom of democratic politics. In precisely this manner, the democratic movement at Cegielski-Poznań in 1956 collapsed into the disorganized upheaval that is now known historically as the "Poznań events." The absence of adequate internal communications within the variegated components of the Poznań working class converted an incipient democratic movement into a street riot.

The form of social organization known as an interfactory strike committee effectively filled this lacuna. It provided the means for the knowledge of more experienced worker activists to be transmitted to the least experienced—from a politicized core of activists to a politicizable mass base of rank-and-file workers. In short, it provided a mechanism to achieve the absolute cornerstone of democratic politics—a self-created environment of information that can lead to coherent collective action.

Thus, two essential tools of democratic movement building were fitted into place by the dramatic and tragic events on the Baltic coast in 1970. How effectively those tools were employed, then or later, depended, of course, on the experience, judgment, and democratic proclivities of the people who utilized them. In this sense, too, the chaos on the coast was instructive, for these events constituted a gigantic schoolroom in which activists could ponder the relative merits of possible future courses of action.

Before entering this "schoolroom," one caveat is offered. In common with other social groups in other countries, the members of the Polish working class cannot be understood abstractly—either individually as "workers" or collectively as a "proletariat" possessing specific kinds of endowed characteristics. In all societies, "classes" (even when they can be identified with apparent precision) are in constant states of change from what they used to be and will never be again to what they are becoming. The only "characteristics" that can accurately be affixed to such social formations are those that their own actions make visible at any one moment of time. All theories about presumed "class conduct" notwithstanding, it is still necessary to ascertain concretely what a group of people are doing before one can determine what they are, what they think, or what their political intentions or capabilities might be. It is to this task, personal and individual as it intermittently must be, that it is necessary once again to return.[19]

By 1970, the regime of Władysław Gomułka had grown so distant from the Polish working class that the first secretary and his associates on the Central Committee seriously misjudged the probable response of workers to a broad increase in the price of food. The angry resentment that immediately surfaced throughout the country stunned Gomułka and the culminating explosions on the Baltic coast shocked and immobilized him. He sent others to the coast and tried to direct countermeasures from his isolated bastion in Warsaw.

Upon learning of the increase on Saturday, December 12, coastal workers spent the weekend in various stages of brooding discontent. At W-4 of the Gdańsk shipyard, Henryk Jagielski summed up the mood: "Among our trusted colleagues, we agreed that Monday we were not going to work." Within an hour of reporting to work on Monday, the early-morning shift was in turmoil. The first workers to organize a work stoppage were the skilled engine mechanics. By 9:30 a.m., over one thousand other workers had begun to gather in front of the administration building. The ensuing dialogue predictably produced nothing of substance, and workers began to talk among themselves about possible courses of action. This was something of a new experience. The Gdańsk shipyard in 1970 was not Cegielski in 1956; four months of intense self-activity had not preceded the moment of possible assertion. At any rate, a march to party headquarters was proposed as a way "to let the town know that we've stopped work, that we're striking." Monitors were picked to keep order. Though hurriedly improvised, the result could scarcely be described as a spontaneous rising. Something less than a thousand (predominantly young) workers filed out of the shipyard in orderly formation and marched to downtown Gdańsk. At party headquarters, a junior official tried to bring calm by suggesting the workers name delegates to come inside and negotiate. The workers turned this down out of fear, born of past experience, that anyone named would be arrested when the confrontation was over. The official suggested that everyone should return to work. This, of course, merely enraged the crowd. It also imposed on them the need to do something, as the pressure they had applied had proven inadequate.

It occurred to some at this point that various prior chores had not been accomplished; accordingly, a group of workers marched to the radio station and tried to persuade the personnel there to broadcast an appeal for all coastal workers to join the shipyard demonstration. It soon occurred to other workers that still another chore had not been done; accordingly, a second group went to the Gdańsk Polytechnic in an effort to recruit students to enlist in their ranks. The sundry comings and goings of newly activated workers became a kind of mobile colloquy that generated occasional new chants: "Out with Gomułka," "The party lies," and, in exasperation, "Out with the party." A sound truck was liberated, and someone, declaring a "democracy," employed it as a popular forum through which individuals wishing to express and opinion could do so. A high point was reached when a worker intoned, "We want butter just like the comrades. We want our children to go to school just like the comrades. We don't want know-nothings ruling the country. We don't want party committees running factory councils and unions. Down with the Red Bourgeoisie!"[20]

The first clash occurred in late afternoon when militia troops appeared and foolishly tried to contain the crowd by dispersing it with nightsticks. Here, finally, was something for workers to do. They surged against the troops. Some rocks were thrown. Windows shattered in a ministry building and then in a bank. Soon after six p.m., the crowd, now swollen with youthful onlookers and increasingly emboldened, returned to party headquarters. Units of the security police fired tear gas into the throng. The resulting dispersal merely spread the violence. Roving squads of demonstrators set fire to newsstands and to official cars, then to the railroad station, and finally to buses and shops. Some looting began. Civic order no longer existed in the city of Gdańsk. It might be added that not much that resembled a coherent "movement" existed either.[21]

In saying that not much that resembled a coherent "movement" existed on December 14, I do not wish to denigrate—any more than was the case for the street demonstrators in Poznań and Budapest in 1956—the authentic popular longing that fueled street actions on the Baltic coast. Far from it. To the extent that the capacity of people to act politically is a function of a specific process of overcoming acquired forms of deference and resignation, impromptu street actions constitute a part of the process. Such activities provide a mode of expression that helps overcome both inherited habits of silence and the self-denigration that silence fortifies. Spontaneous street actions are certainly political and are especially so in the sense that—as proved to be the case on the Baltic coast—they may help generate later actions that become even more powerfully and effectively political. There can be, and often is, a direct causal relationship between extemporaneous street actions and eventual movements. But the relationship needs to be understood in proportion. There is a worldwide tendency to collapse the two forms into one—so that a street action is taken to be a movement. Indeed, for many observers, such dramatic and highly visible activity in the streets is the only or even the ultimate sign of the existence of a "movement." I would suggest that this is a very mystified understanding of popular politics and an even more mystified grasp of the prerequisites for popular democratic politics. A great deal of misplaced historical sentiment has been bestowed on dramatic moments of street violence. Such events can play a role in stimulating political consciousness, but the extent of this stimulation is easily exaggerated—and routinely is.

It cannot be emphasized too strongly that democratic insurgencies are extremely difficult for people to fashion. Given the constancy of deeply felt grievances in every society in every era, there would be far more "movements" except for this controlling but routinely overlooked fact. The point can never be overstressed: popular democratic politics is rare in history—rare because it is hard to fashion and because the process of democratic movement building is not widely understood. This fact merely emphasizes the scope of the organizing achievement Solidarność represents. The most pervasive consequence of these circumstances is that a sophisticated awareness of the process of movement building does not form a conceptual centerpiece of any traditional modern "ideology."

But whatever street actions reveal or conceal about the internal social relations of the rioters, it is historically unassailable that such activity carried the capability of shattering the complacency of ruling elites. Beyond question, the events of December 14 in Gdańsk made a profound impact upon Poland's party elite. In Warsaw that evening, a hopelessly isolated Gomułka verified he had little faith in what his regime had accomplished in fourteen years. He came to the conclusion that the coastal disturbances could only be the result of an antisocialist conspiracy in the working class. Gomułka thereupon dispatched his closest associate in the Politburo, Zenon Kliszko, to the coast with instructions to act forcefully to restore order. Upon arriving, Kliszko promptly verified how isolated he had become. He, too, promptly became satisfied that a "counterrevolution" had begun. Kliszko overruled dismayed coastal party officials, declared a state of emergency, and assumed total direction of the governmental response to the working-class protest.[22]

At the Gdańsk shipyard the next morning, Henryk Lenarciak watched a rapidly escalating situation. A veteran of eighteen years in the shipyard, Lenarciak had engaged in many conversations over the preceding two years with a deeply disgruntled and thoughtful young worker named Lech Wałęsa. Neither had been among those who left the shipyard the day before, though another W-4 veteran, Henryk Jagielski, did go. Indeed, it was the warning of veteran workers like Lenarciak, who had lived through the 1956 street demonstrations on the coast, and Jagielski that induced those who did march to appoint monitors and to take other steps to insure discipline among the demonstrators. But on the second day, the mood among the workers who gathered in front of the shipyard administration building had become even more volatile. The representative of the local party committee was abruptly shouted down by workers. Several thousand joined the new march to the party headquarters. Lenarciak's young associate, Wałęsa, apppeared in the front rank. The chants on the second morning were considerably more insurrectionary in tone. "Gomułka out" became an occasion "Hang Gomułka."[23]

Upon hearing word of the new demonstration, Kliszko decided the shipyard needed to be quarantined to prevent more workers from marching into the city. Police helicopters equipped with loudspeakers appeared overhead to warn workers to remain in the shipyard. Officials in the yard reiterated this advice. Those workers who nevertheless ventured out became instant targets of snipers Kliszko had ordered sequestered in the area. The spectacle of troops firing on workers had a decisive effect on the thousands who had to that point refrained from direct participation in the demonstrations. Blocked from the streets, they cast about for some other less perilous form of positive action. The possibility of an "occupation strike" had been bandied about by groups of workers ever since the price rise had first been announced three days before. As party officials in the shipyard took note of the increasing level of restiveness among the work force, they hurriedly sought and obtained permission to preempt the workers from independent organizational action. Under party direction, the workers named a strike commmittee and declared a twenty-four-hour occupation strike. This presumably would keep Lenin Shipyard workers off the streets. Though the committee was dominated by reliable party people, a gloss of authentic worker representation was achieved by naming prominent activists to the committee. Among those named were Henryk Lenarciak and his younger associate from W-4, Lech Wałęsa.[24]

At the same moment, a huge throng of workers from shops and factories throughout the city joined the core of shipyard workers in front of party headquarters in downtown Gdańsk. Not surprisingly, the party was a bit short of eager representatives to address the crowd. As in Poznań fourteen years before, workers entered the building and discovered in the party dining rooms and kitchens varieties of elegant foods obtainable only in party stores. And as in Poznań, these exhibits served to stoke the indignation of the workers. Soon thereafter, fires were set in the building. At this early hour of the morning—9:30 a.m.—few ranking party officials had arrived. Low-level party workers found themselves trapped inside; the prospect of leaving the building by walking into the midst of massed ranks of angry workers did not appear attractive, nor did staying in a building increasingly filling with smoke. The party newspaper reported the next day their sense of shock. "The silent passivity of the crowd of adults, their indifferent or aggressive attitudes, making it impossible to save the people inside the party building, is horrifying." Party workers waved a white flag and attempted to bargain their way to safety through the crowd without being roughed up. They were not invariably successful. According to an eyewitness, the workers at one point took down the red party flag, raised the red and white Polish flag, and sang the national anthem. A hook-and-ladder fire truck arrived with the possible purpose of rescuing people from upper stories, but it was intercepted by the crowd and set afire. Party functionaries eventually got out, some by helicopter, the last two by wrapping their faces in wet rags and running along fiery corridors. Throughout the city, the younger generation of Poles—in the working class and in the security service—battled one another amid a spreading panorama of fire, wreckage, and death.[25]

At a tense meeting of the provincial party committee in Gdańsk that evening, an extraordinary scene unfolded. An outraged Kliszko simplified matters by summarizing the day as one of "counterrevolution." He was especially infuriated by the effrontery of workers who openly admitted, or worse announced, that they had taken part in the burning of the party headquarters and were proud they did. The provincial party itself had now become divided and immobilized. Many did not attend Kliszko's meeting out of feelings of sympathy for the demonstrators or of hostility for the party's recent actions. Kliszko in turn attacked the Gdańsk party committee, which he accused of "disappearing" in the face of the workers' militance. He decided to call in the army, led by special detachments of armored troops.[26]

By the following morning, after three days of unalloyed alienation and a day of random violence, signs of a coordinated workers' movement began to take form. At the Gdańsk shipyard, thousands of workers arrived to find their workplace surrounded by tanks. They filed past the machines and gathered in a mass demonstration in front of the shipyard's administration building. Well aware of the command structure of the "workers' committee" named the day before—and aware also that the committee had not performed serious programmatic acts after its formation—they discarded that body and elected their own committee. Lenarciak and Wałęsa were once again named, but this time they were surrounded by like-minded worker activists rather than docile party loyalists. Wałęsa was also named to a four-person presidium. Under the guns of surrounding tanks, the new committee declared an occupational strike without time limits and set about the business of drawing up a list of worker postulates.[27]

It is vital to take note of this development. Inside the Gdańsk shipyard, the working class had broken free of the party and had created for itself an institutional democratic space. It was a space to think in, a space in which it became possible to formulate a program for workers—without the "help" of the party. At the center of this worker experience was a 27-year-old electrician named Lech Wałęsa. Ten years later, at a decisive moment of worker restiveness and indecision, he would intervene with a five-word summation of the lesson he had internalized on December 16, 1970, and which became conventional wisdom among worker activists in the ensuing years: "I declare an occupation strike."

While the leadership turned to the business at hand, some of the younger workers decided they could not stand the humiliation of being forcibly confined to the workplace by tanks. They charged out of the shipyard and were promptly cut down by automatic weapons in the open space right outside Gate No. 2 of the shipyard. The massacre on the coast had begun.[28]

The troubles in Gdańsk on December 14 mobilized Gdynia on the fifteenth and the escalating modes of protest in both cities on the fifteenth and sixteenth activated Szczecin on the seventeenth. The dynamics, however, were not quite the same. In the Paris Commune Shipyard in Gdynia, workers staged a mass meeting on the morning of the fifteenth, drafted a set of demands, elected a strike committee, marched downtown, and waited patiently as their committee negotiated with municipal authorities. The chairman of the town council, a party functionary named Jan Marianski, was conciliatory. He settled some issues on the spot and promised to take up others with appropriate authorities. The strike committee, satisfied for the moment, left to consult with the work force. In Gdańsk, meanwhile, the party building was burning, and Kliszko was fighting to control his rage. When the Gdynia delegates returned for discussions that evening, they discovered Marianski had been arrested on orders from Kliszko for having negotiated with "counterrevolutionaries." Three worker delegates were also arrested, an act that ensured the total mobilization of the Gdynia shipyard. The authorities learned the next day that various enterprises in Gdynia had joined the shipyard workers in forming an Interfactory Strike Committee. They discovered this when they received angry demands that delegates to such a body, who had been arrested, be released. With clouds of smoke from Gdańsk blowing across the sister city, Gdynia workers "arrested" several shipyard officials and held them hostage against the return of their elected delegates. On the evening of December 16, army units entered the Paris Commune Shipyard. Across the tri-cities, the shipyards, docks, and supply facilities had become silent. Occupation strikes had engulfed the entire bay area as workers everywhere gathered in angry meetings to draw up steadily lengthening lists of demands. A very broad-based workers movement had begun to form on the Baltic coast.[29]

Late in the afternoon and into the evening of December 16, shipyard workers began receiving diametrically opposed signals from the party hierarchy. First, at 4:30 in the afternoon, the association of the shipbuilding industry, a party management group, sent a telephone order to the Paris Commune Shipyard in Gdynia to suspend work. The shipyard was to remain closed until further notice. That evening, however, Deputy Premier Kociołek appeared on television. He denounced all demonstrators as "social scum" bent on destroying Polish culture, described the worker postulates as "unrealistic," categorically rejected demands that imprisoned worker delegates be released, and ended by imploring all workers to return to work the next morning. Party officials in the Gdynia shipyard were dumbfounded. They went to Kliszko's "command post" and pleaded with him to permit government spokesmen to announce over radio and television that workers were to disregard Kociołek's message and refrain from reporting to work. Mass media had to be used because the shipyard officials could not reach their workers by telephone, even should enough operators be available; only 12 percent of Gdynia workers owned telephones in 1970. Kliszko, however, rejected the appeal. On the contrary, Kociołek's message to resume work was repeated on television at 11 p.m. The same evening, additional armored units were dispatched to Gdynia where they took up positions around the Paris Commune Shipyard. This time, the tank guns were pointing out from the shipyard rather than directed toward it. With a sense of foreboding, Gdynia officials made one more attempt to avert what they were increasingly convinced was an inevitable disaster. They sought permission to order the morning commuter trains of workers to bypass the shipyard station and discharge passengers away from the concentration of armored troops. This request was refused.[30]

And so it was that the massacre at Gdynia was set up the next morning. The Gomułka regime was determined to deliver an emphatic message to the workers of Poland. At ten minutes to six, the first of the commuter trains ground to a stop at the station adjoining the shipyard, and hundreds of workers poured out to discover an arsenal of weapons pointed directly at them. In scheduled increments over the next forty-five minutes, other commuter trains arrived and disgorged more floods of workers. An eyewitness account by a Polish journalist, Barbara Seidler, fills in the details:

> Near the kiosk, the army places a huge loudspeaker. When the people start to pour out of the trains, a hoarse voice calls on them to go back, not to approach the army and the tanks, but the crowd grows and presses closer and closer to the

> cordon. Shots are fired by security forces. There are paving stones beneath the crowd's feet. The bullets ricochet off the pavement. Then the bridge above the tracks begins to burn. Tear-gas cannisters are thrown from militia helicopters into the crowd.[31]

There were now dead and wounded workers laying in the street and a full shift, nearly five thousand workers, reeling in shock and rage. Acts had to be performed to recover one's sense of self, to blot out the humiliation and denigration of having literally been despised to death by one's own government. Some act. Any act. Oil was taken from a locomotive and used to set the station on fire. The throng summoned the defiance to press against the cordon of troops and surround the tanks. More shots were fired and more tear gas. Under this renewed proof of the party's disdain and arrogance, the last connecting strands between workers and the state snapped. The workers advanced, were fired on, scattered, regrouped, advanced again. Columns formed, infiltrated, and drove the forces of order back. It went on for four hours in a morning filled with the fire of automatic weapons.[32]

There is a moment in all this that a photographer with a single click of his camera will etch into the memory of the Polish people and later, through the film artistry of Andrzej Wajda, into the consciousness of people throughout the world. A door is ripped off a commuter train and used to support the body of a dead worker. Six men lift it to their shoulders and lead a surging crowd through the streets of the city. It is a stunning photograph, one that exudes the emotions of horror and massive indignation and embattled dignity that were at the heart of the coastal rising. When Wajda came to make his classic film of Solidarność, *Man of Iron,* he could, in the name of high drama, do no better than re-create with zealous attention to detail the scene fixed by the photograph. It stands as the ultimate worker condemnation of life in the Leninist state.[33]

But the significant evidence of 1970 on the coast has not yet been fully noted. Two hundred miles away, at Szczecin near the East German border, workers watched the escalating situation in the tri-cities and used the period of December 14–16 to establish communication links between factories. When party leaders at the Warski Shipyard refused to talk to the assembled work force on December 17, workers there led the way into the streets and were joined by sizable numbers from other enterprises throughout the city. Confrontation was immediate and quickly erupted into widespread street warfare. Agonized officials went on radio to appeal to the population to cease "demolishing our city." Physical destruction was immense at the party headquarters, at the militia building, and at a number of other public structures.

Amid the wreckage that night, the party decided to try the same tactic on the eighteenth in Szczecin that had briefly worked in Gdańsk on the fifteenth. Party functionaries in the shipyard were enjoined to declare an occupation strike as a means of keeping workers off the streets. The tactic was not wholly cynical, as party activists among the work force, led by Mieczysław Dopierała, were themselves determined to reform the party. A considerable amount of democratic organization within the work force took place in the Warski Shipyard on that day and evening. Demands were collected from every shop section, and three delegates from each were named to a plant-wide committee. The delegates sorted through the postulates and fashioned a set of twenty-one demands that collectively constituted a new plateau of working-class thinking in People's Poland.

Some grievances, while deeply felt, were inheritances from unsuccessful past struggles. Szczecin workers called for "dissolution" of the official trade unions that were described as having "never defended the working masses." The official unions were to be replaced by a worker-centered institution organized for such defense. The old idea of 1946–47 and 1956–57 had made its way out of the kitchen once again, more clearly stated than ever before. Szczecin workers also cracked the party censorship in what for a time was a most profound and dramatic way. Worker militia patrolled the shipyard and shared with army units the maintenance of order throughout the city. Workers also interpreted events in radio interviews, and their thoughts went out over the airwaves unedited. Journalists, too, found they could, for the moment at least, describe events as they saw them. Lateral communications became widely established in the working class; no less than ninety-four enterprises in Szczecin were listed as participating in the municipal strike committee that presided over what effectively became a citywide general strike.[34]

But the most important feature of worker self-activity on the coast was that the breadth of intershop and interfactory cooperation manifested itself intellectually in a broad-gauged set of twenty-one demands that addressed a number of central features of life in a one-party state. Workers simply believed that it was debasing for the party to think that people who spent half their income for food could live with dignity under a food price increase of up to 36 percent. Therefore, workers had to protect themselves against the results of such thinking by creating unions outside the framework of the official trade unions. But beyond these two basic and interrelated considerations, worker postulates extended to the privileges of the party apparatus (both administrative and material), the excesses of the political police, and the class-based narrowness and general unfairness of one-party state planning. Present in all the demands was a simple longing for dignity for themselves in the workplace and for their families in society. The tactical and strategic culmination of all these activities was the formation in Szczecin of an Interfactory Strike Committee.

Because of its strategic importance as a means through which to achieve the basic objectives of Polish workers, this final development merits careful appraisal above all others. No organic democratic magic attaches to a name, whether it be "interfactory strike committee" or something else. A new conceptual plateau had been reached, but not a body of experience in how to function within its framework. Though ten enterprises in the Szczecin area were initially affiliated with the Interfactory Strike Committee, the new entity reflected the presence and purpose only of the Warski Shipyard leadership and that of a sister shipyard at Gryfia.

Nor did the new institution yet constitute in any ideological sense a "free space." Dopierała and his two closest colleagues, Józef Fischbein and Józef

Kasprzycki, were party members—party reformers—whose ultimate goal was a restructuring of the party rather than a structural liberation of the working class from the party. It may perhaps be said more precisely that they wavered between the two. Dopierała himself participated in the writing of the demand for his shop section which called for "independent trade unions." But beyond mere words (however potentially relevant those words might be), Dopierała's personal career was enmeshed in the party. He was the kind of activist who would push the party as far as it would go and, in the end, would acquiesce in the party's ultimate course of action. This reality of course decisively circumscribed his usefulness to workers, as the party intended it should. Such was the "Leninist norm" which Dopierała stretched but did not violate. But his own life was transformed by the experience. During a moment of highly energized self-activity, he discovered a new sense of possibility both for himself and for the working class. Unfortunately, when the party reconfined both and demonstrated in so doing its relationship to popular aspirations, some circuit within him shorted out. He was never again able to perform with the élan that characterized his actions in the middle of December 1970. After the events of 1970–71 had run their course, he took a job in the bureaucracy and disappeared from Polish history.[35]

In any case, Dopierała soon got at least the appearance of the party reform he sought when Gomułka fell on December 20 and Edward Gierek came to power. The development ushered in an intense new war of maneuver between coastal workers and the new party leadership, a struggle that was to last throughout the months of January and February and continue in modulated form through most of 1971. It is essential to record that the worker portion of this equation was fundamentally altered by all that occurred during the epic December week of initial worker self-activity. From Gdańsk to Szczecin, workers gained priceless experience in conducting occupational strikes. Concretely, they learned how to survive and supply themselves inside their places of work while planning under extreme pressure from the state their own future course of action. The occupation strike, like the interfactory strike committee, was not merely a name for something. It provided people with a means of redefining power relationships.

While coastal workers pioneered new institutional forms of self-organization, the regime of Władysław Gomułka stumbled through its final crisis. Gomułka himself reached a stage of genuine panic on the night of December 17 and sent a frantic message to Brezhnev in Moscow pleading for Soviet troop intervention to put down "the counterrevolution." Unimpressed by this description of events, Brezhnev refused. In Warsaw, Gomułka, after much stalling, was forced to call together the Politburo on December 19. In the course of the seven-hour marathon meeting that ensued, he suffered a stroke. From a hospital to which he was taken in the early hours of the twentieth, Gomułka reluctantly submitted his resignation. Brought to power in the wake of the Poznań rising in 1956, he was brought down by the coastal rising of 1970. He understood Leninist forms, but he never understood the Polish working class.[36]

In the final days before Christmas, Edward Gierek assumed control of the ruling party. He promptly presided over an eerily familiar replay of the politics that Gomułka himself orchestrated when he came to power in 1956. Now it was Gierek who exonerated the workers and placed blame for the country's social upheavals upon the deposed party leadership. The working class, said Gierek, had been "provoked beyond endurance." In a similar reprise, the party propaganda machine announced that under Gomułka "the picture presented by propaganda was far from reality." The head of the central trade union was sacked, and his replacement dutifully uttered the obligatory announcement that the official unions had "drifted away from the working class." But with coastal strikes spreading to Poznań, Warsaw, Wrocław, and Katowice, something more than words and different from tanks was needed to restitch the gaping rents in the social order. Gierek gambled and withdrew the army from the coast, suspended Kliszko's emergency decrees, fired Kliszko, and released seven billion złotys "in consultation with the workers" for aid to low-income families. Poland passed an uneasy holiday season with an expectant and restless working class gearing itself to renew the struggle for shop-floor democracy.

In the first week of the new year, thousands of Gdańsk shipyard workers struck again, demanding the release of some two hundred imprisoned workers. The action inaugurated three weeks of active unrest in a coastal working class seemingly determined to demonstrate the extent to which it had found its voice. Local party officials could manage nothing better in reply than inept evasions and awkward silences. Occupation strikes thereupon erupted in Gdańsk, Gdynia, Sopot, Elbląg, Słupsk, and Szczecin on the coast, but produced nothing beyond cosmetic changes achieved through decisions made in Warsaw to sack local party secretaries on the coast. Still another occupation strike in Gdańsk in mid-month was temporarily pacified when the party in Warsaw agreed to receive a coastal delegation, but anger flared anew when it was discovered that authentic worker advocates were systematically kept off the delegation. Meanwhile in Szczecin, the party-led committee headed by Dopierała gradually lost credibility with the work force. The committee was finally replaced on January 22 after another round of insurgency from below that brought to the fore a tier of worker activists unconnected to the party apparatus. Within twenty-four hours, a general strike loomed again in Szczecin, even as still another gathered force in Gdańsk.[37]

It may be observed that the reciprocal relationship of the party to the working class in Poland reproduced a scenario of causation in 1970 that was astonishingly similar to the drama of 1956. The command economy lurched toward crisis, and the government tightened up on work norms throughout the society; the workers rebelled, and the party promised reforms; when changes did not materialize, the worker mobilization enlarged, and the party attempted new evasions. By the third week of January 1971, the party found itself in the same tactical position that prevailed in November 1956. At the earlier time, some creative party official had fashioned an elaborate "experimental program" to contain both the specific Poznań insurgency and the general outbreak of working-class self-activity that Poznań had generated. In 1971, however, the degree of organizational autonomy achieved by coastal workers far exceeded anything constructed in Poznań in 1956. With general strikes threatening in two cities, Gierek was deprived of the maneuvering time necessary to fashion a systematic containment operation. Accordingly, he was forced to go personally to the coast to face the working class. The result was an unprecedented dialogue in a communist country. The head of the regime was forced to listen for hours to bitter worker complaints about the arrogance of the party, the corruption of the economy, and the humiliation of the nation.

In an adroit piece of theatrics, Gierek deflected criticism by agreeing with it; he emphasized the extent to which past party errors had carried the economy to the brink of disaster and then skillfully pled with workers to aid him in rescuing the situation. "Will you help me?" he concluded. Workers inquired as to whether their sacrifices would eventually be repaid. They were assured they would. In Szczecin on January 24 and in Gdańsk on January 25, the workers agreed to help the first secretary. In this way, the crisis seemingly passed. But in Łódź, low-paid textile workers simply could not accept the continuance of the December price increase. Within weeks of the coastal standoff, Łódź workers had positioned themselves for a general strike. With more strike threats poised elsewhere in Poland and the coast still restive, Gierek reluctantly gave up. He rescinded the price hike for which Gomułka had imperiled his regime. In anticlimax, the country returned to the level of unresolved social tension that passed for normality in People's Poland.[38]

Such were the historical events which appeared on the surface of Poland during the crisis of 1970–71. But beneath these events, another reality lurked. To make it visible, it is necessary to shift from broad historical description to a more tightly focused analysis of the new worker leadership that emerged from the crisis.

In Szczecin in January, the party committee that directed the first occupation strike was deposed and replaced by a tier of nonparty advocates, actively led by a worker named Edmund Bałuka. Joined by his comrades, Bałuka had replied in the affirmative when Gierek asked "Will you help me?"

In Gdańsk in December, the party committee that directed the first occupation strike was deposed and replaced by a tier of nonparty advocates, including Lech Wałęsa. Wałęsa, too, said yes to Gierek's plea.

But all was not the same. To understand why the Polish August of 1980 centered itself in the Lenin Shipyard in Gdańsk rather than in the Warski Shipyard in Szczecin, it is fruitful to trace in tandem the wildly divergent careers of Edmund Bałuka and Lech Wałęsa. Beneath the surface of historical description of working-class self-activity repose two distinct dynamics: one of indignant anger, repression, defeat, and exile; and the other of indignant anger, repression, survival, and victory. In addition to embodying structural disjunctures, the fracture opened in communist hegemony in Poland in 1980 has a human dimension. Indeed, it is a dimension whose practical and theoretical components must be isolated as prelude to an overall understanding of the dynamics of the Polish democratic movement and of such movements generally.

In his personal manner, Edmund Bałuka was what U.S. workers would describe as a "tough guy." He "didn't take any lip from anybody" and had, in fact, once socked a policeman. He had also tried unsuccessfully to escape to the West. He spent some time in prison for each offense. Bałuka did not like the party, he did not like the state ideology, and over a beer he would say so. His story in the 1970s is rather quickly told.[39]

The party strategy of "regaining touch with the masses" in 1971 involved official acquiescence in workers electing their own activists to minor leadership roles at the point of production. Thus, Bałuka became an official of the Warski trade union and participated in the new program of worker self-management that the party proclaimed. It took awhile for the new pattern of governance to become fully visible to him. He discovered in fairly short order that he was outnumbered in the plant council and that ordinary workers had no more autonomy in the new trade-union structure than in the old one. As the party gradually refashioned its network of control on the floors of Warski's far-flung shop sections, Bałuka rebelled and was in due course sacked from his official position. When he continued speaking up, he was fired from the shipyard and blacklisted so that he could not find work elsewhere. In 1973, fearing for his life, Bałuka escaped to the West. There he stayed until Solidarność arranged for his return in 1981.[40]

In Gdańsk, Lech Wałęsa also became an official in the spring of 1971. His soft-spoken and more cautious ally Henryk Lenarciak became head of the workers' council in W-4, while Wałęsa was elected by the work force to the job of safety inspector. It was an organizer's position, one that permitted him to roam throughout the shipyard and escape close surveillance by party functionaries. Wałęsa, too, tested the new techniques of "worker self-management." He watched carefully the manipulative mechanisms the party employed to control the shop councils. With difficulty, W-4 remained worker centered; sometimes, even the whole shipyard council did, though the party routinely employed a mixture of intimidation and cajolery. But the party's ultimate weapon was the integrated structure of hierarchy that extended from the trade union in the shop section to the Gdańsk regional committee and thence to the central apparatus in Warsaw. Battles lost by the party at one place could easily be won at a higher level. For workers in a plant to gain a measure of autonomy, a plant-wide mobilization of workers was required; citywide autonomy required a prior mobilization of all enterprises in a city; regional autonomy required regional organizing. Thus, when Wałęsa, like Bałuka, satisfied himself that little room for maneuvering authentically existed for rank-and-file activists in the Lenin Shipyard, he told Lenarciak that shop councils, as such, were a conduit to a dead end.[41]

This understanding came to Wałęsa gradually through the months of 1971. At the climactic January meeting of shipyard workers with Gierek, he had listened carefully to the first secretary's ready acceptance of the party's multiple blunders in the past. Along with Lenarciak, Wałęsa had met personally with Gierek as a member of the shipyard presidium. He had taken Gierek at his word and had entered the reformed shop council with the intention, as he told all who would listen, of keeping the council "in close touch with the people."

In the ensuing months, he watched for the first time from the inside how Leninist norms worked in People's Poland. He intermittently expressed his own views in meetings, sometimes quietly and briefly, sometimes at length and with passion. These interventions succeeded only in making him more of a marked man in the eyes of the party than he already had become after the December events. At the end of his term, Wałęsa declined to run for reelection, even though implored to do so by Lenarciak and his workmates. For his part, Lenarciak decided to give the council one more try. Wałęsa, meanwhile, began to search for new ways to assert the workers' cause beyond the party-controlled shop floor. In what might be called a quest for modes of extra-party programmatic action, Wałęsa in May 1971 played an active role in a memorial ceremony for the fallen workers of December 16. He went beyond this to join in a call for the erection of a monument to those who had died while contending for the rights of Polish workers. He repeated this role on the occasion of a second ceremony on December 16, 1971, and again on December 16, 1972.[42]

But as a closely observant party quickly came to realize, what can be understood as "monument politics" went quite a bit beyond a memorial service. Though it was masked in the trappings of a quasi-religious ceremony (who could oppose the laying of wreaths at the spot where comrades had fallen?), monument politics had the effect of keeping alive the worker politics of 1970. The party, therefore, endeavored to co-opt this meaning by officially participating in the ceremonies. The results, however, were not satisfactory. Monument politics enhanced worker self-activity rather than confirming party hegemony. It had this effect in 1971 and again in 1972. In 1973, the party forcibly quashed all further demonstrations at the site of the December events of 1970. In the same year that Bałuka was forced into exile, party cadres in the Lenin Shipyard, once again confident in their control of the shop floor, moved to purify the work force by sacking the activists of 1970. Some eighty-eight militants were fired, and Wałęsa was targeted to be one of them. But by then, he was a five-year veteran of the shipyard and had earned a wide reputation among production specialists as a superior electrician. Foremen and supervisors saddled with production quotas did not relish the loss of highly skilled specialists. Beyond this, it turned out that Wałęsa was almost uniquely popular with his workmates. They respected his skills both as a craftsman and as a worker advocate. They were attracted by his independence and responded both to his personal warmth and to his impudent, sometimes boastful humor. He was a worker's worker. Led by Lenarciak the workers' council in W-4 threatened to strike if Wałęsa were fired. The supervising manager, never keen on the sacking in the first place, managed to prevail. Wałęsa was given a strong warning to settle down and, in the end, was allowed to keep his job.[43]

Throughout 1973 and into 1974, the regime's techniques of repression reached their most sustained intensity of the decade. Intentionally, the crackdown accompanied a planned economic expansion. After riding out the initial crisis with the working class, Gierek embarked on a bold economic gamble to employ Western credits to modernize Polish industry and to increase exports sufficiently to finance repayment of the loans. The force-fed economy was further stoked with increased imports of consumer goods that served the highly relevant purpose of quieting working-class discontent. Yet even during the high tide of the economic expansion, the working class in Poland remained restive. Coastal longshoremen caused repeated work stoppages from the end of 1973 through the summer of 1974. In the same year, trouble once again developed in the Paris Commune Shipyard in Gdynia as well as at Łódź and at enterprises in the Warsaw area.[44]

Among other problems, such never-ending tension at the point of production reflected the poor working conditions that prevailed throughout the country in the 1970s. It was a subject to which knowledgeable Western trade unionists were especially sensitive. They noticed, for example, the sustained argument between Gdańsk workers and the party over the dangerous practice of sand-stripping old paint from ships. Prolonged worker agitation and slowdowns had finally succeeded in getting the practice officially banned. Nevertheless, it continued at the point of production under orders of low-level party foremen. An English trade unionist, Denis MacShane, summarized: "As a Catch-22 consequence, the Gdańsk workers could not get the necessary protective clothing because in officialdom's eyes the practice of sand-stripping no longer existed." Similar non sequiturs described production relations in industries across the nation. Thus, even during a planned expansion, unreformed bureaucratic habits kept workers in a state of more or less constant unrest.[45]

A final grievance took on ominous meaning in the first half of the decade. As consumption mounted, Gierek moved to merge a loosened economy with tightened political control in a new grand strategy of communist stability. The security police were ordered to clamp down on all "antiparty agitation." This scenario expanded almost geometrically throughout 1972–73 and was poised for even further expansion when the OPEC crisis and the resulting leap in world oil prices in 1974 wrecked Gierek's long-term plans. When—at a much later date—the postmortems were all in, it developed that state economic planning in the Gierek era had been monopolized by a tiny sector of the party's technical apparatus in Warsaw, with the result that some elements of the expansion had been badly conceived. A considerable amount of money had been invested in inadequately productive capital improvements and such other totally nonproductive items as equipment for the security services. Still other funds were lavished on vacation houses and other frills for a bloated party bureaucracy. The whole strategy, in any case, was a classic example of mismanagement by command planning. The ultimate effect was the imposition of an austere retrenchment.[46]

This altered the Polish scene dramatically. If the worst features of Stalinism were to be avoided, the unanticipated economic decline could only be safely counterbalanced by a softening of political controls. Beginning in 1975, worker activists gradually discovered through their own self-activity that they had more room for maneuvering than in the darkest days of 1973–74 when, as Wałęsa understated, "the [police] interrogations were rather brutal." By the time the Gierek-led ruling party had backed off, activists had made themselves unusually well equipped to undertake judicious experiments on the edge of legality. By mid-decade, they learned a whole series of necessary lessons. The first concerned the party's national leadership. When the confirmed evidence of daily life was added up, it was apparent that Gierek had lied to Bałuka and Wałęsa personally and to the working class generally. First came Gierek's purge of some 100,000 "marginal" party members. While some were certifiably incompetent, the only deficiency of many others was their genuine alarm over the way the party treated the population. There was the further matter of the blacklisting of shipyard activists, many of them conscientious advocates like Henryk Jagielski and Edward Nowicki, whom men like Lenarciak and Wałęsa respected. Transparently, the party leadership endeavored to keep its promises to consult with the working class by doing so essentially with a gun.[47]

The second lesson concerned the "reformed" trade unions. The shop councils simply were not interested in "staying in close touch" with workers; party functionaries were concerned only in acting in whatever way was necessary to retain party privileges. Nothing more really needed to be said.

The third piece of wisdom concerned the security services. Here the details were a bit more complicated. The level of astuteness in the security service was not, on the whole, very high, and this folkway sometimes extended surprisingly far up the chain of command. While it was true that a pedestrian mentality armed with great power could be a very dangerous commodity, activists slowly came to understand—sometimes through brutal experience—that some modes of "dialogue" with policemen worked better than others. Outright defiance while under arrest was almost certain to evoke outright violence. "Tough guys" got brutalized. On the other hand, it was quite apparent that those workers who showed excessive fear or indulged in sniveling deference also seemed to invite macho excesses by policemen. Quiet dignity was the best posture, particularly when coupled with an easy poise that conveyed familiarity with police methods without indicating approval. One learned the applicable nuances over time through on-the-job training in detention cells.

Such were the insights one gained from sustained self-activity. One learned about authority as it functioned in the shop section, in the party, in the police station, and in the state. One accepted the occupational hazard of arrest but did not brazenly seek it. The art of movement politics. It was a far cry from Bałuka's politics.

Still, one had to press to encourage others to press, and it was hard to probe the outer limits of permissibility without risking going over the line. Early in 1976, Wałęsa went too far and paid a heavy price. The occasion was a meeting of the trade union in the Lenin Shipyard. With the economy in increasingly obvious distress, the party's platitudes seemed to have exhausted Wałęsa's decreasing store of patience. Lenarciak had finally given up on the shop council and was retiring as its powerless and confined chairman. He intended to go quietly, as Wałęsa had before him, but the young man now said too much had since happened. It was time to press. "You are too soft," he told Lenarciak, "you ought to be more decisive."[48]

And so Wałęsa spoke to the workers and to the party—and he spoke for a long time. In fact, he delivered himself of a summary review of the Gierek administration. The nation's number one Pole had misled the people; he had not kept his promises; he had abandoned the working class. The speech was so popular that the workers promptly elected Wałęsa as their representative to a forthcoming union conference. Unfortunately for Wałęsa, a ranking party official attended the meeting. He did not relish the trouble such an outspoken worker could cause at higher-level meetings; he also did not relish the passive attitude of party personnel who let the worker talk so long; and he did not relish the worker. A few days later, the director of the Lenin Shipyard received orders to fire the man.[49]

The irony was that Wałęsa was not only popular with workers, he was liked by some of his superiors because he was such an outstanding producer and also because his obvious sincerity was augmented with such engaging humor. He raised morale in the whole shipyard. The section manager declined to fire him. The order was repeated, more threateningly, and again refused. The manager himself was fired. Lenarciak completes the story: "the new manager signed Wałęsa's dismissal form the day after taking the job. He was not a bad fellow, just cowardly."[50]

Wałęsa was not the kind of brinksman who philosophically believed that the genus of brinksmen were fated to "win a few and lose a few." He was, in fact, bitterly disappointed: "What's the matter with them?" he exclaimed angrily to Lenarciak. "I don't drink. I'm honest. I come to work on time. Why are they doing this to me?" In the privacy of his own mind, Lech Wałęsa lived in a moral universe. It was not an easily verifiable proposition, but it was inherently a spur to insurgent behavior.

Wałęsa put the party through every possible extremity, appealing his case, pressing the issue. The final decision by a labor court merely stoked his indignation for future battles. In criticizing party officials at a union meeting, said the judges, he had provided the Lenin Shipyard management with sufficient grounds for dismissing him.[51]

In the two months that followed his dismissal, the young electrician learned new details about the power of the state and the limitations of that power. With a police dossier that tracked his activism, Wałęsa had trouble finding employment. But his skills were quite varied. He was now an experienced electromotor mechanic who, as the saying went, could fix anything. He survived outside the state economy until a building company that needed a fixer hired him in May 1976. The company was a shipyard supply enterprise named ZREMB, and it had not been schooled to oppose frontline labor activism. "Just do your work well," the manager told him. "I won't bother about all the rest." What Wałęsa proceeded to do, unsurprisingly, was undertake to organize the work force at ZREMB.[52]

But the most sweeping organization plans in 1976 were being made at the national level. Throughout 1975 and into the spring months of 1976, planners in Warsaw attempted to evolve a new strategy to avert what increasingly loomed as a national economic disaster. The first big round of debits that financed Gierek's failed gamble were now falling due. Poland could not repay its loans, could not, in fact, repay the interest on its loans. There was no more money to siphon out of capital outlays and shift to consumer subsidies to ward off working-class unrest. Something had to give in Poland, and it had to be the working class. The long-delayed Gomułka price hike of 1970 simply had to be implemented. Despite all the internal party discussion and the variety of plans that materialized, the final product had an ironically Gomułka-like ring to it: one quick devastating price hike of 30 to 100 percent for food and 69 percent for meat. The plan was submitted formally to the docile Polish parliament, the Sejm, on Thursday, June 24, prior to a scheduled "consultation" with the nation's working class on Friday. Final passage by the Sejm was to occur on Saturday, in time to take effect on Sunday, June 27.[53]

The Sejm did its part on Thursday, and the next morning the "consultations" began. By noon, the party knew it was in serious trouble. Reports of work stoppages came in to Warsaw from every part of the country. Worse, in many cases they took the form of orderly occupation strikes. Precise figures are impossible for Poland, but a conservative estimate would place over half of the major industrial plants in the nation out on strike the very first day. Occupation strikes spread across the Baltic coast; textile plants in Łódź were out; Cegielski-Poznań was out, as were the mining districts of Silesia, Żerań, and Nowa Huta. The list must have seemed to Warsaw officials to be astonishingly and depressingly long—hundreds of occupation strikes in every province of Poland. But at Radom, some forty miles from the capital, the work force was more than "out." It had exploded.

As at Cegielski in 1956 and Gdańsk in 1970, all had seemed quite orderly in Radom at first. One of the enterprises regarded by the party as possessing a singularly "reliable" work force was the General Walter Ordnance Factory. The workers there emerged in the streets in columns carrying red flags and singing both the "Internationale" and the Polish national anthem.

The scene unfolded in the present tense like an old movie that had premiered in Poznań, enjoyed a revival in Gdańsk, and was showing belatedly in Radom. The workers arrive at the party headquarters to protest the price increase, but the party's first secretary refuses to speak to them; the first workers to break into the building are shocked at what they find; "Look how these bastards live!" shouts one who bears samples of expensive foods; in short order, files are tossed out of windows and the place is set on fire; the resulting street battle lasts from noon until midnight; destruction is enormous. "The riot at Radom."[54]

Even before nightfall, the government backtracked, Gierek himself personally assuring a delegation of women textile workers who surged in from Łódź that the price increase would be promptly withdrawn. Meanwhile in the suburbs of Warsaw, workers at the Ursus tractor factory had stopped a Warsaw-to-Paris train by sitting on the tracks. They quickly cut the power and erected barricades to render further traffic impossible. International passengers were informed of working-class grievances in Poland and then were assured they could resume their journey when the price increase was rescinded. The decision to revoke the increase was announced over Warsaw radio that evening.

At the Lenin Shipyard in Gdańsk, an occupation strike materialized. The same was true at Gdynia, Elbląg, Słupsk, and Szczecin. Edmund Bałuka had said at the end of the troubles in Szczecin in 1971 that "we now know how to strike; we don't know how to win one." After 1976, the second point was not so certain. If the working class had yet to demonstrate its ability to force the government into structural changes, it had definitely illustrated its increasing potential to veto unpopular policies. The ground that divided the party from the working class in Poland had become sharply contested terrain.[55]

In the immediate aftermath, the party was shocked, subdued, and less confident than ever. The events of June had uncovered whole layers of questions that seemed to demand answers. How could the party have misread the temper of the working class so totally? The central bureaucracy in Warsaw had conferred with all but one of the forty-nine regional party secretaries on the specific formula of the proposed new policy. Not a single one had indicated that the price increase needed to be trimmed to avoid serious repercussions among workers. Could it be that the regional secretaries had grown remote from the very social realities they were supposed to help shape—or at least report on? And how about the trade unions in the plants and in the regions? Where were they in the midst of this giant mobilization? In retrospect, the absence of detailed information from that sector was positively alarming. Both in the party and in the trade-union structure a vacuum seemed to exist where a rich lode of social information should have been. The questions loomed.[56] But no one in the leadership pursued them. Instead, a generalized reflex action prevailed. The hooligans who had gone on a rampage in Radom and, to a lesser extent, in Ursus needed to be taught a lesson. The security service and the court system were instructed to hop to it. The party flexed its muscles while its brains further atrophied. And the economy slipped further into chaos.

In seeming paradox, the Polish working class had struggled to its feet in the 1970s. Not only were workers successfully reclaiming old shop-floor traditions, they had begun to pioneer new and broader forms of self-organization. Most tellingly, the breakthroughs on the Baltic coast developed in the presence of a shocking massacre that solidified the permanent alienation of the working class from the workers' party.

The central cultural reality in Poland was that no one outside the working class saw what was happening. To those who lived their lives securely removed from Poland's centers of production, evolutions in working-class sensibility could not be detected. The actions of workers, viewed from afar, sent an entirely different signal. In 1956, the workers had risen in Poznań and had destroyed party buildings in an orgy of street violence; the same thing had occurred on the Baltic coast in 1970, and it had happened again in Radom in 1976. To those who could see workers only when they did something spectacularly abnormal, the verdict was clear: stasis, not development, characterized the Polish working class. It was a settled perception shared by observers of Poland who agreed on little else. Quite unwittingly, the observation provided a powerful indication of the narrowness of "consciousness" that persisted in educated circles—in Poland and throughout the world. Undetected by political theorists of various sanctioned persuasions, this condescending myopia would prove to be self-confident, tenaciously cherished, and therefore capable of feeding an enduring illusion. Indeed, uncontested as it was at the level of high culture, these remote ways of seeing mass society would ensure that when Solidarność appeared, it would be similarly misread.

4 ▪ The Movement Experiments in Self-Activity

A whole nation cannot emigrate, so it undertakes a migration . . . It regains security in customs so old, and therefore so sacred, that authority fears to combat them.

Ryszard Kapuściński, on "Iran"

AFTER THE PASSAGE of thirty years, the promises of the party were in ruins. In the last half of the 1970s, the once gleaming factories had lost their social meaning. Industrial production seemed only to feed more industrial production even as the lines in front of food stores got longer, the housing shortage grew more severe, and Poland's trading relationships with the rest of the world became grotesquely unbalanced. The government media trumpeted a propaganda of success in the face of incontrovertible evidence that the nation was getting poorer. The party had become transparent in ways it did not intend, and disbelief had become endemic. To participate actively in Polish communism was to live a lie. The signs of crisis were everywhere.

The very rigidity of the party elite constituted its principal contribution to the dialectics of democratization soon fated to envelop People's Poland. The contrasting flexibility—and the central thrust for change—came out of the realities of daily life. There, in the social world populated by Polish workers, could be found the building blocks of a new Polish society.

As in other countries, new entrants to the work force in Poland generally started at the bottom of the social and industrial pyramid, performing unskilled manual labor, often in undesirable locations. In the Lenin Shipyard, these unfortunate neophytes were likely to find themselves dispatched to the hull sections, where through summer heat and winter storms they worked out of doors to erect hulls of ships. Not much in the way of "career advancement" was readily possible in the hulls; should one master a higher skill, one still applied it out of doors in all kinds of weather. Accordingly, the turnover rate was high. Hull workers came to be known as the shipyard's foremost center of "flyers"—workers who moved from job to job.

But workplace difficulties were not restricted to the depths of ships. Throughout the shipyard, one of the most depressing aspects of daily life on the job was the chronic shortage of proper materials when they were needed. The shipyard was a traffic jam of criss-crossing workers endeavoring to scrounge materials. Something on the order of 25 to 40 percent of work time was spent that way. Such structured inefficiency lowered morale, for it was self-evident that no "new Poland" could be built in the midst of such ingrained customs of production.

But living conditions were as exasperating as working conditions. Young bachelors were quartered in dilapidated dormitories, old hotels, or barracks-like apartments—three or four to a room. When he first came to Gdańsk in 1967, Lech Wałęsa lived in such circumstances. A year later, he and a 19-year-old woman named Danuta were married. Like Lech himself, she had recently come to the city from the Polish countryside. In 1972, Lech, Danuta, and their two children moved up in the world—to a tiny one-and-a-half-room apartment on Wrzosowa Street in the working-class community of Stogi. Like every worker district in the cities of the coast, Stogi was a socialist community and a Catholic one. A number of cooperative social activities were available, which was fortunate since Stogi families had little money to spend on anything besides necessities. The parish church played a much more active role in the life of Stogi's residents than churches did in most other worker quarters throughout the world. This, of course, was a Polish phenomenon rather than merely a coastal one, and it shall be necessary in due course to explore its dimensions. The Wałęsas, in any case, participated fully in matters religious in Stogi. As children kept coming in the decade of the 1970s, the Wałęsa family, five, six, seven, eight in number, was a predictable and even an impressive sight walking through Stogi to Mass on Sunday. On feast days, the windows of their flat could be expected to display pictures of saints. The Wałęsas were unprepossessing, down-to-earth, and warm—a good Catholic family at home in the workerish world of Stogi.[1]

The Stogis of Poland constituted a world of skills and crafts, one populated, in Wałęsa's term, by "the boys." Working conditions and shop-floor problems assaulted all workers, and it was generally understood, without thinking about it, that if anything were to be done to change things, they would have to do it. There were gradations, of course—workers who were "strong" or "honest" (the terms in Poland tended to mean the same thing) and others who were "shaky" or "cowardly." The adjectives, favorable or otherwise, betrayed the fact that the shop floors of Poland were places of intense coercion. Nowhere was state censorship more consistently compulsive in Poland than at the point of production. A worker's primary identity was largely determined by how this fact was confronted on an individual basis. By the mid-1970s, virtually all shipyard workers, save the very newest arrivals, understood the party for what it was. The last working-class connections to the United Workers' party on the coast had expired under the fire of automatic weapons and tanks in 1970. There were almost no believers in the party within the coastal working class anymore—nor very many, if the truth be known, within the party itself. Party members within the working class were seen as weak and "shaky" or simply "dishonest"—mere careerists.

In severing connections with the party, however, workers carried with them one core ingredient from the official doctrine. With its drab four-story apartment blocks, Stogi was, to its occupants, an essentially egalitarian place, full of essentially egalitarian people. Indeed, to the extent that many of the ideas both of the Church and of socialism harmonized with these currents, such ideas took on added life. And the Church's deep-rooted connection to the nation, to the very idea of Poland, gave it additional meaning.

But, in the world of "the boys," egalitarianism carried implicit gender implications. Intuitions about female equality that have taken fragile form in the West have not found a ready environment of acceptance in Poland. With its big steel economy and overconcentration in heavy industry, the nation's workplaces reproduced traditional divisions of employment rooted in nineteenth-century patterns of industrialization. Women worked in textiles, in the health professions, and as secretaries in offices. Through its hierarchy of priests, the Church unconsciously and profoundly reinforced male assumptions of status and authority. The party, too, both in its interior style and in the public posture of the official media, was almost as traditional as the Church. In the long years of movement building that culminated in Solidarność, agitation for women's rights did not acquire a visible form, either in the working class or in the nation as a whole. Nevertheless, in terms of worker militancy, some of "the boys" were women. But whether they were male or female, and whether they were socialists or Catholics or Catholic socialists, few Polish workers believed in the party anymore.

They also did not believe in being repressed out of existence. They had watched many a sacking and many a blacklisted "leader." They understood why the Edmund Hulszs of the coast had been banished to the hinterland and why the Edward Nowickis had trouble finding a job. The one incontestable accomplishment of the party in its three decades of rule in Poland was its success in imprinting on the nation's psyche the controlling power of the party police. Any worker who talked straight in a shop-floor meeting could routinely draw the overwhelming support of coastal workers. The difficulty was that so many such speakers soon got silenced through intimidation or, if not silenced, dismissed. Repeated regularly enough, as it regularly was, such a development confirmed among workers both the power and the illegitimacy of the party-state. Even more to the point, summary dismissals of worker spokesmen confirmed among the rank and file the need to be personally careful in expressing publicly one's disbelief in the party. The organizing task, in short, was not to convince workers the party was "wrong" or illegitimate. The "consciousness" of Baltic workers was quite a bit more than adequate on that score. The task, rather, was to find a way to act publicly in opposition to the party without having one's life ruined. In practical terms, the organizing dilemma facing would-be activists was quite real: how to perform the acts necessary to encourage workers without undercutting oneself by getting fired and inadvertently proving once again the futility of all such activism.

In the 1970s, coastal activists attended to this duty. In so doing, they learned to seize upon available symbols, both national and cultural, and invoked as well the specific working-class symbols generated on the coast. Wałęsa, typically, came to this strategy gradually, one experience at a time, learning the intricacies of organizing by doing it, learning to protect himself from police excesses while being interrogated, learning to debate with the state by debating in his own trials, learning what workers responded to by observing what left them unmoved, learning to evaluate his peers by testing his evaluations over time.

Because insurgent democratic organizing is a comparatively rare political experience, there are several common ways to misread this political experience. One is to impose on it other and lesser objectives. The Polish writer and social critic Ryszard Kapuściński has written with understanding and insight of societies trapped in alien totalitarianism. He has brooded over the plight of a population so confined that many thousands and even millions contemplate desperate options—including leaving. "But a whole nation cannot emigrate, so it undertakes a migration in time rather than in space. In the face of the encircling afflictions and threats of reality, it goes back to a past that seems a lost paradise. It regains its security in customs so old, and therefore so sacred, that authority fears to combat them."[2]

In Kapuściński's words, an afflicted people had to "dig in" to preserve its cultural and psychological essence against the ravages of the alien power. This intuition expresses a literary sensibility, an insight into ways of preserving the living meaning of inherited cultural experience. Kapuściński was ostensibly writing about Iran, but most of his readers knew he was writing about Poland. His message is not without its uses for any beleagured culture, but it also is one that is inherently passive politically. One digs in when no serious political act seems possible. Kapuściński's tactical solution was essentially spiritual. As such, it was not useful to Polish activists whose spirits were holding up but whose public lives were imprisoned. They wanted to dig *out* from under the leaden weight of the party, not dig into a spiritually nourishing cocoon. They wanted to create their own space in a public terrain controlled by an alien power, not hide from that power. Kapuściński's words might help keep one spiritually warm on a long totalitarian night, but they offered no encouragement to creative activity in the light of day.

Nevertheless, some of the cultural components of Kapuściński's formula for passive survival could prove quite useful in the service of an unresigned vision. He suggested the utility of "customs . . . so sacred that authority fears to combat them." This nicely summarized the instinct of many Poles. Indeed, it explained the strategy that undergirded "monument politics" in the early years after the December tragedy of 1970. In 1976 on the coast, monument politics was resumed under the aegis of student activists who, in due course, would create an institution known as Young Poland. The monument ceremony of December 1976 was small, but it was repeated again on May 3, 1977 (the anniversary of Poland's Constitution in 1791) and on December 16. The two dates were to become the twin high points of movement mobilizations during the last four years of agitation that culminated in the Polish August of 1980. May 3 was democratic day on the coast; December 16 was working-class day.

To the discomfort of the party, the symbols employed became inextricably mixed. The celebration of the ancient democratic constitution on May 3 took place at the site of the fallen workers of 1970. Conversely, the formal commemoration of the dead of 1970 on the December 16 anniversary of the massacre became a ceremony of democratic assertion. Wreath laying and public speeches honoring the worker dead were at one and the same time sacred customs that were overtly political. A semiannual irritation to the state, monument politics was also awkward to repress without subjecting the state to ridicule.[3]

A relevant factor in this context was the deteriorating economic climate of the late 1970s. Poland urgently came to need Western credits, yet President Carter had quietly but effectively linked U.S. funds to reasonable conduct by recipient countries in the field of human rights. He had, in fact, publicly orchestrated such linkage on a visit to Warsaw. Financially strapped as the Polish economy was, it was scarcely in Gierek's interest to crack down on wreath laying. The ruling party decided to take a poised view of monument politics, doubtless consoling itself that such activities did not presage an imminent revolution. Sacred customs—but not quite the way Kapuściński envisioned.[4]

Lech Wałęsa, for one, also found a way to put his highly prized electrical and mechanical skills to work in the cause. By some miracle of labor, ingenuity, and scavenging of parts, he was able to resuscitate an ancient Warszawa automobile, obtain a driver's license, and plaster copies of the democratic constitution of 1791 over its windows. He thus created a kind of mobile democratic exhibit that signaled the presence of self-activity down every street he drove. Public display of the constitution was not, *per se*, illegal; it was, after all, a historical document. But in the social climate the party's police strove to maintain, its public flaunting violated proper deferential form preferred by the authorities in People's Poland. Wałęsa's relic of a motorcar thus became a mobile symbol of opposition in Gdańsk. It also represented one more small but tangible step toward taking back some of the public space that had long been enclosed within the party's monopoly of civic expression.[5]

But recruitment to self-activity was not a simple matter in an insecure state with an overpopulated security apparatus. In the prevailing atmosphere, the practical demands of democratic construction were severe. To recruit, one had to risk entering the public space where the police found the requirements of surveillance considerably easier to manage. The complete avoidance of detection and arrest was, in fact, impossible. Gradually, a workable set of priorities sorted itself out. First, one had to endure the near contradiction of recruiting as much as possible while undergoing as few arrests as possible. The sites of recruiting actions always changed, but the forms remained the same: one recruited on trams and buses, in shipyard locker rooms, on hulls of ships and in similar out-of-the-way places, and before, after, and sometimes during Mass. Catholic feast days offered modest opportunities, and so did anniversaries of Polish insurrections, of which there were many. Two favorites emerged—an occasion and a site. The occasion, of course, was December 16: monument politics. The site was the dawn commuter train from Gdynia that carried the early shifts to the Lenin Shipyard. It ran so often and made so many stops in Sopot and Gdańsk that activists could move through and on and off with the kind of mobility that helped keep people out of jail. On the eve of important actions, veteran organizers tried to make certain that the commuter train was adequately staffed for leafletting. If the organizing challenge facing Polish activists was to create methods to take serious democratic politics out of the kitchen and into society, worker activists on the coast had clearly found some ways.[6]

They were not the only Poles to do so in the first eighteen months following the 1976 crisis. In Warsaw, the sheer violence and lawlessness of the government's crackdown on Radom workers moved a small group of intellectuals, led by Jacek Kuroń, to step publicly into view as a "Committee for the Defense of Workers" (KOR). Originally composed of fourteen and later of thirty members primarily of the literary and technical intelligentsia, KOR set about collecting money for families of imprisoned Radom workers, for medical help for the victims of police beatings, and for legal help for those charged with state offenses. Beyond this, KOR began to publish scrupulously researched and rigorously accurate descriptions of doubtful state evidence and inappropriate trials. KOR's *Information Bulletin* was born in this way, not as a *samizdat* publication but as an open one complete with names, addresses, and phone numbers of its creators. KOR thus nominated itself to address two highly relevant tasks: to speak for civil society, which had no voice previously, and to demonstrate intelligentsia support of the working class. In the spring of 1977, the organizers of KOR, unsurprisingly, found themselves in jail. Their personal survival and the subsequent expansion of their activities are a centerpiece of the next chapter. Here it is necessary merely to insert a brief summary of KOR's utility between 1977 and 1980. While most of the group's activity was centered on a vital contest with the state to diminish the censorship and create at least a bare minimum of room for public discussion of politics and social life in Poland, three of their number, Kuroń, Jan Lityński, and Henryk Wujec, attempted additionally to take the lead in connecting directly with the working class.

After the expenditure of considerable energy, courage, time, and funds assisting worker victims of police repression in Radom, KOR tried to capitalize in the autumn of 1977 by forming a nucleus of a "free trade union" in Radom. KOR managed to locate one worker who would publicly associate with the idea, but he was unable to recruit anyone else. In a repressive atmosphere, the initiative died—a result that seemed to impress upon Kuroń the perils of premature worker activism. In the spring of 1978, a provincial initiative—in Katowice by Kazimierz Świtoń, the owner of a radio shop—succeeded in producing a publicly announced free trade union with four members. Soon thereafter, Świtoń met Kuroń in Warsaw and an argument ensued over basic insurgent strategy. Kuroń asserted that the public unveiling of an independent union without a mass base would doom the group to immediate police repression that would destroy them all. According to Jan Józef Lipski of KOR, "the argument left Świtoń with such a dislike of Kuroń that he later attacked him and KOR in the official government press during the period of Solidarity." Świtoń joined ROPCiO (the Committee for the Defense of Human and Civil Rights), another newly formed opposition group.[7]

In founding their union in 1978, the four activists in Katowice called on

Polish workers elsewhere to do the same. With one exception, no one did. The exception was a worker in Gdańsk named Kryzsztof Wyszkowski. A well-read carpenter who devoted part of his spare time to preparing the work of Witold Gombrowicz for *samizdat* publication, Wyszkowski was a nonparty socialist who despised the governing regime. A possessor of a subtle understanding of the party's mechanisms of social control at the workplace, Wyszkowski together with two other activists, Andrzej Gwiazda and Antoni Sokołowski, joined in the founding statement on April 29, 1978. Gwiazda met Bogdan Borusewicz, the lone member of KOR on the coast, and was put in touch with Jan Lityński in Warsaw. Lityński opposed the initiative, repeating the advice Kuroń had given to the Katowice group a few weeks earlier. "We opposed this move. We did not want to have weak free trade unions, without strong roots in the milieu," said the KOR spokesman.[8] The Baltic workers went ahead anyway, though their immediate fate seemed to verify the accuracy of Lityński's ominous forecast. In addition to the three signers of the founding statement, the new independent union counted in its core group Borusewicz, Gwiazda's wife Joanna, Wyszkowski's brother Błażej, Wałęsa, and Anna Walentynowicz. The latter was a veteran worker at the Gdańsk shipyard. A cataclysm immediately hit the group. Wyszkowski was beaten up by the security police, and Wałęsa was saved from being sacked at ZREMB on fabricated criminal charges only by a threatened strike of his co-workers. Gwiazda, too, was jailed and fined, and Błażej Wyszkowski was relentlessly persecuted. It looked for awhile, however, that the most savage blow would come from within. One of the founders, Antoni Sokołowski, was unable to stand up under the harassment of the security service. He was threatened and bribed into denying his association with the free-union movement. The party press published a forged letter bearing Sokołowski's name denouncing the new unions. This was immediately followed by Sokołowski's grief-stricken recantation. His further value to the movement was, of course, destroyed. The episode merely verified the perilous status of working-class organizing in a police state.[9]

Under such circumstances, the principal service for workers inherent in the formation of KOR lay in the elusive area of the human spirit. For worker activists, the appearance of KOR in their town, if it happened, would generate the kind of boost in morale that came with the knowledge that one was not alone, that one's goals were deemed by outsiders to be important, and that these unknown others were willing to prove the authenticity of their belief by providing material help. Such psychological support might not count much as a blueprint for systemic movement building, but it was fully as vital as the money and printed matter that KOR people hoped to provide.[10]

Certainly for a Baltic activist named Lech Wałęsa, the formation of the "Free Trade Unions of the Coast" in 1978 was a genuine shot in the arm. After years of going it alone and being forced to recruit his own potential allies, Wałęsa's initial meeting with the organizing cell of the free trade union offered what he called "my first taste of genuine human solidarity." The date of the group's formation—April 29, 1978, right before the May 3 date that signaled another round of monument politics—perhaps provides a clue as to why Wyszkowski moved into action when he did. As previously noted, the accelerating economic crisis had caused the regime to ease political controls in the interest of maintaining contact with Western sources of credit; the short-term social result in Poland was the appearance of little grouplets of human-rights advocates. The founding of KOR in the autumn of 1976 was followed by four other small opposition groups: the largely middle-class Committee for the Defense of Human and Civil Rights (ROPCiO) in the spring of 1977; a student off-shoot of ROPCiO which soon evolved in Gdańsk into the Young Poland movement; the scholarly based "Flying University" in January 1978; and a patriotic sect called the Confederation of Independent Poland (KPN) in 1979. After the fashion of dissident factions, these groups sometimes had their sectarian disputes (particularly KOR and KPN) when they jostled against one another in various Polish cities. In Gdańsk, however, a workable degree of cooperation was generally achieved, at least to the extent that the group directed themselves exclusively against the regime rather than against each other. Young Poland tended to merge courage and sensible public actions with rather cosmic private discussions. This combination generated spirited internal morale if not always a good grip on realpolitik.[11]

Under the thoughtful stewardship of Aleksander Hall, the student group in Gdańsk grew and had achieved by 1980 a genuine presence. Young Poland was instrumental in reviving the prewar National Day, May 3, as a second fulcrum for monument politics in Gdańsk. Monument politics had originally been the property of the workers' movement, and the formation of the "Baltic free trade unions" on April 29, 1978, had the immediate effect of reclaiming that presence in the monument demonstration of May 3. On the other hand, the motive of Wyszkowski, Gwiazda, and Sokołowski may have simply been, as they said in their founding announcement: "While remembering the tragic events of December 1970 and acting in compliance with the expectations of numerous groups and milieu of the Baltic Seacoast, we wish to follow the lead of our Silesian colleagues by organizing independent trade unions."[12]

Wyszkowski was a politically sophisticated activist with a strong anticlerical tendency and an effective speaking style. Though no one in Poland was Wałęsa's superior in developing organizing tactics, Wyszkowski broadened Wałęsa's understanding of the existing social forces in Polish society. Wyszkowski's brother, Błażej, was also a thoughtful militant and possessed the extraneous but interesting talent of being an Olympic-class sailor of small boats. Andrzej Gwiazda and his wife, Joanna, were both trained engineers who, along with Bogdan Borusewicz, a graduate of Catholic University in Lublin and a member of KOR, brought a good measure of formal education to the core group. All three had a strong strategic sense of the centrality of the working class in the struggle with the state. The final founding member, Anna Walentynowicz, was a 53-year-old crane operator who brought to the group the experience of years of activism plus an undiluted sense of moral indignation. She was widely known in the Lenin Shipyard.[13]

After the free union declared its existence and took its public baptism in the monument demonstration of May 1978, early repression focused on Wyszkowski, Wałęsa, and Borusewicz. But all of the core group as well as their most activist recruits were routinely harassed. The ability to resist time-honored techniques of police intimidation was really the first occupational requirement of members. Wałęsa learned to appraise the potential value of recruits by how well or poorly they assimilated training in this art form. Like the natural activists who were their counterparts in thousands of factories across Poland, the militants of the coast knew the ways of the party in society and in the trade union on the shop floor. Also, in common with their counterparts elsewhere, they became periodic residents of police detention cells.[14]

In the autumn of 1978 came the creation of a poorly stenciled, badly printed news sheet called the *Coastal Worker*, for which most of the founders wrote. Though produced by makeshift methods, its editorial content, workerist in tone, was always germane. Distributing the paper became a possible means of recruitment and a source of arrest.[15]

The evolving status of dissent in Eastern Europe over the long years since the end of World War II was once explained to a British trade unionist by his Czechoslovakian counterpart: "Listen, my friend, twenty-five years ago, these people could have been tried and shot. Fifteen years ago they would have been put in prison. Now they simply lose their jobs. That's progress under Socialism." In Poland in the late 1970s, the organizing challenge was to avoid even the last-named sanction. Monument politics, however, made this very difficult. Despite another concerted defense in his behalf by his co-workers, Wałęsa was finally sacked from ZREMB following the December commemoration in 1978 and sacked again from Elektromontaż following the 1979 demonstration. In both 1979 and 1980, Wałęsa had long months of unemployment during which he managed to become a full-time organizer. Financial help came from KOR, ROPCiO, Young Poland, Catholic sources, and from collections taken up among co-workers at ZREMB and Elektromontaż.[16]

Life in the little cell of free-trade unionists was a complicated mélange of public and underground activities. Wałęsa and Wyszkowski were the most effective organizers, though Borusewicz and Walentynowicz were also helpful. Young recruits like Bogdan Lis, Jerzy Borowczak, Alina Pieńkowska, and Andrzej Kołodziej brought along other young workers. Lis was instrumental in organizing Elmor, where he and Gwiazda worked. Strategy sessions could rarely be conducted around the kitchen tables of the core group since Borusewicz, the Gwiazdas, and, before he left for Warsaw, Wyszkowski were often under surveillance and Wałęsa was followed wherever he went, even before his third dismissal in January 1980. But Walentynowicz's apartment sometimes served as a meeting site, as did those of Bogdan Felski, Lech Sobieszek, and Kazimierz Szołoch. Szołoch was a tough and experienced veteran of the shipyard Strike Presidium of 1970. Sobieszek, chairman of the 1970–71 workers' council at Siarkopol, was a specialist in providing steady verbal inoculations against anyone believing anything the party said.[17]

Wałęsa's eclectic approach to recruiting got him in trouble with the Gwiazdas and with Borusewicz. The strategic differences between the militants were sometimes sharp and reflected a fundamental disagreement as to what constituted legitimate movement politics. Recurring contentions with security police while distributing leaflets, driving his Warszawa, or simply walking around Gdańsk gradually convinced Wałęsa that the movement needed some militants who had at least a slight disguise—that is, people who did not look, dress, or talk like workers. Too many routine but nevertheless vital chores in the area of communications were systematically short-circuited by police interrogations and forty-eight-hour arrests. To Wałęsa, the movement had only one absolutely essential purpose—to mobilize the working class. Failing that, any "success" was meaningless. Achieving that, any failure could be overcome. Consequently, sectarian disputes between the various sects on the coast bored him. When the workers were mobilizing for struggle with the state, he felt that all the activists—whether KOR, ROPCiO, Young Poland, or KPN—could provide needed cadres for communications between the far-flung working-class constituencies of Poland. What the working class movement needed was activists, and particularly activists who did not look like workers and could more easily evade the security police.

However, to the militants in the opposition groups—none of which had or could get a mass base outside the working class—their very identity was grounded in their particular ideology; it determined why an activist was a member of one sect rather than another. For Wałęsa, a worker who got into an argument with another worker over sectarian differences was simply putting his sectarianism ahead of the working-class movement. This was self-defeating, he believed. There was no ideology that had the potential transformative power for Poland possessed by an independent workers' movement. It would change the very contours of daily life and liberate all the groups to contend with one another and the state as well.

Consequently, when not recruiting workers, Wałęsa attended meetings of all the activist groups in Gdańsk and thereby became acquainted with the militants of each. In the aggregate, they brought a considerable amount of talent and courage to the opposition cause. The driving force behind Young Poland in Gdańsk, Aleksander Hall, presided over the group's uncensored publication, *Bratniak*. It was widely read among students. Maryla Płońska and Ewa Osowska were other Young Poland militants who could always be counted on at the two annual climaxes of the coastal movement—monument politics on May 3 and December 16. Young physicians, like Adam Słonimski, and medical students, such as Dariusz Kobzdej, were also dependable and increasingly experienced. Young Poland was feisty; one of its militants, Jan Samsonowicz, blithely referred to the party as "a class of swindlers" and to communism as "an experiment which failed." It went over big in the halls of Gdańsk Polytechnic.

Borusewicz came out of this milieu and still dropped in from time to time, but Young Poland was too narrowly nationalistic and geopolitically fanciful to suit his tastes. Wałęsa, it seemed, could endure any amount of rhetoric as he counted the house and measured the commitment of individual rhetoricians. He also moved through the civil-libertarian world of ROPCiO, sparked on the coast by a dedicated human-rights advocate named Tadeusz Szczudłowski. But Gwiazda complained loudly when Wałęsa took a deep breath and ventured into the precincts of KPN. In the Confederation of Independent Poland, the lyrics of romantic patriotism reached such heights that the militants of Young Poland seemed in comparison to shrink to the size of spineless internationalists. KPN was too much, said Gwiazda, way too much. It was a sentiment that was echoed down the line to the centers of KOR in Warsaw. Wałęsa, bent on recruiting activists to serve his own cause, ignored the complaints.[18]

The differences between KOR and worker activists like Wałęsa were more than tactical. They reflected the differing understandings of politics that developed between movement ideologists and movement organizers—between democratic theorists and democratic activists. It is a schism that threatens all social movements. Occupying one perspective are those who see activism as the product of an intellectual discovery that transforms the meaning of prior experience; occupying another are those who see the creation of activists as the product of an experiential discovery that transforms the meaning of prior belief. While the two are not mutually exclusive, the first primarily can happen in universities where, among students, societies are analyzed; the second can happen in society where among workers various relations of production materialize. To "happen," both require self-activity, though of a very different kind. One begins in seminars, the other on shop floors. One understands experience as interpreted through ideas; the other understands ideas that develop out of experience. One produces the Jacek Kurońs of movements as well as the more passive intellectuals who appreciate his insights but do not relish the costs of participation in his cause; the other produces the Lech Wałęsas of movements as well as more passive workers who grasp his purpose but do not choose to accept his risks. In their respective understandings of movement politics, the influence of Kuroń depends upon his ability to convey his interpretive insights through the use of sophisticated verbal skill; the influence of Wałęsa depends upon his ability to convey his sophisticated experiential knowledge through the timely use of familiar expression. When they fail to persuade, traditional habits of conduct prevail: no movement materializes.

Because movements seek to correct large imbalances of power, the challenge is severe, the demands of performance high, the tolerance for error low. The political odds, it seems, are quite high. This helps explain a recurring historical anomaly: while the occurrence of injustice in human societies is constant, the number of large-scale movements in history is quite low. In most times and places, the demands of movement building are simply too great. People live their protoinsurgent lives in the privacy of their kitchens and never find a way to connect with the larger society. Given the enormous power imbalances in Poland, the creation of such a large-scale movement as Solidarność is therefore a striking historical event. These dynamics, however, leave unexplained whether, at a given moment, the Kurońs or the Wałęsas are "correct," or whether either is. It does help explain, however, why such people disagree from time to time while coexisting as devoted proponents of the same cause.[19]

In Warsaw, Jacek Kurón was the resident in charge of democratic theory; in Gdańsk, Wałęsa was the resident in charge of democratic organizing. Because they shared—or appeared to share—essentially the same strategic conception of what could be done to democratize Poland, they generally appeared to agree on matters of fundamental importance. This circumstance eventually would change. But as events were to demonstrate, Wałęsa persuaded worker activists to accept qualified and committed intellectuals as secondary advisors in the Gdańsk negotiations, while Kuroń tried to persuade the Warsaw intelligentsia that the working class was necessarily the central pivot of the struggle. Both, needless to say, had a strong sense of self. There was a distinction; because Wałęsa understood the concrete demands of recruitment that constituted the centerpiece of movement building, he concentrated on the task at hand. Unfortunately, Kuroń and other Warsaw intellectuals (both the small group comprising the democratic opposition and the far larger number watching safely from the sidelines) did not fully grasp the arduous demands surrounding this central aspect of the struggle against the party-state. They therefore profoundly underestimated the relevance of Wałęsa's organizational sophistication. They underestimated as well the sense of tactical and strategic prerogative such knowledge instilled in him. A huge communication gap opened, the causes of which no one elected to ponder. KOR intellectuals assumed they would lead when—through the mysterious laws of history—the movement would spontaneously appear. One day the workers would rise and the moment of insurgency would suddenly "happen." The historical and theoretical literature of the past on which they relied said so. On the other hand, Wałęsa felt he had to be present or there might never be a "happening." He had not read the literature.[20]

Wałęsa, therefore, prevailed in the dispute with Gwiazda and the KOR activists because no one could deflect him from his seemingly random pursuit of activists. The puzzlement about Wałęsa's stance extended to precincts beyond KOR. A ROPCiO-connected lawyer who frequently represented Wałęsa, Lech Kaczyński, attempted to take a benign view of the young electrician's blasé response to the differences among opposition groups: "Wałęsa was very active but he was less interested in politics in general than in the concrete problem close to his heart. He never ceased emphasizing worker solidarity." In two sentences, the well-meaning movement lawyer succeeded in revealing both the depth of his political innocence and his rather robust capacity to patronize workers. Alina Pieńkowska, who was close to Borusewicz and KOR, merely complained that Wałęsa "does not know how to submit himself to rules."[21]

The feeling took root in KOR that Wałęsa was uncontrollable. To remedy this, Borusewicz and Gwiazda arranged for most of the core group from the coastal union to travel to Warsaw for what they regarded as necessary counseling from Kuroń. The meeting, however, was unsuccessful. Though Kuroń adopted the quiet mien of a sage, Wałęsa was not responsive. To observers, the conversation took place between "two adversaries." Such tensions, however, were not decisively important in 1979–80. The essential struggle in Poland remained between the party and the working class, not between Wałęsa and KOR. Recruiting remained the overriding task of the movement, and it was a problem that could be concretely addressed only on the ground in Gdańsk. Because ROPCiO and Young Poland were less severely repressed by the security police than were the free unions of the coast, they had far more militants at their disposal. This was particularly true of Young Poland. So, Wałęsa stayed in close touch and paid his dues; Young Poland's militants made the monument speeches on democratic day, on May 3, every year while Wałęsa got arrested for distributing leaflets. When he lost jobs because of his monument speeches on December 16, working-class day, he could take satisfaction in knowing that the increasing crowds each year were partly due to the zealous reciprocal leafletting of Young Poland's troops. Cooperation between workers and students clearly helped the movement.

In terms of the dynamics that produced the Polish movement, there is a potential misreading implicit in this evidential review. Because the Baltic movement eventually produced Solidarność and Wałęsa became its highly visible symbol, many observers (and Wałęsa himself in his ghost-written autobiography) were moved to comment approvingly or critically on his actions. As a result of this outpouring of commentary—by fellow workers, party members, Church officials, Polish intellectuals, and Western journalists and scholars—the world knows far more about Lech Wałęsa's actions in the 1970s than about any other Baltic worker in the decade between the December rising of 1970 and the Polish August of 1980.

This information is more than suggestive, it is exceedingly helpful in delineating the contours of the evolving social relations between the Baltic population and the party-state in the period that produced the final building blocks of Solidarność. But it would be a cardinal error to see the achievement of these building blocks as the work of one man, however imaginative and persistent he might have been or however constructive his role might have seemed in helping to shape a nascent popular movement into full and self-conscious expression. Human agency is a fundamental component of history. But while people can act, and must act, for movements to happen, the range of possible activity is limited by compelling social constraints and by the actions of relevant "others"—in short, by fortune as well as by conscious self-activity. In this overriding sense, Wałęsa is best understood as an individual who affords a window into the complex collective experience of Baltic workers. We can see them acting, often through the observations of witnesses who consciously focus their principle attention upon Wałęsa, and we can also see them acting when Wałęsa is not present. The ultimate point is that the sequential evolution of these collective actions constitute the movement; Wałęsa's actions, in themselves, do not. He offers a relevant view into many of the operative dynamics of the movement; he is not the movement.

In any case, Wałęsa's last dismissal before the Polish August was accompanied by tragedy. The murder of an activist from Elektromontaż in January 1980 provided the Baltic opposition with an emotional bond. Out of the agony it generated, the party ironically became the foremost victim. It also generated within sectors of the coastal working class yet one more reason to support with zeal the first Gdańsk demand later in the year.[22]

When Wałęsa was fired from ZREMB for his part in the monument ceremony of 1978, he remained unemployed for months under a police blacklist.

But the tactic backfired. The police discovered that Wałęsa not only somehow managed to feed his family, he used the new freedom to labor as a full-time organizer. He was constantly watched and periodically interrogated and jailed on forty-eight-hour detention, but the treatment seemed only to make him function at a more intense pace. It also tied up considerable personnel in the security service. After three months of this, two unidentified functionaries one day paid a call on the Połmozbyt works and informed its manager that he was to hire Wałęsa. The authorities had decided it would be easier on all concerned just to tie him down to a job somewhere for eight hours a day. But Wałęsa turned down the job offer at Połmozbyt and the next day signed on at Elektromontaż.

In-plant surveillance of Wałęsa, though constant at Elektromontaż, was apparently ineffective. A fellow electrician, Florian Wiśniewski, marveled at Wałęsa's ability to build a shop-floor movement. "He had an incredible facility for establishing contact," said Wiśniewski, by way of explaining why so many workers at Elektromontaż took part in the monument action of May 3, 1979. Wałęsa had escaped arrest the day before the demonstration by the expedient of not reporting for work and spending the night away from his apartment. On December 16, he further enraged the authorities by getting himself smuggled into the shipyard through a police cordon and suddenly emerging before seven thousand people to make what turned out to be a prophetic call for the erection of a monument by December 16, 1980. Security police blanketed the demonstration site opposite Gate No. 2 and took many photos. Elektromontaż workers were active in the throng, particularly a 19-year-old driver named Tadeusz Szczepański, who energetically attended to the loudspeaker equipment and refused to be intimidated by toughs in the security service.[23]

On January 15, 1980, Wałęsa and fourteen other workers were dismissed from Elektromontaż. Thirteen of them were new recruits to the free union who had been photographed at the monument demonstration. One of them was Szczepański. The next day he disappeared. More than four months later, his body was found in the Motława River, his head badly battered, his fingernails pulled out, and his feet cut off. Opposition forces throughout Gdańsk were deeply affected. The funeral at St. Brigida's became an extraordinary event. Opposition militants from every milieu showed up at Wałęsa's apartment on Wrzosowa Street. Holding a large wreath, trimmed in red and white and bearing the inscription "To a Union Man" and the signature "From the Free Trade Unions of the Coast," Wałęsa attempted to lead the throng from his residence. Security agents tried to snatch the wreath, and a struggle ensued. Wałęsa was overwhelmed, though not before Maryla Płońska tried to restrain the agents. She took a blow to the stomach while Wałęsa got an elbow to the head. Both were arrested, and the wreath was confiscated.[24]

There was a singularly Polish touch to this grotesque action. Security officers—party members but also Catholics—brought to St. Brigida's a plain wreath minus inscriptions and the colors of Poland as replacement for the one they confiscated from Wałęsa. Some 3,500 people packed into and surrounded the church and were themselves surrounded by a vast array of security troops. The work force of Elektromontaż, now deeply politicized by events, attended *en masse*.

Wałęsa's firing from Elektromontaż four months earlier had proved both tumultuous and educational. The plant's workers formed an independent Workers' Commission in December, and 168 of the 500 employees signed the new union's founding announcement. Under strike pressure, management at Elektromontaż agreed to take the union's president, Wałęsa, back. The party promptly intervened and countermanded the decision. The management at Elektromontaż was able to ride out the resulting turmoil with considerable help from the security police. Repression of Elektromontaż employees was widespread, a fact that helped explain the near unanimous turnout for Szczepański's funeral four months later. But at the time, Wałęsa's dismissal was confirmed, and the independent union movement was destroyed at Elektromontaż. The lessons of 1956–57 and 1970–71 were once again reaffirmed: plant-sized independent unions had no real chance against the state apparatus. The free union had to be bigger in scope.

During the entire last half of the 1970s, as Wałęsa was forced into moving from the Gdańsk shipyard to ZREMB to Elektromontaż, Danuta and their growing brood of children continued to live in the small flat in Stogi. Wałęsa had, in the course of these years, acquired a name for himself. He had leafletted on and off the commuter train at all the stops through Gdynia and Sopot. Police interrogations and forty-eight-hour detentions gradually became so common that Wałęsa and the security service had a reciprocal interaction that increasingly seemed to tilt in Wałęsa's favor. He made a judicial proceeding out of every attempt at intimidation, keeping KOR, ROPCiO, and government lawyers in gainful employment and leading his own defense with or without counsel. Every act was a political act; every trial, a recruiting opportunity. The government took to scheduling his trials on Holy Saturdays to diminish attendence and when that did not work, to filling the courtroom seats with security agents. Trial venues were changed at the last minute and even moved out of town. The dismissal proceedings from ZREMB were held in Gdynia.

When worker activists were arrested, Wałęsa appeared at their trials and once tried to defend his close friend, Błażej Wyszkowski, by loudly demanding whether the court proceeding was "another railroading." His casual and open manner masked some inner drive that caused him to probe the limits of the possible. He inevitably made errors in sometimes pressing too hard, but he developed over the years an uncanny feel for the dynamics of confrontation. He also acquired an insight into the psychological drives and needs of his adversaries in the official trade union, in the party, and in the police.[25]

This political sensibility gradually extended to the Church as well. In religious terms, he was a kind of deeply believing mystical Catholic who went to Mass to think and to take on new energy. But as surely as if he had read Kapuściński, he exploited with growing skill the political uses of such "sacred customs" as monument politics and prayer services. He attended a private service every day for five months until Błażej Wyszkowski was released from prison. He first met Dariusz Kobzdej, who became one of Young Poland's tireless militants, at a prayer action for political prisoners. The rituals of the Church often had definite uses, though benefits depended upon the surrounding context. When priests came with candles, programmatic forward motion could on occasion be dissipated into static meditation. But this discovery, too, was useful. On those occasions when workers were goaded by police provocations into an angry mood that threatened to precipitate an intemperate and self-destructive response, a timely moment of prayer or the singing of the national anthem could help restore a poised consideration of the task at hand. The uses of the Church were like anything else—timing and purpose determined utility.[26]

By the summer of 1980, the coastal police had taught Wałęsa almost all a citizen could learn about the multiple ways of judicial repression. These lessons dated back over the years. During 1974 and 1975, one of his oldest and most intimate relationships—with Henryk Lenarciak of W-4 at the Lenin Shipyard—had been gradually derailed through a clever police eavesdropping operation. Conversations each thought were private had a way of becoming discretely public, so that each began to doubt the loyalty of the other. They stopped speaking to each other for months, and the spirit of the movement in the shipyard sagged noticeably as a result. Eventually, they discovered what had happened and restored their relationship. In 1978, the police decision to take away Wałęsa's driver's license was merely part of a long-running psychological campaign aimed at destroying the jaunty confidence that was such a telling aspect of his recruiting ability. His response was to turn his new immobility into a recruiting tool. When he emerged from a forty-eight-hour detention, he would get on a bus without money and borrow złotys from total strangers while telling all on board the details of his false arrest and his activities in the free-union movement.[27]

He knew workers everywhere: in shipyard supply enterprises, in the repair shipyard, on the docks, in the transport unions, in Gdynia and Sopot, and in all the opposition milieus. He knew the old militants—Walentynowicz, Sobieszek, Szołoch, and his old comrade Lenarciak. He knew the "flyers" from the hulls, how to take an interrogation, and how to lose a surveillance team in the streets of Gdańsk or the byways of Stogi. He was, as the world would discover, a man of unusual gifts. He had an acute and retentive mind and a brilliant strategic sense of high politics. All this would become apparent later. But even before the Polish August, his comrades in struggle on the coast understood his organizing talents.[28]

Borusewicz and Gwiazda knew them, too. Each brought considerable strengths to the movement, but some things are not fair in life and politics, and one of them surely concerns the uncertain connection between commitment and organizing talent. Andrzej Gwiazda had a commanding, almost imperial presence, a quick mind, and an iron determination to free himself and the rest of the Polish nation from as much of the party apparatus as was geopolitically possible. He could speak with power and clarity, but—except in his own plant—it necessarily had to be before a gathering that someone else had organized. He possessed none of the essentials of the craft. He was never on time, did not know how to put unsettled people at ease, and, not least, simply could not generate a sense of camaraderie among those he knew casually. He was not one of "the boys," nor could he use the term with the affection and communitarian meaning it seemed to have when Wałęsa used it. He was the kind of political activist who was good at protecting the palace from swindlers and the parliament from the ravages of demagogues. But he was not a tribune of the people for whose cause he so passionately labored.[29]

Borusewicz was more of an Indian scout than a guerrilla general. He could survey terrain, geographically for a leafletting campaign and politically for booby traps. He had a fine feel for the subtle ways the looming Soviet presence did or did not impinge upon movement possibilities, and he could edit an underground journal. Borusewicz was adroit in checking the fine print on a party promise and adept at looking up in time to detect pickpockets from the ruling apparatus trying to slip softly among the ranks of the faithful. But while he enjoyed the attention he had earned, he did not relish the blinding glare of a center spotlight.[30]

For Andrzej Gwiazda and Bogdan Borusewicz, whatever one did to recruit for a movement was a craft whose most subtle components lurked in some unseen corner just over the horizon. After the "Free Trade Unions of the Coast" had first proclaimed their public existence in 1978, the initial hard core of half a dozen members soon discovered they were not to be stampeded by eager militants seeking a proper home. Wałęsa told them they were going about it all wrong and tried to set a proper example. In reciprocity, he attended the seminars on history and politics that Borusewicz organized, though he had little to say. He primarily utilized such occasions to measure the movement uses of the participants.

For his part, Gwiazda, acknowledging his own shortcomings as an organizer, began speculating on the possibility of KOR sending someone from Warsaw who possessed the necessary organizing magic—whatever that magic might be. Gwiazda enjoyed Kuroń, learned from him, respected him as all did who were privileged to watch the range of his dialectical mind at work. But the world of docks and shipyards, supply factories, and shop floors, was a world of social relations that was not the accustomed turf of Jacek Kuroń, who therefore had no concrete craftsmanship and experientially grounded advice to pass along on this score. No proven organizer of workers came from KOR, because KOR had none to send. The best the coastal movement had, it soon became apparent, was Lech Wałęsa. So when he was sacked at Elektromontaż in early 1980 and it became apparent he would be blacklisted, Borusewicz and Gwiazda implored him not to support himself, as he easily could, by working in small private garages. Such places necessarily contained few employees to recruit. The means were found through KOR and sundry coastal sources to keep Wałęsa going at full-time organizing.[31]

The decision, in any case, was not some sort of desperate expedient. The pace of organizing had begun to pick up in the second half of 1979 as the increasingly disastrous state of the nation's economy continued to open ever larger fissures in the social fabric of Polish society. As the rate of deterioration noticeably accelerated in early 1980, almost everyone—in the party and outside it—sensed the reckless speed of the roller coaster that Poland had become. As the summer began, the opposition movement had never seemed more relevant.

On July 1, 1980, Edward Gierek, the first secretary who had come to power on the wreckage of Gomułka's abortive price increase attempt of 1970, instructed the party media to announce a broad new price hike in food. As the particulars were a bit awesome—increases ranged from 90 to 100 percent on some types of meat—an attempt was made to more or less whisper the necessary information to the Polish public. The party's economic plan was announced after the fact on July 2 in an interview with an official of the state retail organization. The following day, the Polish Interior Ministry's information agency, Interpress, informed foreign correspondents in Warsaw of the existence of "consultations" between workers and the government. When the spokesman commented that "what is economically justified is not always approved by the people," the possibility occurred to journalists that strikes might in fact be underway across Poland. One was confirmed at a vehicle parts plant near Gdańsk at Tczew, and there were unconfirmed reports of others in the textile center at Łódź, the Huta Warszawa steel plant in Warsaw, and (perhaps most significantly because of the 1976 troubles) at the Ursus tractor factory. As the days passed, blips continued to appear on government radar focused on the working class: work stoppages at a petrochemical plant, a television tube factory, a Miełech aircraft plant, the Żerań auto complex, the Świdnik helicopter factory, and more rumblings at Ursus where seventeen thousand workers, 40 percent of the work force, were thought to be on strike.

In a speech to the Central Committee, party leader Gierek emphasized that the higher meat prices were permanent and that broad wage increases were unacceptable. The speech was broadcast on television on July 19 and augmented by a second broadcast in which the head of the party's media division appealed for "civic responsibility." Meanwhile, it appeared that local officials across the country had been given a degree of latitude in pacifying strikers. Pay increases of up to 10 percent were granted to over thirty factories.[32]

By July 15, strikes which had first erupted in Lublin on the tenth had begun to reach citywide proportions. By the twentieth, a strike of Lublin locomotive engineers had blocked the rails for four days, affecting traffic to the Soviet Union. A story that began to spread like wildfire along Poland's railroads and thence out to the population told of strikers in Lublin using welding torches to freeze the wheels of railroad cars to the tracks. Supplies of meat for the Soviet Union were not moving. The story proved to be true. Gradually, sensitive observers of the Polish scene tumbled to the fact that something approaching a general strike was materializing in Lublin. A newspaper in the city printed a back-to-work call that appealed to patriotic sentiment and added an oblique reference to "anxiety among our friends"—a reference to the big Soviet brother. Since all services in Lublin had come to a standstill, government vehicles were sent in to ensure distribution of essential supplies.[33]

In Warsaw, KOR's Jacek Kuroń began keeping score for the benefit of foreign correspondents. By July 18, Kuroń's tally listed seventeen factories on strike; no less than fifty-one others had settled through pay increases to compensate for the price rises. But Kuroń's telephone relay system was not simply directed outward to the West; it also extended laterally across Poland. Every incoming caller bearing new local information for Kurón was told of the spreading wave of strikes in other industrial areas. Thin as it was, Kuroń's network of activists across Poland had now been placed in the service of the working-class movement. In a revealing admission, a Warsaw radio commentator conceded that the troubled situation indicated that the party and trade-union structures had failed to stay in close touch with Polish workers. While this was a familiar invocation of an ancient propaganda ritual, the more relevant fact about the official union was that its historical role as a mechanism of shop-floor censorship had been breached by the Kuroń network. As the month wore on, the scale of workers' demands broadened in scope and deepened in structural meaning as immunity for strike leaders and calls for new trade-union elections plus other revisions in the role of the official unions increasingly appeared as points of grievance. After a thirty-minute warning strike by Gdańsk transport workers and additional difficulties reported in coastal shipyards, the party seemed to react with a historically based awareness. Strike prevention committees were established at the Cegielski engineering works in Poznań and at plants in Wrocław, while unsolicited pay increases suddenly were announced at Radom and Toruń, with similar unsolicited settlements reported for some shipyard workers in Gdańsk and Gdynia. The Ursus workers, representing almost as much "historic" memory as those at Cegielski, Gdańsk, Gdynia, and Radom, had been placated earlier in the month, so the party deemed it unnecessary for new overtures there.[34]

Near the end of July, the party decided its frantic fire-fighting activity had proved adequate to the challenge. The government's chief trade-union troubleshooter, Deputy Premier Mieczysław Jagielski, had quieted the storm in Lublin with across-the-board wage settlements. On July 27, Edward Gierek decided he could safely take his summer vacation and departed for a three week sojourn in the Crimea. Ten days later, amid much fanfare in the government media, meetings of the long-moribund Council on Worker Self-Management were announced for cities across the country. Shop-floor concession and jail-house repression became the order of the day, however, as the security services tightened the screws behind the scenes. The most visible leader of striking Warsaw garbage workers was arrested. With a detailed specificity that history generally fails to record, the same thing was happening to worker advocates across the country. Except within the working class, the names of most of these victims of repression remained unknown—historical hostages to the government's control of social information.[35]

By mid-summer 1980, whatever thirty-five years of working-class self-activity had taught Polish workers now existed as an invisible store of knowledge locked in the minds of the workers themselves.

Since it provides the essential causal dynamics that produced Solidarność, it is prudent to review this experience as it existed in its final pre-August form. In terms of organizing, the knowledge that was acquired over thirty-five years was quite concrete. It therefore did not appear to be "ideological" as that word is generally used to describe political horizons. But this worker knowledge was in fact organically ideological in the deepest sense: it constituted the building blocks of autonomous conduct. In keeping with the rhythms of human experience, these understandings were acquired experientially; that is to say, episodically or haltingly as individual workers and work forces engendered various kinds of protests over many years. There had been an intense struggle between Polish workers and the party in 1945–47. As the victorious Polish United Workers' party imposed its own trade-union structure over a subjugated work force, workers were forced to accept the fact that their shop-floor traditions had no place in the new People's Poland. For their part, young workers simply never experienced these traditions at work. After the long Stalinist night of Bolesław Bierut in Poland, the word "strike" became too dangerous to use and passed out of common usage. When workers became so provoked that they tried to move to organized protest, a whole range of new terminology came into play: *awarie* (breakdown); *przestoje* (bottlenecks); *przerwy w pracy* (breaks in the work); or simply *zebrania* (meetings).[36]

Though workers at Cegielski-Poznań could through the use of such euphemisms mount a series of more or less silent strikes in 1956, the plant-wide occupation strike of prewar fame was not employed. This is where matters stood in the late 1950s: a structurally confined working class, bent on reclaiming its traditions under close daily supervision, rousing itself through self-organization, and glimpsing for the first time, amid the wreckage of Poznań, the idea of a union free of the party.

As the party contradicted and discredited itself in the aftermath of October 1956 and in the Gomułka years that followed, Polish workers slowly initiated an almost totally unseen struggle to reestablish some sort of collective identity for themselves in the new industrial Poland. The absence of lateral lines of communication within the working class was exposed and the regional organizing imperative made a bit clearer. On the Baltic coast in 1970, the occupation strike was recovered—under the rifles of the militia and the guns of armored troops. However tenuous they had always been, the last connecting links between the party and the coastal working class were irretrievably sundered by the December massacre of 1970. Most of Poland's major industrial enterprises conducted repeated occupation strikes in the 1970s, and did so with increasing efficiency and frequency as the decade wore on. The self-awareness generated within the working class by this prolonged conflict became a social component of life in People's Poland. The occupation strike became the rule; unfortunately for the workers, so did the ruthless sacking and blacklisting of those shop-floor militants who had become the foremost on-the-job repositories of working-class experience. But group experience could not be "sacked" or "blacklisted." It merely persisted as hard-won experiential memory, a part of received knowledge that became the accumulated social wisdom of the Polish working class.

The most immediate piece of wisdom was the discovery that every assault by the party upon a worker advocate—every intimidation, every dismissal, every blacklisting, every imprisonment—was a fundamental assault on workers as a whole. The demands put forward during occupation strikes, therefore, increasingly included a proviso for immunity from reprisals for all strikers. This reciprocal embrace—trade-union discipline and shop-floor response, party repression and organized riposte, grievance and deceit, assertion and evasion—became the received experience of both the Polish working class and the Polish party. It was not a balanced contest fought on familiar terrain with identical weapons and predictable results. Rather, it was one in which strategic weight always rested with the party, but also one in which the tactical armory available to workers constantly expanded. For this reason, the terrain was constantly in a process of shifting. The maneuvering room of the party narrowed as the tactical and psychological space possessed by the working class inexorably expanded.

By 1970, there was enough of this experience in a part of working-class Poland—the Baltic working class—that it could be codified in an institutional form: the interfactory strike committee. Workers in one plant had to have acquired enough of a strategic sense to want to know the names and capabilities of worker activists in another plant, and then they must have further endeavored to meet before an effective interfactory relationship could be established between them. The first Interfactory Strike Committee, established on the coast at Gdynia, was arrested within twenty-four hours of its formation. The second—at Szczecin in 1970—was neither internally democratic nor, in direct consequence, effective. A functioning interfactory strike committee between five, ten, or twenty enterprises could not adequately be comprehended simply as an idea; it was the culmination of a veritable ensemble of organizing skills that instructed workers in the specifics of self-activity against the state.

Similarly, an occupation strike involved far more than the simple idea of calling one. It involved an array of communication, logistical, and security prerequisites, all of which became vastly expanded and complicated in the course of an action of prolonged duration. By the mid-1970s, most large enterprises in Poland, a nation of large enterprises, had acquired a measure of expertise in the practice. The significant development in 1976 was not, as so many nonworkers believed, that Radom and Ursus had erupted in violence but rather that scores of other enterprises across Poland had protested Gierek's price hike by mounting orderly occupation strikes. By the end of the decade, the occupation strike had become something of a settled craft in most places across Poland.

But on the Baltic coast, something beyond this had been achieved. The pioneering and employing of the interfactory strike committees by coastal workers was the decisive structural ingredient in the historic emergence of Solidarność. It also was one that intellectuals in the universities or in the opposition did not fully understand or, indeed, recognize as existing. This negative relationship was rather stark and was, in fact, reinforced by a striking fact: neither in the two weeks of intense confrontation between the Gdańsk Interfactory Strike Committee and the government during the Polish August nor in the six weeks of labor unrest that preceded it did interfactory strike committees or demands for independent trade unions emerge from Radom, or Ursus or Żerań—that is, from the Warsaw region where KOR's "consciousness-raising" efforts were presumably most powerfully at work—and none came from Lublin or Kraków, where university influences, Catholic oppositional influences, and KOR influences were also said to be at work. It did not emerge for the elementary reason that the necessary sequential achievements of movement building that produced Solidarność were not part of the heritage of self-generated organizing acts within either intellectual or working-class circles in those regions.

This social fact pointed to an intellectual one. Participants in all existing intellectual tendencies in Poland, whether partaking of oppositional activity or not, employed an extremely abstract and experientially remote language when it came to the matter of thinking in print about large-scale democratic forms. All were therefore quite tentative in their understanding of the structural components organic to an empowered democratic polity. That is, they could think about "civil society" as an idea, but they possessed no social experience upon which to base a language of politics that might concretely lead to the creation of such a liberated community. When that community somehow appeared without their knowledge, they literally did not have the words to describe its components or to characterize the specific organizing ingredients that brought those components into existence. At the climax of the Polish August, the completed edifice of an empowered civil society was visible for the world to see. But what its building blocks were—how it came to exist—remained a mystery.

The process by which Baltic workers constructed the building blocks of the house they called Solidarność was not mechanistic. Although in retrospect it is relatively simple to array the organizing pieces aside one another in sequential order so that (1) an occupation strike may be seen as the necessary step toward (2) an interfactory strike committee dedicated to (3) the achievement of a self-governing trade union independent of the party-state, the decisive prehistory of the Polish August was not set in place in such an orderly manner. Each component was a product of knowledge acquired through collective assertion over thirty-five years in an ever-changing social setting that continued to generate distressingly persistent social and economic problems.

Though the result possessed what might be called a culminating logic, this is an understanding that is apparent only after the fact. When enough specific knowledge had been gained by sufficient experiential struggle, the result could retrospectively be seen as "culminating." The Polish August was absolutely culminating in this sense. But the purposeful order that became visible in the Gdańsk shipyard was a very long time coming, and the route to it, traversed as it was by humans, was quite disorderly. When all the pieces were in place the resulting building block—the first Gdańsk demand written by an experienced MKS presidium protected by a sea of workers self-organized in an occupation strike—merely looked orderly. The less one knew about the preceding thirty-five years of struggle, the more orderly it all seemed. Indeed, if one knew little or nothing, it could be described rather simply as "spontaneous," "not entirely spontaneous," "the work of antisocialist manipulators," or as a "miracle."[37]

We may pass quickly over such scholarly, journalistic, and administrative non sequiturs. The occupation strike of the Polish August was scarcely an ephemeral dropped from the sky. Not only had coastal working classes from

Gdańsk to Szczecin lived through sustained and repeated occupation strikes through December–January 1970–71, the same tactic had been used again in Gdynia in 1974 and in Gdańsk in 1976, as well as in Elbląg, Słupsk, Szczecin, and a score of other plants on the coast throughout the 1970s. The interfactory strike committee was a continuing part of the memory as well. But beyond this, the organizing experiences that preceded these mobilizations and the organizational constructions that were fashioned during mobilizations had become a part of the received historical memory of thousands of workers on the Baltic coast.

And yet, and yet . . . despite the overwhelming nature of evidence which had accumulated over three decades of striving, despite the momentum it embodied and the passionate hungers to which it testified, despite its implicit imperative that when the Polish working class moved it would move first on the Baltic coast, things simply did not work themselves out that way. Perhaps, with a kind of inverse logic, it is proper that this was so. For all those who are skeptical of the presumed universal benefits of science, history has a wonderfully disorderly quality about it—inevitably so, given the unscientific character of humanity's social relations. Everything that the Baltic working class had taught itself was in place on the coast on July 1, 1980. It was present, if anywhere at all then in that tinderbox of working-class self-activity, the Lenin Shipyard in Gdańsk. And yet in July, in the Gdańsk Shipyard—nothing happened. The revolution knocked, and nobody answered. The Polish August did not happen in July.

They tried, of course But Szołoch from the 1970 presidium worked in another enterprise; Wyskowski had left the coast two years earlier; the Gwiazdas worked at Elmor; Wałęsa and Borusewicz had quit working anywhere and were full-time activists, but without credentials to enter the Gdańsk shipyard. Of the core group of the free unions, only the crane operator Anna Walentynowicz worked in the shipyard. Nevertheless, they told themselves, there were enough young militants. So they tried in July. But it did not happen—and it was the second failed mobilization of the year. In January, Walentynowicz and a 20-year-old activist named Andrzej Kołodziej had tried and failed. Kołodziej had been fired.[38]

What did happen in Poland in July was that Żerań had stirred and gotten a 10 percent raise, Ursus had achieved a huge mobilization, Łódź engineered a mass stoppage, and Lublin had pulled off a citywide general strike. Something over one hundred enterprises throughout Poland had confronted the authorities and had gotten some relief, though admittedly without achieving any structural breakthroughs. But while there had been rumblings in Szczecin, in Gdynia, and in Gdańsk, nothing happened. The coast waited.

So, on August 7, the party started it. On that day, supervisory personnel in the Gdańsk shipyard settled a score that had been gnawing at them for two years. They fired Anna Walentynowicz. It dated back to the summer of 1978 when Edward Gierek had honored the shipyard with a visit. A massive refurbishing and repainting operation had preceded the visit, and a number of functionaries had seized the occasion to install truckloads of new furniture for themselves. The concluding banquet for the first secretary was sumptuous, sumptuous beyond belief, beyond reason. As the shipyard rumor mill had it, the sustained foolishness over the visit of the first secretary had cost over a million złotys. When it was over, Walentynowicz had written about it in the *Coastal Worker*. It was an effective piece entitled "A Sincere Discussion Over a Piece of Meat." Not only did she make the shipyard administrators look like potentates, but they came through as groveling potentates. It was a critical tandem against which there was no plausible defense. So, the deeply offended functionaries reassigned her from her crane operator's job; she wormed her way back in; they harassed her; and finally on August 7, 1980, they fired her.[39]

The cranes in the Lenin Shipyard are huge. They tower over everything and can be seen from any point in a city of half a million people. Anna Walentynowicz had been handling one for as long as people could remember. Additionally, she was personally well known for her years of activism. She was, in short, a highly visible person in a highly visible job. When fired, she was just five months short of retirement, after having worked in the Lenin Shipyard for twenty-eight years. It was not the party's finest administrative moment.

The militants of the free trade unions met among themselves over the next two days and decided, at Wałęsa's suggestion, to go for it. They originally planned the strike for August 13 but moved it back a day to accommodate Wałęsa, whom they had agreed should lead the strike. Because Danuta had just given birth to their sixth child, he said he was needed at home. To avoid arrest, neither Wałęsa nor Borusewicz slept at home on the night of August 13. The logistical plans, while routine, were important in their details. The basic leaflet simply reviewed the history of Walentynowicz's career and of her firing. The simple facts could scarcely be improved upon to achieve any greater impact on readers. A movement printer in Gdynia ran off several thousand copies. Strike posters were also printed—to be stowed in the locker rooms of selected shop sections until the appropriate moment. The leaflets were to be distributed by long-tested means on familiar tram and bus routes and on the morning commuter from Sopot and Gdynia. The streetcar and bus drivers and the personnel on the railroads were all dependable. Henryka Krzywonos would lead the transport workers out in a solidarity strike as soon as it was certain the occupation strike at Lenin had settled in. Gwiazda and Lis would do the same at Elmor; Sobieszek, at Siarkopol; and Lewandowski, at Gdańsk Port. The leafletting teams were young in age but experienced in the crafts of the movement: Bogdan Felski, twenty-three; Jerzy Borowczak, twenty-two; Lucjan Prądziński, twenty-one. In due course, Wałęsa and Borusewicz would make their unauthorized appearance in the shipyard. With the July failure close in memory, it was to be a maximum effort.[40]

On Thursday, August 14, they pulled it off. The sequential details will never, in retrospect, be perfectly set in proper order. Like regimental reports from the American Civil War, the matter of which regiments led which divisions into the battle lines first is a bit clouded by the demands of ego and the pride that goes with organizing. K-5 was out early, and so was K-3 and K-1, but the mechanical and engine shops were full of militants too. At W-4, following Wałęsa's dismissal in 1976, Henryk Lenarciak had found another younger colleague—Stanisław Bury. Together, Lenarciak and Bury had things well under control in W-4. The first ranks of workers out of each shop joined with the others, and they marched the length of the yards, picking up workers by the hulls and the engine shops, and then back again as their ranks grew to thousands and the more cautious ones joined in. By nine o'clock, some eight thousand had assembled in front of the shipyard administration building. They chose a strike committee by a militant criterion, if not quite a democratic one: "We need people we can trust, who have authority in the shops, in the work brigades," said Jerzy Borowczak. "Let them come forward."[41]

The shipyard director, Klemens Gniech, an amiable functionary when he was not mad about something, put up a bold front but achieved nothing other than contributing to the general uncertainty. It was sometime before 10 a.m. when Wałęsa suddenly appeared on the improvised platform atop an escalator and confronted Gniech with an angry interruption: "Don't you recognize me? I worked in the shipyard for more than ten years. Did I steal anything? Was I a thief?" The sudden intervention, purposeful, authoritative, and angry, stunned Gniech and the surrounding throng as well. Though it left Gniech speechless, the workers quickly began to respond to the familiar story of blacklisting they knew so intimately. And of course, many knew Wałęsa as well. He was, in fact, the best-known activist in the shipyard. He went on: "It's been four years since I've had a regular job and I still consider myself to be a real shipyard worker." They cheered now, restoring his status by instant plebiscite as he demonstrated once again how to stand up to the party.[42]

As future events would prove, Wałęsa was not adept at giving prepared speeches. His natural milieu was the give and take of an informal setting. But his words that morning, while extemporaneous, had been well planned. He intended to lead the strike, and the decision was not a recent one, nor one merely assigned him by his fellow militants. Rather, it was a conviction that had come over him gradually in the past several years as he reflected on how many nuances of power he understood that others had not quite yet learned. He would lead because he knew the most. It was a settled conviction. "The one who acts the best takes the lead," he had once told Gwiazda. The performance with Gniech, in part, was to remind everyone of his shipyard credentials. That accomplished, it was time to be programmatic: "I declare an occupation strike. I have been given the trust of the workers. We are occupying the shipyard. We aren't going anywhere until we're sure we've gotten what we wanted. We're staying. This is an occupation strike. I'll be the last one to leave." It was vintage trade unionism, except for the last sentence. The final seven words came out of the experience of the movement. It was a matter of knowing what people needed before they would commit themselves. "I'll be the last to leave," was a talisman from one of the members of the collective at the moment of its re-formation. It was the sort of statement that journalists writing on deadlines would try to characterize and, in so doing, use words like "charismatic." Hyperbole aside, it was a statement that came out of experience. The power of the seven words lay in the fact that they were germane to the moment.[43]

If the Polish August can be seen as a route marked with one hundred milestones, Wałęsa helped everyone travel something like the first fifty of those milestones on the first day of the strike. He had climbed over the fence to make certain that the aimlessness of July would be overcome, that the proper moment should be seized and the strike declared in authoritative tones. But beyond this essential starting point, he had learned over the years—in the shipyard in 1970–71 and again in 1976, at ZREMB in 1978, at Elektromontaż in 1979–80, and in the free unions—that the key to solidarity was communications. Uncertainty and weakening resolve appeared when rumors appeared, and rumors came when people did not know what was happening. Wałęsa addressed this problem on the first morning.

His name had been added to the list of the strike committee that was handed up to him on the escalator platform. "Will you accept me?" he asked. "Even though I'm not a worker in the shipyard?" In the momentum of events, approval was automatic. The experiences accumulated over the years by the shipyard workers then produced a spate of programmatic suggestions that carried the strike to its first plateau of accomplishment. Someone in the crowd said that demands should be drawn up, and this process began. Someone else wanted to know where Walentynowicz was, and another worker moved that her name be added to the strike committee, an act that in itself might help restore her employment status. "Bring her to the shipyard," a voice added; "in the director's car," suggested another. While this was being done, Director Gniech and the strike committee moved toward their first negotiating session, whereupon Wałęsa made his move: "It must be done so everyone can hear—over the loudspeaker." Gniech was taken aback. A negotiation conducted over loudspeakers? It could not be done, he said. The room was not properly connected. Unfortunately, Gniech was talking to an electrical specialist. Over Gniech's continuing protests and with Wałęsa's promises that no negotiations could proceed without a connection, the necessary arrangements were made.

Thus in one negotiating sally, and with technological finality, Wałęsa solved the entire problem of internal communications in the shipyard for the duration of the strike. The most anxiety-ridden worker in the last rank of the shipyard would be as informed as the most active militant. In one stroke, the rumor factor had been reduced to an absolute minimum. If the worker leadership could now keep both its poise and its programmatic militancy, the entire work force would support them. They would be politicized by events. They always were—when they could get information on events. To anyone who understood the worker milieu on the coast, the incorporation of the loudspeaker into the movement was an important step toward binding the community.[44]

As all the world discovered in the next seventeen days, the Strike Presidium performed with resolve and sophistication throughout the long confrontation with the government. The strike leaders had to deflect a considerable amount of nervous advice from various quarters, but they did this, for the most part, with economy and dispatch. Stratigically more important was the fact that they had to demonstrate to the government that they knew precisely what they wanted, that they had their priorities in carefully considered order, that they were familiar with every government delaying maneuver, rhetorical ruse, and tactical deception, and, not least, they they could hold together the vast array of delegates from striking factories and the diverse rank and file of the shipyard. Above all, however, they had to stand up to raw power as the security service tried to destroy their organization by severing their communication links and as the party tried to sap their morale by controlling what the people of Poland heard about the struggle.

The thrust and parry had begun on the first morning when Gniech discovered that his conciliatory manner merely seemed irrelevant. With grace, he agreed that the strikers "certainly" had concerns that merited attention, but there seemed "little point" in bringing Anna Walentynowicz back to the shipyard. She was, after all, getting on in years and was in bad health. Wałęsa did not argue. He just shook his head. Finally, he said the negotiations could not proceed without her. Gniech could find no alternative to the acceptance of this demand. The workers had no cars; Gniech reluctantly agreed to send his. The woman he had fired a week earlier was going to return in style. It was around noon when the director's gray sedan brought Anna Walentynowicz through the main gate, inching through masses of cheering workers. She emerged a bit stunned and tried to piece together what must have happened to account for such an unexpected turn of events. A huge bouquet of flowers was presented to her as she was taken by a worker escort to the strike committee. She learned that she was a member of the Strike Presidium. It indeed seemed probable that she was going to get her job back. She surveyed the sea of workers who had done this thing for her and brushed back tears. The Polish August had begun.[45]

When self-activity moves from something that one person does—something that can be understood or misunderstood simply as a function of character or personality—to something done through the collective activity of a group, and when the resulting achievement extends beyond the reach of anything a single individual might manage, the personal relations of people within the group undergo a dramatic change. When individuals who have suddenly acquired an enhanced sense of self gaze upon others who have been a part of the joint effort and who have acquired the same enhanced sense, a new kind of connection occurs. The feeling is distinctly one of personal achievement, but it also is organically a product of collective action.

Beyond this, the new sensibility has an emotional significance that can yield surprising political breakthroughs. Scholars have long recognized this sensibility, or this bonding, but have experienced great difficulty in finding adequate words to describe it. The phenomenon can, of course, be abstracted and presented in a way that drains away most of its vitality. It can, for example, be described as the emergence of "class consciousness." In its vagueness, this abstraction unfortunately succeeds essentially in the way it conceals the social reality it purports to describe. Precisely what did Anna Walentynowicz feel standing in front of cheering workers that she did not feel in her apartment an hour earlier? A new sense of self, a new self-respect, and a new connection to those who cheered her? And what did the workers feel? Was there now some form of collective self-confidence, a tangible quality to fuel a new social action still to come? Is some sort of emotion being described? Perhaps what is visible here is also a new piece of knowledge to the effect that a united work force can restore lost jobs to crane operators. If so, then social bonding that produces political confidence is a merger of knowledge with emotion, a conjunction forged by lived experience. These ingredients took a tangible form when cheering workers escorted Anna Walentynowicz and her bouquet of flowers to the rostrum of the strike committee. Surely at that moment, she felt the connection more vividly than anyone else. But everyone there felt something. However it is described, it was the product of self-activity, and it constitutes the very soul of social movements.

In its grip, people do things they have long thought of doing but have not done because such actions would violate their understanding of sanctioned custom. It happened to Stanisław Bury that same afternoon. The citadel of electromechanical technicians in the Lenin Shipyard was W-4, Henryk Lenarciak's old home. Under Bury's guidance, the electronic workers of W-4 decided to mount antennas so they could pick up Radio Free Europe that afternoon. They had reason to believe, they told themselves with high good humor, that the Gdańsk shipyard would be in the news that day. The department manager appeared and inquired (though he doubtless did not use this terminology) just what all this self-activity was about. Bury explained that because Radio Free Europe was routinely jammed, the workers were erecting a tall antenna so they could hear more clearly. Bury spoke these words out loud where the ruling party's representative could hear them. The manager could find nothing to say. After that, everyone in W-4 felt a little taller—especially Bury. Together with another worker, Bury secured all the acetylene, gas, and electrical equipment so that no party hooligan could come in and set off an "accidental" explosion as a pretext for bringing militia into the shipyard. The same thought occurred to other coastal workers with long memories. A worker militia was formed, given distinctive arm bands, and put in charge of security throughout the shipyard. The strike committee also banned all alcohol from the shipyard. Experience—the occupation strikes of the 1970s—informed action.[46]

But experience was not confined to the shipyard. When word of the successful occupation strike was confirmed, militants throughout the tri-cities readied themselves for the morrow. On Friday morning, the Paris Commune Shipyard in Gdynia shut down under the leadership of the same 20-year-old Andrzej Kołodziej who had failed in his strike attempt in Gdańsk in January and been fired for his efforts. The other large port facilities came on line with an impressive display of solidarity—the repair shipyards and the longshoremen at Northern Harbor. Next door to the Lenin Shipyard, at the shipbuilding supply enterprise Elmor, Andrzej Gwiazda, and Bogdan Lis demonstrated what an elaborate internal organization developed over two years could do. Elmor workers simply came to work Friday morning and struck. Meanwhile, a large number of militants led by Florian Wiśniewski shut down Elektromontaż by way of paying their compliments one more time to the security police who had killed Tadeusz Szczepański. And ZREMB promptly came on line with its solidarity strike. With the memory of the coastal massacre driving him, as it had for ten years, Lech Sobieszek led Siarkopol out, and Stefan Lewandowski did the same at Gdańsk Port. In no case was this a private feat. Each "leader" was surrounded by militants whose actions were a product of prior experience. Measured by any of the standards previously established by working classes anywhere in the world, the Friday mobilization in the tri-cities was an impressive achievement in precision, order, and speed.

The city of Gdańsk had come to a standstill as the buses and streetcars stopped running. Henryka Krzywonos of the transport workers directed the logistics. It was the transport workers' second job action of the month, and the gears of internal communication ran smoothly. The first red and white flags began to appear from office windows. Generalized support for the strikers appeared among white-collar workers who, as witnesses to December 1970, had a certain built-in militance of their own. The display of the national colors expressed their sense of the connection. They and the workers were the nation; the party was not.

Both in Gdańsk and in Warsaw, party functionaries frantically tried to keep up with the pace of events. By the end of the second day, enough bits and pieces of fact and rumor had filtered out of the coast that the first foreign journalists began to look into the situation on the Baltic. Whatever was happening there seemed big.

As drawn up by the strike committee on Thursday afternoon, the workers' demands had acquired quite a bit of muscle beyond the simple matter of the rehiring of Anna Walentynowicz. Wałęsa's name had been added to the rehiring agenda, but beyond these personal items were matters of structural centrality: the formation of a union independent of the party and immunity from reprisals for all strikers. Also demanded was a 2,000-złoty raise, erection of a memorial to the victims of the 1970 massacre, family allowances for workers equal to those received by the police apparatus, abolition of the special retail stores for party members, earlier retirement, improvement in food supplies, and publication of the strikers' demands in the government media. Given the fact that each demand was necessarily circumscribed in scope by the necessity of being organic to the specific strike in the Gdańsk shipyard, the demands effectively opened up for review an interesting number of the structural components of the Leninist state: the confinement of the working class in the official trade unions, the ruling-party tradition of repressing working-class advocates, party censorship over social information, the special privileges of the governing elite and of the security apparatus, and finally, the low priority ascribed by the party to the material condition of the population, including food supplies as well as pay. It was not bad work for one afternoon of labor on the first day of an occupation strike. None of these postulates could be called "creative," a product of some late-forming consciousness. Each one was a product of experience; each one had been demanded in the past.[47]

But all was not effortless efficiency in the interior of the worker mobilization. On the third day of the strike, Wałęsa and his associates on the strike committee made a seminal error, and it almost wrecked everything. It came on Saturday morning, August 16. Since his opening joust with Wałęsa over the issue of the loudspeaker system on Thursday, shipyard director Gniech had learned a great deal about the maturity of the strike leadership and the overall dynamics in which he was caught. Soon after being forced to retreat on the Walentynowicz issue, he had been boxed in on the question of the monument. He had tried to explain that the area outside Gate No. 2 where the workers had fallen in 1970 had been set aside for a new supermarket. Strike committee members were unmoved by this piece of intelligence, and one of them walked up to the microphone and addressed the massed thousands listening in the shipyard: "Do you want a monument?" The thunderous answer not only reduced Gniech's maneuvering room, it impressed on him the extent to which the public nature of the negotiations robbed authorities of the strategic advantages normally adhering to power itself. After consulting with the Gdańsk party committee, Gniech was able to return with an affirmative answer; the workers would get their monument. But instead of mollifying strike committee members, the concession only seemed to embolden them. It was at this juncture that Wałęsa chose to emphasize the issue of new unions, adding the further admonition that the strike would not end until all their demands were met.

That night, the shipyard director gave his superiors in the Gdańsk party committee a sobering report on the temper of the strike committee. He did not like the dynamics. It appeared to him to be the kind of committee that near the moment of settlement might suddenly escalate demands. This was almost unprecedented in People's Poland. The party committee worked out a plan, and on Saturday morning, Gniech presented it to the strike committee. Before the party could feel secure about the reliability of any settlement that was reached, the shipyard director said, the unrepresentative nature of the strike committee needed to be corrected. He suggested that each shop section name four delegates, a move that would construct a very large negotiating group. Wałęsa and the committee could find no grounds for opposing this suggestion (it was democratic, was it not?) and reluctantly agreed; the party promptly and effectively used its muscle and its knowledge of the work force to elect a "soft" set of delegates. When they met, Gniech made a great show of first rejecting and then appearing to reconsider a 1,500-złoty a month raise. A number of new committee members engaged on this level—the 1,500-złoty level—and the entire negotiation seemed to lurch into an area of possible overall agreement. Wałęsa made an impassioned speech, stressing the necessity of a 2,000-złoty raise, of better food distribution, of rolling back the meat price increase, and—a sudden thought—of full compensation for workers during the period of the strike; above all, he reiterated the need for free trade unions.[48]

Gniech had been granted almost total flexibility by his superiors: do whatever necessary to end the stoppage at the Lenin Shipyard before the whole Baltic coast became the site of a vast occupation strike. Gniech agreed that Lenin Shipyard workers should have a free trade union; he assured them that the government was reconsidering the price increase; new attention was being devoted to food supplies, and he would prove his good faith by agreeing to the 1,500-złoty demand. It was adroitly done, and the majority of the new strike committee went for it. The formal vote had the effect of ending the strike. Gniech pointed out that it was now Wałęsa's turn. As chairman of the Strike Presidium, it was his duty to announce the end of the strike. Unhappily, Wałęsa cast about for some expedient but could find none. The workers had voted, and he was bound by movement experience but also by principle and simple logic to respect their wishes. He did the best he could to preserve things for future struggles: "We have done a great thing," he announced, "but we shall do still bigger things for the good of the shipyards and our motherland."[49]

The audience immediately around him was sprinkled with the most anxious workers in the place—delegates from all the smaller and therefore weaker work forces in the striking factories of Gdańsk. They would be crushed if the shipyard workers settled. Henryka Krzywonos of the Gdańsk transport workers, a militant who, like all the coastal militants, knew Wałęsa, said: "You can't do it, Lech. If you abandon us, we'll be lost. We can't fight tanks with buses." She was aware as she spoke that both of them understood that the entire city of Gdańsk had been paralyzed by the solidarity strike of the transport workers in support of the shipyard. Her moral authority at that moment was overwhelming. Her point, in any case, was instantly reinforced by the other delegates of striking factories who surrounded Wałęsa. Here was a way out, and Wałęsa took it with dispatch. It was, of course, a thoroughly undemocratic way out, though he found a way to exonerate himself as a person if not as a worker representative. Speaking in a loud voice he said:

> We must respect democracy and therefore accept the compromise, even if it is not brilliant; but we do not have the right to abandon others. We must continue the strike out of solidarity until everyone has won. I said I would be the last person to leave the shipyard. And I meant it. If the workers who are gathered here want to continue the strike then it will be continued. Now, who wants to strike?[50]

This was the most awkward moment in the Polish August, so awkward, in fact, that contemporary chroniclers of the event in newspapers around the world and in subsequent scholarly articles and books have stepped gingerly around its implicit undemocratic dimension. And with understandable reason. Wałęsa's inquiry "who wants to strike?" could have only one honorable answer, for it was the ultimate in loaded questions. The moral authority of the delegates from the small supporting factories created an unresolvable political dilemma: to honor the long-term democratic objectives of coastal workers as a whole, Wałęsa would have to ignore the short-term democratic objectives of the officially elected shipyard committee—objectives that had been affirmed in an honest vote.

All movements at some juncture are forced to confront contradictions of this kind. For understandable reasons, such moments tend to be immobilizing. Things either fall apart or, to avoid that, a potentially crippling deception is introduced into the structure of advocacy the movement represents. At such decisive junctures, its helps if the movement in its natural evolution has genererated indigenous leadership possessing deeply internalized, long-term strategic objectives. The Baltic movement had. Wałęsa's reaction to Krzywonos was prompt and his explanation to the group almost effortless. It also logically stated the democratic options about as well as they could be stated. Of course, all who wanted to strike shouted their approval, and all who wanted to go home were silent. The politics of the situation impelled both. And so, for the moment, the strike was saved, and the mobilized movement was—however shakily—intact.

The difficulty, however, was that all who remained silent promptly began voting with their feet to go home. From the standpoint of the militants, it was even worse than that. All those who had not heard Wałęsa's second speech and did not know the options were also going home. In an instant, it seemed, the party possessed all the cards. The moment Wałęsa's announcement ending the strike was completed, Gniech shut off the loudspeaker system. He had suffered all the democracy he needed. He then hurried back to his office and picked up another microphone. While Wałęsa was making his second speech to the comparatively small group of workers who could hear him, Gniech went on the shipyard loudspeaker to repeat at regular intervals the announcement ending the strike. Workers began leaving in droves. They had won a 1,500-złoty raise and some sort of free union. Their own spokesman had said they had done "a great thing." The loudspeaker said it was over. It was late Saturday afternoon, they had not been home for three days, and they had a lot of things to do. They left.

The militants did all they could to stop the flow. Alina Pieńkowska made an impassioned speech at Gate No. 3 and managed to stem the tide somewhat. Afterward, she joined Wałęsa and a Young Poland activist, Ewa Osowska, riding through the twilight on an electric trolley, exhorting the workers to disregard Gniech's announcement. Through these and other efforts, the militants were able to hold the structure of the strike action intact. Some thousand or so workers stayed the night.[51]

The heritage of coastal workers now exerted a decisive influence. The militants of the Lenin Shipyard and the activists presenting the twenty supporting enterprises in Gdańsk gathered that evening and formed an Interfactory Strike Committee. This broad-based entity—one that insured lateral lines of communications within the coastal working class—grew logically out of the very structure of the collective self-activity that Gdańsk workers had achieved. Having cooperated closely for two days, the representatives of the twenty affiliated enterprises formally organized themselves as a single umbrella structure, one designed to coordinate the strike as it spread. They had informally created their own interfactory structure in the course of fashioning the strike; now they gave it a name. Both moves—what to do and what to call it—came naturally out of the coastal heritage.

Composed as it was of delegates selected for their prior experience in leading job actions in their respective factories, the presidium of the new committee was laced with militants. In addition to Wałęsa and Walentynowicz, Andrzej Gwiazda and Bogdan Lis came on board representing Elmor, Joanna

Duda-Gwiazda appeared for Ceto, Lech Sobieszek logged in for Siarkopol, Henryka Krzywonos and Zdzisław Kobyliński for the transport workers, Florian Wiśniewski for Elektromontaż, and Stefan Lewandowski for Gdańsk Port, among others. In the aggregate, the committee represented something over one hundred years of working-class assertion on shop floors and in society. The lesson of 1970 loomed large—very large—but the lessons since then were important, too. Working into the early hours of Sunday morning, the newly formed MKS drafted sixteen points and on Sunday afternoon and evening completed what the world would know as "the twenty-one Gdańsk demands." It featured, of course, the first demand that would result in the creation of Solidarność: "Acceptance of Free Trade Unions, independent of the party and employers in accordance with Convention 87 of the International Labor Organization concerning trade union freedoms, ratified by the Polish People's Republic." After thirty-five years of postwar struggle and a decade of sustained activism on the coast, the first demand was scarcely a subject of prolonged debate.

The Gdańsk demands of the Polish August reflected the central structural objectives set down by coastal workers in their demands of 1970 and 1971. The relationship was direct and causal, for the memory of 1970–71 was both emotionally powerful and ideologically instructive. The demands of the Lenin Shipyard—drawn up during the occupation strike on the fateful December 16, 1970, when the first workers died outside Gate No. 2—had specified that "trade unions on all levels should be made up of nonparty members." Similarly, the postulate of Gdańsk Port in 1970 had read: "Trade unions should be nonparty and must support the working class." These were not isolated expressions. Among Baltic workers in Szczecin and Gdynia as well as Gdańsk, the prominence of the demand for independent unions was massively self-evident. It also was absolutely controlling. Ninety-eight sets of demands had been collected from coastal enterprises and shop sections during the demand-gathering campaigns of 1970–71, and by far the most important structural change sought by a substantial majority of coastal work forces was trade unions independent of party influence. The leading authority on the coastal crisis of 1970, the political scientist Roman Laba, conducted a comparative study of worker demands from 1970 to 1980 and was struck by "the continuity of the demand for free trade unions." Laba concluded: "The workers' devotion to this demand is so consistent that, in reading the documents, it sometimes seems as if nothing changed in Poland between 1970 and 1980. On this demand, there is no evolution, no turning point."[52] In essence, the experiential knowledge of 1970–79 provided the structural and ideological basis of the Polish August of 1980.

But more than the first of the twenty-one demands of 1980 represented the aspirations of 1970–71, however centrally important that demand demonstrably was. A clean way to cut through the blanketing censorship imposed on the working class by the official trade unions was proposed in the 1971 demand of the Lenin Shipyard that not only called for unions independent of the party "on all levels" but specified the need for "the publication of an independent newspaper of the trade unions." Similarly, the 1980 demand for media publication of full information on the formation of the MKS and details of its demands reflected the 1970 insistence on "official publication of workers' demands." Both in 1970 and in 1980, the workers demanded abolition of the special privileges of the party, militia, and police apparatus; immunity of strikers from reprisals; and pay for strike periods. Clearly, the drafting hand of the working-class heritage was evident in the Polish August.[53]

These historical imperatives helped to put into perspective the contributions of the intellectual community to the Gdańsk demands. They were three in number, visible in one demand and as adjuncts to two other demands. The sixth demand called for the government to take "real steps to extricate the country from the crisis situation by providing full public information on the socioeconomic situation" and "giving all social groups and strata the possibility of participating in discussions on a reform program." In addition to the workers' historic demand for the restoration of the rights of strikers dismissed from work for the crime of publicly articulating worker grievances, the fourth Gdańsk demand contained a clause calling for the restoration of rights of "students expelled from universities for their convictions." This clearly was a contribution from the intellectual community. (It also reflected Bogdan Borusewicz's personal history.) Finally, in the second demand that the government live up to the constitutional guarantees on freedom of speech was a call for freedom from repression of "independent publications." The existence of such publications was a product of the opposition movement from 1976 to 1980; there were no "independent publications" to be defended in the Baltic demands of 1970–71. So this clause, too, can be seen as a product of the efforts of the democratic opposition rather than of the workers' movement.

But none of these contributions from the intellectual milieu were remotely as important as the package of demands inherited from the original strike grievances of August 14–15, which, in turn, were traceable to the Baltic heritage of 1970–71. When the historical record of the coastal working class is set down beside the detailed history of the democratic opposition, the accumulated evidence is overwhelming: like the Baltic movement itself, the Gdańsk demands of the Polish August had their origins in working-class self-activity.

Before turning to the evolution of the democratic opposition, perhaps one aspect of the Polish August may be commented upon by way of illuminating the centrality of the workers' movement to the workers' movement. In the months and years following the Polish August, KOR enthusiasts zealously pushed on Western journalists and on scholars the "influential" role of the December 1979 "Charter of Workers' Rights" in "shaping the consciousness of workers" for acceptance of the Gdańsk demands. This claim will be subjected to historical scrutiny in the next chapter, but for the moment, it is probably much more historically justifiable (though doubtless just as inappropriate) to say that the working-class demands of 1970–71 were "influential" in shaping the consciousness of KOR intellectuals for acceptance of the "Charter of Workers' Rights" in 1979. An obvious problem of the claim of the intelligentsia is that KOR's worker-oriented left wing did not respond well to the first Gdańsk demand when it appeared on August 16; nor did the organized intelligentsia when it arrived in Gdańsk and began lobbying on August 23; nor did the unorganized intelligentsia, which, beyond having opinions, did not play a role in the confrontation. Whatever the Charter of Workers' Rights did or did not achieve with workers, it clearly had little impact on the intelligentsia or on KOR itself. As it had for thirty-five years, the Polish intelligentsia saw politics as pivoting centrally around the censorship, not around an organized popular constituency that could protect such freedoms when they were rhetorically granted by the government. This latter goal, as the pages of *Robotnik* made generally clear and as Jacek Kuroń himself made specifically clear, had to be approached gradually.

These fundamentally differing emphases between workers and intellectuals had their origins in the instructive potential of daily life—or if one prefers, in one's relationship to the means of production. For intellectuals—novelists, economists, poets, journalists and scholars, people whose life gained meaning by the expression of their understandings of life—the crippling moment came with the censor's blue pencil. Things ended not at the typewriter but at what the typewriter produced; creativity stopped dead at the party's censorship office. But for workers, censorship came at the very point of production itself—on the shop floors where incoherent production relations made for waste and inefficiency, where party disdain made for dangerous safety conditions, where worker creativity about organizing production ran afoul of "the plan" or the prerogatives of those in charge of the plan. There was no "bottom drawer" where workers could hide their creative thoughts from the party. Their very creativity itself was consumed by the blanketing control that the party's trade union ruthlessly imposed on the shop floors of Poland. It was for precisely this reason—verified by the experiences of daily life—that workers understood the censorship in structural terms that went beyond the understanding of the intelligentsia. This was the reason workers understood the centrality of the first Gdańsk demand in ways intellectuals could not. It represented the very essence of their struggle for free expression.

But no element of the Polish August was more difficult for the world's intellectuals to accept than this central causal relationship. To those on the left, the fact that workers could at some moment "know more" or have a "higher consciousness" than intellectuals, seemed to violate all they had internalized about the proper role of intellectuals in politics. To those on the right, the same fact seemed to violate the most elemental presumptions of social class. To both groups, the Solidarność experience seemed to define "politics" itself in a way that did not harmonize at all with received understandings of the functioning processes of social life—either in the world they lived in or in the world some of them longed to create. Nevertheless, sustained inquiry into the interior of the democratic movement in Poland, both within the working class and within the intelligentsia, would yield compelling evidence that made boldly clear these political and cultural contradictions. Either received conventions of analysis would have to be readjusted, or Solidarność would remain a mystery.

What the Polish movement exposed was both the vitality and the innovative potential of a self-constructed civil society when the people occupying it were in effective communication with one another—that is, when democratic social relations had been licensed among disparate people. Within a voluntary social formation responding to such democratic imperatives, long-cherished assumptions about historical causality, narrow and confining as many of such assumptions were, melted away. Both in terms of interpretive logic and in terms of social theory, these practical and theoretical traditions simply could not stand in light of the social evidence that appeared. The entire fifteen-month life of legal Solidarność would lay before the world the dynamics of popular democratic politics with a clarity that made these organic contradictions awkwardly visible.

The first thing that became clear was the variety of the sources of innovation available to proto-democrats who had freed themselves from the grip of formulaic thinking. The twenty-one Gdańsk demands were not a product of a tiny cabal working in a smoke-filled room on a long Saturday night. The rebirth of the interfactory strike committee of 1970 represented the creation in the working class of a kind of popular forum that offered democratic space to anyone who cared enough to occupy it. As it happened, a number of opposition militants of the coastal movement desired fervently to occupy it.

They were all there in the Lenin Shipyard on this historic Saturday night—ROPCiO and Young Poland and even an activist or two who, like the more passionate advocates of KPN, nursed as their most cherished immediate hope the complete sovereignty of Poland as a nation. So constructed, the meeting was a classic confrontation between realpolitik and what individual intellectuals thought was real in politics. Tadeusz Szczudłowski of ROPCiO was a passionate democrat who had demonstrated his willingness to pay for his beliefs with police beatings and jail terms. He saw politics as a set of formal rules, the attainment of which was seemingly a product of demanding them publicly. Therefore, as if the Leninist state did not exist, Szczudłowski proposed the total abolition of the censorship. Before this appealing suggestion could get very far, Bogdan Borusewicz, a worker activist with KOR connections, injected a bit of history: "you know what happened when they abolished censorship in Czechoslovakia in 1968?" The enthusiasts quieted down a bit at this sober reminder, and Borusewicz drew a black line through the proposal. Another middle-class militant had his eye on the legislative process. He wanted free elections. Aleksander Hall of Young Poland intervened to say that it was not strategically wise to destroy all the party's maneuvering room. So that suggestion, too, got a quick black line. In the political environment pervading the Eastern bloc in 1980, the popular movement simply was not yet strong enough for such ambitious goals.[54]

It would be wrong to suggest, however, that the impassioned political dialogue in the shipyard that night—the first public democratic debate in Poland in over forty years—was a cacophony of extraneous and fanciful yearnings. The majority of people in the room understood from experience that democratic prospects in Poland did not at the moment extend one millimeter beyond the strength of the organized working class itself. The sheer numerical weight of the eighteen-person worker presidium and the multiplied years of shop-floor militance they collectively represented insured that a certain strategic realism would be present in the room. Because the meeting constituted a self-generated parliament of the working class, the Interfactory Strike Committee could, if its organizers were so inclined, have tossed all of the representatives of the opposition groups out of the room and proceeded to write their postulates without benefit of middle-class advice. But as Wałęsa's own organizing career had vividly demonstrated, such was not the self-conception working-class militants had of their movement. They saw the Interfactory Strike Committee as the entering wedge into Leninist Poland for the benefit of the whole population. The objective was to pressure the party to come to democratic terms with society and thus transform the style of governance in the country.

Men and women who had spent their working lives under the shop-floor tyranny of the official trade unions and who had been harassed and even deprived of their jobs because they dared to breach the censorship by specifying the duplicity of the party's unions knew in their bones that the forthcoming struggle with the state turned on finding ways for the MKS to fashion as much institutionalized protection for itself as was strategically possible. Internally, the independent union movement needed to solidify its fragile mass base in the Lenin Shipyard when the working day began on Monday; in the meantime, it needed to augment as rapidly as possible the number of enterprises on the Baltic coast that could generate shop-floor meetings and elect delegates to serve on the Interfactory Strike Committee. In ideological terms, it needed to be understood that anything that worked at cross purposes to these internal prerequisites—whether free elections or free expression at police stations—was conceptually counterproductive. Even the most self-important spokesman for an opposition grouplet had to be conscious of the strategic orientation imparted by the simple location of the meeting room—in an industrial enterprise protected from a police raid by organized workers conducting an occupation strike just outside the meeting room.

What would happen later in Poland when this environmental influence was not so apparent was anyone's guess. Should the workers prevail in their effort to create a substantial arena of democratic social space in Polish society, one the entire population could enter, it was an open question as to how long these central strategic understandings would remain predominant. There was not much in the daily comings and goings of middle-class life—either in the white-collar bureaucracies or in the universities—that habituated people to think in egalitarian terms or to hone their capacities for structural analysis of democratic social relations. If ROPCiO's Szczudłowski was an example of middle-class preoccupations—and he surely gave voice to a number of common presumptions as to what constituted the priorities of democratic politics—the organized workers might soon face an unexpected and frustrating imbroglio with the opposition sects to array alongside the decisive struggle that loomed with the police-empowered functionaries of the party. Time would tell. For the moment, the twenty-one demands hammered out that Sunday night were both germane to the future course of the workers' movement and sensible in terms of Poland's geopolitical situation.

The original shipyard demands of two days earlier served as a firm basis for the night's work, of course. Beginning with the demand for free trade unions, they neatly codified the heritage of 1970–71. Borusewicz and the Gwiazdas worked well with Wiśniewsksi, Kobyliński, Krzywonos, Sobieszek, Lis, and Wałęsa in fashioning a well-conceived set of postulates. In sixteen words, the second demand built neatly upon thirty-five years of worker experience with police abuse: "Guarantee of the right to strike, and the security of strikers and persons who help them." The latter phrase "who helped them" acknowledged the appearance of opposition groups in Poland in the past four years; prior to 1976, there had not been anyone beyond the workers themselves who had "helped." The third demand wrestled with Szczudłowski's preoccupation; it was a judicious attempt to impart genuine meaning to the existing constitution on the matter of "freedom of speech, print and publication" by specifying an end to "repression of independent publications." The third demand also smuggled in the question of religious freedom by calling for "access to the mass media for representatives of all denominations."

The only structure that could guarantee any of these basic civil liberties, of course, was ratification of the first demand for independent unions. Without that protection, everything else would collapse. The third demand also had the merit of being a bit more discrete, especially in the sensitive area of Soviet-Polish relations, than ROPCiO's blunt call for an "end of the censorship." But if the third demand was pleasantly general in its impact, so were most of the rest of the demands. Because they dealt with food distribution, health, personal security, old age security, and housing, the demands reflected not only the hopes of workers but those of all Poles. It was precisely this reality, of course, that provided the theoretically justifying substance behind the idea of a popular movement grounded in the nation's workers. In directing attention to their own needs at the bottom of society, the activists of the working-class movement insured an egalitarian thrust to the agenda for all Poles. And it was precisely this reality that made the creation of Solidarność such a milestone in mankind's democratic development.

At the moment in the Lenin Shipyard on August 16 when worker self-activity moved from a local occupation strike to a regional mobilization under an Interfactory Strike Committee, the weight of this long experience rather effortlessly asserted itself, as it had on the coast ten years earlier.

It would be inappropriate, however, to leave the impression that worker activists were the sole source of creativity during the exhausting drafting session on Saturday, August 16. The chief benefit of free debate, of course, is that good ideas can come from anywhere. It was the relentless Szczudłowski of ROPCiO who came up with the suggestion that a wooden cross be made to honor the dead of 1970, a cross that would be blessed and dedicated by a priest when he came to say Mass on Sunday morning. The eyes of every creative organizer in the room lit up at the suggestion. Mass on Sunday morning in the shipyard centered around the dedication of a cross to the victims of a state-sponsored massacre? Of course! It would be the ultimate in monument politics, sanctioned and carried out by the Church itself—and in support of an occupation strike for free trade unions. Absolutely! Borusewicz got busy on the suggestion at once, as did a coterie of Young Poland militants. Unfortunately, Father Henryk Jankowski of St. Brigida's did not feel he could conduct a Mass on state property after the strike had been officially ended. A flood of piety overwhelmed the militants. Did not the priesthood realize it was a mortal sin to miss Sunday Mass? Jankowski, wavering, said he would need higher authority. Such was not immediately forthcoming. The activists did not give up, however. They poked around in the Catholic hierarchy, pressing and cajoling, until finally the bishop of Gdańsk, 71-year-old Lech Kaczmarek (the memories of 1970 vivid in his mind) agreed to present to party authorities the proposition that bringing the strikers under pastoral care on Sunday might have a settling effect on the situation. The coastal party, which had been as shaken by the results of Zenon Kliszko's militarism in 1970 as everyone else, agreed.[55]

By this circuitous route, the Church in the personage of Henryk Jankowski entered upon the Polish August. As a religious service, Mass that Sunday was merely a bit out of the ordinary, but as a political event, it was singularly helpful to the workers' cause. The legitimacy of the strike, in question since Wałęsa's double announcement of the previous afternoon, seemed a bit restored amid the flowing vestments of the tall and almost imperially resplendent Jankowski. Some five thousand or more took part in the open air ceremony at the shipyard, including two thousand or so townspeople who stood just outside the gate. The world saw photographs of Polish shipyard workers kneeling to take Communion while engaged in a struggle against the state. The image was both striking and puzzling. Non-Poles were forced to ponder the meaning of a militant occupation strike against the state being blessed by the Church . . . at services blessed by the state . . . at which a cross honoring workers killed by the state . . . was blessed by the Church. If outsiders tried to ascertain whether the strikers were radical socialists or pious churchgoers, they had an equal right to wonder to what extent the communist party had endorsed the Catholic liturgy and to ponder the depth to which the Church of Rome had become infected with some form of syndicalism. Only in communist Poland could the Sunday morning scene in the Lenin Shipyard on August 17 have taken place. In the era of Brezhnev, one can guess what the Soviets thought.

At the meeting of the MKS that first Sunday evening, the strikers took stock. They had a powerful twenty-one-point agenda, though it rested on a shaky base of one thousand shipyard strikers and a thin array of cooperating factories. The big test in the shipyard could not be addressed until the first shift started Monday morning, but the matter of recruiting the rest of the coastal working class was a topic of immediate relevance. People needed to be informed about how to select delegates to serve on the MKS! Orders were dispatched to all underground printers to get to work on copies of the twenty-one demands and on leaflets detailing the structure of the MKS plus the urgent need for all factories to organize immediate solidarity strikes and elect delegates. But all telephone and telex facilities had been shut down by the government on Friday morning; not only was the coast effectively closed off from the rest of the country but each coastal town was itself isolated from every other. Scores of couriers were needed and vehicles to carry them.[56]

But the really crucial signals that first Sunday night were sent by fifty thousand striking workers from two dozen enterprises in Gdańsk to the thousands of other workers who lived around them. The first leaflets were out, and they helped, but even better conduits were available. The city was alive with stories of the formation of the MKS and also stories of the Sunday Mass in the shipyard which seemed to have brought a measure of sanction to the entire effort. There were even more pointed developments to ponder. The shipyard had stood by the transport workers—had stood by everybody in fact. Now clearly was the time for all other coastal workers to respond by presenting a united front to the state. That night, militants in virtually every uncommitted factory, assisted in many cases by militants from factories already striking, made plans for Monday morning.[57]

In one sense, the entire environment of the coast during the last half of August constituted one long crisis. But within this general context, heightened moments of crisis occurred. The most critical began Saturday afternoon, with the near collapse of the strike itself, and extended through the opening hour of the workday on Monday. As many militants saw matters, Monday morning was the decisive moment of the entire strike. Either some of the work force went back on the job, thereby undermining the basic integrity of the strike and placing a stamp of confirmation on the Saturday settlement, or the entire work force joined the occupation strike and thus solidified the movement for the duration of the confrontation.

Over the weekend, the government media had steadily chipped away at the integrity of the strikers, implying that they had forcibly restrained honest workers who wanted to go home and suggesting in a dozen ways that the holdouts were the flotsam of the shipyard, the loafers, malcontents, and hooligans with which any society had to contend. On Monday morning, the voice of shipyard director Klemens Gniech greeted incoming workers over the loudspeakers even before the gates were open: "On Saturday we ended talks with the representatives of all shipyard workers. We came to a full agreement. Let's remember that we and we alone are responsible for the shipyard and for everything that happens here." The message was repeated every few minutes as the work force filed up to the gate, strikers on the inside, the remainder outside.

Wałęsa suddenly appeared, climbing up where all could see him. Many things had happened to him during his years of self-activity, and he called on that experience now. He surveyed the crowd and then, alone, in his loud distinctive voice, began singing the national anthem, "Poland Has Not Yet Perished." The workers joined in on both sides of the gate. With a measure of solidarity to build upon, Wałęsa then reviewed the shipyard director's claims, refuted them briefly, and moved to the central issue—the situation of all the striking workers throughout the tri-cities. The strike, he said, was for the defense of all workers of the coast and, indeed, of Poland. Turning to the throng waiting outside, he said: "Don't hesitate to come in. We have to fight for what is rightly ours. Come to us, shipyard workers. There's nothing to be afraid of."

After all the counterclaims and anxieties of the weekend, Wałęsa's surprisingly short but emotionally effective speech was simply a quiet call to the working class to come together as one. There was a pause—the gates were open now—and some strikers called to the others to come on in. A vanguard of mostly younger workers came, to cheers of encouragement; then they all came, to prolonged cheering. Klemens Gniech took one look, retired to his director's office, and was scarcely seen for the duration of the struggle.[58]

They were together. The movement once again had its mass base in the shipyard. And the MKS was protected against the militia. Nothing short of tanks could now break the strike. A vital step had been taken.

The Polish party and perhaps all Leninist parties learned that Monday that it is impossible in an industrial society to close off entirely a politicized working class that possesses a compelling need to communicate with itself. Beginning early in the morning, trucks rumbled to the Lenin Shipyard bearing building materials, prefabricated parts, electrical assemblies, and the thousands of other components required by modern ships—components built by the supply enterprises ringing the Baltic shipyards. Drivers left carrying dozens of vivid scenes in their minds created by the spectacle of an occupation strike in the largest shipyard in Poland. In the immediate Gdańsk area, information flowed essentially along a network of workers. But up and down the coast away from the tri-cities, worker couriers were augmented by couriers from the Gdańsk opposition—militants from ROPCiO, KOR, KPN, and especially from Young Poland. But also involved were people who were stirred by the drama and importance of the moment and whose one connection was the underground printer who had run off strike leaflets or the driver of the vehicle that was carrying them. Sunday evening and all day Monday, they fanned out across northern Poland.[59]

So it was that by mid-morning on what can retrospectively be understood as the decisive Monday of the Polish August—the day of consolidation of the movement—another group of trucks began lining up at the Lenin Shipyard. They did not carry maritime supplies; they carried delegates representing newly striking factories arriving to take their place on the Interfactory Strike Committee. The entire shipyard that Monday began to take on the appearance of a theater being readied for opening night. Banners went up; posters were made; tents and tables were set up; and mattresses, kitchen utensils, and food were hauled in literally by the truckload. At the gates, official welcoming committees quickly got into the spirit of things. Each incoming vehicle carrying MKS delegates was announced with studied formality over the shipyard loudspeaker, which, like everything else within view, had become the property of the workers. "Mr. Tadeusz Kowalski of the Mopot Furniture Company, Mr. Ryszard Kochanowski and Mr. Anka Smolar of the Gdynia Metal Bearing Combine . . ." An approving foreign journalist reported: "Each new arrival was announced with the flourish that might have been accorded by a tail-coated footman to a Count Potocki or a Prince Radziwill."[60]

Of course the occasion in the shipyard had a historical meaning no court ball ever attained. By early afternoon on Monday, August 18, the count of affiliated factories had passed 100. By nightfall, it had reached 156. No one knew or could find out the precise number, but probably a quarter of a million coastal workers had affiliated themselves with the Gdańsk MKS by nightfall on August 18. The barrier to lateral communications had unequivocably been breached, the historic censorship of the official trade unions shattered, and the regional organizing imperative attained.

In all the dimensions that bore on movement building, the enduring historical barriers to large-scale self-organization in Poland had been overcome. Every basic objective had been achieved: a massive occupational strike at the shipyard, instantaneous lines of interior communication by loudspeaker to the entire work force of fifteen thousand strikers, the creation of a poised and experienced MKS, a broad structural agenda led by a demand of vital centrality, and a far-flung mobilization of regional enterprises so that the strike weapon of the MKS could be brandished with genuine authority and stark economic power. The process of self-education through self-organization had continued to the very brink of the strike in the shipyard: Henryka Kryzwonos and Zdziław Kobyliński had led a dramatic stoppage of the Gdańsk transport workers as late as August 2, and the solidarity strike of the same workers with the shipyard had been effectively organized on August 15. There was not a single activist in any enterprise in the tri-cities who did not have ready access to the organizational details needed to elect delegates and affiliate his shop with the Interfactory Strike Committee.

It was this thick web of information, transmitted by and received by knowledgeable activists in many scores of enterprises ringing the Bay of Gdańsk, that lay behind the gigantic mobilization of August 18. All members of the MKS presidium understood the centrality of communications, not only through loudspeaker with the thousands of shipyard workers whose occupation strike protected the committee but also with the tens of thousands who had affiliated with the MKS and with the hundreds of thousands beyond the tri-cities who still needed to be contacted and briefed. With the phone lines down, there was much work for couriers to do, though the solid base of the mobilization was now beyond question.

Since the end of World War II, no working class in the Soviet bloc had been able to travel the long and perilous road that carried Baltic workers to the Polish August. All the pitfalls that had spelled defeat in Poland in 1945–47, 1956–58, 1970–71 and in 1976 had, this time, been overcome. In five days of purposeful effort, one moment of frantic improvisation on a Saturday afternoon, and an extemporaneous appeal on a Monday dawn at a shipyard gate, Polish workers had been able to draw on the storehouse of experiences their own efforts had generated. They had passed every test that had suddenly appeared during the five days. For the moment at least, the organizing crisis in the Polish working class was over. As a direct consequence, the crisis of the Polish state had begun.

The final giant stride to this strategic plateau came on the fifth day of the strike—before anyone in Warsaw or anywhere else in the world had caught more than a hint of the organizing achievement. On Monday, August 18,

1980, in Gdańsk, Poland, there occurred one of the great mobilizations in the history of the world's industrial working classes.

Ironically, the achievement would pass virtually unnoticed in the mountainous literature that would soon begin to accumulate around the phenomenon called Solidarność. All the experientially grounded methods of internal communication that coastal workers had fashioned so arduously over so many years—without which the events of August 18, 1980, could not have occurred—would disappear under a global cultural presumption that mere citizens (and workers at that) could not conceptualize and consciously carry out such an achievement.

The demand for free trade unions, though faithfully tracking the demand made in December 1970 and again in January 1971—a demand made, reflected upon, and made again—would be seen not as the logical culmination of Polish workers' long and often bloody combat with the state-owned trade unions but rather as an import smuggled into the minds of workers by distant intellectuals in Warsaw. Nor would the astonishing mobilization of August 18, 1980, be seen as the capstone of years of oppositional labor to develop internal communications links—new forms tested in oppositional experiences in many workplaces and integrated into a multi-factory structure of assertion. Rather, credit for the mobilization would go to a distant radio station, Radio Free Europe. In time, other causative analyses, also unencumbered by research, would uncover an even more elemental dimension—the Pope's visit to Poland in 1979, which could be seen to have fashioned a kind of moral rebirth and a mass public bonding in Poland that produced what was called the "climate" that made the Polish August possible. In these crude ways, deeply sanctioned cultural beliefs and social suppositions would shape a collective view from afar that would seriously minimize a number of extremely significant Polish democratic achievements. The wonder of it all was that so many in Poland and abroad who repeated these fanciful and partisan clichés had no wish to minimize anything. In fact, a great many possessed a palpable hunger for more democracy in their own countries.

As has often been said, history is not "what happened," it is what people persuade themselves happened. Deeply sanctioned political traditions shape this process of persuasion, directing attention to certain kinds of events and away from other kinds. Class presumptions and individual psychic needs harness received habits of observation (in the process warping them in unpredictable ways) so that harmonious and reassuring explanations of causality may be formulated. Nothing is more historically certain than this intricate tension between idea and action, between event and its perception, and finally and most politically galvanizing, between the reality of causation and the future political uses of partisan explanations of causality.

By nightfall of August 18, 1980, the Interfactory Strike Committee of the Lenin Shipyard had organizationally put itself into position to speak for 200,000 or more Polish workers in behalf of a cherished dream—the creation of a civic space of their own that was structurally independent of the ruling Leninist party. Soon, well-meaning intellectuals would come to the shipyard and oppose this bold aspiration on the grounds that it was politically beyond reach. Soon, the Primate of Poland's Catholic Church would tell the workers to stand down. Soon, the leaders of the democratic opposition in Warsaw—the leading theoreticians and organizational tacticians of KOR itself—would also tell the workers to back away from their first demand. None of this earnest counseling would emanate from an experientially grounded understanding of the coastal workers' organizing achievement.

In consequence, during the two weeks of crisis that remained, KOR intellectuals and Poland's Primate were to repeatedly illustrate their own lack of any political grasp of the transformed dynamics of power on the Baltic coast. They remained in the grip of conventional political expectations because they remained physically distant from the coastal environment in which the inherited basis for such constrained expectations had been organizationally shattered. In revealing contrast, the academic advisors who came to the coast and experienced firsthand the coastal mobilization, quickly laid aside long-held understandings about the state and society and, newly instructed by what their eyes beheld, took station in support of the worker agenda.

In this highly energized crisis environment, Radio Free Europe continued, as it had virtually every day since July 1, to broadcast news items about industrial unrest in Poland. Though the distant news reporters of Radio Free Europe had no more grasp of the structural dynamics underlying the coastal mobilization than KOR's partisans in Warsaw did, they were able to inform their audience in Poland that something big appeared to be occurring on the Baltic coast. They were even able to report that an "Interfactory Strike Committee" had been formed in Gdańsk, Szczecin and Elbląg, though it was clear from the absence of organizational detail that the newsmen of Radio Free Europe had no idea what such an unprecedented entity was or what it concretely did to enhance its assertive authority.

In any case, it scarcely mattered. In the days immediately following the Monday mobilization, hundreds of reporters from around the world came to the Lenin Shipyard to report with greater precision than Radio Free Europe could manage the surface detail of social life in the shipyard. One thing seemed to be clear, at least from the vantage point of the shipyard: Polish workers had somehow achieved some interesting methods of communicating with one another.

They had, indeed. It was a talent that had required thirty-five years of effort and suffering to fashion. Indeed, the communications links were now so deep that they were beyond the vision of all those who had grown accustomed to conventional hierarchical politics. No one saw, and therefore no one reported, the courier war.

The Strike Presidium had certainly done all it could to make the matter visible to the government. They had stressed that they did not even want to begin talks until the couriers were released from prison and the telephone blockade was lifted. The government nominally agreed to these preconditions in order to get negotiations underway but then blandly argued on the first day that "no political prisoners existed in Poland." The Strike Presidium replied—repeatedly replied—with a meticulous list of couriers who had been arrested and beaten and made it clear to Jagielski that substantive discussions about the twenty-one demands could not proceed until both preconditions were met.

The deputy premier's dazed response—"Are the talks to be suspended on this one issue"—became the first tangible indication on August 23 that he had begun to sense the new dynamics of state-society relations that had been created by the Interfactory Strike Committee (see Chapter 1, pp. 9–23, esp. 21). Like other intelligent members of a centralized hierarchy, Jagielski, as the party's chief labor negotiator, knew precisely the role of the official trade unions in enclosing Poland's workers within an organizational blanket of silence. He knew why the telephone blockade was necessary and why the workers' response to the blockade—couriers armed with the organizing details as to how other workers beyond Gdańsk could act in order to affiliate with the MKS—had to be suppressed. He understood that the government was in a race against time, that workers from Silesia to Kraków needed to be kept as confused and uninformed about the organizing components of the Interfactory Strike Committee as they so far had been. As a negotiator, his task was to get past awkward "preconditions" and move toward a quick settlement, even at high economic cost. Jagielski had been able to do so in Lublin, but he faced a different structure and a different set of programmatic demands on the coast. But he did possess an intimate understanding of the workers' strategic objective to free their couriers and break the communications blackout. Unfortunately, the meticulous naming of couriers by the Strike Presidium made it more difficult for him to lie his way out of this impasse. If the couriers were not stopped, Silesia would inevitably learn enough to join the strike, and then the government would be reduced to the awkward and unwinnable choice between another 1970 military solution or else a fundamental structural alteration of the party's relationship to society. His task was to deflect presidium members from their obsessions about couriers and telephones. They seemed to have an ominously accurate grasp of the communications requirements of movement building.

But visiting observers, both journalists and scholars, were not as sagacious as the party's leading labor negotiator. Despite the specificity of courier names listed by the strikers in defense of their "helpers," journalists decided the workers really were talking about Jacek Kuroń and the other KOR activists who, since August 20, had been in jail in Warsaw. The prevailing assumption in middle-class circles, both foreign and domestic, was that no "helpers" could be more germane to the cause of shipyard workers than the activists of KOR.

The ironies infusing the Polish August are many, but the decade-long misreading of the courier war is probably the most stark. At the very moment the Strike Presidium aggressively pushed its preconditions, Jacek Kuroń was doing everything he could to deflect shipyard strategists from their first demand. Indeed, the entire trajectory of his oppositional career underscored his settled conviction that the postulate for independent unions was a classic example of worker "impossiblism." In the week following his first reading of the twenty-one demands—on August 17 in his apartment in Warsaw—Kuroń was to pursue a consistent strategy to persuade the strike leadership to alter course. The purpose of Kuroń, arrayed alongside the purpose of the couriers, tracked diametrically contradictory courses of action.

None of these underlying strategic tensions were visible to casual observers of the shipyard scene. Jacek Kuroń was the most visible "helper" in Poland. The courier war was unseen. Nevertheless, despite misconceptions about operating dynamics that were spreading and deepening among outside observers, the crisis of the Polish state had, in fact, begun.

5 ▪ Tributaries

I thought it was impossible, it was impossible, and I still think it was impossible.

Jacek Kuroń on the Polish August

VICTORY, AS THE saying goes, has a thousand fathers. Unexpected victory, however, has ten thousand. To anyone who understood the postwar history of the Polish working class, the five days of aggressive organizing that bracketed the first weekend of the Polish August produced a social fact that was heavy with historic meaning. The party would either have to pay the very large price necessary to crush the Baltic movement with force or else accept fundamental changes in the very structure of Polish life. Actually, however, after a mere five days, neither the premise nor the fateful option was at all evident because no one outside Gdańsk understood the organizing achievement of coastal workers, and thus, no one understood the nature of the crisis.

Party functionaries—whether located in the overbloated bureaucracies that sprawled across Warsaw or in the cloistered confines of the Central Committee itself—could do no better than react with ancient tactics and apply them through ancient rituals. They told their underlings on the coast to negotiate hard, concede where necessary, and get people back to work. Similarly, within the intricate structure of the Catholic Church, neither parish priests and bishops nor lay intellectuals had yet sensed the true dimensions of the emerging confrontation. The episcopate was content to issue pious reminders of the need for prudence on all sides. And finally, within the intelligentsia, there existed three identifiable strata: a traditionally passive mainstream of several hundred thousand people who watched but did not act; a progressive intelligentsia centered in the Clubs of the Catholic Intelligentsia, but extending to small grouplets that floated in and around the party; and finally, a more activist minority that comprised the overt democratic opposition. The latter group was centered in KOR but included other small sects that had developed subsequent to KOR's appearance in 1976.

It is necessary to specify—as of August 18, 1980—that none of the participants in these various sectors of the Polish intelligentsia grasped the structural implications of the organizing achievement coastal workers had already set in place. The tiny handful of Warsaw intellectuals who had heard about the Gdańsk strike could do little more than focus with some trepidation upon the 1970 massacre on the coast or the violent repression at Radom in 1976. From all quarters, reactions took these classic forms because none of the groups had gained strategic command of the central reality that had come into being over the five days: the truly beguiling chance for victory that the workers' organizing accomplishments now afforded.

When the Polish August reached its triumphant climax during the ensuing two weeks of tension and the nation awoke to discover itself to be living in a radically different kind of September than anyone could have remotely foreseen, and when sober reflection upon this circumstance conjured up a possible Polish future brimming with social vitality—when all *that* became evident—a great deal of retrospective historical reconstruction immediately began on all sides. Suddenly, everyone in Poland, it seemed, was responsible for the independent trade union Solidarność.

Interpretations quickly appeared that demonstrated the centrality of the Church's role in establishing the proper climate for Solidarność. The emotional and pageant-filled visit of the new Polish Pope to his homeland in 1979 was interpreted after the fact as a mystical moment when Poles came together in a "transformation of consciousness" and saw themselves as a unified nation that would go on to reassemble as Solidarność the very next year. Appropriate homilies from 1976 would be found that seemed to point to the time when the Church spoke out for workers and thereby enabled them to acquire the sense of themselves necessary to fashion the twenty-one Gdańsk demands.[1]

As with the political advocates of Catholicism, so, too, with the promoters of the cause of the intelligentsia. Once Warsaw intellectuals had assimilated the astonishing dimensions of the social earthquake that had transformed Poland, activists within the various streams of thought in the democratic opposition similarly began rummaging through their private archives to locate the appropriate initiatives that seemed to bear on the decisive struggle in the Lenin Shipyard. The right-wing Confederation of Independent Poland (KPN) soon began reminding one and all that Leszek Moczulski, the organization's founder and guiding spirit, had published in 1979 a book, *Revolution Without Revolution*, that called for the creation of citizen institutions independent of the party. KPN strongly implied that Moczulski had been the one to lay down the blueprint for independent unions that the workers had taken up on the coast.[2]

Doubtless to KPN's dismay, two other groups, The Society for Scientific Courses (the "Flying University") and the Clubs of the Catholic Intelligentsia had been ably represented on the committee of advisors in Gdańsk that worked with the Interfactory Strike Committee. Their claims seemed to have a geographic pertinence missing from KPN's scenario. When Bronisław Geremek and Tadeusz Mazowiecki, along with Waldemar Kuczyński, Tadeusz Kowalik, Bogdan Cywiński, and Andrzej Wielowieyski, returned to the capital after the historic signing of the Gdańsk agreement, they were hailed as national heroes in Warsaw's intellectual circles. Indeed, in view of the limits of the contributions of this group in fashioning the twenty-one demands, the unrestrained enthusiasm with which accolades were bestowed proved sufficiently embarrassing to the recipients that reconsiderations became both immediate and imaginative in these quarters, too. The more some of them thought about it, the more the panel of advisors could be seen to have played a definitive and even creative role in wresting mighty concessions from the government. In time, this view expanded so that one of the advisors, Tadeusz Kowalik, came to see the meaning of their "working group" with the government's experts as the centerpiece of the negotiations—relegating the "public" meetings between the workers and the government commission to a lesser role. Kowalik managed this claim through the following language: "An old shipyard worker . . . told us that people were worried by the shortage of information and hence open to rumors. Of course we agreed with him. But our position was difficult. Should we have not agreed to the working groups? Should we perhaps have admitted they were to some extent used as a substitute for plenary sessions under public control."[3]

In this circuitous and suggestive manner, Kowalik "admitted" his own role was central—an authentic substitute for the role the world (erroneously?) thought that Wałęsa, Gwiazda, Wiśniewski, Kobyliński, and other workers on the presidium were playing. In light of what actually happened in Gdańsk, and the true power relations between the workers and their advisors, this rather strained description by Kowalik was a surprisingly aggressive sign of the hunger for political relevance that long years of party monopoly had engendered in sectors of the Polish intelligentsia. After all, they told themselves, Poland had been transformed by Solidarność and the advisors had been in Gdańsk when it all happened. They had been significant in historically causative ways. All of this rhetoric was quite understandable; nevertheless, it was all quite overblown.

In time, the relentless assertion of self-promoting claims by representatives of various factions in Poland had an inexorable impact on those who wrote books on Solidarność. A sophisticated English journalist would write a generous and essentially fair interpretation of the transformation of the postwar Polish Church from a narrow-minded, inward-looking, anti-Semitic, and medieval institution to a more broadly tolerant and judiciously progressive influence in Polish life. His summary conclusion, however, constituted a breathtaking leap in historical causation: "It is hard to conceive of Solidarity without the Polish Pope." Similarly, an American-trained sociologist, writing on "patterns of social inequality and conflict," congratulated the intelligentsia for abandoning its previous isolation and joining up with the working class after 1976: "Only after the 1976 insurrection . . . did [the workers] affect the 1980 change in Polish society. No force other than the working class can challenge the powerful political elite in the People's Democracies, but this class must be provided with the leadership of more enlightened people." This comforting judgment reflected what soon became the conventional wisdom within the Polish intelligentsia following the creation of Solidarność.[4]

To comprehend the rise of Solidarność, it is essential to sort through this maze if the operative causal dynamics are to be clearly illuminated. Toward this end, a sober review of the activities of Polish intellectuals in the postwar years is clearly in order as a precursor to any summary conclusions of the role of the intelligentsia in the rise of Solidarność.

For Westerners, the first step is to bear in mind that in Poland, the word "intelligentisa" is employed to describe a very large group of people; it sometimes is extended to include virtually everyone who performs any kind of mental, as distinct from manual, labor. While it occasionally proves useful descriptively to refer to the intelligentsia as a distinct social formation possessing an inherited worldview that can be fairly broadly applied, this characterization can easily be carried too far. It is nevertheless clear that several strong group tendencies existed. The intelligentsia, traditionally anti-Russian, had a predisposition to be hostile to Soviet-style socialism, a predisposition that had grown over the years, so that by 1980, it ranged from strong to total. These inclinations extended through the upper and middle echelons of the party which increasingly since 1945 had been supplied by the intelligentsia. This awkward state of affairs did not merit the label of a "hidden truth" about socialism in Poland because it was a fact too ubiquitous to be hidden, and the Soviets were painfully aware of its depth. Since the party, in any event, was not an institution bent on socialism, this circumstance did not constitute a burdening contradiction in the day-to-day administration of the country. A much bigger problem for the functionaries who ran the party was the task of finding some way to connect with Polish society while functioning under the constraints imposed by Soviet strategic requirements. No Polish party leader had been able to cope with this contradiction for long, and all in the course of trying had been forced to increase the rigidity of the censorship the longer they stayed in office. This dynamic ensured that the bulk of the intelligentsia would be more or less constantly alienated from the party.[5]

In general, the Polish intelligentsia had been badly battered by the party throughout the postwar years. Centralized rule was inherently humiliating to all except the powerful few at the center, and the morale and self-respect of the intelligentsia had over the years been a special target of a succession of interior ministers and heads of state. As a group, intellectuals proved tenacious in resisting their own brutalization and developed a remarkable range of social and psychological defenses—all in the name of maintaining some workable space beyond party control in which lives could be led and ideas could be thought about. There was some measure of help. Residents of the capital could get the news by radio on three stations; whenever possible, most listened to the BBC or Radio Free Europe—known as "Warsaw IV."

Inevitable losses had to be absorbed in the process, however. A government that could force the citizenry to get topical political information only from another country was effectively reminding all concerned of its power. If two generations of suffocating rule orchestrated through the "leading role of the party" had provided any instruction at all, it was that the idea of the leading role of the party was not easily questioned. In the course of committing a truly distressing number of blunders, the party had taught intellectuals that Leninist norms produced a style of governance that was crony-ridden and enormously wasteful, but it also profoundly instructed one and all that the party could not as an entity be challenged. The constant and unending din of government-controlled information had a powerful and subtle effect. The droning of semi-sophisticated hacks in the print and electronic media actually proved more mind numbing than the occasional adroit piece of internal propaganda that the party was sometimes able to foist upon the population. The "daily news" in Poland was so transparently false that its level of hypocrisy insulted the intelligence of the entire population. The daily drippings of officialdom simply washed into all the public spaces of daily life. One could not simply pretend to be above it all as a psychological gambit necessary in avoiding a permanent state of rage. For that would outmaneuver the censorship only at the cost of remaining ignorant of all human affairs beyond the circumference of one's own daily life. Therein lay the maddening power of the censorship; even when it was transparently ridiculous, it prevailed at a strategic level by robbing people of the kind of intellectual power that only an informed sensibility can mobilize. To read, observe, and learn without becoming contaminated—that was the challenge of every minute of every day in Polish life.[6]

This puts the matter broadly. A number of courageous individuals tested the constraints in Poland, and some of them performed ingenious creative feats, politically as well as culturally. But the personal price they and their families paid was inevitably high, and the object lessons thus dispatched through society were quite sobering. One had to be very strong to be creative in People's Poland.

Yet for all this, there persisted a powerful countertradition traceable to the nation's vulnerable geographical location and the long history that the resulting geopolitical realities had imposed. Fate placed the Poles on a parcel of Europe squarely between two numerically superior nations, the Germans and the Russians. Poland's history has consequently been a centuries-long saga of invasion, partition, foreign domination, and a corresponding litany of popular risings and insurrections, almost always against desperate odds. It is a tradition that can too easily be dismissed as "apocalyptic" or "romantic," for such terms conceal the desperate yearning for autonomy that is the great connecting thread of the Polish national experience. For the past three centuries, authentic nationhood existed only during an eighteen-year period between the two world wars. The two generations of Soviet domination since 1945 are merely an extension of the long-standing national tragedy.

The cultural result of this tortured past is a national literature that is held together by the leitmotiv of heroism and guilt in the face of tragedy and a national calendar that is pockmarked with patriotic anniversaries that commemorate passionate insurrections. Most of the risings, of course, were against Russian czardom, and this fact, despite the ideological non sequiturs involved, increasingly energized modern-day party censors to find "antisocialist" intentions at work on a great many occasions in seventeenth-, eighteenth-, and nineteenth-century Polish history. Indeed, the grotesque distortion and outright suppression of the nation's historical experience through such mindless censorship became the single most intellectually humiliating aspect of life under the centralized bureaucracy. The cultural devastation that began in 1945 had become a cataclysm within the first generation. In a rare moment of public candor in the Polish Writers' Congress in 1968, some of the nation's most prominent artists and intellectuals escalated one another into what turned out to be a remarkably free-wheeling assessment of the impact of the censorship on cultural life. In a magisterial summation, the philosopher Leszek Kołakowski touched upon every major theme of that memorable evening:

> I repeat, and we will repeat endlessly, the most banal truths: that cultural life requires freedom . . . to reflect on culture, its values and possibilities. But even this reflection is not possible, for every discussion is fortified with prohibitions and is inevitably systematically falsified . . . We have reached a shameful situation where the whole of dramaturgy, from Aeschylus through Shakespeare to Brecht and Ionesco, has become a collection of allusions to People's Poland . . . The theatres are rendered unfit out of necessity when it appears that nothing is neutral, for where it is necessary to return to the most elementary values, everything becomes suspect . . . Let us consider the degradation of the Polish film, let us consider the appalling and miserable system of information in the press, let us consider the restrictions and harassment practiced in Poland in the humanities, in current history, sociology, political science and law. Let us consider the poor, deplorable discussions in which no one ever dares say what is really the matter, for everything leads to the forbidden fruit . . . The monopoly leads to the degeneration of the same twig that enjoys the monopoly . . . Every evaluation is called antisocialist and viewed as aiming at the reestablishment of capitalism. *But this is blackmail* . . . In point of fact, if we were to acknowledge that socialism could exist *only* in conditions where culture was trampled upon, we would have to acknowledge that socialism is impossible or that it could be only a parody of one's own premises. Yet, I can imagine a kind of socialist life in which this unbearable and destructive state of affairs . . . shall be abolished. We want the abolition of such a situation in the name of socialism, not against it.[7]

Kołakowski had been expelled from the party for a similar speech in 1966. As a result of the declaration in 1968, he was deprived of his chair at the University of Warsaw and was forced into exile. The scope of the censorship remained unchanged. It therefore continued to be the focus of reform efforts by the artistic, technical, and academic intelligentsia into the 1970s.[8]

On a more mundane level, daily life within Polish society was afflicted by a maddening ingredient that had the effect of concentrating people's attention upon the possibility of short-run improvements rather than upon more organic matters of structural change. This had to do with the Olympian level of incompetence that had been attained by the party's central-planning bureaucracy in Warsaw. An astonishing number of things did not work in Poland, particularly very important things. State planning was at the head of the list. Increasingly over three decades, financial and material resources were routinely misallocated, disorganizing production and demoralizing middle and lower levels of management as well as workers at the point of production. Leninist theory played a hand in a number of these persistently recurring disasters, particularly in agriculture where modern equipment and fertilizers were funnelled into malfunctioning state farms while more productive peasant plots were intentionally starved of capital input. Spasmodic efforts to alter this policy could not overcome the party's prior momentum in the opposite direction and the peasant hostility that momentum had generated.

Leninist customs extended to industry as well, of course, and especially to heavy industry. Party centralizers were wedded to the idea of "economies of scale." The conviction that large units of production are inherently more efficient than small or midsize units has gained much sanction in the modern world (despite mounting countervailing evidence in a number of fields ranging from agriculture to banking), but nowhere on earth does an enshrined faith in large-scale production have greater currency than in countries enclosed within the Soviet orbit. The Polish party regarded as one of its greatest achievements the creation from scratch of an enormous steel complex at Nowa Huta (New Forge) near Kraków. In architectural design, the housing at Nowa Huta achieved a monotony of such august proportions that it almost seemed to represent a conscious effort by state planners to diminish the self-esteem of the work force, largely comprised of peasants drawn from the surrounding countryside. Permanent pollution problems were structured into the production process, and the predicted economies of scale vanished in a thicket of supply, product, and marketing inefficiencies. Nationally, the government achieved some tangible successes in the field of technical education, but ironically, the routine inefficiencies of the prevailing system merely served to push newly trained mechanics and technical specialists to higher levels of frustration.[9]

Daily life under such stultifying constraints generated an immense longing among the Polish intelligentsia. Thousands of people came to understand that small changes in decision-making processes could engender fairly substantial improvements. The thought kept at least some measure of hope alive, even in the worst of times. But even small changes in the functioning of government ministries, basic industries, research and technical institutes, or in the educational apparatus could be vitiated by a sudden return to traditionalism at a higher level of the bureaucracy. The mechanisms of authority, vertically organized through the party and sanctioned with the theoretical dress of Leninist democratic centralism, drove Polish intellectuals to distraction. By 1980, two generations had grown bitter under its yoke. But prospects of attacking the concept of democratic centralism itself dissolved under the heavy hand of the central bureaucracy in Warsaw and its ubiquitous police apparatus. These strategic heights had to be conceded. But smaller foothills conceivably remained within reach. The technical intelligentsia came gradually to focus its attention—necessarily discretely—upon ways to ameliorate the censorship and to open the technical journals to more creative analyses and perhaps even to creative planning. At propitious moments, such as the coming into power of Władysław Gomułka in 1956 or of Edward Gierek in 1970, the inevitable "thaw" offered the prospect of relaxed control over cultural affairs as well. Unfortunately, normal patterns of authoritarian constraint habitually returned in fairly short order, as each incoming party mandarin came to terms with the sundry self-interests of an overstaffed and police-driven bureaucracy. In response, the intelligentsia tried to remain adaptably flexible and ready to capitalize on any temporary opening. Over time, the sheer narrowness of the available opportunities tended to compress people's sense of the possible, so that even marginal double entendres successfully slipped past officialdom were cherished. These little coups—in popular songs, in short stories, in theater productions, and in scholarly papers—and the jokes that accompanied their retelling, became the buoying staples of intellectual conversations that otherwise stood in more or less permanent danger of being uniformly depressing.[10]

The accumulated impact of thirty-five years of life under the Polish United Workers' party taught the intelligentsia to see prospects for future democratization in terms of the ebb and flow of the censorship. In the 1970s, Polish writers vented their rage in passionate lines, some of which were later printed after Solidarność had pried open the censorship or published abroad after martial law. Some came from the distinguished pen of Kazimierz Brandys:

> Someone from the outside, who came here from, say, Venezuela might have the feeling that Poland's cultural life was on fire. What could possibly be lacking here! There's everything, and everything in its proper place: creative work produced here, translations of literature from the three worlds, analyses of contemporary civilization, sociological polls, structuralism, human rights, civil rights, new terms—paradigm, trauma, trends . . . Everything, the entire spectrum! How many years would a person have to live here before he could decipher this system of shams and smell the hovering stench. Here everything conceals the corpses hidden under the floor. A book that is being written up is supposed to drive the book that is not published even further down. Buried between the lines of every text printed are the names one is not allowed to mention, the facts about which one remains silent. The places and dates blotted from history create a dead zone whose silence is filled with the noise of artificial polemics and where semipoets give interviews to semijournalists and express their semitruths to their semireaders. There is no lack of anything here except for the half that has been amputated.[11]

Beyond question, the censorship strangled the national culture. It was an area in which a breakthrough at some point was an unassailable prerequisite for serious improvement in the quality of Polish life. But *it was a dependent prerequisite*. Liberalization within the structure of the censorship could not, in itself, be culturally self-sustaining; the permanence of any relaxation of state control was dependent upon the existence of *other forces in society* capable of providing some measure of protection against the powers of the central political bureaucracy to reclaim, whenever it wished, any component of its authority that had been temporarily relinquished.

As a simple problem in political logic, this connection was self-evident. Yet most Polish intellectuals rarely thought about such social relationships as prerequisites to an effective democratization of their society. The subject simply was not within the mainstream of discourse. The dependence of cultural liberalization upon the appearance of some unseen "other force in society" which could help protect the population against the bureaucracy constituted a proposition that was psychologically difficult to focus upon. This was an absolutely crippling circumstance, and it will be necessary to explore the subject in detail in the context created by the arrival of Solidarność. But it is first necessary to pencil in the dimensions of the cultural iron curtain that walled off the Polish intelligentsia from the great bulk of the nation's citizenry.

It needs to be stressed at the outset that the Polish intelligentsia was not alone in its self-absorption. The trait is present to a greater or lesser degree among self-conscious elites the world over. But it was fortified in Poland by a long-standing tradition of conceptualizing the idea of the nation and the idea of popular aspiration in terms of the intelligentsia itself. Here could be found the defenders of the democratic idea for the long centuries of foreign domination. As it understood these events, the intelligentsia had led the risings, provided the political rationale for the insurrections, articulated the democratic manifestos and proto-constitutions that occupied such a conspicuous space in the national memory. The intelligentsia was the nation, the conscious nation.

A final dimension was at work within the intelligentsia, one that in some ways was the most tenacious of all. Apart from whatever claims it made for itself on the basis of historic service, contemporary occupation, or intellectual merit, a significant sector of the Polish intelligentsia possessed a deeply internalized conception of itself as a class anchored in historic bloodlines and noble lineage. The Polish word applicable here is *Szlachta*, denoting a petty nobility dating from the late Middle Ages and often visible in contemporary Poland through distinctive endings of surnames. As one authority has summarized matters, "*kultura szlachecka* (the noble ethos) has become one of the central features of the modern Polish outlook." The Polish intelligentsia, seeing itself as embodying the noble ethos, was the *Szlachta*'s modern manifestation, frequently as identifiable by name and occupation as by noble forebears. The point was not that the Polish intelligentsia was wholly of *Szlachta* origin—which it was not—but that the ethos dominated its worldview.[12]

"Other forces in society?" As an operating cultural assumption, there were not any other forces that readily came to mind. In the milieu of the intelligentsia, political analysis began with a certain global sweep which tended to channel discussion toward lofty geopolitical realms far removed from the anxieties and tensions afflicting Polish workers. Simply enough, thoughts about workers did not intrude with sustained regularity into the discussions of the intelligentsia. Indeed, in the prevailing worldview, Poland possessed two historic problems—Germany, which had, temporarily at least, faded to marginality, and Russia, which was at the height of its oppressive power. Because of the Soviet Union, the nation was saddled with an all-powerful party of proconsuls which represented on Polish soil the domestic priorities of another country. The modern struggle for the soul of the nation took place within the intelligentsia: the opportunistic sector, which staffed the party with proconsuls and their footmen; the heroic sector, which formed the democratic opposition and intermittently had to assert its views from prison; and the mainstream of the intelligentsia, which endeavored to survive with dignity while judiciously maneuvering against the censorship. Given the political realities which welded so many Soviet divisions to Polish borders, the struggle had always been desperately unbalanced, but it was perceived to be the only one that could be waged. The contested terrain revolved around the form and scope of the censorship which, in its pure Leninist form, sapped the vitality of all institutions and poisoned the culture itself. Understandably, the Polish intelligentsia thought about civil liberties when it thought about politics.[13]

In a political world so conceptualized, little room remained for "other forces in society." Little room remained, consequently, for reflection upon the weaknesses inherent in a political strategy that depended upon the presence of allies who could not be seen. When workers were thought about at all, it was with a measure of distaste. They were believed to be so deeply preoccupied with economic issues that their attention span on issues of freedom was that of a child. One Polish intellectual summarized the relationship: "As a thinker, I am an alien in a nation of nonthinkers." Others, less in the grip of complacent self-pity, nevertheless also preferred to see Polish workers primarily as the bulwark of a massive and faceless citizenry ignobly warring against the best features of the human spirit. The writer Kazimierz Brandys, for example, was so offended by what he took to be the thoughts of the average Polish citizen that he experienced great difficulty, as a communist and later as an ex-communist, getting close to life in the popular mainstream. He was aware of no debilitating effects upon his descriptive precision, however, as the term he habitually used—"the masses"—was perceived by him to be fully adequate to meet his literary needs. Though the psychological distances involved between subject and object were patently enormous, Brandys wrote as if he had an intimate knowledge of the popular milieu. It was not a place of variety and tension, but rather one of pathetic monotony that was easily described: "When one speaks of freedom or justice, they only listen with reluctance. Those are threatening words, they interrupt the humdrum of daily life and the consequences they entail are unforeseeable. Nor do the masses like to hear talk about moral courage, because they see it as an expense beyond their means which could ruin them."[14]

After the Polish August, Brandys wrote about Solidernośċ with surprising brevity considering its transforming impact on Polish life. His style was a bit leaden. Perhaps, in light of his distance from popular life, such was the only style he could summon: "At the time, I had argued that the masses in Poland were passive and in danger of being Sovietized. I was mistaken. I underestimated their hidden reserves. We awoke to a new society from one day to the next. As for me, it has suddenly become possible to travel. I got a passport . . ."[15]

Brandys had long since left the party, where his condescending views had harmonized rather nicely with prevailing attitudes. But he did not move to the democratic opposition. He simply took his place in the mainstream of intellectual discourse. His view of the party had changed, but not his views of society. His old class outlook, therefore, transferred rather easily to his new surroundings. Condescension toward "the masses" was an energizing feature of both his old and new environments.

Such were the presumptions within the mainstream of the Polish intellectual world. But there were a number of Poles who, with varying degrees of thoroughness, transcended the limitations of condescension in the name of making themselves more relevant to the nation's political future. One of them was Jacek Kuroń, a founder of the opposition group KOR. Early in the Polish August, Kuroń was among those who quickly recognized the strategic and symbolic significance of the Gdańsk strike. He used his Warsaw apartment as a communications network to spread word of the strike, via long-distance telephone, throughout Poland. Unfortunately, he was arrested along with twenty other members of the opposition on August 20 and was thus prevented from playing an active role in the negotiations between the workers and the government that began on August 23.

Nevertheless, it is appropriate, at this junction in the sequential evolution of the Solidarność movement, to review the work of Kuroń and his associates in KOR, from its formation in 1976 through the eruption of the Baltic coast in 1980.

By the standards of Poland or any other country, KOR's unrelenting four-year campaign for civil liberties in Poland constituted a courageous chapter in the annals of democratic advocacy. The militants of KOR had all been arrested many times, brutalized, harassed relentlessly in large and petty ways, jailed on trumped-up charges, and subjected to harrowing psychological persecution. At a time when part of the regime's ubiquitous power had lain in its ability to force its victims to suffer in obscurity, isolated from the world, KOR had borne witness of state persecution countless times—in Radom, Ursus, Kraków, Silesia, and elsewhere. Repeatedly, they had gone to jail for bearing witness. But they had done more than offer their bodies and their personal freedom as hostage to their dream of a democratized Poland. They had helped create half a dozen journals and had reached out to every sector of Polish society. KOR's *Information Bulletin* had begun the attempt to open up Polish society by publishing details of police beatings and the names of victims and by otherwise challenging the government's monopoly over social information. KOR's Intervention Bureau put indicted students, workers, and intellectuals in touch with lawyers and the victims of police beatings in touch in physicians. KOR had raised money to pay for both. KOR activists had held dramatic fasts to put public pressure on the government and developed a worldwide network of contacts through Helsinki Watch, PEN, Amnesty International, and a dozen or more socialist, liberal, and communist political institutions.

Though the members of KOR went to prison, they also got people out of prison, including, as their status grew, each other. They helped found, stimulated, or subsequently associated with a remarkable range of free, if embattled, institutions: the "Flying University," the Independent Publishing House NOWa, free trade unions in Katowice and Gdańsk, and student groups in half a dozen university towns. They cooperated with Russians the world had heard of, such as Helena Bonner, and many others principally known only to opposition groups in Eastern Europe. They worked with their equally courageous Czech counterparts, including Vaclav Havel, in Charter 77. Virtually every such joint action with opposition groups in other Eastern bloc countries, carried out under the most extreme kinds of police surveillance and harassment, was paid for in jail sentences.[16]

With pride and irony, Jan Józef Lipski wrote that in many areas of opposition activity, "KOR did not suffer from competition." Attempting to do so much and to cover so much ground, the militants of KOR could not, of course, be uniform in the depth to which they pursued each initiative. Some of their creations, however, had great depth, such as the underground publishing house NOWa. Among the victims of the tyranny of the party was "a constantly persecuted Polish culture." Under the brilliant and inventive leadership of Mirosław Chojecki, NOWa set out to make available to the population the great names of émigré literature—Gombrowicz, Miłosz, Wierzyński, and others—and "to offer Polish writers of all generations the possibility of reaching readers inside the country." Everything came from NOWa—lyric poetry; novels from the Soviet Union, Germany, England (inevitably George Orwell); and volumes of economics, political science, and history. *Zapis*, *Puls*, and *Krytyka*, underground journals of literary and political opinion that rolled from NOWa's presses, alone accounted for thirty volumes of free commentary. All was done on the run with inevitable breakdowns caused by arrests and confiscations of equipment. Though Lipski sighed, "There were not too many people who really mastered the printing techniques," a kind of professionalism in printing eventually emerged.[17]

The militants of NOWa, particularly the younger ones, became pirates, second-story men and women who lived shadow lives in the name of free expression. Some publications were printed after hours on government equipment, a practice the militants considered "consistent with the Constitution which guarantees that printing facilities will serve the working people of towns and villages." A special satisfaction attached to the use not only of government presses but "the proper utilization" of government-owned paper and ink. The exploits of Chojecki became legendary. He was regarded by many foreign correspondents as the most successful buccaneer in the ranks of NOWa. A paragraph from Lipski's definitive history of KOR may perhaps serve to summarize this aspect of life within the world of the democratic opposition.

> Not only the duplicating machines but also stencils, printer's ink, binding gadgets, and other equipment traveled secretly across the border making the work of supply extremely bothersome and difficult. But paper was never imported; often it was even bought retail. All this material, including the reserves, had to be stored somewhere. New locations for print shops had to be sought out constantly, and they had to meet at least certain minimum requirements. Even housing was not a simple task for the printing services. The paper, the equipment, and the printed editions had to be transported from place to place. Sometimes bags and knapsacks were sufficient; usually cars were needed. Thus, there emerged a circle of specialized associates of KOR, NOWa, and the individual periodicals, whose job was precisely to take care of all this. Distribution started from storage places, and then the transports were gradually broken down into smaller and smaller parcels, until they reached rank-and-file carriers. Even the best known KOR members and associates took part. It was rare for entire editions to be confiscated by the police, but losses "on the lines" in the last or near the last stage of distribution were not uncommon, and were the most severe.[18]

In certain ways, KOR became absolutely unique as an institution of political assertion. Since the invention of the printing press, underground groups dedicated to political, religious, and social restructurings of one kind or another have intermittently appeared and tried to persist in the face of brutal repressions at the hands of a variety of emperors, kings, generals, and republics. Inevitably, the lives and the very values of people engaged in insurgent activity became disfigured by the forces of the authoritarianism they challenged. Radical despair could lead to a kind of authoritarianism among the insurgents themselves, turning them into mirror images of that which they opposed. The political practice set in motion by V. I. Lenin's *What Is To Be Done?* provides the most sobering example of this mutation, of course. The great challenge, one that many besides Lenin failed to meet, was to remain democratic while at the same time remaining insurgent.

KOR set a new standard for the world in this respect. KOR's founders were ideologically diverse, an amalgam of socialists and liberals, Catholics, Jews, and nonbelievers. The binding ideology was a simple and profound belief in human freedom. In carefully measured words, Lipski described the milieu of KOR as one "fundamentally conditioned by Christian ethics," but it was even more precisely descriptive of KOR that this characterization was not pressed very far or very often. Like coastal activists, KOR learned to use "sacred customs" for political purposes, including a collective fast at St. Martin's Church in Warsaw early in KOR's life, one that seemed to lead to an emotionally bonding experience for the participants. But KOR systematically avoided religious display, either for its own sake or as a recruiting device. Another attribute common among the activists of KOR was their sophisticated understanding of the critical difference between patriotism and nationalism—a shoal upon which many a Polish cause had historically foundered. The group came down against the use of patriotic props and the "overuse of symbols," such as Polish crowned eagles and signs of Fighting Poland. Such practice was regarded as "patriotic exhibitionism." It was not merely a matter of aesthetics; KOR's fear was that acts which encouraged popular participation in the "antique shop of Polish patriotic phraseology" would divert attention and energy away from the essential task of building an independent social life. The purposes of the democratic opposition, said Adam Michnik, one of KOR's most articulate strategists, were more serious and more demanding than mere gesturing and flag-waving: "The duty of the opposition is to participate continually and systematically in public life, to create political facts through collective action, and to propose alternatives. All the rest is just literature."[19]

To the ancient axiom "all that is necessary for evil to triumph is for good men to do nothing," KOR's answer was to do something—to put the glare of the *Information Bulletin* on hidden police fabrications and star-chamber legal proceedings through which the government brutalized the population, to challenge the cynical hypocrisy of the judicial system with intelligent writs and forthright appeals made openly and legally, to mobilize Amnesty International, Helsinki Watch, and PEN so that the weight of world opinion could be brought to bear on the ruthless pettiness through which large crimes were routinely committed by the state apparatus in Poland. Activists from KOR showed up at court trials where they were routinely harassed, threatened by police authorities and thugs, and sometimes beaten. Still, they kept returning. While the kangaroo proceedings inside Radom courtrooms revealed one dimension of party justice, the bizarre events outside revealed yet another. The future director of NOWa, Mirosław Chojecki, was cornered in the courthouse and kicked and beaten by a large group of "unknown perpetrators." A professional blow to his head caused a subcutaneous hemorrhage over half his face. When the thugs manhandled Chojecki to the top of a flight of very steep stairs and prepared to toss him down them, he managed to save himself by breaking free and executing a daring jump to safety. All this took place in broad daylight and in the presence of scores of witnesses, while KOR's attorney, Andrzej Grabiński, banged futilely on the closed office door of the policeman on duty. Two weeks after this incident, ten KOR observers were pelted with dozens of eggs in the courthouse corridors, and one of them, Kuroń, was beaten and kicked. At the time, eggs were unavailable for popular consumption in Radom. KOR lawyers filed formal complaints, but the prosecutor refused to take action "in view of the minimal social harmfulness of these acts."[20]

There can be no question that such arduous efforts served the basic moral objectives that were part of the rationale for the formation of KOR put forward by its founders, including Jacek Kuroń. But Kuroń had an even deeper purpose. The intellectuals who had sat on the sidelines while the working class challenged the state in 1970 had to redeem themselves by taking a stand in 1976 when the workers protested once again. Kuroń said he and other activist intellectuals organized KOR because "we were ashamed the intelligentsia had been silent in 1970, and '71, and we wanted to restore its good name." To do so effectively, Kuroń felt, intellectuals had to go beyond moral stands and social aid, fundamentally important as both were, to proceed to organize an effective worker-intelligentsia coalition. For a time at least, it seemed KOR's initial work in Radom and Ursus might effectively achieve that end, for the linkage became quite palpable very early. For example, the well-known Polish actress, Halina Mikołajska, transported KOR teams to Radom several times, even though police agents tampered with her automobile and slashed her tires. The first time her car was towed to a repair shop in Radom, the actress was recognized and promptly learned that everyone there knew why she regularly came to the city. Her car was taken out of turn and repaired immediately with payment refused. Future free service was offered to all KOR people by the Radom mechanics.[21]

Attempting to capitalize on such links, members of KOR tried to persuade workers in both cities to sign letters to the Polish parliament protesting the false imprisonment and blacklisting of so many hundreds of people. The Ursus letter eventually gained over 1,100 signatories while Radom, a deeply intimidated city in 1976–77 produced less than 200 signers. Moreover, Radom workers who were encouraged to write individual letters protesting their own brutal treatment were further harassed and beaten by security thugs. Despite such instructive social information as to the differences in the two work forces, the activists of KOR pushed forward the idea of creating a free trade union in Radom in the autumn of 1977—probably for the elementary reason that KOR had helped more people in Radom and, thus, had more contacts there. It was a seminal error. The founding of the free trade union was proclaimed by the union's sole member, Leopold Gierek. The sustained persecution of Gierek that subsequently ensued in Radom brought the project to an abrupt halt. Gierek apparently was unable to recruit a single person to associate with him publicly in the free union.[22]

In Warsaw, the three KOR members most deeply committed to the project of an active intellectual-worker coalition—Kuroń, Jan Lityński, and Henryk Wujec—drew a fundamentally negative conclusion from the episode, one that guided their strategy for the rest of the decade: worker activists who wanted free trade unions could not go public until they had recruited large-scale support from their co-workers. After the Radom failure, KOR got out of the business of initiating any new free trade unions in Poland. The two that appeared anyhow—in Katowice and in Gdańsk in 1978—did so on workers' initiatives that were immediately opposed by Kuroń and Lityński when they heard about them.[23]

To understand the sharp limits to KOR theorizing, it is necessary to review briefly the trajectory of speculations generated over time by both Jacek Kuroń and Adam Michnik. Michnik was an essayist and a student of politics. His thoughts on the subject of worker organizing, never fully developed but never totally abandoned either, turned on some sort of Polish variation of what Michnik called the "Spanish model." As Michnik understood the Spanish effort, workers, previously identifiable to each other through their ideological beliefs as socialists or communists, created covert independent structures outside Franco's state-run trade unions in Spain. But the tactical ingredients of shop-floor organizing were beyond Michnik's experiential grasp, and he was never able to suggest an organizing strategy to apply this "model" to Poland. The problem was that the organized structure for workers in Spain was fundamentally different from the social circumstances of workers in Poland. In Spain, at the end of the Civil War, the communist network (a social formation already recruited) remained. The Spanish effort thus concerned the redeployment of a movement that already existed. In Poland, the problem was to *recruit* a movement. "It is hard to tell when and how more permanent institutions representing the interests of workers will be created and what form they will have," Michnik mused. "Will they be workers' committees following from the Spanish model or individual labor unions or mutual aid societies?" When opposition intellectuals could do no more than suggest courses of action that ended in question marks, they were clearly not engaging in the political act of movement building.[24]

In a more politically engaged manner, and much more relentlessly, Jacek Kuroń also attempted to address the organizing dilemma. For Kuroń, the controlling political experiences came early in life—namely, his early education into Marxism and his private and public break with that tradition. The latter came in 1964 in his "Open Letter to the Party," written with Karol Modzelewski. The evaluative criteria Kuroń applied were derived from his own negative experiences as a party member, richly corroborated by his independent observations of the condition of the Polish population. His conclusion was unequivocal: the party had failed, and Poland needed another revolution.

The structure of logic in the "Open Letter" was Marxist and, as the force of the prose indicated, no less telling in its effect because of it. But it was also blurred and incomplete in its response to the Leninist principles informing the party's methods of rule. To Kuroń, the party had failed to serve all society because it was so remote from the working class that it could not function in a democratic manner. But given its bureaucratic class basis, the party could not reform itself; it had to be overthrown by workers in order for the necessary democratic restructuring to take place. In this language, the party remained in Kuroń's structural universe. Though he was to spend the next quarter century of his life in forthright and demonstratively courageous public opposition to the party, serving eight years in prison in the process, he never fully expunged the party from his thought processes. In one sense, of course, this result was merely dialectically coherent—one had to think about the party in order to think of coherent ways to oppose it. But this necessity also impelled taking due note of the party's continuing power, a consideration that gradually encouraged Kuroń to see a diminution of party authority as a political prospect that was necessarily tied to the number of party reformers who could be encouraged to exist.

Imprisoned for three and a half years for the "Open Letter," Kuroń emerged just in time to defend the students engaged in the "March events" of 1968. He received another three-year prison term for his troubles. He, thus, was socially quite distant from the Baltic mobilization of 1970–71. Though he generally read political events with a keener eye than most Warsaw intellectuals, Kuroń did not do so with respect to the long-term implications of the structural achievements embedded in working-class assertion on the Baltic. Rather, he joined the rest of the nation's intellectuals in not understanding these developments. Unlike most, however, he was disillusioned if not exasperated that the Polish intelligentsia had remained politically passive through the long months of working-class agony and frustration on the coast. He resolved that it would not happen again.

And thus it was that when Gierek's ill-conceived economic program culminated in the price restructurings of 1976, precipitating a wave of strikes across the nation and violence at Radom and Ursus, Kuroń played a critical role in the organization of KOR as a public institution to assist worker victims. Over the next four years, Kuroń increasingly came to think about "social movements" as the structural base of systemic reform. KOR itself was understood as such a movement in embryo, and it, like any others that might somehow appear, necessarily existed "outside" the party. Indeed, part of the theoretical rationale for KOR turned on the understanding that public space in Leninist societies had to be created by the conscious acts of a consciously active citizenry.

For a time, Kuroń believed this generalized strategic necessity extended to the creation of trade unions independent of the party. It was for this reason that he attempted to help create the beginnings of such an institution in the course of KOR's efforts for the worker victims of the police violence at Radom and Ursus in 1976. But the repression was both prompt and overwhelming.

In terms of ongoing democratic theory, this denouement became a decisive juncture for Kuroń, for it persuaded him that the public unveiling of free trade unions was an inherently self-destructive act. Unless some kind of extended preliminary organizing work had been achieved that could provide an effective working-class defense of such structures, independent trade unions would be destroyed by the police at the moment they became public. It was for this reason that Kuroń and Lityński opposed free trade unions when they were organized from below by activists in Katowice and Gdańsk in 1978. The disastrous experiences at Radom and Ursus were cited as the compelling evidence.

But precisely what form "extensive preliminary organizing work" could take became a political challenge that the KOR activists never solved. Since the party in 1976–77 had once again so convincingly demonstrated its repressive power over workers, Kuroń, reduced in possible options, resumed his speculations about reforming the official unions. This pursuit harmonized with his inherited beliefs, never completely transcended, that party reformers were a necessary wedge into the authoritarian state. Though this possibility did not substantively materialize in Poland, the eventual emergence of Gorbachev in the Soviet Union demonstrated that speculations about such a development were not wholly illusionary. In any case, to Kuroń, efforts to increase the number and also the resolve of reformers in the party, perhaps coupled with a genuine worker presence in the official unions, might coalesce to produce some sort of effective democratization of the party.

In this way, KOR's organizing failure of 1976–77 at Radom and Ursus caused a basic shift in long-term strategy. When *Robotnik*, the new KOR journal aimed at workers, began publication late in 1977, its editors took the position that authentic worker advocates needed to go into the official trade unions in order to begin the process of making those moribund and state-owned institutional shells truly responsive to the work force. An early goal would be to try to get enough worker advocates inside the official trade unions to shape policy there. As an eventual goal, a free trade union was a legitimate objective, but its premature public unveiling could only result in the destructive repression of the organizing core of worker militants at the hands of the party's police apparatus. The organizing failure at Radom had instilled in *Robotnik*'s editors an appreciation of the merits of caution.

Unfortunately, as Edmund Bałuka, Lech Wałęsa, and scores of Poland's most persistent worker activists knew, a profound innocence surrounded this strategy. The party used the official union as a suffocating blanket of censorship and control. The party *invited* worker activists into the official unions, the better to watch them, the better to control and limit the range of their shop-floor influence, the better to intimidate them into passivity. By the late 1970s, only relatively inexperienced worker activists could be persuaded into the party-controlled confinement of the official trade-union structure. By the time KOR was formed, veteran worker activists like Wałęsa knew better. They had received a convincing personal education on the subject.

Nevertheless, this assumption guided Kuroń and *Robotnik* from the organizing failure at Radom and Ursus in 1976–77 down to—and through—the Polish August of 1980. Kuroń saw KOR as an exemplary pressure on party reformers, a way of combatting deference in both the party and in society. One demonstrated a free way to act by publicly commenting, judiciously but seriously, on the party's authoritarian habits, and if one was occasionally sent to prison, that fact was also exemplary in that it demonstrated that jail was not the end of the world. One came out, resumed civic life, and continued the pressure. As Michnik put it, everything else was "literature."

Here was the organic politics of KOR. It was a model of civic courage. But for workers, KOR's message had no tactical clarity. In fact, the specifics of the message did not extend beyond the assumption that public formation of free trade unions ensured instant repression and that reform through the official unions was necessarily the only survivable short-run course. The essential preliminary for independent unions—prior organization of strong worker formations to defend the idea when it became public—was a task whose organizing prerequisites left KOR theorists baffled. *Robotnik*, therefore, was silent on this fundamental level of politics from 1977 through the Polish August of 1980.

The principal flaw in *Robotnik's* strategic conception of infiltrating the official unions was not necessarily that it charted an inherently tangential approach that channeled organizing energy into a state-controlled cul-de sac—which it did—but that the strategy itself was so vague and generalized that it failed to address the concrete problems of organizing. The dilemma facing worker veterans on the coast like Wyszkowski, Gwiazda, Borusewicz, Walentynowicz, Wałęsa, Sobieszek, Lenarciak, and Szołoch was a very specific one. The task at hand was to find a way to persuade the rank and file that they could perform an act of self-assertion—any act—that was germane and at the same time not personally dangerous. Workers did not need to be told that the official trade unions utterly failed to serve their interest. A bit of time spent on a production line anywhere in Poland was enough to drive this lesson home. But workers who acted out in the open—to criticize an action or nonaction by the official unions, for example—became instant objects of harassment. Under the circumstances, the organizing dilemma had to do with finding a way to agitate without being repressed; workers who watched activists being arrested merely had their fears confirmed and their own passive stance verified. On the other hand, the simplest way for militants to avoid repression was to perform no public acts of assertion. This kept activists out of the clutches of the police, but it also left the workers unorganized and the official unions unchallenged. Kuroń's suggestion that the official unions be "infiltrated' did not really address this dilemma. The simple fact of the matter was that one could not "quietly" infiltrate. For workers to know who was on their side and who was not—either inside the official unions or outside of them—some sort of issue had to exist upon which people could see a visible division of opinion. Any activist who raised such an issue was performing a public act that made him instantly vulnerable. Even more important, the internal structure of the official unions made those organizations extremely difficult to "take over" from the inside. There was always a higher level that had not been "taken over," leaving one who had infiltrated far enough to raise an issue still exposed. All the worker efforts since the end of World War II in Poland conclusively pointed to the invulnerability of the official unions to this kind of "reform." And the same efforts revealed the futility of "worker councils" as long as such plant-sized structures were forced to try to survive at the bottom of the state's unionized structure. The organizing lesson of the postwar years in Poland was that no worker activist with any experience at all could put faith in either shop councils or trade union reform.

It is remarkable that the *Robotnik* editors did not know this—that is, did not know either the extent or the causes of rank and file alienation from shop councils as well as from reform of the official unions. Workers in Poland repeatedly organized themselves in order to try to prevent a sacking (or, of course, to protest a government food price increase), but it was a gross error to see this as an indication of some latent hunger for worker councils. Thirty-five years of self-activity on the shop floors of Poland had taught some lessons. Similarly, Michnik's idea of organizing Spanish-style "worker commissions" reflected a dismaying lack of concrete knowledge either of shop-floor conditions or the history of Polish workers in combating those conditions.

Unfortunately, such tactical and strategic subtleties at the point of production were beyond the ken of the editors of *Robotnik*. Kuroń was a committed democratic theorist, but he had little exposure to a working-class milieu beyond what he might have experienced in prison in the mid-1960s, and he had had no exposure at all to the day-to-day environment that existed at the point of production. That world was alien to him, and he could not think about it with experiential specificity. Nothing he ever wrote, before or after Solidarność, dealt in a sustained manner with this basic organizing problem that faced Polish workers at the point of production. As with Kuroń, so with his less experienced associates. Lityński was a young, college-trained computer scientist. He knew those things about life and power that one could learn growing up in an intellectual household and going on to college. This environment was not noticeably instructive as to the subtleties of working-class self-activity. Similarly, Wujec was a young physicist who, after graduation from Warsaw University, had honed his ideas about the party and the state as an activist in the Clubs of the Catholic Intelligentsia. The specifics of shop-floor organizing were as distant from his world as they were from Lityński's. Both were disaffected intellectuals who were attempting to learn how to oppose the state by learning how to mobilize the nation's intelligentsia. This they did through exemplary (and personally dangerous) public acts, followed by public meetings, letters, manifestos, contacts with world public opinion, and by contributing to opposition publications. All such activity was necessary, as well as being heroic, but it did not prepare its practitioners for the politics of the shop floor. From the day it was founded in 1977 to the day the Gdańsk agreement was signed in 1980, *Robotnik* was unable to address this central working-class problem with sophistication or precision. The journal, published semimonthly with a national press run of twenty thousand, eventually succeeded in conveying to an undetermined number of workers the information that somebody in Poland was sympathetic. The information was heartening to workers (a political value not to be dismissed), but it also possessed minimum practical utility. Additionally, most of the strategic advice *Robotnik* offered about infiltrating the official trade unions was unsuitable for the Polish situation and had been disproved by prior worker efforts. Finally, what *Robotnik* did say about free trade unions—that they were an essential long-term objective—constituted an idea that had gotten embedded in the Polish working class before most of the *Robotnik* editors had reached adulthood.[25] (See Chapters 2 and 3.)

Did KOR "prepare the consciousness" of anyone in Poland? Absolutely. The activities of KOR were relevant to the heightened awareness of three groups of Poles. In the two years between 1976 and 1978, KOR played a remarkably successful role in reorganizing the consciousness of the Polish United Workers' party. It forced the party to choose between standing before the world as an administratively incompetent and socially ruthless police state and acquiescing in the permanent existence of a creative and unintimidated intellectual opposition that was both watchful and articulate. As the economy worsened and Poland's need for Western credits became increasingly imperative, Gierek chose the second option. KOR made President Carter's human-rights program look good in tangible ways, and Carter's program and his country's checkbook proved timely for KOR. Additionally, KOR helped activate other opposition groups, and all of them, with KOR in the lead, helped overcome the inherited passivity of increasing numbers within the mainstream of the intelligentsia. KOR's exemplary role in this regard was demonstrably more central than that of any other group in Poland. Finally, the activist intelligentsia, in all its sundry raiments, had a measure of impact on the Catholic Church, though gains here were less pronounced, particularly at the level of the Primate himself.

But KOR did not "prepare the consciousness of workers for the strikes," because the understanding of worker militants as to what they could or might be able to do was far more experientially varied and tested than was KOR's. This circumstance explained why the Polish intelligentsia, including KOR generally and Kuroń specifically, was taken by surprise by the events of the Polish August. It explained why Kuroń was aghast at the bluntness of the first Gdańsk demand, just as it explained why the academics who went to Gdańsk had such "grave doubts" about the same postulate that they called it "unrealistic" and "irresponsible." It explained why, at the height of the crisis in the Polish August, Kuroń would go to extreme lengths in his effort to persuade the Strike Presidium to back off from their first demand. It explained why the Catholic Church, and particularly the Polish Primate, Stefan Cardinal Wyszyński, failed impressively to read the dynamics at work on the Baltic coast as of August 18, 1980. None of these three groups either theoretically or experientially comprehended the organizational troika that produced the Polish August—the occupation strike plus the Interfactory Strike Committee plus the preeminent demand for free trade unions. None anticipated the (different but related) experientially based communications networks that held the troika together: the united force of workers in the shipyard, linked in their understanding of events by loudspeaker, through which fifteen thousand workers physically protected the Interfactory Strike Committee; the hundreds of supporting factories which sent to the MKS the delegates that armed it with economic authority sufficient to force the government to the bargaining table; and the fact that these first two prerequisites constituted and explained the voluntary social formation upon which the demand for free trade unions rested and from which it derived its authentic political power.

In the course of sundry self-justifying rationales, various intellectuals in Poland and abroad would subsequently point to the workers' "complete linguistic inadequacy" and their "semantic shame and incompetence" as the reasons why "this class must be provided with the leadership of more enlightened people."[26] Except for those few who had matured within the coastal workers' movement itself, there was not a single representative of the party, the episcopate, or the intelligentsia who understood the workers' organizing achievement and the political realities it created. The power of cultural condescension to blind its possessors has perhaps never been demonstrated with more devastating specificity than it was on August 18, 1980, in the Lenin Shipyard in Gdańsk, Poland. The implications that flow from this historical fact war frontally against a number of sanctioned epistemologies that provide the structure of contemporary sociology, political science, and political theory.

With this as background, it is now possible to begin the task of placing the labors of Polish intellectuals into a coherent framework with respect to the structured emergence of Solidarność in Poland. But to do so, it is essential to trace not merely the ideas but also the actions of intellectuals as they functioned within the insurgent context created in Poland by the Baltic movement.

Prior to the emergence of KOR, the history of postwar protest activity by Polish intellectuals centered on three distinct moments—the immediate postwar years, when a number of intellectuals signed onto the Polish United Workers' party while others joined workers and peasants in the Home Army (AK) in opposing the Stalinist takeover; 1956–57, when the philosopher Leszek Kołakowski led a number of young communist revisionist intellectuals in creating the short-lived journal *Po prostu*; and, finally, 1968, a brief period of student protest known as the "March events." The latter occurrence implanted a deeply felt memory in the 1968 student population that had two subsequent effects: it energized a number of young people who later would associate themselves with KOR, and it provided for others a somewhat tortured explanatory rationale of postwar Polish history that would assist intellectuals in fashioning a post-facto causative role for themselves in the formation of Solidarność. The bizarre "March events" thus merit a brief review.

The affair in 1968 began with the closing by the government of the Warsaw revival of a drama entitled *Forefather's Eve* written by the nineteenth-century Polish poet and playwright Adam Mickiewicz. Set in an era of czarist occupation, the play was full of anti-Russian lines that proved highly popular with the twentieth-century Poles in the audience. The response night after night was so exuberant, in fact, that the Ministry of Culture, in a typically blockheaded invocation of the censorship, closed the play, thus precipitating a series of protests by Warsaw students. At one juncture, the party transported truckloads of security thugs, dubbed "workers," to the campus of Warsaw University to break up with physical force a student demonstration. The action seems to have confirmed the opinion of many Warsaw students and undefined sectors of the older generation of intellectuals that Polish workers were both doltish and brutish. In any case, after the Polish August of 1980, a four-step explanation for the origins of Solidarność became conventional wisdom within the Polish intelligentsia: (1) the intelligentsia protested in 1968, but the workers did not join them; (2) the workers protested in 1970 but the intellectuals remained passive; (3) KOR created an intellectual-worker coalition in 1976; (4) workers and intellectuals joined forces in 1980 and Solidarność was the result. Though relentlessly advanced by the intelligentsia as a post-1980 explanation of causality, this scenario was politically primitive. For participants in insurgent coalitions to act in concert, prior communication between the component parts of the coalition are essential. Neither the students in 1968 nor the workers in 1970 fashioned the necessary contacts with the other prior to the moment of protest. Indeed, after the students of 1968 and the workers of 1970 began organizing themselves, the thought belatedly occurred to each that allies would be helpful. Thus, the marching workers of Gdańsk in 1970 spontaneously decided to parade to the Gdańsk Polytechnic on an *ad hoc* recruiting mission, and the students of 1968 belatedly addressed a plaintive mimeographed appeal to workers. The protest agenda in each case was parochially restricted to student issues or worker issues. Indeed, in the case of 1968, the students failed to mobilize not only workers in their protest of the censorship but also the surrounding Warsaw intelligentsia. Adam Michnik, then a student, protested vigorously and was dispatched to prison, and Jacek Kuroń, a few years Michnik's senior, also protested and was jailed. But the intelligentsia as a whole, unorganized as it was, remained generally passive in the presence of student activity. The essential conclusion that emerged within the intelligentsia, inappropriately enough, was that Polish workers were unresponsive to the issue of censorship and that the Polish working class as a whole had somehow failed the students in ways the intelligentsia as a whole had not.

The role of experience in determining consciousness was vividly demonstrated by the varying responses to the events of August 14–18 of three cooperating Polish oppositionists. The three were Bogdan Borusewicz of Gdańsk, Jacek Kuroń of Warsaw KOR, and Konrad Bieliński of NOWa. All three men were college-trained intellecutals, all were victims of the government repression of student protests, and all were engaged or became engaged in pursuing a worker-intelligentsia coalition. Additionally, all three were associated with KOR, and all eventually came to feel that the achievement of free trade unions constituted the only ultimate solution for the plight of the Polish population. But the three militants had a radically different experiential relationship to the Baltic working class and, as a result, had radically different understandings of what was happening in the Lenin Shipyard. All, therefore, had differing insights into what was possible for the nation during the Polish August.

Borusewicz was an 18-year-old youth at the time of the December massacres on the coast in 1970 and, thus, played no role in the succession of occupation strikes or in the struggle over worker councils that followed in the 1971–72 period. But along with everyone else on the coast, he acquired the essential "memory of December" and demonstrated that fact by participating in the 1977 monument demonstration outside the Lenin Shipyard. Borusewicz was one of the core group of the Baltic free trade union in Gdańsk and suffered the repressions that were soon visited upon each of its members. He participated in the planning of the 1980 mobilization and closely followed the evolution of the strike committee of August 14 into the Interfactory Strike Committee of the sixteenth. He shared—effortlessly—in the commitment to the first Gdańsk demand that was so characteristic of the coastal milieu. Indeed, his strategic conception of precisely what helped and what harmed the workers' cause was so thoroughly in hand that he had no difficulty at all during the drafting session on the twenty-one demands in taking the floor to refute Tadeusz Szczudłowski's attempt to burden the independent union movement with a crude demand for total abolition of the censorship. As well as anybody in Poland, Borusewicz served as an intellectual who worked intimately over a prolonged period of time with workers.

Though Borusewicz could move within the atmosphere of the Warsaw intelligentsia, it would be an exaggeration to say that he could inspire the kind of rank-and-file confidence in his leadership that Wałęsa could summon among workers. But along with Gwiazda, Sobieszek, Szołoch, Lenarciak, Wałęsa, and the other veteran worker activists on the coast, Borusewicz understood the essential prerequisites for shop-floor democracy in Poland. As the challenges of August appeared, he was as ready for them as the range of his talents, which were considerable, permitted.[27]

In the years between 1977 and 1980, Kuroń did not familiarize himself with the working-class culture of self-activity that surrounded Borusewicz on the Baltic coast in the 1970s. Though Kuroń sometimes met Borusewicz when the latter made trips to Warsaw, Borusewicz's intellectual style and personal manner did not succeed in conveying a sense of the shop-floor organizing history embedded in the coastal working class. When Kuroń met Wałęsa for the first and only time prior to the Polish August, he was apparently unable to grasp the implications of Wałęsa's confident assertiveness or the range of prior experiences that undergirded that style.[28]

Thus, in the first week of the Polish August that preceded Kuroń's arrest in Warsaw on August 20, his response was broadly what one might expect of a worker-oriented metropolitan intellectual who was not grounded in the social formation that was at that moment emerging in a distant part of the country. Kuroń was delighted on August 14 to add the symbolically significant name of the Lenin Shipyard to the list of Poland's striking enterprises. And two days later, no intellectual in Poland appreciated more immediately and fully than Kuroń the strategic implications of the creation of the Interfactory Strike Committee in Gdańsk. Because of the telephone blockade, he did not learn of the formation of the MKS on the coast until the night of August 17. But when he did, he promptly told the foreign press (a source he had cultivated over the years) that the mobilization on the Baltic "created an essentially new situation" in Poland. Nevertheless, while remaining publicly enthusiastic, Kuroń was privately alarmed to learn of the first Gdańsk postulate. The unequivocal demand for a trade-union structure independent of the party was, said Kuroń, "in no way reasonable." The demand itself constituted a systemic challenge to the basic precept of a Leninist party as the controlling social force in society. As such, the demand would drive the party to order a violent repression.

In no sense was this stance a spur of the moment response. Indeed, behind Kuroń's reasoning was a lifetime of strategic brooding about the seemingly intractable problem of conceiving a coherent strategy that could bring democratic forms to Leninist Poland. Unfortunately, the analysis Kuroń developed following the organizing failure at Radom and Ursus was tactically immobilizing and strategically contradictory. But Kuroń held to his course as the nation's economic crisis deepened in 1978–79 and maintained it down to and through the Polish August. The practical effect was functionally central: it placed the democratic opposition at cross-purposes with any independent workers' movement that might be organized. Kuroń's analysis ensured that if and when such an organizing moment occurred, the specific groups of workers engaged in the mobilization effort would be able to extricate themselves from KOR's self-conceived cul-de-sac only by charting a different course of their own. Since the organizing dynamics necessarily transpired publicly where all the elements were highly visible, the structural impediments intellectuals threw up in the presence of worker activity were quite plain to see.

KOR's faulty reasoning had its origins in a fundamental misreading of the self-activity of Polish workers over the long time span from the coming of communist authority in 1944–45 down to the Baltic mobilization of August 1980. Over this entire period, Poland's workers, confined within the intense shop-floor control of the party's trade unions, engaged in a prolonged institutional escape attempt, a collective effort to break the barrier of the party's iron censorship over their working lives. Throughout the postwar era, the party responded in a consistent manner; it made promises it did not intend to keep and did not keep. Though the evidence of this dialectic was quite plain, it involved dynamics of worker organizing that Kuroń, Michnik, and other Polish intellectuals did not grasp. Indeed, these relationships play no role in the worldwide interpretation of Solidarność, even though they have exerted a controlling force both on the way the Polish intelligentsia reacted to the Polish August at the time and how that decisive climax has been subsequently portrayed.

After the defeat of the independent shop councils by the security services in the 1944–47 period, workers endured the bitter years of Polish Stalinism before a four-month organizing effort at Cegielski in 1956 opened a new chapter of struggle. Cegielski workers exposed and immobilized successive layers of party duplicity in W-3 shop sections, in W-3 as a whole, in the Cegielski plant as a whole, in the provincial party in Poznań and, finally, in the Ministry of Machine Industry in Warsaw. This prolonged effort set the stage for a monumental public deception by the party when, on June 27, 1980, in a mass meeting at Cegielski, a high party official named Roman Fidelski publicly contradicted what he had promised to the Cegielski delegation on June 26 in Warsaw.

What happened next illuminates the first specific misreading by Polish intellectuals of the democratic self-activity of Poland's workers—setting in motion the interpretive dynamics that would produce the nationwide misreading of the Polish August. Having exhausted every internal mode of assertion available to them, Cegielski workers took their organized expression "out-of-doors where it could be seen." This forthright effort unfortunately exposed a prior organizing lapse, the absence of thick lines of internal communication between the Cegielski workers' movement and the rest of the enterprises in the city (see Chapter 2). The Cegielski workers lost control of their own movement, and the riotous "Poznań events" were the result. While the rising set in motion the complex intraparty struggle that culminated in Gomułka's accession to power in the "Polish October," the nation's intellectuals generally failed to make this causal connection—one that made publicly visible the potential structural power of the nation's working class under Leninist confinement. A second and more subtle interpretive lapse was the failure to see the elaborate organizing effort that preceded the Poznań rising. Polish intellectuals saw only June 28, not the months of intricate organizing before June 28. They saw, in short, a one-day carnival of violence by the unlearned. The third interpretive failure was to overlook the growth in consciousness at Cegielski generated *after* June by these organizing experiences. Fidelski's gross deception, massively fortified by the police and judicial repressions that followed, convinced Cegielski workers of the need for unions independent of the party. An entire work force of nonbelievers surrounded a shaken coterie of party apparatchiks, some of whom themselves had become covert nonbelievers.

Meanwhile, in their maneuvering against the Stalinists, the so-called "Polish faction" in the Central Committee moved in July 1956 to broaden its popular base by authorizing workers' councils. While the Cegielski movement was by far the most experienced and therefore the most autonomous of the social formations that emerged in Poland's industrial working class in 1956, even the party-organized councils were sites of worker assertion against the power of the state-owned trade unions. In the months of worker self-activity that followed, the expedient of the "experimental program" was thrown into the breach by the party to redomesticate the shop councils under party control.

Precisely the same dynamics—but involving still additional levels of organizing achievements—functioned on the Baltic coast in 1970 and 1971. When Gomułka's December restructuring of food prices found workers once again securely confined within party-controlled unions, coastal workers tried to break free by taking their struggle out-of-doors and into the streets. Once again, the dialectics of organized assertion and consciousness-raising inexorably transpired. To contain the workers in the Lenin Shipyard following the rioting on December 14, the party's shipyard activists organized an occupation strike on December 15. The very next day, nonparty workers reorganized the strike under their own leadership. Not only did similar worker-controlled structures emerge in Gdynia and Szczecin, but, even more important, the conceptual breakthrough to interfactory organizing emerged. However, the people in such expansive structures were not experientially grounded in knowledge of how to conduct cooperative assertion within them. Nevertheless, they learned a great deal in December 1970 and even more in the opening months of 1971 about the essential dynamics of interfactory organizing.[29]

Once again, as at Cegielski, the sheer violence of the repression in 1970 fortified worker alienation from the state-owned trade unions. This time, however, the memory of the "December massacre" was so profound that it worked its will on nonworkers as well, so that party members, the clergy, and white-collar intellectuals all shared in different ways in a generalized conviction that a new route to civil peace needed to be created in Poland. So widespread was this belief on the coast and so intense was the memory of the December massacre in the Baltic working class that popular self-activity in the 1970s reached levels of cooperation and experientially based sophistication unmatched anywhere else in Poland.

On the docks, in the shipyards, in the transport services, and in the maritime supply enterprises, occupation strikes increasingly materialized as the decade proceeded. This occurred even in the period of "success propaganda" in the first part of the decade, but it became even more visible as the economic crisis deepened in the second half. The workers who participated in these self-organized job actions and the activists who led them all internalized the experiences—and the social knowledge that accompanies such experiences—of movement building. Collectively, what Baltic workers were experiencing in these years can be understood as the structural prerequisite for popular democratic politics; namely, the acquired knowledge that effective popular assertion was necessarily tied to controlling one's own organizational space.[30]

As the eighteenth-century American revolutionary Thomas Jefferson had understood the precise practical and theoretical relationship that must exist if political forms are to work in a democratic manner, a functioning civil society depends upon organized "elementary republics." The precise application of the analogy to Poland was sharply limited by the fact that in modern Leninist states, small "elementary republics"—plant-sized worker councils—were easily overwhelmed by the multilayered vertical structure of the party-controlled central trade unions. To be practically capable of social self-defense, "elementary republics" in Poland needed to be very large.[31]

Baltic workers, however, were not consciously pioneering democratic theorists. They came to interfactory organizing as a practical evolution in their prolonged struggle with Leninist trade unions. The point is that they reached this organizing plateau as a direct experiential product of their organizing efforts over time. By the end of the decade of the 1970s, the occupation strike was an experientially tested skill of the coastal working class, one confirmed not just at Elektromontaż, for example, but also at a number of enterprises in the 1970s—and in 1976 at literally scores of enterprises throughout the maritime provinces.[32]

But Kuroń, in company with the Warsaw intelligentsia as a whole, did not know this. Kuroń did not "see" the occupation strikes across Poland in 1976, the ones in the years before 1976, or the ones after. He saw, rather, the violence at Radom and Ursus. Reinforcing this impression was the ruthless party repression he, his KOR associates, and the workers had suffered in the aftermath. The Warsaw intelligentsia generally and Kuroń personally also had a different memory and understanding of the December rising on the Baltic in 1970. The immediate evolution of occupation strikes that began to materialize *after* the first day of street politics and the long strike-marked struggles and demand-gathering campaigns that followed in 1971 also constituted events that did not register within the world of the intelligentsia.

In short, Jacek Kuroń did not see the experiential progression of working-class self-activity in Poland from 1945 through 1980; he saw, instead, a consistent thread of worker politics, one that seemed to characterize itself again and again as a violence-prone social formation. Though Warsaw intellectuals fully shared this presumption, there were elements in Kuroń's own worldview that exacerbated his individual misreading. Because of his own fixation on structural change through reform elements in and close to the party, Kuroń found it difficult to make a structurally serious distinction between a party-organized shop council and a worker-organized one. Thus, Kuroń, along with many subsequent students of Poland,[33] saw, as the capstone of worker achievement in 1956, Lechosław Goździk and the party-organized shop council at Żerań, not the independent organizing achievement of Cegielski workers at Poznań. About the latter, he knew nothing. About the organizing experiences and resulting political evolutions within the Baltic working class in 1970, in 1971, and in the years thereafter, he similarly knew nothing (or at least nothing that he understood). What he did know—and he knew it with experiential power from Radom—was the capacity of the state to repress independent shop councils.

This combination of empirical ignorance on the one hand and highly selective experiential knowledge on the other led Kuroń to a fatal conclusion. He developed a structural analysis for the democratic opposition in Poland that was programmatically untenable. He shared with Adam Michnik the strategic insight that the working class was the one social formation that the party really feared. Kuroń had also been able to convince younger oppositionists, such as Jan Lityński and Henryk Wujec and others, in KOR, of this reality. Though the editorial staff they had put together for *Robotnik* was heavily dominated by intellectuals (one worker was on staff) and though *Robotnik* additionally was unable to chart an organizing strategy that concretely addressed the social circumstances prevailing on Poland's shop floors, the journal nevertheless represented visible evidence that at least some intellectuals in Poland sought an authentic worker-intellectual coalition against the state. This was a genuine signal, but it is important to specify the sharp limits to its practical meaning in organizing terms.

In concluding that workers could not successfully organize outside the framework of the party-controlled trade unions, Kuroń set a strategic course that unwittingly gave license to his greatest fear. The temporary rampages in the streets in 1956, 1970, and 1976 had emerged from local work forces possessing genuine and deep-seated grievances that they were desperately seeking to express. It was precisely because they could not assert themselves within the party-controlled unions that they took to the streets. In the absence of independent worker structures, the streets became the only remaining channel through which workers could perform an act of protest. If this energy were to be harnessed for coherent autonomous activity, it needed its own self-generated instrument of expression. Independent unions might be dangerous (no question about *that*), but they were *less* dangerously explosive and self-defeating than no independent unions at all.

The Achilles heel of Kuroń's vaguely defined strategy for infiltrating the official unions was that no informed worker activist in Poland had any respect for that course of action. The Matyjas, Bałukas, Hulszs, Goździks, Nowickis, Lenarciaks, Sobieszek's and Wałęsas of Poland's working class had all tested that route, individually and in company with fully organized independent shop councils. They had all learned the structural power of the vertically organized party trade-union structure. To all of them, the idea was hopeless—an illusion potentially persuasive only to the inexperienced and the innocent.

Poland's Baltic workers had especially learned this unwanted truth inside the party-controlled unions at Szczecin, Gdynia, and Gdańsk in 1971–72 and had confirmed it again and again throughout the rest of the decade. Still, the most energetic activists kept experimenting for some formula for autonomous activity outside the party unions—at Gdynia in 1974, on the docks in 1975 and 1977, at a score of enterprises in 1976, at Elektromontaż in 1979, and in the transport system down to August 2, 1980. While they had not yet found a formula for breakthrough, they had confirmed one elementary fact: the official unions were not a site of hope.

What coastal workers did accomplish, however, proved to be one of the organizational centerpieces of the Polish August. In the years of experimental thinking and acting between 1970 and 1980, the decisive ingredient that Baltic workers constructed through their own efforts was a thick web of internal worker communications outside the official trade-union structure. The shipyard activists knew which workers could shut down public transport in the tri-cities; Henryka Krzywonos and Zdzisław Kobyliński. They knew who counted with the workers at Gdańsk Port: Stefan Lewandowski; and at Siarkopol: Lech Sobieszek. If Wałęsa so much as blinked, Florian Wiśniewski and a small army of militants at Elektromontaż would move into independent action—to honor themselves as well as to pay their respects to the murderers of Tadeusz Szczepański. And the activists knew that Bogdan Lis and Andrzej Gwiazda could probably shut down Elmor any time the word to do so went out.

Experientially, something was in place on the Baltic coast that Kuroń had never heard of—an interfactory structure of cooperation. Moreover, the activists possessed interior lines of communication that could make such a structure of cooperation function under pressure. Jacek Kuroń drew the wrong lesson from Radom and Ursus in 1977. As a result, he oriented himself to a course that was diametrically in conflict with the workers' course.

These differing perceptions—within the democratic opposition in Warsaw and among the working class on the coast—asserted a controlling force upon the decisive events of August 17–18, 1980, in Poland. The resulting contradiction was to prevail throughout the remaining days of the Polish August and was to exert an elaborate influence on how these events would later be perceived and written about. Indeed, these contradictions were to ensure that internal tensions within the Polish movement would persist through the remainder of the decade.

Kuroń, KOR, and the entirety of the Warsaw intelligentsia on the one hand and the Baltic working class on the other lived in different experiential worlds; they would pursue radically different paths through the Polish August. These divergent trajectories were quite public and easily visible in the twenty-four-hour period between nightfall on August 17 through the climactic day of mobilization that followed on August 18. To view this decisive moment from the inside is to uncover the contradictions that played themselves out on the critical day of the Baltic mobilization.

When the phone lines were cut off on August 15, Kuroń knew only that an occupation strike had been organized in the Lenin Shipyard. He learned about the formation of the MKS and the drafting of the twenty-one demands on the evening of August 17, when young Konrad Bieliński, who had come upon the coastal strike in the course of a Baltic vacation, returned to Warsaw with a copy of the twenty-one demands. Though Bieliński was full of enthusiasm about what he had seen, and Kuroń himself duly noted the existence of something called an Interfactory Strike Committee, the older man was not pleased with the programmatic result of that organizing achievement. The first demand for independent trade unions struck Kuroń as transparently excessive. Still, the Lenin Shipyard was a famous name, one totally enveloped in the mystique of December 1970, and it represented a notable addition to the list of sites of industrial discord that had punctuated the Polish scene since the beginning of July.

Kuroń's tactical decision on the night of August 17 was to try to capitalize upon the Gdańsk strike through his carefully cultivated network of foreign journalists but to do so in a way that put the best possible explanatory shape upon the political meaning of the workers' demands. The details of information packaging on August 17 paint in rather brilliant hues the contours of all that followed.

The call for free trade unions was part of the demands of the original strike committee of August 14, reflecting the heritage of 1970–71 on the coast and the demand-gathering campaigns of those years. The same objective was the cornerstone of the sixteen demands fashioned by the newly formed MKS on Saturday night, August 16, and fleshed out to a full twenty-one demands on Sunday. But in his carefully crafted interpretive analysis drafted for the foreign press corps, Kuroń bypassed the first Gdańsk demand and instead introduced the third demand, which he emphasized as being the first of the issues that were "on the more political side."

Kuroń had a well-earned reputation and attendant status as a news source in Warsaw. Western correspondents had a rather large and predictable brand of skepticism through which they viewed official government press releases on almost any topic—and particularly on the subject of protest activity in Poland. Kuroń's fact-gathering network was a great boon to journalists imprisoned within the party-controlled news environment. Information provided by Kuroń had proved accurate time and again and, indeed, served as effective raw material for questions journalists put to the Politburo's propaganda chief, Jerzy Lukaszewicz. More than once, Lukaszewicz would deny facts contained in Kuroń's press advisories only to have subsequent events force him to recant. Interestingly, the relative reliability of Kuroń's press summaries paradoxically set up a new level of competition among foreign journalists. Since all got the same information from Kuroń's press releases, the more aggressive journalists personally sought him out by phone for additional bits of news to give their stories a character distinct from those of their journalistic rivals. Kuroń, needless to say, enjoyed this competition for his views and was quite happy to oblige.[34]

Radio Free Europe, untroubled by hard deadline imperatives and possessing access to all Western news dispatches, was able to summarize the totality of Kuroń's messages, conveyed both in his press releases and on the phone to individual journalists, and the station could do this with a thoroughness that often went beyond that which any single reporter could achieve. One of the station's staff writers, Ewa Celt, had much from which to work in the daily press reports of August 18. Drawing on the Warsaw dispatches of the *New York Times*, Associated Press, and United Press International, each, in turn, based on Kuroń's interpretive summary of the Gdańsk demands, the Radio Free Europe staffer reported that "on the more political side, strict observation of freedom of speech and the press is demanded, including ending censorship and reprisals against independent publications, as well as access to the media by representatives of all religious dominations." Still tracking the KOR interpretation, Radio Free Europe went on to list other demands pertaining to the excessive privileges enjoyed by the party apparat and the security services. A new topic, described as "in the labor area," was then introduced in the following language: "The committee demanded the right of all workers to strike and official assurance that those involved in any labor disputes would not be victimized." Buried deeply in both Kuroń's press release and the Radio Free Europe report was the only reference to the first Gdańsk demand, a brief sentence to the effect that the MKS "also called" for "full freedom of unions in the country as guaranteed by the 87th International Labor Convention signed by Poland." This vague allusion was immediately followed by two obfuscating sentences to the effect that "all interference on trade-union activities by administrative agencies must be stopped," and, "moreover," the official media "should henceforth provide accurate and adequate information on recent developments in the labor dispute as well as the formation of the MKS." In its concluding interpretation, Radio Free Europe passed on KOR's privately gathered "assurance" from "circles close to the party" that the government "had no intention whatsoever of using police or military forces to quell the unrest."[35]

Two elements that bore most directly on the dynamics of the Polish August became clearly visible in the Radio Free Europe report. First, it was possible to read the KOR press release—indeed, it was structured to encourage precisely such a reading—to indicate that the workers' demand for labor reforms pertained to the structure of the existing official unions. The opening twelve words of the Gdańsk document, calling for "acceptance of free trade unions independent of the Polish United Workers' party," is nowhere visible in the KOR analysis; nor is it visible in the subsequent Radio Free Europe summary. The strategic substance of the demand for free trade unions was so blurred in content and minimized in its centrality that it virtually disappeared.

Kuroń's instinctive focus on reform groups "close to the party"—a perspective that was to dominate his understanding of the strategic meaning of the Baltic strike—was also foreshadowed. It was significantly coupled with another overriding concern, active since Radom, about police repression. Indeed, the entire subject of the Gdańsk demands "in the labor area" was introduced by a reference not to the first demand for free trade unions but to the demand for the right to strike and an accompanying reference to the "official assurance" by Polish authorities that those involved in labor disputes would not be victimized. To underscore this emphasis, Kuroń connected both themes in his conclusion that "circles close to the party" had given assurances against repression. A week later, at the height of the Baltic crisis, these themes would dominate Kuroń's futile advice to the strikers.

But the most immediate relevance of both the Kuroń interpretation and its faithful reproduction by Radio Free Europe concerned its relative unserviceability as an instrument of movement building. This central political aspect is most usefully considered in the context of events in the interior of the movement on the Baltic coast.

While Kuroń placed what he regarded as a much-needed protective veil over the workers' demand for independent trade unions, activists on the coast employed precisely the opposite emphasis on August 17. Communications between the tri-cities and the rest of Poland may have been cut off, but the phones were in working order within the metropolitan area itself. They got unprecedented use that Sunday night. Not only were the activists of the Strike Presidium, representing the twenty enterprises already affiliated with the occupation strike in the Lenin Shipyard, busily organizing, but veteran activists in other coastal enterprises were on the phone to each other and to their respective rank-and-file workers. That same Sunday night, the red and white flag of Poland was unpacked from scores of closets, and plans were made to display the national banner the next day—on the front of buses and trams, from shop-section ceilings, from enterprise gates, and from the windows of office buildings. All would help instill the necessary "movement spirit."[36]

But the most inspiriting element in the entire coastal panoply was the opening twelve words of the first MKS demand, calling for "acceptance of free trade unions independent of the Polish United Workers' party." To the citizens of Gdańsk, Gdynia, and Sopot, the twelve words had a wonderful clarity about them, the kind of clarity that could move people to perform a political act that long had intimidated them—public opposition to the party-state.

A central reality concerning the cultural components of democratic politics, as distinct from the familiar fabric of elite politics, here comes into full view. For ordinary people remote from power to be encouraged to collective action, clarity of purpose is needed. For widespread resignation, carefully instilled by centralized power, to be transcended, a clear and patently worthwhile objective is needed. For nagging and immobilizing fear, carefully cultivated by ruling authorities, to be overcome, evidence that the goal is worth the risk is needed. "Free unions independent of the party!" In Poland, the opening words of the first Gdańsk demand passed this test in a way unrivaled by any public document since the end of World War II.

On Sunday, August 17, as Kuroń crafted his carefully worded interpretation for the foreign press, movement printers in the tri-cities worked overtime running off leaflets of the twenty-one demands. Under the pressure of time and paper availabilities, some shortened the list. Small broadsides were created, restricted to the first three demands—on the free unions, the right to strike, and the censorship. A slightly longer broadside also appeared, listing the first six demands. In whatever detail of specificity, in whatever quality of reproduction, the demand leaflets were snapped up by waiting activists who, that night, knew they possessed the organizing opportunity of their lives. If the quiet workers—the family men and the intimidated ones—were ever to make their own private statement against the party, now was the time. The message from the activists was reassuring; all any worker had to do was stand with the Lenin Shipyard. In so doing, they were standing by the bus drivers and the tram operators, standing by the supply enterprises that had already joined the MKS—with many more sure to follow. The movement's message was clear: get ready to meet in the shop in the morning and elect delegates to the Interfactory Strike Committee. Unfurl the red and white flags. Unions free of the party, now or never.[37]

On Monday, August 18, after the moment at dawn when the militant named Wałęsa made his short speech at Gate No. 2, the word swept through the tri-cities that the whole shipyard had become one big occupation strike. By nightfall, no less than 156 enterprises in the tri-cities had joined the Gdańsk MKS. In so doing, coastal workers consolidated the structural base of the Polish August.

In its initial phase, the mobilization was the product of the intimate personal associations that coastal activists from dozens of enterprises had made with each other in the years since the December massacre had first bound the people of the coast in a passionate—and angry—memory. At the outset, these associations constituted the internal communications network of the coastal movement. It was soon augmented by the loudspeaker system that connected shipyard workers to each other and to the strike leadership. But in the early afternoon, when the number of affiliates to the MKS had already passed one hundred, another ingredient was added. When it became apparent that the mobilization would be successful in the immediate bay area, it also became clear that couriers were needed to carry the necessary organizing details to the rest of the coast and to the rest of Poland. But ominous news was also widely distributed about the roadblocks that the security services had thrown up around the Gdańsk region. Cars driven by workers were being routinely turned back.

This was the movement moment for the middle-class activists of ROPCiO, Young Poland, and KPN—people who did not look like workers and therefore might have better luck getting through police blockades. But these activists, too, needed an effective "cover." On Monday afternoon, a physician left Gdańsk to perform consultant work on surgery scheduled in a distant city. A supply truck left Gdańsk loaded with arcane pipefitting components ostensibly needed for a Soviet ship docked in Szczecin. A technician at a coastal institute departed for a research project near Elbląg. A student from Gdańsk Polytechnic set out to take some lab equipment to Wrocław. The supply truck to Szczecin got through; the physician and the student did not. The courier war had begun. In due course, the first announcement went out over the shipyard loudspeaker listing the license plates of unmarked automobiles of the security services that were being used to trap couriers. What might be called the preconditions for what would soon become the negotiating sticking point of the workers' "preconditions" were being experientially established.[38]

That Monday afternoon the broadcast of Radio Free Europe, with its strange, Kuroń-contrived emphasis on every part of the twenty-one demands except the most important demand, was heard by those Poles in the Gdańsk area who were listening. But the information provided was silent on the mechanics of movement building. To those distant from the coast who did not know what an MKS was or how to join it, Radio Free Europe could provide no information and neither could Jacek Kuroń. It was not merely that they were structurally uninformed; the key recruiting tool—the union free of the party—was missing from the message. The most relevant subsidiary informational tool—what to do to prepare a factory for affiliation with the MKS—was also missing.[39]

By Wednesday, August 20, the MKS counted over 250 affiliate enterprises and the chairman of the Strike Presidium began telling visiting journalists that the strike could go on for five years because the fear was gone and "the shipyard workers are not the same people anymore." Late in the evening on Friday, August 22, the first of the academic emissaries, two men from Warsaw named Geremek and Mazowiecki, arrived in the shipyard bearing both a letter of support and anxious worries about the first Gdańsk demand. Within three more days, Wałęsa individually, the Strike Presidium as a whole, and rank-and-file shipyard workers would all participate in the task of convincing the advisors that the first demand did not require amendment or modification. The advisors, instructed by this, but, perhaps more profoundly instructed by the organizing scene visible in the shipyard itself, willingly took their station in the Polish August. The date was August 25, 1980.

But in Warsaw, Jacek Kuroń and KOR were unable to make a similar transition. Possessing no experiential knowledge from the shipyard with which to augment their prior strategic understanding, the theorists of KOR fidgeted in prison from August 20 through the final resolution of the crisis on August 31. Information they did receive was garbled, contradictory, and (whether from police sources or party sources) variously suspect. But nothing that Jacek Kuroń heard dislodged his prior conclusion that the workers' demand for independent unions was a classic example of "impossiblism." By insisting on this structural transformation, the coastal militants were therefore endangering the real gains already won—or at least the gains that were rumored to have been won.

By August 25, one thing was certain. The workers were still insisting on their preconditions, and no talks had been held with the government since the opening session on August 23. During the two-day test of wills between the government and the Strike Presidium, the government had staged a bit of bureaucratic theater by replacing a number of party functionaires. But the gambit had left the strike committee unmoved. On the central issue of free unions independent of the party, the government was also unmoved. Gierek was content merely to speak favorably about free elections within the existing trade-union structure. After two days of this, Kuroń could stand still no longer. On August 25, his views, long held and fulsomely orchestrated as recently as August 18 to the foreign press, were dispatched one last time through the Warsaw correspondent of the Stockholm daily, *Svenska Dagbladet*. Kuroń's strategic interpretation was relayed through a (necessarily) unnamed KOR spokesman who had escaped the police dragnet on August 20. The Swedish correspondent Boho Sheutz conducted a long interview with the KOR representative and entitled his subsequent report "Call Off the Strike." He quoted KOR to the effect that "the strikes have now achieved more than anyone could have imagined when the conflict started, and the strike committee ought now to decide to call off the strike after getting all the regime's promises properly in writing." The Swedish reporter specified that the KOR analysis came on the heels of the Polish Politburo's reshuffling of high-level party functionaries and after the government had "promised workers free trade-union elections." Correspondent Sheutz went on to report that "through its channels in Gdańsk, KOR is trying to persuade the strike leaders that they must break off the strike in order to make sure of the successes that have already been achieved."[40] KOR reasoned: "The promise of free union elections with a limited number of candidates could be the seed of a reform movement within the system, but what is important is to see to it that the promise can be put into effect, so the trade unions do not become the regime's tools again in a couple of years. Even though it is too early to say what changes in the government and the Politburo will mean in the long run, everything indicates that the reform groups within and close to the party now have a better opportunity to make their voices heard."[41]

Such reasoning reflected Kuroń's strategic conclusion as to the limits of possible change in People's Poland, one grounded in an acceptance of the centrality of "reform groups in and close to the party." If properly and prudently encouraged, such groups might "make their voices heard" so as to plant "the seed of the reform movement within the system." This approach, of course, constituted a far different understanding of the social base of reform than one conceptualized as a mass-based social formation—a trade union—existing independent of the party.[42]

While Kuroń remained isolated in Warsaw, the young mathematician, Konrad Bielińksi, underwent a profound political metamorphosis during the last days of the Polish August. Bieliński had never been a part of the worker-oriented faction of KOR. He had cut his teeth in opposition activity at Warsaw University during the student protest of 1968 and had deepened that experience in another round of confrontations in 1973. After his dismissal from the university, he became an early participant in KOR activities and had soon gotten caught up in the slippery world of underground printing and publishing. With his eyes fixed firmly on the intelligentsia, Bieliński drifted quite naturally into the exciting and conspiratorial realm of NOWa, where he became a close associate of Mirosław Chojecki. His editorial interests were focused in the intellectual journal *Krytyka*, not on the less-elegant journal *Robotnik*. By 1980, he had become a seasoned oppositionist, sophisticated in the ways of police surveillance and experienced in the art of underground publication. Bieliński was one of KOR's second story men, a tactician of high culture trained in stealth. The theoretical and tactical world of *Robotnik* and of Kuroń, Lityński, and Wujec—where planning around kitchen tables considered the seemingly intractable problems besetting the formation of a worker-intelligentsia coalition—was not his natural milieu. Chojecki and the world of NOWa was. But the molecules of historical causation have certain affinities. As luck would have it, a new duplicating machine for NOWa arrived in Poland from the West early in August 1980. Chojecki and Bieliński left Warsaw and drove to the place where it reposed, ready for pickup. They decided to combine the chore with a seaside vacation. And so it was, on the weekend of August 16–17, that the two NOWa operatives found themselves in Gdańsk on the Baltic coast.[43]

What Bieliński and Chojecki encountered was the activated insurgent culture of the Baltic coast. Gdańsk was a city in which party functionaries were grumbling about one-hour warning strikes by the municipal transport workers, a city in which enterprises that were on strike were being cheered and those that had not done so were making ready to earn some cheers. It was a weekend in which two thousand townspeople joined five thousand shipyard workers in a special outdoor Mass to bless a cross dedicated in memory to the victims of the 1970 massacre. In short, on the weekend they spent on the coast, Bieliński and Chojecki encountered worker politics, sacred politics, and organizing politics of a kind they had never seen before. In politically activating ways, this experience seemed to make up for any number of missed sessions around Jacek Kuroń's kitchen table. One did not have to be a theorist of intelligentsia-worker coalitions to sense the importance of what had happened in the Lenin Shipyard. On Sunday night, August 17, Bieliński took the duplicating machine and his new awareness back to Warsaw and immediately put both to work. After showing the twenty-one demands to Kuroń, Bieliński began making copies while the *Robotnik* circle set about drafting a statement of support for the coastal workers.[44]

The same day on the coast, of course, marked the organizational consolidation of the Polish August. From August 18 forward, everything in Poland accelerated geometrically. The party, awakening from its torpor, created a special commission to deal with the Baltic troubles and appointed a functionary named Tadeusz Pyka to head it. He immediately left Warsaw for the coast and began calling in the party troikas from coastal factories as a prelude to fashioning a strike-ending paper agreement that could be trumpeted by the government media. Meanwhile, within the opposition, even as security forces arrested more than a dozen KOR activists, Konrad Bielińksi went not to jail but back to Gdańsk. Chojecki had been recruited to courier duty and had set out by automobile on the nineteenth for the industrial city of Elbląg on the upper coastal region near the Soviet border. He was soon intercepted by security forces and sent back to Warsaw. But that same afternoon, worker couriers arrived in the shipyard from Elbląg. They proved to be more than mere couriers seeking information or even delegates joining the MKS. The information they carried was truly uplifting: Elbląg, too, had formed an Interfactory Strike Committee and enterprises in the region were at that moment affiliating with it. And from Szczecin, at the other end of the Baltic coast on the East German border, came word that an MKS had also been formed there. The phones were down, but Szczecin had gotten early word about Gdańsk anyway—from couriers dispatched from Gdańsk. The ethos of the coast was something of a marvel; the memories of 1970 were very long.

Here was something that no theorist and no tactician—inside the working class or outside of it—had ever thought through. A way did indeed exist to form public committees and make public demands in People's Poland without subjecting the organizers to instant police repression. Surround the organizers with an occupation strike of thousands of workers and augment the strike committee with solidarity strikes by hundreds of other enterprises. It was not Kuroń's idea nor Wałęsa's nor Borusewicz's nor Wyszkowski's. Rather, quietly it had simply grown out of the accumulated experience of the coastal working class itself, a kind of collective imagination which, step by step, seemed almost as if it were orchestrated by collective wisdom. But it was not orchestrated. It simply grew, logically, out of the coastal workers' own experience, as they improvised to meet the successive challenges to their organizing effort. It could be understood, after the fact, that the Polish August had a certain democratic ring to it.

Such, of course, was not a sanctioned way for persons of authority in Poland to analyze popular movements. The party retrospectively saw—was already announcing that it saw—the conspiratorial hand of antisocialist forces, outside agitators, who were manipulating the honest postulates of the working class. In a fit of unintended self-mockery, the party would prove it thought these manipulative pseudoworkers were named Lech Wałęsa, Anna Walentynowicz, and Andrzej Gwiazda. The party announced that it would agree to negotiate with the MKS if these three subversives ("hard ones") were eliminated. With even sharper unintended self-mockery, the party would soon specify that the manipulators were, in fact, Jacek Kuroń and his antisocialist cabal in KOR.

Confusion about causation was by no means limited to the party. The office of the Primate—in an unfortunate moment of myopic pique at Wałęsa for his repeated demonstrations of independence from high church advice—would eventually pinpoint "Trotskyite" influences manipulating Wałęsa and the workers' movement. By this, the Warsaw functionaries of the Church meant to pinpoint Kuroń. With disarming candor, Kuroń undercut everybody (including the enthusiasts of KOR) by saying blandly that he had opposed the first Gdańsk demand when he heard about it, opposed it upon reflection, and still opposed it while under police detention. "I thought it was impossible, it was impossible, and I still think it was impossible." All was said in joyous, triumphant, and candid good humor. Kuroń was not maneuvering for a personal place in history; he was trying to help democratize Poland in his own lifetime.[45]

Despite all this confusion as to who generated the workers' movement, almost everyone tried to contribute to it—especially workers. Gdańsk bakers brought bread; farmers brought food. A giant poultry co-op announced it was disastrously short of fodder; along the Northern Harbor, longshoremen promptly unloaded a grain ship and truckers took the fodder to the co-op; the co-op sent chickens to the shipyard, to the longshoremen, and to the truckers. It all seemed a lot more intelligent than the party's mode of planned underconsumption. The militants of Young Poland, those not serving as couriers, helped Alina Pieńkowska oversee the gathering of food and medical supplies outside the shipyard. Many other students who had no prior political experience showed up to help. What internal communications network sent for them? The word-of-mouth milieu of the coast itself.[46]

The party helped, too. The claims of the Pyka commission—that seventeen factories had agreed to back-to-work settlements—were destroyed in speeches made to the MKS by delegates from the same factories who provided intimate details of Pyka's worker agreements made without benefit of conversations with workers. The lateral lines of communications the workers had constructed undermined the party's power to lie and thus the party's power. The fact that delegates from the seventeen factories openly cast their lot with the MKS improved everybody's morale and confidence. On August 21, the party recalled Pyka to Warsaw and sent in his place a deputy premier of People's Poland, Mieczysław Jagielski. He was instructed to reevaluate the entire coastal situation and, if necessary, to negotiate with the MKS. Too many enterprises on the coast had been shut down for too long.[47]

The Church helped. Individual priests and bishops found ways—supplies, people, food, money, suggestions. This last was quite mainstream: everybody had suggestions for the strikers.

The intelligentsia helped. On the twenty-first, Lech Bądkowski, a Gdańsk writer with a record of past difficulties with the censorship office, organized a meeting of the Gdańsk Writers' Union and presented a manifesto of support for the twenty-one demands. Their endorsement in hand, he went to the Lenin Shipyard where, after weathering the credentials check at Gate No. 2, he was permitted to approach the presidium, on which a small man with a moustache presided. Wałęsa greeted him with a smile and the words, "At last our literary men are here." The presidium chairman introduced Bądkowski and invited him to read his statement over the microphone. It seems that Bądkowski had been censored one time too many, and he was determined to show what he could do when released from bureaucratic restraint. His lyrics pulsed with the rhythms of freedom and his melody rang with national hope. The manifesto earned a roaring ovation from the MKS. Some of the delegates—quickly joined by the others—spontaneously rose to sing the national anthem. It was a moment of unity and joy, and Wałęsa seized it. He proposed Bądkowski's immediate election to the Strike Presidium and, in the generosity of the moment, secured it unanimously. After thinking it over, Gwiazda and Borusewicz decided Wałęsa had been too cavalier in his willingness to tamper with the nerve center of the workers' movement. As Wałęsa saw matters, he was merely engaging in the movement's oldest activity—recruiting. He felt it should be self-evident to everyone that he was simply broadening the base of the working-class movement. He did not give the matter a second thought.[48]

Thirty-six hours later, Wałęsa met for the first time the two arriving intellectuals from Warsaw who would soon lead the team of advisors to the MKS. In view of the long-term ramifications of this meeting—in 1989 Mazowiecki would become prime minister, Geremek would become leader of the popularity elected deputies in the Polish parliament, and Wałęsa, of course, would stand as the worldwide symbol of Solidarność—it is prudent to review the immediate background that informed their first meeting. The intrusion of the Warsaw intelligentisa into the worker milieu of the shipyard was the sequential product of actions initiated by one intellectual and one worker—the editor of the Catholic journal *Wież* Tadeusz Mazowiecki, and the chairman of the Strike Presidium, Lech Wałęsa.

The first initiative was by Mazowiecki. On August 20, as the police rounded up KOR activitists, the Catholic editor recruited to a meeting in Warsaw a number of people from the party's reform faction, others from the Warsaw Club of the Catholic Intelligentsia, and still others associated with the Flying University. With memories of 1970 on the coast vivid in their minds and in a crisis heightened by the volatile nature of the workers' first demand, it seemed imperative that Poland's progressive intelligentsia draft a statement on the situation on the Baltic coast.

The result was the "Appeal of the 64." The intellectuals were unequivocably direct in fixing responsibility: "As a result of years of ill-considered economic decisions and the authorities' assumption that they were infallible, the crisis arrived." The sentence directly reflected the pent-up frustration of the nation's technical intelligentsia and its academic economists who had been systematically ignored by the central bureaucracy in the fashioning of national economic policy. The statement continued: "It was a result of broken promises, of all the attempts made to suppress the crisis, of disregard for civil rights. Once again it has become clear that it is impossible to rule the Polish nation without listening to its voice. With determination and maturity, Polish workers are fighting for the right to a better and more dignified life. The place of all the progressive intelligentsia in this fight is on the side of the workers. That is the Polish tradition and that is the imperative of the hour."[49] In offering public support and solidarity at a time when the Baltic workers were standing against the state, the declaration endangered every one of the signatories whom Mazowiecki gathered in his apartment on August 21.

Reference has been made to the "divisive elements" embedded in the respective histories of the Polish intelligentsia and working class that hovered over their first meeting during the Baltic crisis of 1980. It is now appropriate to detail specific components of these diverse histories that came into play amid the shipyard crisis.

There were historical reasons that induced committed oppositionists like Kuroń to view the first Gdańsk demand as "impossible" and, similarly, induced Geremek and Mazowiecki to have "grave doubts" about the same postulate. The simple, overwhelmingly relevant but culturally elusive fact is that the long history of brooding about freedom on the part of Warsaw intellectuals did not encompass an understanding of the complex process of movement building that Baltic workers had experienced through years of insurgent activity. The capstone of this experience was the workers' settled understanding that the negotiating crisis in Poland—one that turned on the basic question of whether the nation was to have a new social contract between the party and society—had been precipitated by a coordinated, large-scale, multifactory strike of production workers. To the militants of the Lenin Shipyard, what gave the prospect of a "union free of the party" genuine social meaning was the *already existing* evidence of organized workers acting free of the party.

It is essential to spell out the difference between "freedom from the party" as an intellectual construct and the social meaning of the same idea when it was already in being as a product of an organizing achievement grounded in lived experience. The first was a political idea existing in social isolation; the second was the idea successfully projected into social life. The first had intellectual meaning; the second, the idea in action, had transforming social potential. The practical political result—in terms of consciousness—was that, at the strategically decisive moment of confrontation, the workers were "prepared" for the struggle for independent social space and the intelligentsia was not.

The *idea* of independent trade unions, long since brought into being in capitalist societies, was, of course, not conceptually difficult for the intelligentsia to grasp. In 1975, the "Memorandum of the 59" had attempted to make the proper case: "workers should be allowed freely to elect their own representatives in the professional field, in order to make them independent of the party and the state." Unfortunately, the pronouncement unwittingly incorporated a critical misunderstanding of how power worked at the point of production in Poland. It made the achievement of "independence" contingent upon the form of the election, as if free elections would somehow guarantee free unions. Under circumstances so defined, the militants of the union would in actual practice have to win *every* election, win them at every level of the local, regional, and national structure of the union, and, finally, win them under conditions and procedures prescribed by the party which possessed prior control of the entire vertically integrated structure.[50]

Free elections, far from producing free unions under these constraints, could not even effectively mitigate, much less fundamentally alter, the party's iron control. Indeed, though intellectuals in Poland were unaware of it, this was precisely what had happened on shop floors across the country after the war, again after the abortive "Polish October" of 1956, and still again after the collapse of grass-roots worker councils in 1971–72. These events proved concretely and with sobering certainty that workers could not set themselves free from the party with a simple wave of the hand and a "free election." There was a great deal more to the struggle than the simple electoral principle of a free ballot.

Two years after the "Memorandum of the 59" was written, a group of Warsaw intellectuals tried again. The "Declaration of the Democratic Movement" in 1977 stated that "free choice of their own professional and trade representation, independent of the State or party authorities, should be guaranteed to all employees." The statement obviously represented no analytical advancement. The two messages of 1975 and 1977, in fact, boiled down to an appeal to authority to do the proper democratic thing. Neither reflected a political understanding of how an authentic democratic social formation necessarily came into being in an industrial society or how it had to be structured to protect itself against manipulation from above. Neither formulation put the matter as it needed to be put—a popular democratic structure *created* from bottom to top as an institution independent of the party. The union structure needed was one that could survive any kind of election and remain structurally independent of the party—an institution, in short, born in free democratic space. Realpolitik, then, became the defense of that space—to win a legitimate place for it in society.

The "Appeal of the 64," drafted by the Mazowiecki-Geremek group, like the pronouncements of the progressive intelligentsia that had preceded it over the years, was a thoughtful and heartfelt act of social and political concern that exposed its authors to possible police abuse. It had demonstrable political value in its own right. But it possessed strategic limitations because it turned on ideas about politics that were remote from the political realities existing at that moment in the Lenin Shipyard. The first members of the Warsaw intelligentsia to see the difference were, appropriately enough, Mazowiecki and Geremek. They grasped the essential strategic implications of the Baltic mobilization within hours of their arrival in the Lenin Shipyard, and they had internalized their new insight fully within forty-eight hours. In short, their political understanding of what was possible had undergone an experiential transformation. In the inappropriate language of description widely favored in intellectual circles, their consciousness had been "raised." That of the KOR activists still isolated in Warsaw had not.

Wałęsa, of course, did not know any of this history of struggle by Warsaw intellectuals. He knew only that he and other worker activists on the Strike Presidium did not have "grave doubts" about the possibility of free trade unions, as the academic advisor originally did, nor did he believe, as Kuroń and other KOR activists felt, that the goal was "impossible" in the short run. Wałęsa listened to the view of the two distinguished visitors and explained to them his own reasons why the first demand was necessary.

Of long-term importance was what happened next. Instead of withdrawing in disagreement upon hearing Wałęsa's views, Mazowiecki and Geremek performed two strategically vital acts of coalition building. They listened to the reasoning of the workers' spokesman and they stayed in the shipyard to help. That is, they put aside their doubts and took their places as cooperating members of the democratic movement. Because of their considered manner in this conversation, Wałęsa concluded that his two visitors from Warsaw might prove helpful. Subsequent history indicates he was not in error. Later that day, after providing convincing arguments that a panel of advisors would be helpful and also free of decision-making power, Wałęsa was able to secure the approval of the presidium for the creation of such an advisory panel.

Meanwhile, young Konrad Bieliński was busy laboring at a craft at which he was undoubtedly the shipyard's most experienced practitioner—printing and editing for the opposition. Bieliński, together with another young Warsaw oppositionist Mariusz Wilk, had made the necessary arrangements to produce a strike bulletin. It acquired a descriptive identification that had been on everyone's lips all week—"Solidarność." On August 23, the tenth day after it had all begun, *Solidarność Strike Bulletin* no. 1 rolled off the presses. The movement, it seemed, had finally acquired a name. Galvanized by what he had seen in the shipyard, Bieliński, it seems, had in less than a week moved quite a bit beyond the strategic limits envisioned by Kuroń, Michnik, Lipski, and other prominent KOR activists in Warsaw. Within another week, he was to move much, much further—to the consternation of the Strike Presidium.[51]

On Sunday, August 24, Father Henryk Jankowski, assisted this time by other priests, held his second Mass of the Polish August. It was not as timely or as tactically helpful as the first one that had come while the movement was frantically trying to consolidate itself a week earlier. But it was far more publicized. The world press had settled in at the Gdańsk shipyard and photographs of thousands of militant strikers and their families outside the gate receiving Communion in a communist state soon appeared on front pages around the world.

Monday, August 25, was a day of eerie quiet before the ultimate crisis. Deputy Premier Jagielski had taken back to Warsaw from the first negotiating session some ominous news for the Politburo. The Strike Presidium was laced with militants who seemed both articulate and patient. Worse, the MKS now bulged with delegates from something over four hundred factories from across the whole of northern Poland. The site of the talks was surrounded by a veritable army of workers interspersed with literally hundreds of reporters and photographers from dozens of countries. Worst of all, the MKS seemed absolutely immovable on the demand for a union independent of the party. In all the years since the United Workers' party's accession to power, the party had never had to face such an organized social formation. The party leadership in Warsaw agonized in indecision.[52]

In the shipyard itself, the academic advisors whom Geremek had recruited by long distance over the shipyard's only functioning telephone gathered for a brainstorming session. Despite admonitions from both Mazowiecki and Geremek as to the firm resolve of the strikers and the limited prerogatives of the advisors, the newly arrived academics were deeply worried over the first demand. It was not only that the workers' insistence on free trade unions was irresponsible and fanciful, it seemed more than that; it seemed beyond everyone's imagination. Where had such a demand come from? In all the years of opposition activity in which they had engaged, none of the academic progressives had fitted free trade unions into their assortment of initiating strategies for confronting the Leninist state. An independent union was something that could come, if it came at all, only at some later stage. No one in the Warsaw intelligentsia had ever postulated such a move as a possible opening step. The first Gdańsk demand was both breathtaking and frightening. The Warsaw intellectuals prepared that day their own plan for a revised first demand and gave it a name. This was the process through which Variant "B" briefly intruded into the negotiations in Gdańsk.

Poland's Primate, Stefan Cardinal Wyszyński, shared fully in such concerns. He prepared a homily that would soon get national exposure on Polish television and in the government press. The previous week he had tried to calm things down:

> I want to remind you of what this country needs so that calm and reason might reign here: in the first place we must work honestly, with a sense of responsibility and conscientiousness; secondly, we should not squander, we should not waste, but economize, because let us bear in mind that we are a nation still on our way to prosperity; thirdly, we must borrow less but also export less, and on the other hand, better satisfy all the needs of the people, that is their moral, religious, cultural and economic needs.[53]

Despite advice from the bishop of Gdańsk, Lech Kaczmarek, the Polish Primate proved unable to benefit in awareness from the tumultuous events of the succeeding week. The sermon he delivered at the famed monastery of Jasna Góra at Częstochowa on August 26 did not help the cause of Poland's workers. Its full meaning can be conveyed only through a lengthy excerpt. The Catholic Primate of Poland said:

> The better we perform our daily duties, the more assured we can be that the help will come. What is more, the more aware we are of our responsibility for the Nation in performing our daily duties, the more justified or substantiated our rights will be, and in the name of those rights we can advance demands. But not otherwise! . . . The responsibility, then is joint. Why? Because the guilt is also joint. No one is without sin, no one is without guilt. It can manifest itself in various forms. It can be guilt for violating the personal rights of man, which implies a moral and social deformation of man. It can be guilt for not defending our rights, to which we are entitled as we fulfill our duties. This could be the result of a lack of social awareness, a certain passivity and insensitivity to the common weal, social weal, the good of the family, Nation and State . . .
>
> Let us remember how difficult it was to regain our independence after 125 years of subjugation. And as we devoted much time to domestic arguments and disputes, so a great danger threatened us and our independence . . . Today we often complain about the inefficiency of various social and vocational institutions. One branch of work blames another. We know how to enumerate adverse effects of work and management, we know how to talk about various unsuccessful undertakings. We know how to condemn them and how to devolve the responsibility for them on to other vocational groups. But let us think, how we ourselves perform our work? What is our contribution to social life and the national economy? True

> enough, there may be stumbling blocks everywhere. I looked at the fields on my way from Warsaw today. Some of them have been reaped and ploughed. In others, the grain was still standing in shocks or even unreaped. Sure enough, this may be the effect of the weather in that area; nevertheless, diligence and honesty in work do play a part. We know that where there is no honest work even the best economic system will fail while debts and loans multiply . . .
>
> I think that sometimes one should refrain from too many demands and claims, just so that peace and order may prevail in Poland; this is hard, especially as demands can be justified and they usually are sound; but there never is a situation where it would be possible to fulfill all the demands at once, right away, today. Their implementation must be gradual. And that is why we must talk: we shall first satisfy those demands which have priority, and then others. This is the law of everyday life . . .[54]

To equate the party and the workers as "vocational institutions" was hardly a step toward a new society. The party was ecstatic with the Primate's homily. In an event without precedent in People's Poland, the party printed a summary of the sermon on the front page of the Central Committee's daily newspaper, *Trybuna Ludu*. Excerpts were played on nightly telecasts throughout the nation.

The Gdańsk workers were bitterly disappointed. Wałęsa waved away reporters' questions and talked of other matters. He perfected a new quip or two as a means of changing the subject. The MKS intended to stand on its program, irrespective of Wyszyński's untimely advice. Shipyard workers, meanwhile, decided to outrank the Polish Primate. A new sign went up on Gate No. 2: "the Mother of God is on Strike."[55]

The bishops of the Church were also dismayed by the Primate's stunning intervention. They met hurriedly and promulgated another document for immediate distribution on the coast: "Among the fundamental rights of the individual must be numbered the right of workers to form themselves into associations which truly represent them and are able to cooperate in organizing economic life properly, and the right to play their part in the activities of such associations without risk of reprisal.[56]

A remarkably consistent causality connected these agonizing tensions that the Polish crisis generated between groups and within each group. It seemed to apply everywhere. In distant Warsaw, for example, the party's leadership found Jagielski's report hard to believe. But Jagielski was not the most effective advocate of his own new insight; namely, that the Baltic movement was authentic rather than a product of artificial stimulation from some easily expunged spot in Poland. The coastal party, extremely well informed as it was, could document such an analysis with copious detail. The party secretary for the Gdańsk region, Tadeusz Fiszbach, and the party's governor for the city of Gdańsk, Jerzy Kołodziejski, argued for the party to come to terms with society. They took the lead in specifying that the coastal party could live with the first Gdańsk demand. The coastal communists were ready to take the party in a new direction.

Similarly, the distinguished contingent of the progressive intelligentsia from

Warsaw could not fit the first demand into the politics of the situation. But a provincial writer from Gdańsk named Lech Bądkowski took it in stride, wrote a ringing manifesto of support, and straightforwardly took his place on the strike committee when the opportunity was offered.

There was also division within KOR. In Warsaw, Kuroń prayed the workers would come to their senses; in Gdańsk, Bodgan Borusewicz, the comrade-in-arms of Gwialda, Wałęsa, and other Baltic militants, easily took his station at the very heart of the movement for independent trade unions.

The Catholic Church faced the same dilemma. The Primate spoke publicly as if he had spent the days of August cloistered on another planet; the bishop of Gdańsk, in contrast, zeroed in on the first Gdańsk demand with quietly authoritative words of support.

To see this sweeping parallelism as proof of some kind of geographical determinism would, however, be inappropriate. Such a stance would—once again—represent a view of movement politics from afar, from outside the social formation that had generated the new political configuration itself. The broad-based consciousness of coastal intellectuals, priests, and party members could be seen as a function of their vicarious experiences during years of living alongside the workers' efforts. December 1970 was the centerpiece of this experience, but the organizing lessons that developed subsequent to that fateful month were decisively important, too. The Baltic workers learned these lessons. No one else did. But party members, intellectuals, and priests on the coast could see the workers had learned something—something relevant to the realities faced at work. They therefore had a basis for understanding the new things the workers said. Lech Wałęsa, Tadeusz Fiszbach, Bishop Kaczmarek, Bogdan Borusewicz, Jerzy Kołodziejski, Andrzej Gwiazda, and Lech Bądkowski could talk to each other and understand each other—as workers, as party functionaries, as priests, as intellectuals—without unnerving psychological or cultural strain. If they could talk with equal effectiveness to their respective counterparts in the party, the church, the intelligentsia, and the working class, Poland might be able to turn a historic corner. Clearly, the coast had created an internal dialogue that was quite a bit in advance of the nation.

Two weeks into the Polish August, Gate No. 2 of the Lenin Shipyard had become a multipurpose junction that was unique in the world. The gate itself was decked with workers' banners, Polish flags, and religious symbols, including pictures of the Polish Pope. It additionally served as an *ad hoc* message board that boasted, among the slogans, a poem from Byron. It was also a rendezvous where workers engaged in an occupation strike could meet briefly with their spouses and children and exchange food, clothing, and information. It was, moreover, a place of authority where workers wearing red and white arm bands checked credentials and where the MKS's official outriders announced over the loudspeaker the names of each incoming delegate dispatched from a factory somewhere in Poland. Finally, Gate No. 2 was the precise site of the coastal memory, a shrine to the Baltic movement's energetic and tragic past and a symbol of its continuing hope for an open public life for all Poles.

At least twice a day, Lech Wałęsa talked to the shipyard workers. But from time to time at Gate No. 2, he met with his wife Danuta. In a matter of days, their world had tilted in an unforeseen direction. Someone had remarked to Danuta that her husband was a great man, a Pole of historic significance. She could live with that, if it proved to be true.

Somebody, not a family member, not a worker, somebody not even from the Baltic coast, was beginning to reach some sort of similar conclusion about Lech Wałęsa—after only three days of observation. He was a medieval historian from Warsaw, and his name was Bronisław Geremek.

Political coalitions, like other outgrowths of self-activity, are not ordered through scientific means. Irrespective of the material and cultural conditions that brought each component of the coalition into being, their only connecting tissue is human communication grounded in candor. Poland's new worker-intellectual coalition was much more fragile than anyone knew. But as something more than a mere political slogan or tactical objective, it had, on August 26, 1980, finally come into existence. If the independent union could be won, many Wałęsas and Geremeks across Poland would have to try to learn how to talk to one another. It was the ultimate democratic test that history puts to social movements. To fascinated reporters from around the world, the remarkable scenes that kept unfolding in the Lenin Shipyard convinced even the most skeptical among them of an emerging possibility: cooperating Polish workers had brought everyone a long distance down a road where, at the end, people might be given the chance to try to say what they really meant to one another. It was this possibility, so elusive yet so seemingly near, that gave the Polish August its quality of exhilaration and its numbing tension. What each would say to the other, at the climax of the crisis, would surprise and alarm them all.

6 ▪ The Movement Consolidates Democratic Space: The Birth of Solidarność

One of the most striking, and also most baffling, features of the Polish Crisis is the enormous gulf between surface appearances and underlying realities.

Oliver MacDonald, editor, *Labour Focus on Eastern Europe*

IN ITS OUTWARD appearance, the Gdańsk shipyard in the second week of the Polish August seemed to offer a rare spectacle of an entire society organized against the state. Polish workers were obviously visible in large numbers, but also present was the team of advisors from Warsaw whose arrival on August 24 seemed to signal the participation of Poland's academic intelligentsia. On the same day, the Catholic Mass in the shipyard testified to the church's presence also. A great convergence of workers, intellectuals, and priests appeared to have taken place. The fact that a strike bulletin had made its appearance—a contribution of two young KOR activists from Warsaw—seemed to complete the picture since it verified that the most militant sector of the democratic opposition had also been integrated into the coastal coalition.

But the portrait of a great convergence misrepresents the substance of the social formation existing in the shipyard. Profound differences in strategic judgment existed among the disparate elements that were attempting to array themselves against the state. The fact was this: the presidium of the interfactory Strike Committee possessed one view, while the Church, the intelligentsia, and the democratic opposition had other views. Allowing for some subsequent alterations of perspective among individuals from all three groups, the strategic differences were to persist throughout the life of Solidarność and into the period of martial law. Indeed, it would be years before a broader common ground of understanding would begin to be discernible among the multiple economic, social, and religious components that existed and overlapped within nonparty Polish society.

Since all modern societies house disparate elements containing differing perceptions of reality, there was little that was particularly unique to Poland about this circumstance. In most societies, anything that might be called "national unity" has on occasion existed an an idea and also as a kind of yearning, but its actual historical appearance has most often been only a momentary happening confined to the outbreak and cessation of national wars.

The distinguishing political feature of the Polish August was not some multiple convergence but rather something much simpler and yet far more profound: the mobilization and consolidation of the working class. As a result of their condition in Polish society, the men and women who represented the coastal working class on the Strike Presidium saw the possibilities of political change in Poland essentially as something that would come from below and as a function of their own efforts. Though some worker representatives held aloof from involving anyone else in the effort, most felt they could benefit at least to some extent from the presence of allies. But even the workers who held the second viewpoint insisted that their allies understand their role to be supportive rather than controlling. As the Strike Presidium understood the situation, there was, in terms of power, only one certainty in Poland—the party-state. To challenge that power, there was one possible opposition—the working class.

This analysis did not really reflect the view of the religious and intellectual groups that sought to play a role in the Gdańsk negotiations. It is appropriate to keep in mind in this connection that the material structure of the Church existed outside the Polish economic system. The Church was sustained by voluntary contributions, freely given by millions of devoted followers. When the episcopate negotiated with the party-state, it did so with its own long-term survival interests in mind. In pursuit of this strategic objective, the episcopate might from time to time judiciously lobby with the state for a relaxation of the censorship (thus pleasing the intelligentsia as much as itself), and it might also lobby for an easing of restraints upon the working class. But within the circumference of Catholic planning, the episcopate had never pushed such secular goals to the point that the effort might fundamentally jeopardize the Church's own long-term position. Indeed, neither the Vatican nor the Polish population expected it to engage in such high-risk advocacy. The Church counted intellectuals and workers among its flock, but strategically, the episcopate kept its eyes on the state. The working class was a constituent part of the faithful, but it was not the preeminent Polish social formation that served as the centerpiece of Catholic decision making; that centerpiece was the Church itself.[1]

Within the intelligentsia, a somewhat more complex and elusive problem existed. As engaged participants in the shipyard crisis, they generated instructive evidence of their relationship to the emergence of Solidarność.[2] The group of intellectuals who accepted the role of advisors to the MKS Strike Presidium included an economic historian, Tadeusz Kowalik; a sociologist, Jadwiga Staniszkis; and a medieval historian, Bronisław Geremek. Their subsequent conduct provided vivid examples of three different ways of thinking about the Polish working-class movement. Their contribution, both to the chaotic tension of the negotiations and to its resolution, cannot be packaged as a single piece, for it was interwoven with developments that unfolded in the final week of the Polish August. It is necessary, therefore, to treat their roles in the context of these other events as they transpired.

Kowalik's approach to Polish politics seems to have been shaped by 1956. After the high hopes of the Polish October, the reestablishment of tight centralization convinced him of the narrowness and incompetence of the one-party state. His professional activities in the economic realm merely confirmed and deepened this analysis. It may be said after 70 years of Leninist central planning that intelligent economists and economic historians were ill-equipped to co-exist in good cheer with one-party command economies. In the realm of agricultural production, for example, lofty theories about the retrograde impact of "peasant individualism" and other such forms of party apologetics have rarely seemed persuasive to agrarian economists. Kowalik, for one, calmly pointed out that roughly half of the disastrous decline in Polish potato production was traceable to shortages in pesticides. Even more fundamentally, the entire system of reporting and fact gathering within the ruling party was so riddled with factional politics and career advancement that the top leadership routinely made decisions based on erroneous data. From very close-hand observation, Kowalik concluded that in the autumn of 1980 no one in the party apparatus up to and including the first secretary comprehended the extent of the economic crisis looming over the country. Kowalik and economist Waldemar Kuczyński were driven almost to distraction by their intimate knowledge of the massive damage to Poland's future that they saw being fully generated in the ministries of central planning in Warsaw.[3]

But life near the seats of the mighty did little to instruct Kowalik in the day-to-day capabilities of Polish workers. He was, in fact, far removed from the workers' milieu and, as his own writing indicates, possessed little understanding of the social and political ideas that were widespread in working-class Poland. Along with most of the other members of the advisory group, he appeared not to have picked up the slightest hint of the workers' deep concern that members of the intelligentsia were easily intimidated by the party and the party's police. It is evident that Kowalik was unaware that Walesa had to mobilize careful arguments to persuade the Strike Presidium to accept advisors in the first place. In his retrospective account of the academics' role in the Gdańsk negotiations, which he entitled "Experts and the Working Group," Kowalik explained: "My impression is that the choice of the term 'experts' was made by the strikers themselves, perhaps influenced by some myth of intellectual 'expertise.' "[4]

The sociologist Jadwiga Staniszkis was unique among the group of seven that Mazowiecki and Geremek assembled in that, as events were to show, she was closely tied, psychologically, to a number of Marxist intellectual constructs of a singularly elite quality. It soon became apparent that she was not equipped to view the scenes in the Lenin Shipyard with poise; these events were, in fact, deeply threatening to the view of the world she brought to the coast. To see the climax of the Polish August through her eyes is to enter a negatively constructed world of abstraction about workers. Staniszkis was not only briefly an actor on the Gdańsk stage (and thus was later written about by others), but she subsequently produced articles and a book of her own about Solidarność. The combination of sources provides what can only be described as a devastating insight into her actions.[5]

Staniszkis participated fully in the deep reservations the Warsaw academics possessed concerning the first Gdańsk demand. Indeed, while most of her colleagues had, with varying degrees of hope and reluctance, acquiesced by August 26 in the workers' insistence on independent unions, Staniszkis had not. The proposal by the advisors for substantial revisions of the official trade-union structure rather than independent unions—a proposal that became known as Variant "B"—remained very much in her mind as a possible option, even after its rejection by the Strike Presidium. She was alone in this stance.

Resistance to independent unions was also very much in the mind of Deputy Premier Mieczysław Jagielski as well. Stunned by the poise and militance he had encountered in his opening negotiations with the Strike Presidium on August 23, Jagielski huddled with the Central Committee back in Warsaw for the ensuing two days. The first thing party leaders did was to engage in the time-honored practice of cosmetic changes in the party hierarchy, a ploy that produced a fundamentally different response from the workers, on the one hand, and Staniszkis on the other. Within the Gierek regime, a flood of dismissals and appointments to high government positions occurred on the twenty-fourth. Edward Babiuch was replaced as prime minister by Józef Pińkowski. Jan Szydłak, head of the official trade unions ("We will not share power with anyone") was also sent packing. In fact, everyone directly associated with the party's initial move to break the strike without negotiations was dismissed. In the nature of party politics, these moves did not mean that the victims were superceded by officials who sought a more broadly democratic relationship with society. Indeed, one of those newly elevated to the Politburo was Stefan Olszowski, a hard-line critic of Gierek's economic policy and a thoroughgoing believer in the absolute supremacy of the party in society.[6]

The regime's penchant for reshuffling Leninists at the top of the party pyramid each time the working class created a governing crisis had, of course, long since become part of the received wisdom of the Polish working class. Polish workers realized that hierarchies did not change, but that functionaries did. It had happened after the Poznań events in 1956 and again after the coastal upheaval in 1970, and now it appeared to be happening a third time. In the Lenin Shipyard, workers regarded the new ministerial reshuffling as something less than a matter of earthshaking interest. As Wałęsa put it when closely questioned by the foreign press: "I don't know these people. Our main problem is free trade unions; and it is not important for us who will meet with us."[7]

To Jadwiga Staniszkis, however, these alterations were political events of considerable magnitude. Accordingly, she was "surprised" at the lack of interest in the shipyard which she interpreted as indicating that "for the workers, the power structure was out of reach, untouchable and invariable." The irony, here, was singular. The party would indeed be "out of reach" and "untouchable" if the workers backed off from the first demand and supported Variant "B," an option they would not consider, but Staniszkis would. Moreover, the significance she placed on the shuffling of functionaries in Warsaw testified to her own continuing preoccupation with the details of high politics—a folkway not shared in by the workers. Had she known anything about the heritage of working-class self-activity in Poland, the sophistication evident on the coast obviously would not have been so baffling to her. It turned out to be only the first of many events in Gdańsk that she did not comprehend.[8]

Jagielski and the Politburo, however, did more during the two days between the negotiating sessions than simply replace ministerial officials. The party continued to wage an intense war of maneuver with the Strike Presidium over the "preconditions" to end the telephone blockade. The party engaged in many evasions, making repeated references to the "unprecedented situation" as justification for being slow in reconnecting the coast to the rest of the country. The Strike Presidium held fast to the original agreement, which was that reconnection of the telephone lines should precede the next round of negotiations on this issue. As the long experience of workers at the point of production in People's Poland had proved, the case in point went beyond an episodic example of censorship; it concerned the basic question of structural censorship: the thirty-five-year government policy of depriving workers of lateral lines of communications so they would be unable to construct for themselves a working-class social formation in Poland. Some of the incoming delegates to the MKS broke down and cried with joy and anger when they encountered the organized reality of the Lenin Shipyard—joy that such a magnificent pyramid of worker organization had been achieved and anger that their own co-workers throughout Poland had been cynically and systematically deprived by the party of such transforming knowledge. What the presidium of the MKS knew (and the government feared) was that the great industrial enterprises of Poland—at Nowa Huta, Łódź, in Silesia, and elsewhere—would mount solidarity strikes of their own in support of such an achievement as had been wrought on the coast. It was because of this palpable social fact that the workers felt the preconditions represented something on the order of half the battle to achieve independent unions. Given the opportunity to communicate with itself, the Polish working class as a whole would unite behind the first Gdańsk demand. This was the realpolitik of August 26. It explained why the preconditions were so vital.[9]

As the argument continued, the government complained that the strikers were "escalating" their demands. The presidium replied calmly that the preconditions had been preconditions established on August 22—before negotiations began. In a final effort, the government proposed that the issue be placed first on the agenda for discussion at the next session. Worker representatives put the proposal not just to the presidium but to the full MKS. The answer was no—no talks unless the blockade of information and the isolation of the coast was ended. This ended the semantic duel. On the twenty-sixth, after three days of stalling and two days without negotiations, the telephone lines to Warsaw and Szczecin were opened.[10]

The academic advisors to the Strike Presidium learned a great deal in the course of the two days about the ground rules for negotiations with the party-state. In the struggle over the preconditions and the telephone blockade, the academics discovered something about the intricate mechanisms of censorship historically employed by the party to prevent the lateral communications so essential to working-class self-activity. They learned as well something about the workers' determination to employ strike-based pressure to remove those structural mechanisms of censorship. After relenting on the telephone blockade, the government, in a sudden maneuver, made a bid to exclude Kuczyński,

Cywiński, and Geremek from the panel of advisors on the grounds that they were connected to "antisocialist" elements in the opposition. Mazowiecki agreed to this exclusion but was overruled by Wałęsa, who spoke for a Strike Presidium that was seeking to make a fundamental point with the government on the entire question of the legitimacy of the strikers' "helpers." Wałęsa's reply could be swiftly summarized: "no advisors, no talks."

Years of experience in contending with the party in the plant were paying off for Wałęsa; the government backed off on this issue, too. Their acquiescence revealed again the precise source of the workers' authority—the strike weapon itself. However central a recognition, it was something that intellectuals and other nonworkers might easily forget in the course of the semantic jousting they tended to regard as the centerpiece of the negotiating process itself.[11]

The second round of negotiations on the twenty-sixth began as a replay of the first. In generalized language that left the government plenty of room for subsequent backtracking, Jagielski repeatedly emphasized his "sincerity" in reference to most of the demands. But he talked about democratization of the trade unions only in terms of official unions.

Wałęsa, calling for "free, independent and really self-governing unions," set a different tone which members of the presidium insured would dominate the discussion. The workers "did not want to disturb the basis of the social ownership of the means of production," said Wałęsa, but rather sought proper management "both of factories and the country as a whole." Such had been "promised time and again. Now we have decided to back our demand with a strike . . . No one wants to come back every ten years to the same point. Something must be done about it. The fact is we don't want political games of any sort." Wałęsa was at pains to demonstrate that the workers were not enjoying the strike: "We don't want more stoppages, more strikes, this is a last resort, a necessity. But we must have this point. We really won't give it up. Even if we get the twenty points but not this one there will be no agreement."[12]

And so began a concerted effort by the Strike Presidium to induce the government commission to confront the nature of the new dialogue between society and the party. Gwiazda offered a straightforward and effective exposition on the economic incoherence of the prevailing system and the inability of the official unions to live up to their "professed purpose" of defending the interests of workers:

> Ever ready to come out hand in glove with the factory administration which blindly followed directives from above, these unions acted against the interests of the workers . . . Crises happen here time and again: 1956, 1970, 1976 and 1980. They recur after shorter intervals. These were not just political crises, they were also economic. All were caused by the fact that working people had no influence over what was happening . . . These unions have lost all trust and authenticity. Trust will not be restored by means of a new law because the public has also lost faith in laws. Nor will the shipyard, whose employees are in this hall, now have faith in any law. Our aim is to create a genuine organization for employees in which people can have confidence . . . Mass organizations cannot operate simply by

> private conversations. They must be provided with publications of their own to convey information from the top down, and from the bottom up. They must have their own press and be able to write the truth in it, regardless of whether this suits the authorities.[13]

Gwiazda's underlying message was clear; the party needed to understand that reforms of the existing trade unions was no reform. Reinforced by ovations from the mounting number of MKS delegates—over four hundred by August 26—members of the presidium again and again emphasized the incoherence of the command economy. Responding to government promises to build two more housing factories for the coast, Florian Wiśniewski of Elektromontaż pointed to the destructive waste inherent in large-scale production facilities, including those in housing: "When you put up such a colossus, such a Moloch, you have to transport what you build long distances, and transport is very costly. Factories which don't require such transportation are preferable." In response to the on-going government propaganda barrage about "wasteful strikes," Wiśniewski added, "bad management costs more than strikes."[14]

The sweep of the workers' intentions was brought home when Lech Sobieszek of Siarkopol focused on government statements that any new trade-union arrangements worked out would be applicable only to the coast: "Statements like this can bring the whole country out on strike" Sobieszek said—prophetically. "This is not a matter of the private welfare of your government representatives," he added, "but concerns the whole of society, which is ours and not to be trifled with." But, most centrally, Sobieszek emphasized that democratization was the real issue, not the sacrosanct "leading role of the party" that the government media constantly brought up as as defense of existing arrangements: "No Pole really cares whether one gentleman or another occupies this high position or that. My friends, that honestly does not concern us. We are concerned that these gentlemen should function as a healthy organism, permitting themselves to be criticized from time to time, whether rightly or wrongly, so that the flow of information is truthful and honest." This judgment elicited a sharp burst of applause from MKS delegates.[15]

Collectively, the presentations by Wałęsa, Gwiazda, Wiśniewski, and Sobieszek took away the government's attempts at evasion while at the same time focusing attention on the central issue of self-governing trade unions. Presidium members made clear they did not want a union that played the role of a political party, did not question the leading role of the party, did not care which party functionaries ran the country as long as they were in some way accountable to social control, and did not wish to tamper with the social ownership of the means of production. What they had to have, however, was a self-governing union.

Needless to say, no meeting of the minds resulted from this colloquy, but it was quite successful in driving home to Jaglielski the rather stark options that the worker mobilization now imposed on the party; independent unions or military repression.[16]

The next day, the government attempted a basic shift in strategy. First, Jagielski, having noted the remarkable lack of deference on the part of the

Strike Presidium, correctly attributed part of it to the protected setting in the shipyards. The normal trappings of power could not be asserted simply through his own facility with words: not in a workerish environment that was organized in echelons of representation with thousands of strikers in the surrounding yard, nearly five hundred MKS delegates in the same building with the presidium, and the presidium itself numerically dominant over the government commission by an official margin of eighteen to four. Two sustained sessions in such a setting had left Jagielski straining to avoid appearing wholly defensive under the barrage of worker assertions. Accordingly, Jagielski suggested that a reduced committee of the presidium meet with him on the twenty-seventh at party offices in downtown Gdańsk. This suggestion, however, was rejected by the MKS. The Polish working class had long ago learned how exposed its representatives were to arrest during negotiations, and the MKS had no intention of sending anyone out of the shipyard.[17]

From the party's standpoint, this was no longer an unexpected result. But Jagielski had a deeper purpose: to create circumstances leading to a moratorium on the main negotiations and shift discussions to the respective sets of academic advisors each side possessed. Though the advisory group was privately regarded (by both the government and the workers) as softer, the workers agreed to a meeting of the academics after taking the precaution of adding three strong-willed worker representatives to their side of the table. As an additional safeguard, the presidium's prior agreement with its own advisors as to their respective roles insured that no unilateral action by the academics was possible.

The atmosphere changed dramatically in the second gathering of academics.[18] Conviviality had vanished from the government side. Speaking for the party, the prefect of Gdańsk, Kołodziejski, announced that "the center" in Warsaw had concluded that the entire issue of independent unions implicitly cast a shadow over "the leading role of the party." The government insisted that a reiteration of the party's leading role was needed as well as an ideological definition of any independent union that might emerge. The workers saw this argument essentially as a smoke screen. Zdisław Kobyliński, member of the Strike Presidium who possessed a fine feel for party mischief making, simply replied: "Why? We thought that such problems would be elaborated in practice, step by step. We do not want to play the role of a political party." The same point, of course, had been repeatedly made by the presidium itself in the formal negotiations. The whole issue of "the leading role" had historically been used by the party to deflect reforms and was being used, this time, to deflect something clearly more structurally serious, namely, independent unions. The workers' task was to return the conversation to this central issue.

Unfortunately, Jadwiga Staniszkis, alone among the advisors, did not cooperate with the thrust of the workers' intent. She repeatedly returned, as did the government side, to the question of the party's leading role. She took the following position, recorded in her publication account:

> My personal position was that we should explain the situation to all the workers in the presidium and leave the choice to them. If they rejected the political formula necessary to get new unions, they should bargain for a different one (for

> instance "social ownership of the means of production and power to the people," plus a declaration that they would not play the role of a political party). If this was not accepted, they should change to the variant of the reform of the existing union structure. The idea of independent unions should be kept as a pure, utopian dream for the future.[19]

It is clear from this conflicted prose that Staniszkis did not have much of a grasp on the precise nature of the confrontation. Her inherently deferential acquiescence in the government's analysis of "the political formula necessary to get new unions" and the speed with which she was willing to abandon any worker position that "was not accepted" by the party left her, willy-nilly, as an advocate of "reform of the existing union structure." From such a party-centered perspective, it was quite natural for her to dismiss independent unions as a "utopian dream."

According to other participants, Staniszkis's position at the time (somewhat different from her own published version) was that no statement about the party's leading role was needed. Her position on this point lent itself either to a maximalist reading or a minimalist one: either independent unions that acknowledged the presence of no other political force in society or a reform of the old trade unions. Independent unions coexisting in society with the party were not a part of either approach. In fishing in such troubled waters, Staniszkis was, in effect, playing the same politics ostensibly from the workers' side as the government group more straightforwardly pursued from its own perspectives. In any event, Staniszkis soon held court before journalists, explaining that she had decided to "leave" the workers' advisory group as a means of protesting their "agreement" to the leading role of the party. This stance was quite fanciful, of course. It was not within the power of the advisory group to "agree" or "disagree" with any government proposal.[20]

Her bizarre behavior, which was to contribute to a last-minute round of confusion and indecision just prior to the signing of the Gdańsk Accords, seem inexplicable on its own terms. A bit of background, in any case, may clarify matters.

Jadwiga Staniszkis brought to the Lenin Shipyard a settled conclusion that Polish workers possessed "an ahistorical perspective" grounded in "the common inability to generalize from their own experience, which is rooted in limited semantic competence." Workers suffered from "semantic shame" which left them mute on the face of "complexity and contradiction." They were politically passive as a result. When they were seen to act despite their shame and passivity, they could be described as "rebellious masses." The achievement of the workers' movement, Solidarność, was the product of an "escapist-type, utopian perspective, characteristic of traditional status-oriented societies, blended with the rejection of politics (due to its compromise-building character) and rooted in a fundamentalist-moralistic mentality."[21]

The democratic opposition fared scarcely better at Staniszkis's hands, for she saw opposition groups as obstructing relations between party members and the rest of society: "The polarizing tactics used by KOR (under the slogan of defending society against the state) hindered communication." The activists of

Solidarność were characterized as suffering from "monistic tendencies linked with a moralist conception of movement legitimacy" and a "one-dimensionality in perceiving the external world."[22]

Whatever lay behind this elaborate retreat into abstraction, Staniszkis, without prior discussions with her academic or worker peers on the advisory committee, suddenly abandoned the cause of the shipyard workers on August 27.[23] In her interviews with reporters as she departed, Staniszkis planted seeds of doubt—about the fidelity of the rest of the advisors and indirectly about the Strike Presidium itself. In the midst of a confrontational situation in which the alternative ranged from peaceful agreement to military repression by tanks, the suggestion that the advisors had sold out the presidium and, by implication, that a naive presidium might soon sell out the working class was precisely the kind of innuendo that provided the substance for anxious rumor.[24]

It took awhile, of course, for these dynamics to accumulate motive force. On the evening of the twenty-seventh, when Staniszkis ended her brief sojourn in the workers' milieu, members of the Strike Presidium reacted with calm to the report of their academic advisors to the effect that the government had professed alarm about the supposed threat to the "leading role of the party." Wałęsa, Gwiazda, and the others had heard it all many times before.[25] The need was to put this bogey out of sight with as few concessions in the fine print as were practical and thus restore the conversation of the issue of independent unions. Of more relevance to the presidium was Jagielski's address that same evening on Gdańsk television. He had made a quick trip to confer with "the center" in Warsaw and was back to report publicly that the essence of the strike concerned the deep dissatisfaction of the working class over the economy—matters the government had decided, he said, to address with boldness. He announced the existence of a "crisis in confidence" in the official trade-union structure, which he termed a "sick body" to which the government intended to devote immediate attention.[26]

In the face of this ancient refrain, the Strike Presidium simply stood fast. With the telephone lines open, time was now on their side. For two days the good news had been rolling in: Polish workers were busily creating new facts of political life for the Politburo to consider. Transport workers were out in Łódź and Kraków, and in southern Poland the auto plant at Bielsko-Biała had shut down. At the enormous steel center at Nowa Huta, workers had begun drawing up lists of grievances. And there were strikes in Bydgoszcz and Rzeszów. But the big news was from Wrocław where thirty factories employing sixty thousand workers had formed an Interfactory Strike Committee. It was the first MKS in Poland to be formed other than on the coast. Better still, Wrocław was the capital of Lower Silesia, a major industrial area which included mining. The miners, while beginning to make threats of a September 1 strike if a settlement was not reached in Gdańsk, had not yet formed any MKS structures. Silesia, Gierek's original power base, was proving difficult to rally. But it was also crucial because a pivotal portion of the nation's hard-currency earnings came from coal production.[27]

Meanwhile, as the miners debated what to do, the coastal hard core became harder. Over five hundred enterprises had affiliated with the Lenin Shipyard and face-to-face coordinating sessions had been held by the Gdańsk MKS with representatives of the Elbląg MKS and the Szczecin MKS. Over one hundred enterprises had affiliated with the MKS in Szczecin. When the government met again with the Strike Presidium on the twenty-eighth, it knew that the productive capacity of at least 400,000 workers was at stake. As Jagielski talked, members of the Strike Presidium were calm. They were waiting for Silesia.

The twenty-ninth was a day of good news and immense hope plus rumors from Warsaw that heightened tensions. The good news concerned the emerging pattern of a general strike of the Polish working class. There were new strikes in Legnica, Toruń, Włocławek, Świdnik and—a magic name—Cegielski-Poznań. And word came that the brave band of Czechoslovakian dissidents, Charter 77, had released a public statement of support of the twenty-one Gdańsk demands. In Warsaw, KOR beat the drums for the release of its activists from jail.[28]

In the capital, the Politburo engaged in a prolonged debate over the option of military repression. A majority seemed in favor, led by hard-liner Stefan Olszowski, who argued for a formal "state of war." But among the holdouts was the head of the army, General Wojciech Jaruzelski, and the head of the security services, Stanisław Kania. It would be hard to mobilize a repression without their support. One of the hidden truths of postwar Poland was that the morale of the Polish officer corps had been almost destroyed because of the army's role in the 1970 massacres on the Baltic coast. Jaruzelski had been patiently trying to lead the army to recovery ever since, and he had no intention, if he could so manage, of negating his efforts just to save another incompetent party first secretary. Kania, too, wanted the working class to be managed rather than repressed. In the final analysis, such was the stance of Edward Gierek also. Every piece of information pointed in this direction. During the evening, Gierek received news that Silesian miners had begun laying down their tools. In the Politburo on the twenty-ninth, the wheels of decision making turned slowly but decisively toward a peaceful settlement. Jagielski made it clear that the price was independent unions. Late that night, Gierek agreed. There would be no massacre of workers in 1980.[29]

The Polish August entered its final phase. Jagielski took the morning plane to Gdańsk, and when talks resumed, it became clear to the Strike Presidium that the government was paying serious attention for the first time to the demand for independent unions.

Two other issues appeared, however, one quite real and the other fabricated. A matter of high relevance was the scope of the new working-class institution—should it cover the Baltic coast or all of Poland? Wiśniewski of Elektromontaż and Sobieszek of Siarkopol had been point men for a once-and-for-all national solution. The position had been pushed by Wiśniewski in the advisory group and by Sobieszek in the main negotiation. Indeed, Sobieszek, an experienced veteran of 1970, had told Jagielski publicly that the party would encounter a national general strike if it tried to confine the new unions to the coast. But as he had for years, Wałęsa subordinated everything in his driving intention to create at least one wholly autonomous working-class institution. The rest could be left to the immediate future, one the entire working class would have to face—with the coast in the lead, of course. Gwiazda agreed with this strategy, and so, with considerable haste, did the academics in the advisory group. The issue was settled without undue internal tension. Both Wiśniewski and Sobieszek had too much respect for Wałęsa's strategic judgment to create a divisive crisis over the issue.[30]

The second problem was another matter. The presidium had not really handled the government smoke screen over the party's "leading role" with the attention it deserved. The reasons were quite human. The party's leading role, like Poland's geopolitical situation next to the Soviet Union, was a fact of life that was both maddeningly confining and tactically unchangeable. One rebelled against both circumstances with every fiber of one's being, but the MKS did not have the means of altering either. So the presidium had chosen to ignore in discussion the issue of the party's leading role, while resigning itself to what it hoped would be a formal acknowledgment that was as unobtrusive as possible. The real issue, however, was nothing less than the basic dignity of the working class. The presidium's position was tactically sound insofar as it kept the principal focus of negotiations on the basic issue of independent unions. But once that hurdle was apparently crossed, the question of the party inevitably, and awkwardly, moved to center stage.

The problem was that the presidium had not been entirely forthright—that is to say, not entirely democratic—in its method of dodging the issue in the three days since the government, with the help of Jadwiga Staniszkis, had first raised the issue. There was irony enough for everyone in the situation that now materialized on August 30. It turned out that Wałęsa had, at an important juncture, violated his own instincts; the presidium, in deferring to him, had violated its own instincts; and both moves disturbed the environment of worker confidence that otherwise prevailed as the struggle with the government reached its climax.

The origins of this final testing of the movement were rooted in events spread over the preceding forty-eight hours. It had all begun on the twenty-eighth when the Strike Presidium, unmoved by the "leading role" issue that Staniszkis had stirred up, gave the Mazowiecki-Geremek advisory group a tacit go-ahead to work out the phraseology of a draft of the first demand that acknowledged the party's leading role. When the draft was approved by the presidium and ready for reading to the full plenum of MKS delegates, Mazowiecki suggested to Wałęsa that a public reading in front of the world press might complicate the delicate maneuvering room available to both sides. This was simultaneously politically inappropriate and a little self-aggrandizing, but it was an accustomed way people who presumed they had authority thought about politics: that they knew better than the people they were supposed to represent. Wałęsa hesitated, agreed, and the presidium, listening, did not intervene. Since the idea of a closed-door session contravened virtually everything Wałęsa had learned about movement building in his twelve years of activism, he moved immediately to contain possible damage by explaining first to the MKS delegates in the hall and then to the shipyard workers by loudspeaker that a closed working session was momentarily practical and useful. "We will remain in control," he assured them, and for a time, there seemed to be no problem.[31]

But there was a problem. Mazowiecki's proposal for secrecy was precisely the kind of political move toward self-promotion that is a settled part of political habit the world over. Anything that diminishes the size of a decision-making constituency enhances the authority and sense of self-importance of the people remaining within the constituency. The drive to control others by maximizing the information one has and minimizing the information others have is a feature common to hierarchical modes of governance in all the world's cultures. Indeed, as in the case at hand in the Lenin Shipyard, the custom rather easily intrudes into the voluntary social formations that self-consciously seek to diminish hierarchy. Democratic behavior turns on the acquired ability to resist such organically aggrandizing tendencies. As of August 28–30, 1980, Mazowiecki, Wałęsa, and the Strike Presidium had not yet learned how to pass this test. It needs to be immediately added, however, that all three were generally doing better in meeting this difficult challenge than is common. But because they failed in this particular instance, they weakened the connecting link to their worker constituency and thus weakened the chain of confidence that is essential to democratic collective action.

Accordingly, through the long wait on the twenty-ninth, as the Politburo thrashed about in Warsaw and the regular flow of information in the shipyard was disrupted, clots of workers began to become uneasy. What was being discussed, if anything, behind closed doors in the advisory group? Were the workers, in fact, being "sold out"? The clots grew in size. Wałęsa and then Gwiazda provided assurances and succeeded in restoring confidence for awhile. But the doubts returned. The Polish working class had never made a "deal" with the party in all the years of repressive politics in People's Poland. After all that had happened, how could the party be trusted? "I don't like it," workers said to one another, and listening workers grew uneasy again. Amidst the vacuum of information, an old folk memory—indeed, a folk wisdom—had returned: the party was not to be trusted. The proposed draft should have been read to everyone when it was first completed. Such a course would have been forthright, democratic, and therefore safest. When it was finally read publicly, after two days, the level of confidence among the workers was not the same. They had been left in the dark too long.

Adding to the anxiety in the shipyard were ominous developments that had been generated by the party. While the Politburo agonized in Warsaw throughout the twenty-ninth and the workers agonized in the shipyard, the most implacably authoritarian elements within the party set in motion a series of actions designed to influence both groups. Police and media display was intense on the twenty-ninth. The press and television throbbed with strident denunciations of "antisocialist elements" directing the coastal movement, and the police did everything they could to heighten tension and suspicion by interrogating and arresting people all over Poland. Under this pressure, what can be understood as "movement poise" cracked in a number of persons associated with the workers' cause. Even within the presidium itself, doubts appeared that the MKS was about to make a fatal concession on the leading role of the party.[32]

In this one area of the Polish August where their role made them genuinely relevant politically, the Mazowiecki-Geremek team crafted one subtle achievement and one ill-advised phrase. The achievement was to acknowledge "the leading role of the Polish United Workers' party in the state." The last three words were significant, at least in legal terms, for they represented an interesting, though untested, step forward. The party's leading role "in the state" was acknowledged, not its leading role in society. Theoretically at least, this wording constituted a formal public acknowledgment of the existence of democratic space in society. Since the workers intended to occupy this space anyway, it was reassuring that its existence was formally accepted by the government. Unfortunately, the advisory group inserted this phraseology as a dependent clause in a sentence that defined the role of independent unions, thus unnecessarily tying the new unions to the party in clumsy juxtaposition. The sentence read: "While acknowledging the leading role of the Polish United Workers' party in the state and not questioning the establishment system of international alliances, their [independent unions] purpose is to provide working people with appropriate means for exercising control, expressing their options and defending their own interest."[33]

To persons in the shipyard who were not brandishing their own agenda, the wording did not seem troublesome. Reporters from around the world concluded easily enough that the first demand was encased in a reasonable sounding statement—especially given the fact that the demand for free unions in a Leninist state did not pass the test of reason in the minds of many foreign journalists. The worker advisors clearly felt they had performed satisfactorily, and so did Wałęsa, Gwiazda, and others in the strike leadership.[34]

But to some nonworkers who had no mass base in society, the wording was a violation of all that was politically sacred. The Confederation of Independent Poland (KPN) was grounded in a conception of an independent Polish republic in which Soviet domination could be removed "only by eliminating the power of the Polish United Workers' party." KPN had not recruited anything remotely resembling a mass movement on this basis, but its handful of militants had sat around many a kitchen table in the eleven months since their organization had been founded. Without a democratic by-your-leave, these militants gave themselves full permission to shape the ideological nature of the workers' movement. The representatives of KPN were present in the shipyard on the twenty-ninth, and with the zeal of true believers, they lobbied with every worker and MKS delegate they could buttonhole. The message of KPN had the power of simplicity, if not political reality: the first demand was a sellout! Tension would rise among groups of workers, and Wałęsa would then restore a measure of poise with a telling speech to the effect that the only real guarantee workers had for the future was their own independent organization. The KPN, increasingly buttressed by suddenly aroused militants from Young

Poland, would stir things anew. As the day wore on, the presidium itself began to be affected by the alarms that were being so passionately sounded.

The connection between the workers and the opposition groups was neither ideological nor programmatic. Rather, it was grounded in the powerful social fact that arguments against the party appealed to widely held convictions that the party was not trustworthy. There was, of course, nothing the workers could do at the moment to alter this underlying circumstance short of a coordinated nationwide general strike focused on the issue of the leading role of the party. Whether opposition militants knew it or not (they did not, in fact, have enough knowledge of the mechanics of working-class self-organization to have thought about the matter with sophistication), the Polish working class was not yet ready for a coordinated national general strike. So that option—later pronounced by KPN's Leszek Moczulski as the valid road that should have been taken—was not possible. Those on the presidium who had thought the matter through—Wałęsa and Gwiazda most prominently—knew that limitation fully and acted on that knowledge. They had the security of knowing that the majority of the presidium and hundreds of MKS delegates understood in a general sense the same necessity.[35]

But KOR's Konrad Bieliński and Mariusz Wilk had gathered around themselves, as they produced the shipyard strike bulletin, a handful of young oppositionists who succeeded in infecting each other with pulsating anxiety over the issue of the party's leading role. Together with Jacek Taylor, KOR's legal representative on the coast, and a small group of like-minded worker delegates, they began drafting a statute through which the MKS could unilaterally transform itself into a self-governing trade union. Through this device, they seemingly fantasized that they could present the government with a *fait accompli* that would somehow achieve what the Interfactory Strike Committee and its 550 affiliated factories was also trying to achieve; namely, forcing government acceptance of independent unions. The agitation had the effect of representing a conspicuous vote of no confidence in Wałęsa and also in the presidium's leadership, which was seen as persisting in working toward an agreement with a ruling party structure that was at that moment arresting people throughout Poland. It represented also by inference—and very soon explicitly—a thunderous vote of no confidence in the panel of advisors that Wałęsa and other presidium members could be seen talking to from time to time.

Some of the advisors did not help matters by protesting the activities of the KOR group, which the advisors elected to see as a threat that, in Kowalik's words, "might erode the government's confidence in the Interfactory Strike Committee as a partner to negotiations." This latter fear betrayed a still shaky understanding of the nature of the confrontation between the working class and the state, one that vastly exaggerated the legal niceties by ignoring the real sources of power the party and the MKS each possessed. The government was worried about a general strike, not about a publicist named Bieliński who had no constituency. By the evening of the twenty-ninth, even as the Politburo stumbled toward its own policy conclusion, the three groupings of oppositionists among the intelligentsia (represented by the KPN–Young Poland coalition and by the young KOR militants) collectively succeeded in escalating one another and some of the observing workers onto a new plateau of confused thought.[36]

An extraordinary meeting of the presidium, considerably augmented by advisors and self-appointed advisors, took place that evening. Konrad Bieliński, fully as convinced of his prerogatives to shape the working-class movement as KPN militants had been earlier in the day, took the floor to make an impassioned speech. As recounted by one of the dazed onlookers, Bieliński reminded everyone of his service on the strike bulletin and his heretofore gracious silence on policy-making. But, since it was now apparent negotiations "were not bringing results," he had decided to take action to ensure the strikers would not leave the shipyard "empty-handed." Bieliński attacked the advisory group for meeting with the government behind closed doors, "replacing the workers," and announced that the time had come for the opposition to understand itself as a series of "social movements" dedicated to activist policies. Tadeusz Kowalik responded with slightly less heat that no "politics" could be achieved "behind closed doors" since the Strike Presidium retained full decision-making power. In any case, Kowalik believed, correctly, that a legal trade union could recruit millions, while an unrecognized opposition union would do well to recruit thousands. The Strike Presidium remained largely silent through this tangential and unnecessary debate. The wrangling resumed the next morning and proceeded to the point where Bogdan Cywiński, a distinguished philosopher and activist in the Flying University, blew up at the insinuations of the KOR militants and announced: "In that case, we have nothing left to do except issue a short statement and leave."[37]

Whatever the reaction of Cywiński and the other advisors, Bieliński's imperial judgments and frantic strategies did not sit well with Wałęsa, Gwiazda, and the other worker militants. The worker organizers had spent years trying to build the voluntary social formation that now sprawled throughout the tri-cities and across all of northern Poland. It was stunning to discover the extent to which people like Bieliński arrogated to themselves the right to risk the fruits of such difficult working-class organizational effort in the name of some extemporaneously conceived "social movement." The sense of prerogative middle-class oppositionists brought to the shipyard was astonishing. They clearly possessed a curiously paternalistic conception of what a working-class mass movement was and their unlimited rights to intervene in it.[38]

Ironically, before it became necessary for activists on the presidium to inject themselves into *L'affaire Bieliński*, word arrived in the shipyard that the Central Committee of the party in Warsaw had agreed to the first demand. Within minutes of Cywiński's outburst of exasperation at Bieliński, the tension evaporated from the room, and both the presidium and its advisors suddenly found themselves liberated from criticism by excitable and querulous young activists from KOR.[39]

After Jagielski arrived back in the shipyard, it quickly became evident that an agreement was in fact in the offing. The government's chief negotiator publicly revealed his desire to tie up loose ends, reach an agreement, and get the coast back to work—that very day, if possible. As the demands were reviewed, he did, at one point, revert to old habits when he attempted to exempt the strikers' "helpers" from the new guarantees of protection both from dismissals at work and from police harassment. But when the hall erupted in protest, Jagielski quickly backtracked and said, "I accept, I accept." Things moved even more swiftly after that. Wałęsa and Jagielski soon began initialing agreed-upon paragraphs. When all was completed, Jagielski took the government's copy and caught a plane for Warsaw for final approval by "the center." The presidium, meanwhile, went over the entire agreement, point by point, with the full hall of delegates—over six hundred officially affiliated—that comprised the Gdańsk MKS. The final version, of course, contained the necessary wording on "the leading role of the party," a point that presidium had never considered contesting, but one that had engendered great strain.

The anxiety had, in fact, not yet run its course. Indeed, on the eve of achieving what was, by any standard, one of the world's great plateaus of democratic self-organization, the rank and file of the new independent trade union-to-be wrung themselves through one final moment of doubt and confusion. Before leaving for Warsaw, Jagielski had congratulated one and all and had praised the members of the Strike Presidium for the "unambiguous" way they had defined the "ideological and political profile of these trade unions." Suddenly, in the sweet afterglow of the unanimous acceptance of the historic agreement by the full MKS and in the aftermath of a tumultuous celebration of Wałęsa's leadership (thousands of voices imploring "May he live a hundred years"), after all of this, the flood of doubts about the party surfaced one final time. Militants from Young Poland burst upon the presidium, loudly accusing them of a sellout. They were "traitors to the working class," no less. A kind of youthful and decidedly middle-class panic coursed through the hall of delegates. Some of the delegates wondered if perhaps the young representatives of the intelligentsia were right. What did the deputy premier mean by praising the new union's "ideological and political profile"? Had they thrown it all away at the last minute? By early evening, some of the workers had worried themselves into a veritable frenzy. "We will not," one insisted angrily, "accept communist unions." When other workers tried to reassure their distraught colleagues, the effort simply caused a new argument and heightened contention. Soon, the evening was punctuated with sharp arguments that extended even to members of the presidium.

A central fact about the hall of MKS delegates on August 30 was that it no longer housed only MKS delegates. More than a thousand people milled around, many of them oppositionists from every stratum of the Polish population. A journalist from KPN burst into a session of the advisory group during the afternoon, accusing one and all of betrayal—her allusion was to Judas—and as the evening wore on, the number of self-appointed experts of high strategy continued to grow. This circumstance obscured the contrast persisting throughout the day in the mood of the shipyard as a whole from the atmosphere in the hall of MKS delegates. After the excruciating tension of the day before, the shipyard was on August 30 a repository of good spirits. Wałęsa's reports to the rank and file during the day were greeted with enthusiasm and cheering. But the hall, in Wałęsa's words, was "boiling." Things had been stirred up to the point that even the safety of the government delegation—due to return for the formal signing of the agreement—could not be guaranteed. To the astonishment and alarm of the Jagielski group, they received word not to return for the final session until the next morning.[40]

This unseemly turn of events, coming as it did on the very lip of triumph, apparently proved sufficiently out of place, or possibly embarrassing, to so many observers that most chroniclers of the Polish August have passed over it as rapidly as possible. Those who managed to give it a modicum of attention did so as a means of criticizing some person or faction inside the shipyard—the advisory group, Wałęsa, or Polish workers as a whole.

Actually, however, this evening of turmoil may be seen as both understandable on its own terms and positively helpful in providing a way of coming to grips with the complex dynamics that produced Solidarność. It also provides a useful clue to understanding the passionate early months of Solidarność in the autumn of 1980.

A way to approach the subject is to focus on a minor incident that occurred in the hall of MKS delegates fairly early in the struggle. In his comprehensive study, *The Polish Revolution*, Timothy Garton Ash summarized the incident with swift efficiency:

> A nervous-looking man in his mid-thirties walked up to the podium and begged attention for a "historic announcement." Challenged by the chairman, he identified himself as Irenuesz Leśniak, deputy head of the personnel department. Then he read a ten-minute-long prepared statement, clogged with pathos, concluding in a plangent appeal to Edward Gierek "who is for us like the Pope" to come to the shipyard as he had in 1971, "you, Edward Gierek, who alone we trust, because you are to us like a father." Astonishingly, the delegates crowned this peroration with resounding applause.[41]

A moment later, the veteran activist Anna Walentynowicz took the microphone, and the sky came crashing down on the poor personnel director. "I know, Mr. Leśniak," said Walentynowicz. "He has persecuted me for years, and two weeks ago it was he who sacked me." As ashen-faced Leśniak suddenly found himself surrounded on the podium by outraged delegates seemingly bent on lynching him. Wałęsa intervened to bring order and personally escorted the distraught functionary to safety.

How should one interpret this bizarre affair? To some, it could be alluded to as fulsome proof, in Jadwiga Staniszkis's phrase, of the "limited semantic competence" and "self-hatred" of the Polish working class. To Timothy Garton Ash, in contrast, it revealed "something disturbing and something fine about the large assembly which the MKS had become"—"Disturbing" because it showed how unclear many delegates were about their goals, and "fine" because of the "dignity and restraint which, after Wałęsa's intervention, the *provocateur* was complimented out of the yard."[42]

Balanced as such a judgment is, there is need here to acknowledge the complexity of the social formation that is being described. It constitutes a fundamental error in political analysis to see the hundreds of people in the delegate hall as "the delegates" who alternately generate "resounding applause" and cave in to "self-hatred." The only intellectually defensible way a descriptive sentence beginning with the words "the delegates" can be finished would be with a phrase of diversity such as "are variously inclined." Six hundred people transparently had many different reactions to Personnel Director Leśniak. It would be quite incorrect to assume that had Anna Walentynowicz not been in the room, Leśniak could have bowed out of the hall with applause ringing in his ears. A great many workers in the hall besides Walentynowicz had the experience to see through his defense of the first secretary.

In the autumn of 1980, scenes were to take place throughout Poland that corresponded in instructive ways to the scenes that transpired in the Lenin Shipyard during the volatile days of August. On the eve of this nationwide flowering of popular political activity, it is appropriate to establish some ground rules for interpreting such a rare mass-democratic experience that Solidarność, in fact, constituted. The task is to grind an observer's lens with sufficient fineness that distinct human beings rather than a blurred mass can be seen on the stage of politics set by Solidarność. Some sense must be made of the "shipyard workers," the "Baltic workers," and the "Polish working class." The three groups of people are not quite the same, whether viewed with the balanced goodwill of a Garton Ash or with the impulsive contempt of a Jadwiga Staniszkis.

The Lenin Shipyard was a product of the technically incompetent planning that was a routine component of administrative conduct by the Polish United Workers' party. In its production methods, the shipyard was a jerry-built blend of modern and outdated tools and tool systems. A certain pragmatic improvisation explained this disjunction. Because of the certainty that the arrival in the shipyard of materials and parts would be uncertain—a result traceable to the rhythmic structure of "storming"—the degree of automation that was possible was sharply limited. One result was that the shipyard was an artisan's paradise, conceivably one of the last and certainly one of the biggest such bastions of highly skilled labor remaining anywhere in the industrial world. The production methods that came to be the norm in the Lenin Shipyard were highly dependent upon a broad mass of extremely skilled mechanics of all descriptions. Shop sections were well populated by machinists, electricians, and engineering specialists who had matured through successive levels of technical training. The one book-length study of Solidarność that has taken time out to comment in some detail on this phenomenon is Lech Wałęsa's autobiography, *A Way of Hope*. His affectionate and precise description of "the hierarchy of castes" in the shipyard possesses the power and authority that workers can bring to analyses of the work process:

> The highest ranking workers, the elite, in our yard were those in machine shop M-5 who handled the final assembling of the engines. Shop M-4, one step down, dealt with general engineering, and there you'd find some skilled workers such as good lathe or milling-machine operators. The pariahs of the shipyard labored in the hull. But the worst job, in my view, was boiler-making, where thirty-five

> pound hammers were used on sheet metal heated with blowtorches . . . The painters were also regarded as pariahs. They worked in a real hell: seventy percent of them were "temps," and few lasted long . . . With the exception of the engineering crews of M-4 and M-5, the electricians had the best working conditions.[43]

The shipyard, then, was a montage of skills and training and of people possessing the sense of self that went with both. It housed a job-based differentiation of functions, each arranged, in Wałęsa's description, "according to its own self-restrictive laws." It had pariahs, its "temps" in the paint shops, and "flyers" in the hulls. The "flyers" might or might not be able to wage a verbal battle with the Jagielskis of Poland, but the machinists of M-5 seemed well posted to look him in the eye; the "temps" might or might not applaud an emotional speech by an Irenuesz Leśniak, but most electro-mechanics from W-4 had the experience to see through such a political ploy without much strain. Clearly, "the workers" of the Lenin Shipyard, sixteen thousand strong, had lived variegated experiences and had a variegated understanding of work and politics in Poland. But they also shared many beliefs in common—and skepticism about the Polish United Workers' party was one of them.

As for the shipyard, so for the tri-cities. Shop-floor activity had been multiple and varied among the nine thousand workers at Northern Shipyard and the four thousand in the Repair Shipyard, but it took a different form among the five hundred employees of Elektromontaż who had lived through the searing experience of the murder of Tadeusz Szczepański and his emotionally charged funeral. Still other forms of self-activity had been experienced at the smaller enterprises in the metropolitan areas. The MKS itself reflected this variety in the experience and sophistication of the delegates it received from different enterprises. Some applauded Leśniak, and some did not. Moreover, service on the MKS was itself an intense experience, so that "veterans" who had been engaged in the self-activity of the strike since August 16 knew considerably more about the dynamics of the confrontation with the state than those late-arriving delegates who trickled in as late as August 28 and 29. But very few delegates to the MKS could match the political experience of hundreds of militants from the Lenin Shipyard. They had been talking about the politics of self-organization steadily since the December massacre of 1970.

Finally, there was the phenomenon of the coast itself. The environmental experience of coastal assertion over the years had generated a storehouse of knowledge that did not find its equivalent anywhere else in Poland. This reality was verified by the quick formation of an elaborated MKS structure, first in Gdańsk and then soon after the Szczecin and Elbląg. Such rapid structural mobilization occurred nowhere else in Poland. And the remarkable 156-factory mobilization of August 18 in the tri-cities was unmatched in the Polish August. Indeed, in precision and speed, it stands as one of the high points of working-class assertion and cooperation in world history.

A seminal component of shared experience as prologue to consciousness is here exposed to view. The simple discovery that Gdańsk possessed an "Interfactory Strike Committee" did not instruct workers in Ursus in the art of forming such a committee. A theory about a potential organizational structure moved from idea to institutional presence only through the connecting link of internalized experience; hearing about the creation of an interfactory strike committee did not teach one how to form such a structure, any more than reading *Robotnik* on "workers' commissions" taught workers how to emulate communists under Franco. When the first MKS outside the coast appeared in Wrocław, the workers there did not read a KOR pamphlet; they went to Szczecin to get some organizational advice. It was in this sense also that KOR's widely credited claim of "preparing the consciousness of the workers for the strikes" can be seen as intellectually fanciful and institutionally irrelevant. Throughout the Mazowsze region surrounding the metropolitan district of Warsaw, where KOR's influence and organizational ambitions were most visible and where distribution of *Robotnik* was most consistent, not a single MKS appeared during the Polish August. Not at Żerán or Huta Warszawa or less famous enterprises in Warsaw. None of the workers who heard Jacek Kuroń make his pre-August speeches about the need for workers to become interested in civil liberties as well as shop-floor rights went back to their plants armed to form an MKS. In none of the cities where KOR courageously attempted to bear witness and to help the victims of repression did an MKS develop—not at Radom or Ursus or Płock and not in Kraków or Lublin.

No special failing of KOR produced this result. "Consciousness" simply is not "formed" the way intellectuals have come to think of it as being formed—neither through "laws" of production relations nor by reading approved texts or pamphlets. It is not KOR's claim that seems interesting, after the fact; it is the worldwide acceptance of KOR's claim that is most revealing. In fundamental ways, the history of Solidarność contradicts the assumptions undergirding mainstream capitalist and Marxist analyses as to how social movements develop in industrial societies. A basic understanding of what constitutes popular politics is at issue here. The word "politics" itself is poised for redefinition.

At the climax of the Polish August, observers were uniformly puzzled by the seeming volatility of the situation as "the workers" pondered the meaning of a formal agreement with a party-state they did not trust and which was at that moment conducting a public campaign of disinformation and repression. But, in fact, a great many workers and delegates in the Lenin Shipyard held to a firm strategic purpose throughout the crisis. They were, perhaps, hard to see during the hours of the most intense strain, for they were intermingled among people whose actions visibly confirmed they had not retained their poise. But those who thought about social formations abstractly—as "the workers" or "the delegates"—could not hope to penetrate this complexity and analyze it coherently. Such glib generality as is implicit in such descriptive or theoretical approaches has long crippled the social analysis of political movements.[44]

But one additional factor needs to be considered. The political institutions, large and small, powerful and powerless, which succeeded in heightening the level of anxiety present in an already tense confrontation in Gdańsk—the party, KPN, KOR, Young Poland—all shared one transcendent political reality in common. Each of them was spared the burden that is organic to the creation of large-scale democratic movements: none was accountable to the movement's mass base. None, therefore, inherently thought of their relationship to coastal workers in reciprocally democratic terms.

Visible here is an absolutely critical distinction between democratic movements (full of diverse people) and that descriptive monolith, a "mass movement" (comprised of a single entity called "the masses"). It is the latter that is commonly thought of when people brood about or encounter popular politics. Mass movements have "leaders" who are taken to be the necessary objects of careful study, so their modes of manipulating "the masses" can be traced. Leaders are presumed to have goals that are divergent from "the masses," a circumstance that makes manipulation operable and hierarchy inherent. Mass movements are created by "other people"; that is, people not of "the masses," people outside the social formation that comprises the movement. An essential corollary is that movements that appear relatively leaderless at the moment of formation are, perforce, "spontaneous" movements.

By way of illustration, in the Lenin Shipyard the activities of KPN, Young Poland, KOR, and the party may be retrospectively measured for their democratic components. The party scores lowest on a democratic scale, because it tried to influence the workers' movement through public disinformation, private evasion, and the threat of military repression. Next lowest in democratic conduct were the nonworker representatives of KOR gathered around Konrad Bieliński and the shipyard strike bulletin. Scoring best (relatively) were Young Poland and KPN. Both attempted to influence the movement by appealing to its visible mass base in the shipyard and also in the MKS. Bieliński, thinking in hierarchical terms, chose to carry his argument only to the top of the pyramid of worker organization—the Strike Presidium and its advisors. But irrespective of their points of entry, none of the three opposition groups—KOR, Young Poland, and KPN—considered the possibility of a recriprocal relationship with the workers; all forthrightly empowered themselves to impose their own sectarian agenda upon the mass movement, risking the movement's stated objective of independent unions in the process. The exception was ROPCiO, which simply supported the MKS.

KPN's program was grounded in a strategic vision of destroying Soviet influence in Poland by toppling its surrogate instrument of control, the Polish United Workers' party. Independent unions, while not counterproductive to this objective, were not seen as central to it either. Of the two possibilities, either destroying the party or establishing independent unions, KPN chose the former. The militants of Young Poland opted for the same choice. In terms of realpolitik, this choice was logically indefensible; independent unions were an imminent possibility, destroying the party in the Brezhnev era of 1980 was not. But this is merely an important strategic point, unrelated to the task of measuring the presence or absence of internalized democratic values in the various opposition movements. KPN and Young Poland were marginally more communitarian than the nonworker members of KOR (what may be defined for the moment as "young Warsaw KOR") in their approach to the movement. Both groups "saw" more workers than did young Warsaw KOR whose gaze did not extend beyond the leadership group in the presidium. But none of the three groups saw the workers as their democratic allies whose forthright goal of creating democratic space merited collegial respect and therefore did not deserve to be brazenly ignored in the name of some fanciful private vision of even greater democratic space. In this decisive sense, the workers' advisory group, led by Mazowiecki and Geremek, was unique among the groupings of intellectuals in the democratic nature of its relationship with the striking workers. Instead of habitually adopting a patronizing stance toward the workers, the leading members of the advisory group *listened* to workers. Their stance was relational rather than imperial or condescending. Such a political posture toward workers was quite rare in Poland in 1980; as it has been in all other stratified cultures fashioned since industrialization.

When the organizing achievements of the Baltic working class and the close proximity of its goals are considered in the context of the situation actually existing on August 29–30, it is quite instructive to discover that the middle-class militants of KPN, KOR, and Young Poland could, without much hesitation, arrogate to themselves the right to engage in such inherently patronizing relationships with the shipyard workers and their affiliated delegates who by then represented some 500,000 Polish workers. Instructive, transparently indefensible, and possibly even incredible. But very common. Very common indeed. Considered historically, such conduct by class-based opposition groups was entirely predictable. The two features KPN, KOR, and Young Poland had in common were (1) a number of socialists in their ranks, which permitted militants of each to persuade themselves they were not spokesmen for essentially middle-class objectives, and (2) the power of the class-based cultural values that each group unconsciously expressed in its relationship to Polish workers.

The true conflict in the Lenin Shipyard on August 29–30 was a war between the three opposition groups for the right to impose its own sectarian objectives upon the working class. The workers, with some strain, resisted all three.

It is appropriate to emphasize that in this endeavor the workers had some timely support in maintaining the structure of their own assertion in the Gdańsk shipyard. They received it from their advisory panel, which stayed to the end as promised and which never overstepped its agreed-upon authority. Even though Kowalik privately condescended to workers, he publicly associated himself with their goals throughout—as did the entire advisory group once Staniszkis had abandoned the cause. Mazowiecki, Geremek, Kuczyński, Cywiński, and Wielowieski all served the workers' cause with reasoned fidelity.

Members of the Strike Presidium helped in crucial ways, but the most effective spokesman for working-class resistance to ideological takeover was the chairman of the Interfactory Strike Committee, Lech Wałęsa. His most active ally in this endeavor was his longtime associate and personal rival, Andrzej Gwiazda, who, however, was almost invisible in reportorial accounts in the world press.

At one time or another over the seventeen days of the struggle in Gdańsk, Wałęsa won the admiration—for different reasons, of course—of Japanese conservatives, American liberals, English socialists, French Trotskyists, and Italian communists among the ranks of the world's journalists. He talked to the shipyard workers so many times and said so many different things in so many altered situations that his greatest gift—an uncanny ability to explain complicated political nuances in understandable language—became apparent to everyone. By no means did journalistic approval imply that Wałęsa maintained a flawless consistency or that his "understandable language" was always a precisely accurate rendition of his own opinion. Indeed, there was just enough inconsistency for some reporters to conclude that Wałęsa was an "effective leader" but also a "demagogic spokesman." In the fifteen vibrant and volatile months of Solidarność and in the long years of martial law that were to follow down to the resurgence of Solidarność in 1989, far more evidence about the man emerged than was available on August 30, 1980. A full assessment of Lech Wałęsa must be held in abeyance until that evidence is considered.

On the penultimate day of the Polish August, he had one last speech to make under tension, a final self-imposed assignment as a conveyor belt of information to the shipyard rank and file, a concluding tactical mission to carry out in the name of holding the movement together. On August 30, Wałęsa told shipyard workers that they had no institution to stand up to the leading party but their own independent union. "The agreement is clear. The free unions will be those that we elect democratically, those that we decide. Each of us will have the union that we ourselves create." The crowd in the shipyard was immense, a huge jam of people, surprisingly silent. Their needs were great. Wałęsa sensed this as he spoke from an improvised platform near the gate, before thousands in the shipyard and more thousands packed outside. "It is true that not everything is perfect. But believe me, we'll get there . . . We can already see the light at the end of the tunnel. We will get there. I hope that there's going to be no trouble, no outrages. We know that there are provocateurs in our midst; there are people who don't like what is happening. But they will not be given the chance." There was a visible reaction to this and the first shouts of agreement and support. "We must not let slip our chance of winning, because it is just possible." And again, he spoke of the key connection, the point of the strike, the thing they had won: "Listen. We are going to have our own building with a large sign over the door, [the syllables were drawn out for effect] In-de-pe-dent, self-gov-er-ning trade unions." The thought had been in his mind for years; it had floated in and out of conversations around kitchen tables on the coast for a generation; but now, all the "temps" and the "flyers" needed to internalize what it meant. In two weeks of self-education by loudspeakers, many had already done so. In his daily reports, and especially on August 30, Wałęsa tried to increase the number.

The crowd responded. The chants came again, not quite for the last time in August, but certainly never with more tactical political meaning: "Le-szek, Le-szek"—rising in volume each time—"Le-szek, Le-szek." The whole mood changed as Wałęsa, laughing now, began bantering with the crowd, whispering quips about the negotiations, the people's tribune acting out his war against the party's best, a bit of the trickster sending a signal that "they" could not fool "us"—high politics at the grass roots. It was a kind of public discourse that always makes intellectuals uneasy and deeply suspicious, and yet on this night, it seemed so necessary to the supply of information that alone carried the possibility of coping with the rumors and doubts that had spread like a virus among a powerless group of people trying desperately to climb upon the stage of their own nation's politics.

On this night, a journalist of the English Left, recoiling from such scenes, labels Wałęsa a "demagogue." Neal Ascherson is not certain Polish workers have intelligently earned a victory and is not quite ready to regard it as unalloyed progress. In contrast, a journalist of the French Left records the words in careful detail and focuses on the response of the crowd: "Their chants and their hymns warmed the heart," he writes. Jean-Yves Potel is convinced the victory is fairly earned by the shipyard workers, by the MKS, and by the Strike Presidium.[45]

It is a rare historical moment—popular politics in the raw—and all watching bring their own political baggage to the occasion. Poland is not the only country and Leninism not the only ideology that is exposed in new ways by the searing glare of the Polish August.

A poet who (in light of the commercial fragility of his calling) also labored as a shipyard worker noted the anxiety that had built up among his co-workers through the afternoon of alarmed debate in the MKS.

> Through the loudspeaker came voices full of anger, stoking up uncertainty. The frayed nerves of the speakers did nothing to calm ours. The crowd froze in the place where Wałęsa was wont to conduct his evening "vespers." I reached the gate and started to look for someone with whom I could share my feelings of confusion . . . There was Wałęsa, standing up on his cart, talking with the crowd . . . a friendly chat, almost. His evening audience: the atmosphere was more like that at a picnic than a mass rally. He himself had just been through a battle which threatened to break up that beautiful constructed solidarity. Now here, in the evening light . . . Wałęsa could relax. He was at ease. In this crowd he probably found the thing which had been carrying him through those days, those weeks and years, which had made him cling to his belief. He had the crowd's attention, the sense of oneness with it, and the sense of being able to prevail upon it by using the right words, words which would land him on the fertile ground of faith. He was pleased, therefore, to find himself face to face with the crowd which was waiting for him. In a place and time when others might have felt uneasy, when others might have felt tongue-tied or uncertain, he was consumed with joy. "And now we'll go to our homes, take a bath, go to bed and tomorrow, later, our Poland will have more citizens. And now, therefore, let us sing the National Anthem for this country of ours. [And here he started singing the anthem.] Oh, and one more thing: Let us sing a religious song to God because now we can't go any further without God . . ."
>
> I am grateful to him for that Saturday evening, just before the day when peace between Poles was signed. Thanks to him I was able to sleep, even though I was lying on the floor between a desk and a bookcase. I was grateful to him for that peace of mind which, so it seemed to me then, he had rescued at the very last minute . . ."[46]

So the cheers came for Wałęsa late in the evening of August 30 as the coastal movement came back to itself, in a new form, a bit more confident as a result of its new experience. But there was no "one" understanding, no "mass" understanding. The Lenin Shipyard was a much more highly politicized place on August 30 than it had been on July 30. But it remained a place of diverse opinions.

Andrzej Gwiazda also made a memorable speech, short but powerful, in what was his finest moment. Behind the anxiety over the leading role of the party, he knew, was the deeper anxiety about whether the new union would in truth be "independent and self-governing." Speaking to the full MKS, Gwiazda confronted the issue: "Our only guarantee is ourselves . . . We know that hundreds of thousands, millions of people think the same as us. There we have our guarantee. We know the word 'Solidarność' will survive." The phrase stuck in people's minds—"our only guarantee is ourselves."[47]

Throughout the final countdown, indeed throughout the entire ten-day period of intense negotiations that climaxed the Polish August, one calm and thoughtful participant attempted to take note of all that had occurred. By August 30, Bronisław Geremek had become, in his own mind, intimately tied to the workers' movement. Since his arrival in the shipyard toward midnight of August 22; he had offered assistance when asked; otherwise, he had kept his own counsel and generally followed the ebb and flow of events with quiet intensity. His interest was not solely a product of the strike itself; Geremek had spent much of his adult life trying to understand and to fashion democratic structures. He had worked inside the party and outside of it. He accepted the fact that the circumstances surrounding high politics in his homeland forced him into a marginal role.

A medievalist, he became an authority on the Parisian underclasses of the late Middle Ages, a submerged people who attempted to survive with dignity despite all hazards of feudal class and cultural power. In People's Poland, Geremek had had similar problems with the hierarchies of the one-party state. As a young scholar in the 1950s, he poked around in the party's trade union for teachers, but abandoned the party after Poland joined other Warsaw Pact countries in the invasion of Czechoslovakia in 1968. In truth, his remaining connecting links to the party had been severed even before then—by the "March events" of 1968 during which the party launched an inexplicable anti-Semitic campaign in the course of suppressing the student protests of that year. As an orphaned Jewish child who had been saved from the wartime Holocaust by the Gentile parents who had adopted him and given him their name, Geremek was in his politics as in his professional interests always something of a rarity.

He was one of the initiators at Warsaw University of fund-raising efforts for the workers victimized by the 1976 repression, and he subsequently moved with ease and energy into the milieu of the Flying University. He suffered the inevitable harassments—searches of his apartment, interrogations, and occasional forty-eight-hour detentions. But he did not formally join KOR or any other oppositional structure. It could be seen in retrospect that Bronisław Geremek had been waiting all his life for something like Solidarność, something that could offer all Poles—artisan Poles and intellectual Poles—the opportunity to work together for a more open society. He was at home in the worker milieu of the Lenin Shipyard. In his talks with the Strike Presidium, with the MKS delegates in the adjoining hall, and with the enormous numbers of rank and file in the shipyard, Geremek had gotten a profound sense of how deeply Polish workers resented the patronizing attitudes of Polish intellectuals. He had noticed how this resentment extended to the most articulate levels of the workers' movement—and to the Strike Presidium itself. As he also acquired a sense of the subtle understanding of the party that worker militants possessed, he saw how inappropriate much of the posturing by the Polish intelligentsia was in real political terms. It was a matter he intended to take up in a speech he had decided to give to the Polish Academy of Sciences. Members of the intelligentsia had some learning to do before they could be seen as appropriate partners in a coalition with workers.[48]

Geremek had watched the machinations of Jadwiga Staniszkis on the advisory panel, had not grieved at her abrupt departure, had not felt a need to respond publicly to her accusations against the advisory group, and was not drawn by the post-August gossip and intrigue that swirled through the circles of the Warsaw intelligentsia. He was too preoccupied with the workers' movement. Together with his associate Tadeusz Mazowiecki, who generally shared his approach to popular politics, Geremek embodied qualities that were to prove enormously useful to the emerging democratic movement.

Staniszkis, who seems to have seen the strike as a threat to her own politics, held workers in contempt, invariably wrote about them condescendingly, and performed acts that undercut their movement; Tadeusz Kowalik patronized workers while serving faithfully in their cause; Geremek worked in the movement because he understood the organized working class as Poland's key to a democratic future. The differences among the three intellectuals were not small. When Wałęsa asked Geremek to serve as an advisor to the Solidarność National Commission, he readily accepted. The two men, both very private persons, were not intimate associates then or subsequently. They merely shared identical strategic conceptions of the centrality of the working-class movement. In the meaning of the creative Italian democratic theorist Antonio Gramsci, Geremek was as "organic" to the movement as Wałęsa.

On their last night in the shipyard, the Strike Presidium, spurred by Wałęsa, took steps to establish rigorous credentials checks so that only authorized delegates could gain entrance in the morning. The objective was to convene the workers' movement the next day—not a congress of KPN or a rump caucus of the Bieliński group from KOR or a meeting of the adult division of Young Poland.[49]

And so they did. Freed of the brooding anxieties of contending opposition forces, the workers' movement settled down. The final issue, it turned out, concerned not the leading role of the party but the release of political prisoners. The government proved extremely resistant to the idea of agreeing in writing to the release of the workers' "helpers" as part of the Gdańsk agreement itself. Jagielski evaded and finally verbally assured the presidium that the prisoners would be released the day after the agreement was signed. This was good enough for Wałęsa, but not for Gwiazda or, it developed, the plenum of the MKS. On the final day, Gwiazda met no less than four times with Jagielski before this point, too, was advanced a bit. Jagielski read the signed statement to the MKS: "The prosecutor's office will reach a decision on these cases by noon tomorrow; no one will be punished for taking part in or aiding the strike."

At the very end, as the Polish August entered its concluding moments, the final triumph was suspended in a wave of popular emotion that was appropriate to the occasion. Wałęsa addressed the six hundred delegates to the MKS: "*Kochani!* [Beloved], we return to work on September 1. We all know what that day reminds us of [the German invasion of Poland that ignited World War II]. We think about our country and our fatherhood and the family which is called Poland. We have thought a great deal about it during our strike. We haven't gotten all that we wanted, but we got all we could in the present situation." With an eye to the future, he went on:

> The rest we shall also win, because now we have the most important thing: our independent self-governing trade unions. That is our guarantee for the future. We not only fought for ourselves, for our own interests, we have fought for the country as a whole. You all know the tremendous solidarity we have received from the working people. The whole country has been with us. Thank you all for supporting us. In the name of those working-class forces who went on strike, I would address myself to all who have supported us. We have fought together, and we have also fought for you. We have won the right to strike, we have received guarantees of certain civil liberties and, most importantly, we have won the right to an independent trade union. All working people are now able, voluntarily, to form their own unions. They have the right to create independent and self-governing unions . . . I declare the strike ended."[50]

A prolonged wave of applause engulfed the hall—applause and cheers for themselves, for the Strike Presidium, for the MKS advisors, and for the new possibilities for Poland.

Then under the glare of klieg lights (the Polish nation would finally catch a television glimpse of the worker named Wałęsa), they all gathered on the platform. The rostrum was nicely balanced with a statue of Lenin, a cross, and a Polish eagle. Wałęsa turned to Jagielski and thanked him "and all those in power who refused to allow a settlement by force." Turning to the delegates again, he added: "There are no victors and no vanquished; we have settled 'as Pole talks to Pole.' " Jagielski seized this refrain, pleased that they had indeed talked "as Pole to Pole." Then the nation saw Wałęsa sign the agreement with an enormous pen decorated with a photograph of the Polish Pope. After seventeen days of testing, argument, and large-scale cooperation, it was over. Democratic space had been pried open in People's Poland. It housed an independent trade union called "Solidarność."

The next morning, Wałęsa went to the rooms on Marchlewski Street that the government had set aside as the headquarters for the new independent union. The key he had been given would not fit. After some delay, a janitor was found to open the door. With dozens of foreign reporters and photographers crowding in behind him, Wałęsa entered to discover a completely bare office. Reporters looked to him for a cue to the proper reaction to this state of affairs. Wałęsa waved his arms and said: "I am in an empty room, but one full of hope."[51]

The same morning, an engineering works called Hipolit Cegielski ceased production under the pressure of an occupation strike. Also on September 1, Silesian miners in a dozen pits began laying down their tools. By noon the next day, 200,000 were out on strike. The government dispatched the minister of mining, Włodzimierz Lejczak, who was well known to the strikers. He was too well known; the miners refused to talk to him and told the government to send someone else. They would stay on strike, meanwhile. A new spirit, it seemed, was in the Polish air.

In Warsaw, Jacek Kuroń was released from detention. He promptly contradicted Bieliński's entire stance in the shipyard. The workers were right, said Kuroń, to accept the leading role of the party. If the party collapsed, the Russians would come. The important thing was the independence of the new trade unions, he emphasized. But to his surprise and dismay, Kuroń also encountered the immense prestige the academic advisors had suddenly acquired within the Warsaw intelligentsia. The six members of the Mazowiecki-Geremek group were being hailed as national heroes who had wrought a miracle at the negotiating table in Gdańsk.[52]

Kuroń had never been impressed by the capacity for structural political analysis displayed by the mainstream of the Warsaw intelligentsia. He was not impressed on September 1 by the swiftness of their ability to look past the workers' movement. His teeth a bit on edge—he had spent over six years in prison for being forthright in his analysis of the Polish state—Kuroń noted the language of praise for the academic "experts" that constituted the new popular wisdom of the Warsaw intelligentsia. If one had to be an "expert" to gain access to the movement at this late date, so be it. He grimly announced his own new occupation: "trade-union expert." The gesture proved unnecessary. Kuroń received an invitation from Wałęsa to serve as a consultant to Solidarność. In Warsaw, Mazowiecki was appalled at so radical a step in naming the nation's most highly visible oppositionist to such a highly visible role. So, for the same reason, was the Polish episcopate. So was the Central Committee of the party. All quickly understood; the Polish August had ended without bloodshed, but the crisis was not over.

It had, in fact, just begun.

7 ▪ Solidarność Creates a Democratic Culture: Civil Society in Leninist Poland

When I saw it on TV, something happened to me. I felt it inside me, in my stomach. It was physical. I said to myself, "It's going to happen. It's going to happen in my lifetime. It's going to happen to me." I've been a different person ever since.

A Polish machinist, on seeing Lech Wałęsa signing the Gdańsk Accords

THE FIFTEEN-MONTH era of Solidarność embodied social and political ingredients of such contradiction as to be almost unprecedented in the recorded history of social relations applicable to an entire society. On the one hand, the social experience that materialized in coastal shipyards in August 1980 were extended over a four-month period to virtually the whole of Polish society. If not every single Pole in every town and peasant village experienced this transformation with the intensity characteristic of life in the Lenin Shipyard, the broad democratic dynamics set in motion by the Gdańsk Accords nevertheless invigorated daily life in demonstrative ways that affected the entire population. The process took time and involved a number of instructive learning experiences, but these experiences were so vivid for so many millions of people as to constitute in the aggregate an authentic revolution in social relations.

But at the same time, the popular movement could not, and knew it could not, actually take state power. In the Brezhnev era, the continued existence of the Polish United Workers' party was a precondition understood to be essential in preventing the Soviet army from marching into Warsaw. Solidarność, thus, was socially powerful but politically constrained, while the party retained authority but had no social credibility. This bizarre juxtaposition called into question long-standing scholarly presumptions as to what acts were "political" and what acts were not, what prerogatives and potentialities belonged to "society" or to "civil society" and what prerogatives remained solely the property of the state.

It is necessary, then, to review some descriptive terms that have been widely but, I feel, inappropriately applied to the Solidarność era. The confusion has its origin in the unfamiliarity of most students of Solidarność with the long gestation within the Polish working class of the specific organization forms—

and the attendant ideological ramifications of these forms—that produced the very concept of "an independent and self-governing trade union."

Authorities on Poland, grounding themselves in the (untested) assumption that Polish workers had historically been concerned solely with "economic" or "trade-union" issues, have been forced to conclude, as a product of reading the twenty-one Gdańsk demands, that some sort of fundamental change occurred that elevated the workers' horizons from traditional economic demands ("stomach issues") to "political demands." This presumed progression has been abstractly explained as a "transformation of consciousness," variously attributable, as we have seen, to the episcopate or to the intelligentsia. In fact, no such specific ideological transformation from the economic to the political occurred in 1980; as specified in Chapter 3, the key systemic demands, for unions independent of the party and for publications independent of the party's censorship, had been part of the worker demands generated in 1970 and again in 1971 on the Baltic coast. Both were natural instrumental products of the conceptual and organizational achievement of the interfactory-strike-committee structure originally formulated in 1970. The deeper reality was that the entire postwar effort of Polish workers to break out of party control on the shop floor can properly be understood as an organizational attempt to escape party censorship. Additionally, of course, the very existence of massively organized social space independent of the party carried profound ramifications for the Leninist system of governance. Solidarność was "political" from its very inception.

Nevertheless, despite the fact that worker self-activity was perceived to have necessarily evolved from economic to political consideration in order to produce the conceptual sweep embedded in the twenty-one demands, most accounts of the Solidarność era have proceeded under the paradoxical assumption that this newly gained political plateau was immediately surrendered in September 1980. The movement's imperative need to avoid taking state power has convinced students of Poland to see the Solidarność era as consisting of two distinct phases, a "nonpolitical" early period in which Solidarność could be seen functioning "as a trade union" and a later "political" period in which the movement drafted its Action Program aimed at revamping the structure of society. This artificial dichotomy is, to say the least, not helpful—for a number of reasons that go to the very heart of the definition of what popular democracy actually encompasses.

In the late twentieth century, this is an extremely difficult subject to deal with in a culturally comprehensive manner because the global struggle between capitalism and socialism has so politicized the definition of "democracy" that sustained analysis has the effect of making partisans in both camps uneasy and suspicious. Competing loyalties fortified by the Cold War have the effect of making serious democratic analysis appear to be "going too far"—that is, beyond the limits of democratic practice fashioned within actually existing societies around the world. Such analysis gives the appearance of being inadequately partisan to one "side" or the other. It sounds faintly subversive.

It was not always so. This response constitutes a modern anxiety, invested with a disabling culture authority that was not present in earlier eras. A central irony of Solidarność is that its forms harkened back to an earlier time, to historical periods of less-confined democratic horizons. It is our loyalty to the present that obscures this historical relationship and causes modern observers to pass abruptly over the most noteworthy democratic dynamics that gave form and shape to the Polish movement.

To make these relationships clear, it is necessary to review a bit of the history of democratic theory and democratic striving that, in its realized and unrealized modern forms, constitutes the world we live in and the ways we have come to think about that world. It is precisely because Solidarność extended itself beyond the boundaries of modern political convention (though not beyond more expansive earlier horizons) that makes this historical digression necessary.

As a huge popular movement inhibited from taking state power, Solidarność has been seen as a classic expression of mobilized "civil society." It is a term understood to suggest a voluntary association of free individuals collectively forming themselves "outside" the state. It is critically important to note that only since the appearance of the broad political tradition initiated by Hegel and Marx has the term "civil society" possessed such a narrow descriptive range. For example, the ancient Greek *polis* was not only a functioning civil society; its participants were intimately involved in politics. Civil society existed "inside" the Greek state as a functioning political force. As revived in the Italian Renaissance, the idea of civil society was enriched in its social and political dimensions by additional concepts of "civic virtue" before being further elaborated in the commonwealth tradition of seventeenth-century England, a strain of political thought that extended to far-ranging democratic conceptions that materialized within the English Revolution. The idea of a politically active civil society was an animating component of eighteenth-century republican political forms, pioneered in England and expanded upon during the American Revolution. This republican tradition yielded derived organizational forms as a feature of nineteenth-century workingmen's associations, reflecting artisanal egalitarianism, and self-organized rural cooperatives that formed the structural base of agrarian populism. Both of the latter developments, as elaborated in America, were anchored in Jeffersonian conceptions of a properly functioning civil society erected upon, in Jefferson's phrase, "elementary republics."[1]

All such forms of social organization—Greek, Italian, British, and American—grew out of conceptions of a palpably active "citizenship" as a necessary cornerstone of popular democracy. Such an understanding of civil society was inherently political, both in theory and in practice. And "practice" these forms demonstrably enjoyed, for all were reproduced in lived experience within actually existing societies. But none can be fitted into any concept of "scientific socialism" because all such inherently dynamic democratic organizational forms are intrinsically experimental. As Solidarność itself would experience, contingency is necessarily built into the structure of popular democracy. There is nothing remotely "scientific" about democratic forms. Such building blocks as have been fashioned possess value because they have been experientially tested over time, not because they harmonize with some enclosed political theory.

Unfortunately for the cause of democratic governance, the industrial revolution weakened this accumulated experiential tradition. In the first instance, the dynamics of the controlling market that materialized under capitalism inexorably yielded enormous concentrations of economic power that not only narrowed the range of sanctioned political debate but also produced transparent social excesses and systemic human suffering. In recoil, alarmed democrats shifted their attention from a focus on building a democratic civil society to a new objective: harnessing the state as an instrument of ameliorating gross social inequities. The parameters of public discussion therefore gradually contracted in the twentieth century to a narrow argument over the merits and useful extent of state participation in the economy. At the decision-making level, civil society and the accompanying practices of civic humanism virtually disappeared as politics became a truncated discussion among competing or cooperating elites. In this abbreviated political context, recurring economic depressions and other social strains eventually produced the sundry adjustments that comprise the essential ingredients of modern welfare capitalism. These adjustments were brokered by political parties managed by distant elites. The parties themselves constituted a skeletal stand-in for civil society, for they were essentially unoccupied by the citizenry, which merely visited them briefly on election days. Forms of welfare, it was discovered, generated remote bureaucracies that heightened popular preoccupation with rules administered by the state, even as the forms themselves distanced people from each other and, effectively, from basic political decision making.

These modern political customs further weakened the old commonwealth idea of society as a politically active civic community. In its place, the commodified social relations that did materialize produced endless varieties of individual anxiety, occasional private material satisfaction, increasing alienation, and elaborate modes of social escape. Such obsessive preoccupations with the privatization of social life rendered ideas about the structural ingredients of popular democracy—about civil society itself—increasingly difficult for citizens to think about, let along experience in their daily lives. Easily enough, the imperative structural need to strengthen popular democracy by organizing civil society was not a subject that the citizenry of modern democratic states easily understood—Jefferson's writings on the subject notwithstanding. The constrained limits of democratic possibility became a cultural norm of the twentieth century, internalized by elites and nonelites alike. Should individuals somehow find a way to breach these narrow limits, their speculations could be easily dismissed as "romantic" or "utopian." The capitalist response, then, to industrialization produced technologically advanced societies that facilitated social loneliness, reinforced by political resignation. The explanatory rationale for this state of affairs was embodied in an emotionally powerful word of consolation: "progress."

On the other hand, the Leninist alternative dismissed interest in self-organized civil society altogether. In postulating a disciplined revolutionary party to act in the name of society, Lenin completed the final arc in a profoundly undemocratic political trajectory, one that began with Hegel's vision of a truncated civil society unencumbered by either peasants or workers—that is, walled off from most of the population. Marx's settled conviction concerning "the idiocy of rural life" was therefore quite Hegelian. But the social content of Marx's break with his predecessor was centered in his vision of a capitalist-created proletariat as a "universal class" that would produce the revolutionary breakthrough to a society of associated producers. Lenin, giving up on this projection, brought forth his triumph of anti-democratic theory, the vanguard party. To an extent that was unprecedented, the Leninist "leading party" endeavored to occupy every nook and cranny of social space throughout civil society. A forthrightly ideological party should not be content to manage the economy over the heads of a passive and mystified constituency but rather should extract the active endorsement of a properly "class conscious" and therefore aggressively acquiescent populace. The conjunction of "aggressive" and "acquiescent" merely sounds odd, but it is built into the Leninist assumption: only an aggressive (believing) constituency would work hard; only an acquiescent constituency would escape official displeasure.

The twentieth-century societies that have emerged from these competing traditions, though significantly different both in terms of individual liberty and in the availability of democratic forms, share a common tendency. Both have radically reduced the operating space and potential political sanction embedded in the very idea of civil society as a self-organized voluntary social formation serving as the essential foundation of a democratic polity. The reconstruction of civil society is therefore a necessary precondition for democratic renewal in the modern world—a prerequisite to the public recapture of the public sphere of politics.

The era of Solidarność stands as a moment of global historical importance because it generated new democratic experiences that ranged through the whole of society and because, as these dynamics unfolded, they in turn generated pioneering democratic experiments not only in social self-activity but also in economic and political self-management. The Polish movement was not, as some have tried to argue, a "self-limiting" exercise in "anti-politics"; nor was it in its early phases, as others have judged, "nonpolitical." Rather, over a fifteen-month period of intense effort and internal debate, various sectors of the Polish populace successively became earnest participants in an Athenian *polis*, virtuous Renaissance citizens, good commonwealth advocates, zealous republican innovators, aspiring artisanal egalitarians, and, in the end, pioneers bent on scouting out the beckoning frontiers of a self-managing republic. To observers whose eyes are accustomed to hierarchy, such Poles necessarily became, particularly in this last stage, "romantic" and "utopian." This seems an unnecessarily sour and resigned response, one that forecloses serious prospects for democratic achievement by future generations. By reasonable democratic standards, the millions who comprised Solidarność seem to have earned the right to serious inspection for their collective act of generating, under conditions of extreme pressure and incipient repression, social experiences that democratic advocates may selectively emulate and that political theorists may study with profit.

To this end, some distinctions are in order. Solidarność was not a New England town meeting, especially not an eighteenth-century one; nor was it a nineteenth-century workingman's association grounded in an internalized republican heritage; nor was it a populist cooperative self-consciously Jeffersonian in its conceptions of republican justice. It was not a popular expression of the Scottish enlightenment or a conscious manifestation of Chartist aspiration. It was not, in short, the culmination of three and a half centuries of democratic and republican theory or (largely as an oppositional force) political practice in sectors of the West.

Solidarność was Polish, and it drew on two distinctly Polish inheritances: first, the national culture that in all its complexity, passion, and political contradiction had withstood incessant party attempts to obliterate the multiple features that were incompatible with communism; and second, the enormously educational experience of postwar Leninism, which Solidarność consistently utilized as a guiding negative example. For Leninist centralization, Solidarność substituted organizational decentralization; for informational closure, informational openness; for public disinformation, public debate.

But a distinguishing feature of the Polish movement, one that separated it from all other popular expressions of democratic assertion in other countries, was its size. As a movement that came to incorporate most of the adult population of the nation, Solidarność housed competing interest groups with potentially incompatible objectives. For much of its legal existence, the Polish movement was able to maintain the appearance of "unity" because all of its social components were united in opposition to the Leninist party. But during martial law and even more tellingly at the end of the decade, these divergent tendencies would inevitably emerge.

For the entire period from 1980 to the present, any coherent analysis of Polish politics necessarily turns on a close reading of these interior social relations that provided both Solidarność and Polish society generally with the specific political contours that so decisively inform the country's immediate future. Before investigating this complex interior life, it is helpful to engage in a brief sketch "from afar" of the fifteen-month crisis at the apex of Polish society.

It is now evident, after the fact, that the operative strategic reality governing the Polish crisis was the unshakable insistence by the Brezhnev Politburo in the Soviet Union that Solidarność had to be summarily repressed out of existence. The detailed account by the defecting Polish army officer, Colonel Ryszard Kukliński, a CIA agent sequestered in the highest reaches of the Polish General Staff, makes clear the unrelenting nature of the pressure maintained by the Soviets upon their Polish counterparts. If the Soviets felt that the fall of Edward Gierek in the week following the Polish August and his replacement as party first secretary by the head of the security services, Stanisław Kania, signaled an end to party irresolution, they were soon disabused by the persisting aimlessness of the new regime. Twice—in December 1980 and again in March 1981—the Soviets moved to the brink of unilateral military intervention. Soviet nervousness was attributable to what appeared from Moscow as an unending series of ignominious party retreats before "counterrevolutionary forces."[2]

The first round of high politics occurred between September and December 1980 as Solidarność extended its organizing tentacles from the coastal heartland of the movement to the whole of the nation, expanding its rolls from roughly 750,000 on September 1 to something approaching ten million at year's end. During this period, the independent union was officially registered outside the structure of the party's official trade-union apparatus (though not without the pressure of a general strike) and its members protected from police repression (though not without a series of local, regional, and national strikes). These tumultuous events generated enormous tensions within the Polish party, activating hard-line factions in the army, the security services and in the Politburo itself. This condition culminated in police "incidents" that reached a climax at Bydgoszcz in March 1981, precipitating another round of crises that shook the interior structure of both the party and Solidarność. The party's recalcitrance, while inadequately focused and politically ineffective from the Soviet perspective, was sufficiently nonresponsive to the original Gdańsk agreement that it triggered within the party its own insurgency, known as the "horizontal movement." This effort, arising from the lower echelons of the apparatus, sought to reorganize the party along more democratic lines and to array it politically in support of the objectives of the popular movement. Thanks to the vertical structure of the party, this revolution from below was fairly easily contained and formally domesticated in mid-summer 1981. Of greatest strategic importance was the Soviet response to what was viewed as the Polish party's inglorious defeat at Bydgoszcz. Polish General Wojciech Jaruzelski had been elevated to prime minister weeks before the Bydgoszcz affair, and only his firm guarantee that the Polish army would, in fact and in due course, repress Solidarność through military action persuaded the Soviets to refrain from unleashing combined East German, Czechoslovakian, and Soviet armed forces.

The two sessions of the Solidarność National Congress in the autumn of 1981 reaffirmed the creative thrust of the movement, though in a climate of increasing party hostility, planned unresponsiveness, and, in time, planned provocations. Food "shortages," a disrupting factor since late July, were augmented by sharp price increases announced in early October. This move predictably produced yet another wave of strikes across the country, which, on the testimony of Colonel Kukliński, were regarded by the army high command as a justifying setting for the promulgation of martial law. Kania was replaced by Jaruzelski as first secretary, and preliminary preparations for a military coup were set in motion in late October. Solidarność, fighting for its dignity as well as its life, became increasingly alienated and rhetorically radical up to the moment of the December military coup.

Though a number of illuminating details add texture to this brief summary, they can appropriately be considered in the course of investigating the interior life of the popular movement during its fifteen months of legal existence. Since these developments were sequentially related, the period may be seen as pivoting on six evolving dynamics: the politicization of the bulk of the population through their self-recruitment to Solidarność in September-December 1980, culminating in the first crisis with the Soviets; the clumsy but continuous resistance of the party to the implementation of the Gdańsk Accords, climaxing in the Bydgoszcz crisis and the national warning strike of March 1981; the appearance and subsequent evolution through the spring and summer of 1981 of Solidarność's "Network" and its agenda for self-management; the development of the Action Program and its unveiling at the Solidarność National Congress in the autumn of 1981; the proliferation of autonomous activity at the movement's grass roots and its political corollary, "fire fighting"; and the final countdown to Jaruzelski's "State of War" at year's end. It all began, of course, on September 1, 1980, in the Lenin Shipyard.

The era of Solidarność changed so many aspects of daily life in Poland and offered the possibility of changing so many more that the myriad of cultural and political realities that appeared became difficult to sort out and describe. In terms of daily social relations, the explosion of life astonished people. Talent, ideas, and emotions that had been damned up for two generations suddenly washed across the landscape in seemingly unending variety. Daily life itself was transformed into a series of pleasant surprises—intriguing, invigorating, verifying. The outpouring of energy took on so many dimensions that it merited and began to receive foreign scholarly attention purely as a cultural phenomenon.

The graphic arts, always expressive in Poland, abruptly began to demonstrate a remarkable emotional power. All of Poland's long and blood-soaked insurrectionary past seemed suddenly to provide an inexhaustible treasure house of images and symbols through which the promise of the present could be appropriately portrayed. Posters of stunning visual imagery appeared everywhere and in every form: on film and theatrical advertisements, as political pronouncements, or simply as unique private expressions of sensibility. The posters appeared on marquees, in galleries, and on walls and buildings. Poland itself became a gallery. But creative imagination was by no means restricted to professional artists, or even to adults. Art classes in Polish grammar schools revealed such an astonishing vitality that it became possible for creative visitors to suggest a good deal of the social and political meaning of Solidarność simply by assembling and presenting the graphics of children. From one end of the country to the other, the imagination of Poles had been let out of jail. Underground journals of all kinds flourished above ground as a citizenry long starved for information competed with itself in ways to satisfy the general hunger. The private lives of millions became intensely expressive, precisely because one's most personal intuitions could be—and, everyone felt needed to be—publicly expressed and shared.

As in art and private life, so necessarily in politics and public life. The intense experience of Solidarność armed political theorists with distinctly new kinds of historical evidence and, for all who pursued the evidence, opened new democratic vistas for global inspection.

Then, after sixteen months of Solidarność, suddenly on a December night in 1981, everything was violently interrupted. Because of the weakness, incompetence, and greed of the party, the era of legal Solidarność culminated in a new form of state terror within Poland—one that ensured the future appearance of Solidarność in a new form. This was so because everyone understood that the cultural memory implanted by Solidarność would live as long as the idea of Poland did. The memory has a special significance because it was in no way augmented by material considerations. Thanks to the almost incredible immobility of the regime during the entire era of Solidarność, the economic crisis, left totally unaddressed, worsened steadily after August 1980. Poles became more alive and more creative even as they became poorer.

No analysis of such a striking phenomenon can proceed with much coherence through the routine employment of familiar political terminology. The original band of workers that assembled itself in the Lenin Shipyard on August 14, 1980, grew into a mass movement that housed a majority of the adult population of Poland. Solidarność was unprecedented in history—the world's first majoritarian insurgent democratic movement. All previous democratic revolutions, including the many that failed, have been conducted by minorities, sometimes highly politicized, but minorities nevertheless. Other revolutions, including those that attracted millions of adherents, defined themselves as movements of national, ethnic, class, or religious liberation. As it washed against the Leninist state, Solidarność drew from all of those tributaries in the interest of its larger purpose of democratizing Polish social relations. All of which is to say, Solidarność channeled contending currents. Indeed, the movement contained elements of discord that embodied contradictions rather than simple contentions. A great deal of the pioneering of new democratic forms initiated within Solidarność, both in practical terms and in projects that never were able to move beyond advanced stages of planning, were fashioned in an effort to cope with the internal tensions that grew out of the movement's sheer size and broad democratic objectives.

The complexity of these circumstances generates real problems of interpretation. It requires a measure of patience to bring the Solidarność experience into focus. It requires more than that. There is a need to reevaluate prevailing ideological traditions which provide the sundry ways people "see" societies. Failing this, Solidarność remains intractable, a huge lump of people free to think but not to act, a blurred social formation permanently inchoate in the face of party-controlled guns.

It is first necessary to specify, then, that Solidarność tested long-sanctioned assumptions about the dynamics of democratic politics. A certain strategy of approach is required, one that permits a prudent step back from the day-to-day life of the movement so that one can first explore several fundamental assumptions about the unusual forms of popular democracy that Solidarność came to represent. Under the circumstances, a bit of prologue may prove helpful.

Revolutionary leaders as diverse as Thomas Jefferson and Mao Tse-tung have written about a certain human capability they regard as essential to social change. It may be quietly described as the ability to act publicly against sanctioned authority. Once acquired, this capability produces a highly visible result, but its activating ingredient is invisible. The elusive component is something people acquire *before* they gain the capacity to act. It is something that has often been understood vaguely as an "insurgent attitude." It has also been seen as a function of pure "will," as in the injunction, "people must act as if they were free to act." This "something" can also be described as a political stance, as suggested by words like "militant" or "radical." Thus, both Jefferson and Mao can be descriptively characterized, easily enough, as "revolutionaries." Unfortunately, such terms describe self-activity, not the essential prerequisite to self-activity; they describe a result rather than the process necessary to the result. The relevant actors, whether Jefferson or Mao or anyone else, are focused upon and described too late in the process.

Another way to miss the essence of autonomous activity is to focus upon it too early by fixing attention on strategies designed to "activate people." Revolutionary organizers have been told to go out into society and "rub raw the sores of discontent" so that people can be induced to be "angry enough" to act. Such activism can also be misunderstood as a kind of intellectual truth bringing. In its normal state of passivity, the population is "mystified," and it is the duty of militants to teach people to be "demystified." The presumption is that once the activists instruct the population as to the wholesale nature of the injustices being perpetuated by the regime, the people will "rise."

Familiar political perspectives play a major role in how this recruiting activity is described. If the revolution is understood favorably as something that is needed, the militants can be seen as "patriots" engaged in the essential work of "rousing the people" to fight "in freedom's cause." By contrast, if the revolution is frowned upon, a linguistic transformation takes place. The "people" become a "rabble" and the patriot becomes a "rabble-rouser." Much of this terminology, whether overblown positively or negatively, is essentially irrelevant in that it focuses attention too early in the revolutionary process to isolate the ingredient people must have in order to act.

All of which is to say that a certain human capability comes into play after "demystification" and before insurgency, and this ingredient is a central component of social movements. This necessary condition is the conquest of fear and an overcoming of the social habits generated by fear. When a handful of people find a way to achieve this conquest, a political sect appears; when a great number do, a large-scale social movement can form. Since Solidarność was the largest democratic movement in history, to study the Polish movement is to place oneself not only in the presence of a rare social phenomenon but also a rare historical moment—a time when masses of people overcame the binding constraints of life an authoritarian state had instilled in them as ongoing social habit.

One of the academic advisors who came to the Lenin Shipyard on August 24 was economist Waldermar Kuczyński. He watched part of this process unfold in front of his eyes, and he later analyzed the experience he underwent: "The whole stability of the system reigning in the East consists of the accumulation of the people's fear. If it goes on for decades, there is no need for any repression, because people have become afraid of absolutely everything. This does not mean that they are shaking with fear every day. There is a barrier of fear inside them. And it creates very peculiar sensations in people." In time this "peculiarity" describes normal conduct. "For example," Kuczyński further explained, "in Poland three months before everything started, thousands of people went where they were told to go, to the government's annual first of May demonstrations. And some weeks earlier people went to election booths and just put their cards without crossing out any name from the approved list."

It was, of course, precisely such people who performed on center stages during the Polish August. Kuczyński said: "I would treat the shipyard as reduction glass—very concentrated reduction glass. It was almost physical, the way I felt it. And I could feel this barrier of fear go down and grow smaller and smaller from hour to hour. You could feel it everywhere. There was a feeling of liberation from fear traveling from person to person. The social psychology of people was changing."[3]

Another observer was a bit too removed from the center of events to "feel" anything, but he "saw" it. An activist in Young Poland had been on his grandfather's farm when the strike began, and he did not even hear of it for two weeks. He hurried back to Gdańsk and took up station outside Gate No. 2 on August 30.

> It was very, very peculiar when I got back. Strange. In one way, there was a great deal of order. But there were flags all over and slogans everywhere. What I saw in front of the shipyard was something I had never imagined. There were several thousand people, townspeople, standing in front of the Gate from the early morning to late in the night. On the other side were the workers. The moment I recall best wasn't the speeches. Not even Wałęsa. It was the moment when the gate to the shipyard was opened and the shipyard workers came out. People would make a path and would let them go through and the workers would come out. And very different people would come out. And the way the inhabitants of Gdańsk received them was incredible. Everybody would shake their hands and slap them on the back—and singing. There were people crying and people laughing. There was this big determination to construct the new.[4]

Whether one was in the middle and could "feel this barrier of fear go down" or stood outside and discovered that "very different people" emerged from the struggle, whether one sensed a change in "the social psychology" or detected a "big determination to construct the new," the transforming impact of the struggle in the Lenin Shipyard was something that was socially and historically tangible.

Despite the emotionally charged quality of such descriptions, however, it remains unclear precisely what process is being described. To what extent did these alterations depend upon the victorious outcome? Barely a week into the strike, Wałęsa believed the gains were palpable, regardless of the eventual outcome. "Things will never be the same here again," he told Western reporters. "The shipyard workers are different people, now." That is to say, in Wałęsa's view, they were "different people" three days before Waldemar Kuczyński ever set foot in the shipyard and ten days before the Young Poland militant caught his glimpse of "very different people." Is it possible to be more precise than this?[5]

I think so. The task is to construct the concrete ingredients that collectively impart social meaning to the word "consciousness." The decade of experience that people on the Baltic underwent between December 14, 1970, and August 31, 1980, fundamentally altered the social inheritance imposed on Poland by thirty-five years of one-party authoritarian rule. These experiences affected many thousands of people, who, quite naturally, internalized different elements of the process at different times and at different speeds. Coastal activists experienced each stage before the rank and file did and the most persistent and active militants before the less persistent. The totality of experiences also varied in proportion to the degree of involvement of each individual. For great numbers of rank-and-file workers, the most transforming educational experiences came at two moments of large-scale mobilization and of extreme tension: the first during the prolonged confrontation and demand-gathering campaigns from December 1970 through February 1971, the second during the Polish August of 1980.

But it is essential to take full and careful note of the experientially grounded social consolidation that connected these two events—the decade of striving and movement building between 1970 and 1980 that gradually transformed hundreds of thousands of deferential coastal workers in 1970 into the massively organized Baltic movement that generated the Polish August of 1980. Disconnected from one another, these events seem rather mundane, mere "incidents" of discord and the routine repression of discord. But considered both in the sequence in which they occurred and as a whole, they were the concrete building blocks of social solidarity—what the world eventually came to know organizationally as "Solidarność." A vital ingredient—one that connected the experientially grounded knowledge acquired by the most persistent activists to the specific events that ultimately came to involve many hundreds of thousands of coastal people—were the internal communications links that became the lifeline of the Baltic movement. These modes of communication did not fall from the heavens, they were fashioned, one by one, as a product of the experiences veteran activists and newly recruited activists lived through during the decisive decade of worker self-activity.

Major components of this linkage were two mechanisms of large-scale organization: the occupation strike and the interfactory strike committee. But essential to each were smaller components of structural importance: first, the loudspeaker system that connected each shipyard worker to the intimate details of worker assertion and government evasion in the negotiating process between August 14 and August 31; then, the movement's couriers who conveyed the movement-generated information of Gdańsk to less-organized social formations throughout the Baltic region and thereby were instrumental in producing the delegates from six hundred factories who populated the plenum of the Interfactory Strike Committee. The courier system was created by activities predating 1980, during the formation of the middle-class opposition sects that Wałęsa, among others, was so fond of visiting. These groups gathered together the individual Poles the movement recruited as couriers in 1980.

There were many other components that went into the construction of the movement's internal communications network: meetings around kitchen tables in worker apartments, leaflets and the recruiting of movement printers who produced them, the testing of possible public places for effective distribution of information (the trams, buses, shipyard locker rooms, the commuter trains from Sopot and Gdynia), the appropriation of religious and patriotic symbols to the changing needs of the movement as it formed, the discovery of the need for and the subsequent employment of a worker militia to maintain order and security during the length of the occupation strike. Here, then, were the institutional links that made thousands of isolated individuals into the coherent chain of assertion that was the movement. Once in place, these linkages provided a way of communicating to thousands of less-experienced workers the tactical and strategic knowledge gained over time by the most experienced activists.

Collectively, these finite social achievements—each a product of self-activity—had the effect of encouraging the individual self-respect and collective self-confidence necessary for many thousands of heretofore intimidated and silent people to come together as a self-conscious movement of social liberation.

"Consciousness," then, was a product of acquired experience; social knowledge is experiential. It does not occur as a disembodied product solely of the social relations of production; it does not happen merely because "the times were hard" or because "the people got angry" or because the citizenry or part of the citizenry was exposed to dissident pamphlets or manifestos designed to expose the ruling regime or exhort intimidated people to run the risks of public protest. All such explanations describe necessary preconditions for democratic striving. But they are tangential to an understanding of how movements actually happen for the elementary reason that these preconditions are present in most societies most of the time when movements are not, in fact, happening.[6]

In the case of Poland, the coming of "consciousness" can be illuminated in explicitly historical terms. For thirty-five years, the Polish population had "experienced" the leading role of the party. Some, in Lechosław Goździk's phrase of 1956, had been "too full of faith in the party" for some years. But regardless of when this faith began to wane (immediately after the war for some; in 1947, 1956, 1968, 1970 for others), it had withered away to almost nothing by the time Edward Gierek fully consolidated his power in 1973. By then, the Polish working class, particularly on the Baltic, was "demystified" about the alleged benefits accruing from the leading role of the party. But this experiential knowledge scarcely diminished the widespread fear that had been instilled and then reinforced by police terror. That is, workers and other Poles were not yet "different people" merely because they were no longer beguiled by state propaganda. Knowledge of this general kind did not by itself translate into self-generated collective action; nor could it be achieved by reading books and pamphlets. The problem was not complex; it was stark. People were regularly unable to act because they did not know what to do in order to escape the narrow confines of seemingly all-encompassing state power. A political act of self-assertion, any act, had to offer at least a reasonable promise of not being totally self-crippling. Deeply internalized fears that had developed over a lifetime of exposure to authoritarian modes of social organization put a tight lid on any spontaneous free expression. The consequences were too devastating.

What the Polish August gave the Polish people was a kind of Baltic model that could be followed by the rest of the population. It is necessary to emphasize a "kind" of model because, on September 1, few people in Poland knew precisely what it was that had happened on the coast in those weeks. The necessary knowledge was locked up in the minds of the MKS delegates and shipyard workers. They alone had experienced the social dynamics of a mobilized and internally communicating working class engaged in confronting state power. Everyone else in Poland knew only that workers in Gdańsk (elsewhere than Gdańsk?) had "struck" and that they had negotiated with the government for some number of days or weeks. What people outside the coast did possess was a copy of the twenty-one demands as a negotiated component of the Gdańsk agreement itself. As specified in the general agreement ending the strike, the organ of the party's Central Committee, *Trybuna Ludu*, published a full text of the settlement, including the central first demand, and many other journals in Poland followed suit. On September 1, Poland's people, therefore, were permitted for the first time to catch a glimpse of *what* to fight for, but not *how* to struggle for it.

So they came to Gdańsk. Solidarność's little suite of offices on Marchlewski Street became, improbably, a nationwide information center. Within an hour of its opening on the first morning, the office was besieged. By noon of the first day, the line of supplicants wound out into the street. In short order, the new independent union moved to an old seamen's hotel, the Morski, and took over two entire floors. There, the activists of the coast explained to representatives from factories all over Poland the raw mechanics of organizing an independent trade union movement. If it was necessary to secure recognition over the objections of local party officials, instructions were also given on forming an MKS.[7]

In this fashion, a deeply ironic truth slowly materialized in working-class Poland in the autumn of 1980. The MKS had been wrong to seek and the government had been wrong to attempt to deny nationwide recognition of the new independent union. For workers throughout Poland to acquire the sense of self necessary to permit them to behave subsequently in autonomous ways, they needed to earn their own union; to overcome their ingrained fear of the party, they had to struggle with the party. The reverse was also true. For provincial party functionaries across the nation to amend their ways and to begin softening their programmatic arrogance and their administrative condescension toward workers, they had to be forced to confront the organized power of their own provincial working classes. This dynamic worked itself out in every corner of Poland in the first ninety days or so following the historic settlement in Gdańsk. The process constituted the heart of the democratizing experience that Solidarność brought to Poland. More than any other memory, it represented the essence of the democratic legacy that remained within the Polish population after martial law descended.

The old Morski Hotel became the central classroom of a nationwide democratic schoolhouse. The needy came, acquired information, and then, with a kind of tentative confidence, went back to their factories possessing knowledge of the surface appearance, or external scaffolding, of an independent union. Learning began immediately. Once home, they abruptly encountered party plant committees that had no intention of relinquishing their power. Party functionaries evaded, they proposed "steps," they proposed "local alternatives," they even stonewalled. Worker representatives who were still carrying "the fear inside them" made settlements that turned out not to change things much. Some of them returned to Gdańsk seeking further advice. Others turned to their fellow workers, fashioned plant-wide worker committees, and went on strike. At some juncture, the representatives of the workers and the representatives of the party inevitably faced each other in open confrontation. The free union did not materialize until the workers prevailed in that specific confrontation. Only through this engaged experience could fear be burned away. A most essential moment for any local Solidarność lay in the discovery that democracy was more than an idea, it was a way to act.

This learning process was the absolute cornerstone of the new Poland that the population was struggling to attain. It happened very quickly in some places, more slowly in others. But fully functioning independent unions did not occur anywhere until it happened. The formation of a local Solidarność became the pivotal democratic learning experience of the Polish working class—and for Polish office workers, teachers, writers, actors, and any other group that sought to create its own democratic space in People's Poland.

Certain historical anomalies occurred. In Poznań, for example, local efforts to create new structures of independence ran into stubborn opposition from the local party. The lessons of 1956, only partly assimilated by the workers, had not been expanded upon during the subsequent twenty-five years of one-party control. Poznań workers knew how to strike but not how to act after that. Things staggered on for weeks as workers continued to act out inherited patterns of deference in the presence of party officials who had controlled their lives for years. Finally at one aimless meeting, a visiting delegate from Gdańsk, a woman dispatched by the national office of Solidarność, stood up and firmly told Poznań officials what the Gdańsk agreement specified, whether they approved or not. The party's opinions about what the workers could or could not do were, she said, inappropriate. The new independent union would make those decisions, she said. This demonstration of how to act transformed the room. Within forty-eight hours, the delegate from Gdańsk was no longer needed; the workers of Poznań had begun living through a democratic experience on their own. Visible here is the process through which intimidated people learn the prerequisites of autonomous conduct. The coming of collective self-confidence is a political movement of enduring democratic significance.[8]

This experiential juncture contains a political fact that America's Ralph Bunche was once closely questioned about and for which he provided a concise description. Bunche had served as the American ambassador to the United Nations at a time when a flood of new African nations, recently decolonialized, had begun sending delegations to New York. By no means did all of these spokesmen for emerging nations regard America as a dependable ally. Bunche seemed to take this state of affairs with more equanimity than some conservative American journalists found themselves able to achieve. They asked Bunche whether he thought some of the African nations were "really ready" for independence. The racial implications of such questions were rather transparent, particularly to Bunche, who at the time was the highest-ranking black American in the State Department. His reply ended the question and answer period: "People," said Bunche, "are ready for freedom when they are ready to take it." It might be added that, historically, people may well be "ready" long before they have the opportunity to acquire freedom, but they are surely ready by the time they have won it. It is, after all, a quality that can only be learned through experience. Democracy begins with the attempt to have it.[9]

Here then was the central cultural experience surrounding the appearance of Solidarność as an integral part of Polish national life. Encouraged by the example on the coast, which had amazingly been verified by the official press and television, ordinary Poles consulted their most deeply ingrained fears and began experimenting in ways to act, experimenting as if they were *free* to act. Disbelief hovered over the experiment, a feeling grounded in memories of police truncheons and job dismissals that had always awaited all Poles who said what they believed.

The habits of repression immobilize people, and they are habits by no means restricted to police states. One deals here with a folk custom and a cultural fear that is embedded in all humans who necessarily must be socialized from infancy into one or another of the world's hierarchical cultures. Repression is built into the human condition itself, by the helplessness of infants and the necessity for watchful parents who in the course of nurturing acquire a hierarchical sense of their own relationship to their offspring. The child is judged and measured, in schools and later by managers, personnel directors, plant foremen, and other kinds of bosses. One learns which thoughts are permissible and which are not, what behavior is admissible and what is not. In every age group and in every social class, one learns one's lessons and in due course becomes knowledgeable and "mature." Deference is built into socialization. When appropriate conduct is supported by an apparatus of police and judges, detention cells and prisons, the components of deference are deeply internalized indeed.

Disbelief, therefore, contested against the experiment in Polish democracy. The twenty-one Gdańsk demands, the union free of the party, floated in front of everyone's eyes as something apparently attainable and yet, memory insisted, something clearly impermissible. Looking back, after the battle was won, individual Poles could speak lyrically about the moment they reclaimed their own subjectivity. "I was born on September 12, 1980," said a school teacher, "when we [teachers] stood up to the first secretary in a meeting." A small-town machinist confided: "When I saw it on TV [the signing of the Gdańsk Accords], something happened to me. I felt it inside me, in my stomach. It was physical. I said to myself, 'It's going to happen. It's going to happen in my lifetime. It's going to happen to me.' I've been a different person ever since." A local-level Solidarność leader said simply, "I came alive in the Polish August."[10]

It happened so quickly to so many people—ten million in four months—it might be described as a "mass experience." But of course it was not. After watching the brief scene on television of Wałęsa signing the agreement with his oversized pen on the night of August 31, people talked about its meaning in their apartments before they went to bed, talked about it on the trams and buses the next morning, talked about it in the plants and offices of Poland. Cautious people waited for someone to make a move and—that day or the next or the next—"someone" did. People who had "felt something inside my stomach" took the lead, talked to co-workers, formed a group, perhaps a committee. It might be said that people got out of their own private kitchens of repressed subjectivity where they had spent their lives. Groups of workers talked to a shop foreman or a section superintendent or a plant party committee. They talked until they got rebuffed, and then they went back to their starting point and talked the situation over among themselves. They sifted ideas and enlisted more people, and they started the process all over again. The political name given to this activity is "organizing." The term is both descriptively accurate and fundamentally misleading. It passes over the hard part: the time before organizing, the period of months, years, lifetimes, that ends in the very final moment before a subjective act is performed, before people are "born" and "come alive." If consciousness is, indeed, determined by lived experience and if an activated consciousness can find expression only in self-activity, then the operative democratic task of modern politics is to create opportunities for that activity to take place.

On August 14, 1980, democratic space in Poland was occupied by about 8,000 shipyard workers, united both in their tentativeness and in their fragile resolve. On August 17, the space contained 50,000 workers in 20 enterprises in Gdańsk; on August 18, by some 200,000 workers in 156 factories in the tri-cities; and by August 31, by 500,000 to 750,000 from many parts of Poland. The space had a name, Solidarność, which within two more weeks had some three million inexperienced members; within two months, seven million; and by Christmas, almost ten million. These were impressive numbers, assuredly; but the real change was that Poles everywhere had entered into self-activity wherein they could be "born" and "come alive" and through which, in the end, "different people came out."

In ways that people in less embattled societies could scarcely fathom, Poles had spent their lifetimes trying to "come out." In 1944, in the Warsaw rising, 200,000 Poles had died fighting the Nazis as the Russian army lingered on the Vistula, determined that the Red Army, not the Poles themselves, would liberate Poland. Between 1945 and 1947, the Polish "home army," the storied AK, had fought the new communist regime in the countryside, even as the working class waged its bloody battle on the shop floors of the nation. The workers had tried to "come out" in 1956–57; students had tried briefly in 1968, the working class had tried again in 1970–71, in 1976, and again in the summer of 1980.

Over this somber thirty-five years, with or without popular support, individual Poles had "come out" in every year of the nation's postwar existence.

Hundreds of shop stewards had been cashiered or imprisoned between 1945 and 1948, and scores of individual writers, party members, workers, and teachers had, at one moment or another, taken a stand. With judicious impudence, a 29-year-old philosopher, Leszek Kołakowski, had taken a stand as editor of *Po prostu* from 1955 until its suppression in 1957. Each in his own way, a carpenter named Stanisław Matyja and a machinist named Lechosław Goździk had taken a stand in 1956 at Cegielski and at Żerań and had subsequently been blacklisted from their crafts. With guarded daring, Jan Józef Lipski had taken a stand in the quasi-legal Club of the Crooked Circle from 1956 until its suppression in 1962. Two years later, Lipski had organized the Letter of the 34 to the prime minister through which Warsaw intellectuals protested the censorship; the minister acknowledged the letter through the immediate arrest of Lipski. With youthful audacity and relentless intelligence Jacek Kuroń and Karol Modzelewski wrote their brilliant and scandalous "Open Letter to the Party" in 1964 and had gone to prison for three years as a result; Kołakowski spoke out against the censorship in 1966 and was dismissed from the party; Adam Michnik spoke out against the student repression of 1968 and was sent to prison; Kołakowski again denounced the censorship in 1968 and was forced into exile; Edmund Bałuka, Edmund Hulsz, and Henryk Jagielski spoke up for the workers' movement in Szczecin and Gdańsk between 1970 and 1973 and were driven into exile. Hundreds of workers were sacked in 1973 and 1974 after the coastal mobilizations of 1970–1971. The moments of assertion came with increasing frequency: the "Memorandum of the 59," protesting the constitutional revisions of 1975; the formation of KOR in 1976; the "Declaration of the Democratic Movement" in 1977; the formation of the Movement for the Defense of Human and Civil RIghts (ROPCiO) in 1977; the creation of the "Flying University" in 1978; the free trade unions in Katowice and Gdańsk in 1978; the Confederation for Independent Poland in 1979.[11]

For every signer of a letter, petition, memorandum, appeal, or declaration, there were many who wavered and did not sign; for every worker who joined a shop floor insurgency, there were many who thought it over and did not join. There was the case of the two intellectuals who, when asked to sign a protest, replied in contradictory ways. The first said, "I can't. I have a son," and the second said, "I have to sign, because I have a son." Kazimierz Brandys, who took station on both sides of this dilemma, commented quietly: "The two answers express old alternatives, two threads woven through the histories of many cultures. The idea of survival, the injunction of revolt. The preservation of one's existence, the legacy of honor . . . In the end, however, everyone must decide for himself what he fears more—life or himself." It was the pride and agony of Polish history that so many chose to live what they thought.[12]

Here was the particular social poison of the authoritarian state; it forced everyone to pay an intolerable price for the simple preservation of self-respect. No one could speak out all the time; everyone was silent at one point or another. On every shop floor and at every writers' congress, personal humiliation was the enforced cost of getting through the day intact. Yet it was precisely this intolerable condition that inevitably carried with it the germ of emancipation.

So again and again, Poles spoke up. Through all the decades of the postwar years, no working class in the world was more restless, even though it was repeatedly hammered into submission; no intelligentsia was more unconquerable, even though it was humiliated with relentless regularity. Each class counted its victims and remembered its martyrs. The world was told the names of most of the intellectuals; few heard the names of the workers. It had almost become a law of history—in Poland as in every other country.

By September 1, 1980, there were many thousands gone. But there were also many millions who were not gone and who at last saw the opportunity to find their voices. In this sense—the most important sense—Solidarność was not a "self-limiting" revolution. It was, in its deepest meaning, a revolution in social relations. Poland became a place where a majority of the population elected to engage in self-activity. It was the only country on earth where this was true. In consequence, Poland became, for a time, the most democratic society in the world, and Solidarność became a model for people of all nations concerned about democratic governance to ponder.

In 1980 and 1981, foreigners began to do this. Though all visitors to Poland in the era of Solidarność necessarily riveted their attention upon things Polish, they also found themselves thinking about their own countries in a new way. In the spring of 1981, as Solidarność writhed in struggle with a desperate and embittered party apparatus, the American writer Lawrence Weschler found himself comparing Solidarność Poland with Reaganite America and judged the balance to be tilted in favor of the Poles. Everywhere Weschler went in Poland, he encountered passionate conversations about society, about the state, about politics. In America, people talked less about real political issues than they did about the way people in power talked about politics. It was, said the press, the era of the "Great Communicator." It was stylish to talk about the U.S. president's optimistic style as he presided over the most massive redistribution of wealth in the nation's history—a redistribution from the very poorest to the very richest. The president, in any case, seemed to be a devoted rhetorical supporter of Polish trade unions, even as he smashed the air-controller's union in his own country. While political discussions turned on style in America, they were deadly serious business in Poland. The difference was not hard to detect.

But Poland in the era of Solidarność was distinct in many other ways. The most unique feature of the nation's post-August politics was that a powerful movement had come into being that could not, and knew it could not, formally take state power. Located across the Soviet Union's Warsaw Pact communication lines, Poland was an integral part of the Eastern bloc's defense system. The Soviets possessed the military power to topple any Polish government by force, and the Soviet interventions in Hungary in 1956 and Czechoslovakia in 1968 stood as stark reminder of how quickly that power could be used. In the era of Brezhnev, as before, any Soviet guarantee of maintaining the status quo was predicated on the existence of a Leninist party in Poland. Both within the Polish working class and the democratic opposition generally, this fact had long been recognized and was formally codified in the Gdańsk

Accords. In its aims, Solidarność was necessarily a "self-limiting revolution" to the extent that it aimed at democratizing socialist Poland without disturbing Russian strategic interests.[13]

During the sixteen months from the coastal mobilization in August 1980 to the coming of martial law in December 1981, the public appearance of Solidarność seemed to be one of episodic crises: the November confrontation over the registration of the union, the Bydgoszcz crisis in March, the "extraordinary" party conclave in July, and the emotional national congresses of Solidarność that preceded the military coup in December. Though much drama and authentic politics lurked within these visible events, the interpretive tactic of focusing so exclusively upon their public face has the effect of obscuring both the core social experience and the essential democratic legacy generated by the Polish experiment in popular politics.

In Poland, the fifteen months of legal Solidarność saw the unfolding of a nationwide process through which millions of heretofore deferential and resigned citizens recovered their subjectivity by means of self-generated, face-to-face confrontations with the local functionaries of state power. Whatever else is said, this elaborate flowering of popular self-activity—this construction of collective self-confidence—deserves emphasis as the core democratic experience of the Polish people. The process through which rank-and-file Polish citizens actively confronted their received culture of social governance provided the essential psychological precondition and therefore the organic political foundation for subsequent democratic striving.

In September 1980, the first necessity was to export the movement from the Baltic coast to the rest of Poland. While this process proceeded with endless variations over the ensuing four months, a second and equally elemental task quickly took center stage. What kind of internal structure should the democratic social space called Solidarność construct for itself? The party forced Solidarność into immediate consideration of this question by offering a series of often threatening advisories as to what forms—local, regional, or national—trade unions could and could not have. Within two weeks of the Gdańsk agreement, on September 15, the regime put adroit pressure on the workers' movement by officially granting legal registration to a regional—and separate—independent trade union for the workers at Huta Katowice. The move impelled an immediate response by what still remained at that moment an existing but legally unregistered regional union of the coast. If partisans of Solidarność in Gdańsk or Warsaw or anywhere else did not forthwith move to make some structural decisions on their own, the government unilaterally would decide for them. Thus it was that a "national" conference of Solidarność activists met in Gdańsk on September 17 to thrash through the question of what an organized democratic impulse should look like in Poland.

It was a meeting full of the deepest kinds of emotional expression—a kind of coming-out party for worker activists and intellectuals from many parts of Poland—but it was also laced with intense and sometimes strained debate. The circumstance of the moment imposed upon the fledgling movement a choice between two courses of action, both of which seemed necessary but which also seemed to contradict each other. The issue was: should Solidarność, following the coastal model, build from the bottom up a worker-based movement of factories organizationally linked by regions or, conversely and in the name of creating a coherent entity to cope with centralized state power, should the movement immediately create a unified national structure registered in Warsaw under one government statute? In an ideal world, Solidarność could have the time to honor both these conceptions. To be institutionally real, it needed to build itself from below into linked organizations possessing genuine mass roots in society. But to be politically defensible, it needed a structure through which it could speak, or attempt to speak, with a unified voice—as the Interfactory Strike Committees at Gdańsk, Szczecin, and Elbląg had been able to do.

The danger of instant unity, of course, was that Solidarność's planners might inadvertently construct their own "experimental program" or their own "Council on Worker Self-Management" (those large-scale entities that the party had created in the 1950s to outnumber and thereby domesticate autonomous pockets of self-organized workers). This was the underlying fear of coastal activists, given public expression by Wałęsa and elaborated privately by Krzysztof Wyszkowski, the founder of the 1978 free trade union of the coast. But Wałęsa and Wyszkowski were joined by dozens of other coastal militants who understood the intricacy and difficulty of the successive stages of organization they had constructed during the mobilization of August 14–31. As products of this experience, coastal delegates—of all the delegates—were most experientially aware of what precisely had *not* been achieved by other groups of workers represented in the meeting.

Simply stated, coastal militants did not want their own mass-based organization held hostage to an empowered national leadership that represented less-organized groups or, worse, paper groups. Wałęsa announced his fear of a national federation of ten regions "and five which really built from the bottom, what we wanted to build." He added: "And then I fear this defeat because of decisions which are not consulted with the ranks . . . It seems to me that we should divide our roles so that these MKZs [Interfactory Founding Committees] will be stronger and can take care of their own affairs on their own territory." Only in this manner could a strong regional membership be certain, said Wałęsa, expressing his deepest fear, that "no one has the right to tell them what to do."[14]

But the counterargument was also compelling—and it had massive support in the hall. From the most remote reaches of Poland and from newly forming factory groups as well as newly created cadres of groups, delegates had come to Gdańsk to help—but also to get help. Many of these worker militants, some little more than self-selected spokesmen for organized constituencies that did not yet genuinely exist, were thunderstruck by the social environment that suffused the coastal provinces and especially the city of Gdańsk. As one of them put it, when the delegates left the "freedom" of Gdańsk, they would go back to the "barbed wire" of their native provinces. Such workers did not want or intend to complicate the lives of the men and women who had won the great victory at the Lenin Shipyard. What they wanted was succor and protection.

Specifically, they wanted to occupy space under the umbrella that Gdańsk held aloft. To this end, they had a single word that expressed their deepest political need: "unity." The word dominated the speeches of delegates from every corner of Poland. Such an objective authentically expressed the overwhelming short-run political goal of the Polish working class.

It also expressed the considered strategic objective—hurriedly fashioned but deeply reflected upon—of the leading theorists of the democratic opposition in Warsaw. As their spokesman in the Gdańsk meeting, they had an authoritative figure—Karol Modzelewski, the one-time Marxist dissident who as a youth had drafted with Jacek Kuroń the famous "Open Letter to the Party" in 1964. That document had proclaimed in approved terminology the decidedly unapproved conclusion that the party was elite, authoritarian, and antiworker. Young Modzelewski had paid for his courage with three years in prison. But beyond possessing a battle-tested and articulate spokesman, the intellectuals of the democratic opposition had done their homework and—no less relevant politically—they also possessed some additional movement credentials that only a handful of insiders knew about. There was a little-known distinction between the agreement signed with the government by the Szczecin MKS on August 30 and the one signed in the Lenin Shipyard the following day. The workers in Szczecin had kept academic advisors at arm's length throughout their negotiations with the government, and the legal document they signed contained a crippling flaw: no provision was made to base subsequent worker institutions on the self-organized Interfactory Strike Committee. Lacking that guarantee, the legal position of the Szczecin committee was that it had no future organizational place, as an "independent" union, except under the party-controlled trade-union apparatus! In contrast, the Gdańsk MKS had not only enhanced its position with academic advisors, it made up for its initial shortage of legal specialists on August 23 by accepting additional assistance from three of the nation's most experienced opposition lawyers—Jan Olszewski, the legendary Władysław Sila-Nowicki, and Wiesław Chrzanowski. Olszewski had arrived in the Lenin Shipyard on August 25 with a well-crafted preliminary draft that, in due course, became the legal fine print of the Gdańsk agreement. Olszewski's draft carefully preserved the legal independence and autonomy of the Interfactory Strike Committee as the founding committee of the new self-governing trade union. The verified reliability and independence of Olszewski was no less authentic than that of Modzelewski.[15]

On September 17, Modzelewski carried with him a new legal draft—written by no less a personage than Jan Olszewski himself—of a national trade-union structure. Emphasizing the recently unveiled government tactic of registering the Huta Katowice union separately, Modzelewski warned the assemblage with both passion and conviction: "If as many unions arise as speakers in this hall presenting the situation in individual MKZ, then we are finished. Maybe the coast will survive longer but if they finish us off, so will they finish off the coast. It will be the last to go but inevitably it will fall."

With the benefit of the historical knowledge that Solidarność successfully united ten million Poles, it is fairly easy in retrospect to minimize the dangers that concerned Modzelewski. But amid the deeply internalized anxieties of the delegates who filled the Gdańsk hall on September 17, there could be little doubt of the extent of the popular agreement with Modzelewski's conclusion: "This is why I believe that our crying need is for one national self-governing trade union. Ladies and gentlemen, this union must have one statute."[16]

On September 17, the new movement came to its first crossroads. Whatever the eloquence of Karol Modzelewski, whatever the proven legal sophistication of Jan Olszewski, whatever the compelling political drives of the clear majority of the delegates, no person in the hall commanded the prestige and the attendant authority that the electrifying victory in the Gdańsk shipyard had conferred upon Lech Wałęsa; nor, as he had repeatedly demonstrated in the August crisis, was anyone more attendant to the movement's mass base. It is also necessary to add that the outcome of the debate over structure was more significant for Wałęsa in terms of future political standing in the movement than for anyone else. For hours on end, Wałęsa, with timely assistance from the Gdańsk-based chairman of the meeting—Lech Bądkowski—parried and impeded the forces for centralization.

Historians can never divine with certainty the motives of the people they write about. They can only uncover actions and attempt to infer motivation from the evidence implicit in such actions. Wałęsa knew that the strength of the coastal movement was grounded in its mass base. On September 17, he had confidence in the coastal working class and the cooperative experience it had lived through; the status of the rest of Poland, while encouraging, remained problematic. But it is also true that Wałęsa's own preeminent standing in the movement rested upon the loyalty of the social formation he had done so much to shepherd into being on the coast. In this context, Olszewski's statute and Modzelewski's presentation of it represented the organized appearance within the movement of the intellectuals of KOR. Whatever the implications of his own ambition, Wałęsa did not trust the ambitions of KOR members. It was not just that he had observed, unadmiringly, self-appointed authority of a youthful and headstrong Konrad Bieliński to disrupt MKS delegates at the peak of the August crisis (see Chapter 6, p.241), he had also observed, with something less than awe, the earnest, self-confident theorizing of Jacek Kuroń. KOR was a handful—unquestionably dedicated but also conceivably proprietary and paternalistic. Beyond doubt, ample evidence already existed, subsequently augmented by KOR-authored interviews, articles, and books, that the Warsaw intelligentsia had a deeply ingrained habit of patronizing workers.

The internal struggle on September 17 thus flourished on three levels: the strategic problem of the structure of the movement, a class-based dispute between coastal workers and Warsaw intelligentsia, and a clash of well-fortified egos. In this respect, the Polish movement took its proper place among all the world's social movements; it housed diverse human actors with their own experientially empowered views of an orderly future for the movement. It might be added that movements which do not find a way to transcend such inevitable tensions rarely have much of a future.

A final component of what was to become an ongoing feature of the style of Solidarność surfaced in the September 17 meeting. This ingredient was the complex persona of Lech Wałęsa himself. He was not a man who could be characterized in a sentence or even in a brief paragraph. To anyone who saw him perform in public—and all Poland was to see this phenomenon soon enough—Wałęsa had an embarrassingly boyish capacity to celebrate himself. His habit in the middle of a tense debate of assuring workers that "I will fix everything up, don't worry. I will do it" unnerved the rather large number of incipient democrats who filled out Solidarność's legions both in the ranks and in the leadership. And when Wałęsa was assailed by personal doubt—also usually in the midst of policy debates—he demonstrated an astonishing ability to unveil diametrically opposed conclusions in the same sentence. This unique talent moved some of Wałęsa's rivals for leadership to question his basic honesty. Others, less hostile, were left to question his coherence. An early example of this style was provided by Wałęsa's initial response to Modzelewski:

> Ladies and gentlemen, I understand the intentions. And I know that it should be done this way. But after 12 years of struggling with all this, I would not want to lose everything together with you because of a foolish decision. That is what I am afraid of, this defeat, this ignominy. I would also be for this—it is well made. I also am for it. Except a little bit differently. Sure, we need this Commission and we want to elect it, but this Commission does not have and cannot have the right to decide by vote . . . The fact is we cannot permit ourselves to sit down and write a statute now. We would not finish in two years. There will always be doubts. But let us accept that we can amend it [the existing Gdańsk regional statute] and quickly correct it, but only when we have to—till Thursday—we will fix it somehow. And on Thursday, it will be a statute for those who unite with Gdańsk. There is no other way. Except that I would say I would be for the regions having the deciding voice. . . .[17]

In this manner, the hero of Gdańsk informed his comrades that he favored a single union, as long as it was powerless, and a national structure as long as the regions remained autonomous. Olszewski's centralized statute was "well made," and he was "for it" with only "a little bit" of alteration. When this "little bit" was explained, it was clear Wałęsa favored Olszewski's original Gdańsk regional strike upon which the August agreement was based—not the new one calling for a national union! At first blush, it was all a bit dazzling.

But at widely varying speeds, Warsaw intellectuals eventually learned, as coastal militants had begun to learn in August, that Lech Wałęsa could not be understood at first blush. When in doubt or when his instincts clashed with observable and unwanted reality, Wałęsa routinely selected the strongest features of each side of an argument, endorsed both, and bided his time as he groped for a way out. It was a maddening mode of procedure, but it possessed an additional quality that proved a saving feature. While debating with himself in his rambling public monologues, Wałęsa read his audience. It was a silent coupling impossible for his contemporaries—or subsequent historians—to interpret with specificity. The eventual resolutions of conflict that he fashioned seemed to some to be "intuitive" and to others "instinctive." Such descriptions serve only as long as their unspecified meaning is freely acknowledged. In some fashion, he sensed mood or counted the house or felt around for what the movement could do to avoid splitting apart. This last criterion seems a bit closer to the mark. Wałęsa tried to read what the movement needed in order to preserve itself.

A political aesthetic emerges into view—properly Polish and paradoxical. Wałęsa's willingness to arrogate to himself the right to make imperial pronouncements, including mind-numbing and immobilizing pronouncements, verified an intermittent but visible authoritarian trait. Such moments bared him as a Caesar of the incipient Republic. Yet his ultimate measuring stick was always the preserving needs of the popular movement. This guiding focus underscored the authenticity of his democratic drives and revealed him as a servicing tribune of a functioning social democracy. Through it all, the intimacy and candor that surrounded his involvement in each of these personal folkways proved somehow arresting to his most observant critics and emotionally bonding to his millions of admirers. From beginning to end, he labored tirelessly to serve as a reliable voice of the workers and, in so doing, willingly became their hostage as well as their leader. More straightforwardly than most popular spokesmen who have surfaced in history, he measured his constituents and struck a democratic deal with them. If, over the ensuing fifteen months of unrelenting tension, he was sometimes uncertain in the presence of useful options, he was also strongest when the times were most uncertain. On the first day in the shipyard, he ended his speech calling for a strike with the words, "I will be the last to leave." He was to honor this promise through all the days of legal Solidarność—and beyond.

The solution to the structural problem that arrived on Gdańsk on September 17 had its bizarre features. Though it did not appear so at the time, particularly to anxious strategists of the Warsaw intelligentsia, the structure that emerged also possessed an interesting utility. The views of coastal militants essentially prevailed. Power remained at the bottom, in the factories and out of reach of outside influences. But there was also a national "consulting commission" that was established. It was to have an office and one secretary in Gdańsk and was to "exist" on those occasions when it met. In many respects, the internal structure of Solidarność resembled the original Articles of Confederation of the thirteen American colonies, with a weak executive presiding over autonomous structures close to the rank and file of the population. These provisions seemed to indicate a clear-cut victory for Wałęsa and the coastal MKS, and yet Solidarność's national presence—however fragile—premitted the new institution to apply for official government registration under one statute. This was the hope, articulated by KOR, of most of the delegates from the rest of Poland. This solution had the merit of making it much more difficult for the party to play elaborate legal games to divide and control the popular movement. The chances were greatly increased that the responsibility for holding the movement intact would belong to the movement itself.

Lurking within these arrangements, however, was an underlying problem that had more to do with what modern people expect of democratic governance than it did with the reasonable requirements of governance itself. To function as intended, democratic forms require more patience by participants than they have been culturally conditioned by life in modern hierarchical societies to expect. On the decision-making front, Solidarność appeared cumbersome because it was so structurally democratic.

When such underlying cultural considerations are admitted as an organic part of political theory, it becomes quite apparent that there is no structural "quick fix" available to any society in the world desiring to cope with the inequities embedded in its own hierarchical inheritance. Democratic patience is an essential requirement of democratic politics. Despite all homilies mobilized in support of this truth, this contingent understanding of democratic forms is by no means routinely reflected in the dominant political customs that have materialized historically. Rationalizations (such as "efficiency") are constantly being invoked to justify hierarchy. There is no immediate structural panacea that can obliterate this cultural barrier to the appearance and growth of democratic forms in stratified modern societies. The most that can reasonably be expected (it is serious step in the right direction) is the creation of structures of open discussion that people can then test and experience and from which they can learn concrete things both about the forms themselves and about their own individual and collective conduct within them.

One of the chief obstacles to democracy is not merely embedded in the problems of internal structure (critically relevant as those problems are) but also literally in the heads of people, in the received culture of anticipation they bring to collective activity. Specifying these hazards constitutes a simple recognition of the underlying reasons why no culture that can seriously be described as democratic has ever been achieved anywhere or at any time in history.[18] On balance, the incipient democrats who gathered in Gdańsk, Poland, on September 17, 1980, did not do badly. They performed rather well, in fact.

But alas, there was one more strategic consideration that bore on the proceedings, one that went beyond the question of how Solidarność members would relate to one another. If democratic processes require both time and patience, existing forces hostile to democracy can do their best to see that adequate time does not exist. A central difficulty surrounding the entire question of democratic theory—a difficulty that was to haunt Solidarność—is that "political crisis" and "democratic patience" are two entities that can rarely coexist with ease. Before the Gdańsk Accords and after, the party-state in Poland possessed ample power to generate very real and very desperate crisis situations, situations that called for immediate, as distinct from democratic, response by the popular movement. To put the matter in terms of realpolitik, the regime could always create crisis situations that would preclude Solidarność from acting with internal democratic order. It could—and it would.

But before those junctures appeared, the people of Poland participated in the civic culture fashioned by Solidarność. It lasted for many months; it flooded the country with life, ideas, and innovation; it transformed private conversations and public debate; and it brought to Poland a civic vitality unmatched anywhere in the world. It is not too much to say that the period of Solidarność was one of the historic high points of mankind's history of democratic quest.

A marvelous irony attends this fact. Soberly considered, and affording due respect to the looming Soviet presence, the era of Solidarność was a time when very little was possible. Yet applying the identical sober criterion, Solidarność itself was impossible but had happened. Soberly considered, then, concrete evidence existed that—if not everything—much more than routinely recognized was, in fact, possible.

Though it would seem that such conduct might impel at least a minimum of cooperation by the party-state, the existing structure of the Leninist inheritance simply did not house the reciprocal energy and coherence necessary to oblige. The Warsaw ministries dawdled, the media were filled with routine and occasionally strident disinformation, Central Committee members periodically mumbled about the dangers of straying from Leninist norms, and behind the scenes, the vast police apparatus became increasingly frustrated and angry.

Meanwhile, occupying the new social space created by Solidarność, newly minted political observers, social critics, and talented "amateur" journalists filled the burgeoning alternative press with questions and comments about official incompetence, bureaucratic corruption, and—most scandalous of all—with adventurous ideas about social reform. To a proper party functionary, the effect was humiliating—and publicly humiliating at that.

Though party officials and their families could more or less eat, sleep, and shop in a sanitized environment of party stores and party neighborhoods, it was impossible to escape completely the new social realities of Poland. Many entertainment facilities were necessarily public, and whenever one "stepped outside" in Solidarność Poland, the winds of popular expression blew freely, a steady hurricane of cryptic comment and biting humor that stripped away the pretentious façade of official credibility. Public support for Solidarność was so transcendent and mass disdain for the party was so ubiquitous that the children of Poland's elite found themselves objects of ridicule and contempt on the playgrounds, in schools, on public transportation, and wherever else they came in contact with the citizenry. Medium-level and high-ranking functionaries found, as parents, that they had a lot of explaining to do at evening meals. Their offspring had the option of disillusionment at home or abrupt alienation from society as a whole. The tension was real in either case. Both in terms of modern modes of governance and modern habits of thought about governance, Solidarność shattered the parameters of inherited tradition within the Leninist state.[19]

Party members were not the only ones having difficulty focusing on the new realities. Foreign scholars and journalists had their own difficulties. Though observers disagreed with one another in their conceptions of a functioning democratic state—a surfacing of ideology—they shared a tendency to focus only on parts of Polish society. The most widely shared tendency was a pronounced preoccupation with the state. Again, though they possessed differing appraisals both of what the Polish state embodied and the coherence of the route taken to embody it, all carried their own intuitions as to what an intelligent route resembled. In terms of day-to-day social observation in Poland, the popular movement was essentially seen in terms of each critic's underlying cultural assumptions as to what constituted a proper state engaged in modern politics.

The most striking feature that emerges from the mass of social evidence generated by Solidarność in Poland is that the people who come into historical view as highly visible actors—party members, the intelligentsia, the Catholic-episcopate—share a common focus of contention; namely, their differing preferences concerning the "rights" of the Polish working class. But most of these actors cannot "see" workers, and most of those who do cannot relate to what they see in democratic ways. Demonstrably, the only people in Poland who do not have to practice extraordinary cultural dexterity to uncover the complex world of the working class are workers themselves. Western observers cannot console themselves that this malfunction of vision is traceable to some peculiarly Polish problem of optics. Westerners have not seen the self-activity of the Polish working class either.

More than any other historical event in Solidarność Poland, this circumstance suggests the empirical narrowness of received modes of social observation. It also exposes the descriptive by-products of condescension that suffuse the literature of contemporary political theory. But art imitates life; as a thematic ingredient of research and theory, the cultural condescension rippling through historical literature is a direct reflection of the same ingredient at work in society itself. Class-based condescension was a constant destabilizing threat to the day-to-day health of Solidarność itself.[20]

Under the circumstances, it is not surprising to discover that the most adventurous exploration of democratic forms in Solidarność Poland carried their participants into realms of experience that cannot be easily fitted within the constraints of contemporary politics. The most adventurous of these innovations involved the efforts of Solidarność to fashion an intellectually coherent and politically feasible plan of "self-management" on a national scale.

The concept of citizen "self-management" cannot, as a concept, be understood as new. As an idea, it is at the center of much liberal and socialist theory; in practice, however, the idea merely skulks around the edges of lived political reality in contemporary social systems. Its rather tired residue may be seen in New England town meetings and in the bruised remains of worker councils that can be found here and there in Yugoslavia and elsewhere in Eastern Europe. These limp institutional relics of past struggles testify to a sobering contemporary fact: throughout the world, large-scale industrialization, proceeding within increasingly stratified technological societies, has overwhelmed old dreams about "citizen power" and popular "self-government." Important democratic battles have been lost in every society in the world. The elaborate mechanisms of mediation and social control that have matured over time in the capitalist and communist spheres have so diminished people's conceptual space that serious ideas about self-management have now come to seem rather quaint. In many elite circles, it is not considered sophisticated any longer to entertain such thoughts. In this manner, the twentieth-century defeat of shop-floor democracy and of popular democracy is obscured.

But "elaborate mechanisms of social control" did not exist in Solidarność Poland. The party, its prestige in ruins and its leadership barren of initiative, had virtually ceased to exist as a policy-making body. Its only means of achieving anything that might resemble social control was in no sense elaborate. The party had reduced itself to the naked threat of its policing power. While palpably real, this weapon provided very little day-to-day political sustenance to the ruling regime. If the security services went into action, a socially impoverished military dictatorship would emerge; if they stayed home, the party would remain adrift. Neither result filled the vacuum at the center of communist governance in Poland.

Gradually and then with steadily increasing practical meaning, large and specific ideas about *samorząd pracowniczy*—self-management—began to fill this vacuum. Early in 1981, a professor of electrical engineering in Gdańsk, Jerzy Milewski, began to write, and began encouraging others to write, detailed position papers on relevant aspects of work-force control of Poland's enterprises. Particular attention was devoted to questions relating to democratic input from below and management accountability at the top. Led by activists from the Lenin Shipyard in Gdańsk and other prominent industrial enterprises, a new organization bearing an unwieldy name came into being in March 1981: the Network of Solidarność Workplace Organizations of Leading Workplaces. Its name quickly became the "Network" *(Siec)*, and its central goal became the drafting of legislation on "the social enterprise." The structural formulas embedded in the position papers became the Network's recruiting tools in the targeted "leading workplaces." By April, an organizational meeting in Gdańsk brought together delegates from many of Poland's giant enterprises. They created a formal structure. The idea of self-management, given institutional presence by the Network, thus became politically visible in Solidarność Poland.

A central question, of course, was the Network's relationship to the independent union. The new self-management bodies at the factory level saw themselves as wholly independent of the structure of Solidarność—while sharing the same popular constituency. The union would continue to represent workers; the self-management bodies—elected by workers from below—would administer the enterprise.

Here, then, was the emergence of a provisional structural answer to age-old theoretical and practical questions that had haunted advocates of economic democracy. If worker councils democratically controlled a local enterprise, what entity could exist that could connect into national planning in such a way that the local enterprise might logically integrate itself into the national economy? Who was this cooperating manager who would serve as a "friendly adversary"' (See Chapt 2, 80–81) to a democratically empowered work force? The Network offered a practical answer. The workers, organized in a local Solidarność union, represented themselves as a work force; the same workers would elect the self-management body, including the enterprise director, who would sit across the table from them and negotiate all matters concerning enterprise management. These included pay, allocations of income for capital investment, recreational allocations, and all other relevant management-labor topics.

Among other things, central conceptions of the Network clearly warred frontally against the nomenklatura. Merit, efficiency, and, above all, accountability now replaced membership in the nomenklatura as the operating criteria for filling administrative positions in industrial Poland. But this represented only one of the underlying appeals of the Network's plan for self-management. Because of the party's aimlessness, Poland's economic troubles had grown steadily worse in the months following the Polish August. Poles on the street increasingly complained, "No one is doing anything about food shortages!" To a mounting number of Solidarność activists, self-management seemed to provide a way out. As popular interest grew through April and May, the self-confidence of Network militants increased geometrically. By the time of the planning and organizing session in Poznań on June 1–3, self-management had come to be regarded within the Network itself as the absolute foundation stone of the new Poland that Solidarność was endeavoring to construct.

In early June, such an idea had not gained wide currency within Solidarność's leadership itself, particularly among the functioning heads of the union's regional bodies. Many of these union activists viewed the Network nervously. The proliferating local "self-management bodies" seemed to challenge the existing structure of Solidarność itself. Moreover, the Network's proposed draft legislation on the "social enterprise" also carried the movement into heightened and uncharted conflict with the party—an occupation a bit overworked as it was. Nevertheless, the more activists studied the Network's evolving agenda and pondered the implications of its draft legislation, the more sensible self-management seemed. When the party's docile parliament, the Sejm, announced its own new draft law on the subject on June 8—in proposals that made no structural alterations of the party's dominance of the economy—the National Coordinating Commission of Solidarność denounced it publicly. This step had the effect of moving the National Coordinating Commission (popularly known as the KKP) into closer strategic proximity to the Network's position.[21]

Even more important, the Network acquired steadily increasing muscle both from rank-and-file workers and the engineering and technical stratum in enterprises across Poland. Such growth was in large part a function of Solidarność's continuing inability to induce the party to live up to the Gdańsk Accords and to move Poland into the new era. Poles wanted somebody to do something, and the Network was clearly trying to fill the vacuum.

By the time the Network met again in Gdańsk late in June, its activists felt they were ready for a major role in the popular movement. Milewski said bluntly: "It is evident that the union cannot secure an improvement in the conditions of life and that only self-management . . . has a chance of doing so." To charges that the Network had become so aggressive that it appeared to be making a bid for strategic control of Solidarność itself, Milewski replied passionately and imperially: "The union must not surrender the initiative in self-management, but must carry it through consistently in accordance with ideas developed by the Network." Nor was this all. The Network moved beyond Milewski's somewhat self-important declaration to author a direct appeal to regional MKZs to exert additional rank-and-file pressure on Wałęsa and the entire KKP. The regions were asked to bind delegates to the next meeting of the KKP to secure its endorsement of the Network's new draft law.

The government, appalled by Solidarność, was absolutely thunderstruck by the Network's strategic attack on the nomenklatura. Clearing its throat, the party informed the Polish citizenry that the growing cacophony of voices on self-management constituted an "anarcho-syndicalist deviation." But the party had seen merely the tip of a very large iceberg. In the middle of the summer, something called the All-Poland Information Meeting of Workers' Self-Management Representatives brought to Gdańsk delegates from grass-roots self-management founding committees. The arrival of these delegates in Gdańsk marked a striking moment in the history of Solidarność. An understandably preoccupied Lech Wałęsa was stunned to behold no less than one thousand delegates in attendance. A representative from Łódź, Edward Nowak, dramatized the structural implications of the Network's theoretical and practical proposals. In a speech bearing the title "Give Us Back Our Factories," Nowak declared that genuine socialization of the means of production offered "the only way out." He added that the movement needed to pursue this goal even if it required a general strike to do so. What gave such provocative sentiments tangible relevance was the deepening economic crisis across the country. Food was increasingly in short supply in Poland's cities. Workers and the families of workers had begun to miss meals. The lines in front of bakeries and groceries were longer, the results of prolonged waiting increasingly uncertain. The feeling was growing throughout the population that something *had* to be done.[22]

Nowak's speech was dramatic but not out of step with the thrust of the conference. The KKP, including Wałęsa, came under a drumroll of criticism for being unnecessarily vague in its support of the Network's draft law on the self-management of enterprises. Wałęsa's ideas (he was present at the meeting and busy counting the house) were dismissed by delegates as a "cosmetic cure, not a systemic reform." No one in Poland was more experienced than Lech Wałęsa in the art of drawing appropriate political inferences from such an intense dialogue. The passionate words, sometimes fiery but strategically clear, came from delegates elected by workplace committees. The hero of the Polish August was hearing from the movement's mass base.

On July 10, two days after the Network meeting, Wałęsa issued a formal statement that carried Solidarność into new terrain: "In a difficult period when continued conflicts and provocations have limited us to interventionary activity and protracted negotiations, charges have resounded ever more often and loudly that Solidarność is shirking acceptance of even part of the . . . responsibility for stopping our economic slide into crisis . . . This has mobilized a significant part of the union's activists in the direction . . . of radical reform of the economy. Therefore we recognize the initiatives of union activists aimed at organizing workers' self-management groups as undertakings in accord with the needs of society, a most proper way to restore the health of the economy [and achieve] genuine, not formal, national unity."[23]

An underlying democratic dynamic was here made visible. Pushed from below, the movement's leaders were being forced to relinquish the philosophy that had guided them since the Polish August. At that time, the independent union conceptualized itself as essentially a protective institution, one dedicated to "the defense of the working place." The movement had publicly declared it did "not intend to make suggestions that might encroach upon the prerogatives of management." Eleven months later, this stance had simply been undercut by the force of events that followed essentially from the party's sustained resistence to implementing the Gdańsk Accords.

Two weeks after Wałęsa's statement, the full National Coordinating Commission and its distinguished retinue of advisors met in Gdańsk on July 24. The day before (the timing suggests the act was a conscious provocation), the party set a proper stage by announcing plans for a further 20 percent cut in meat rations and 200–300 percent price increases. Understandably, a mood of desperate bitterness surged through the whole country as Solidarność's national leadership assembled in Gdańsk.[24]

It seemed self-evident to the great majority of Poles—including a now silenced minority in the party—that no rational economic policies could be fashioned unless the party enlisted the support of the population by making a deal with it. The debt crisis fueled inflation and sharply curtailed imports of industrial parts and raw materials. This simply translated into declining productivity throughout the country. Things were getting worse, much worse. Poland was in a desperate tailspin, and the first major hunger march—in Kutno—had already taken place.

The popular movement literally lived in crisis. The workers and their advisors on the KKP fully reflected the ever-heightening sense of politics-on-the-edge-of-economic-doom. In the prolonged and intense discussion that ensued, some of the most famous personalities in Solidarność began to sound, in tone if not quite yet in substance, like the activists of the Network. The imperative need for a program of self-management tended to dominate the discussion. Things reached the point where the Solidarność chairman for the Bydgoszcz region, Jan Rulewski (whom the party regarded as a flaming radical), was moved to try to slow everyone down: "I get the impression," he said, "that I'm in 1917 and we're taking power." But another militant, Solidarność's regional chairman from Wrocław, was not moved by such cautions. Władysław Frasyniuk said: "We're pickled in problems, like cabbage in a barrel. There's no one to solve it. Then suddenly here's the initiative coming from the people for whom we've all been waiting." By the middle of July, the Network had succeeded in forging an organizational apparatus linking more than three thousand enterprises across the length and breadth of Poland; less than a week later, it had engendered the dominant momentum within Solidarność's national commission itself.[25]

The political progress of ideas about self-management within Solidarność both gained strength from and were impeded by the surrounding economic crisis. The tactical threads running through the Network's approach to self-management had emerged in response to the Leninist realities that had all but destroyed the Polish economy. In the name of restoring sanity to production, the objectives of the Network had the overriding short-run purpose of ending the food crisis by ending central planning. In contrast, the Network's strategic aims were sharply circumscribed by the need to conform in practical ways to these tactical imperatives. The result was a program that contained a number of pioneering ideas in the area of economic governance coupled with what was as yet an incomplete program of national planning. This unusual conjunction of innovation and constraint yielded bizarre and instructive effects.

From an orthodox Leninist perspective, the very thought of empowering work forces to elect enterprise directors and otherwise participate democratically in (or control outright) economic policy-making constituted the rankest kind of revisionism. As a direct attack on the nomenklatura, it assaulted the very foundations of the party's internal structure and external authority. "A careful reading of the Network's documents," intoned a report of the Central Committee in Warsaw, "leads to the conclusion that Solidarność . . . is engaged in a violent escalation of aggressive activities" and signals a "sociopolitical situation that is getting sharper in large and small workplaces." Whatever this was supposed to mean concretely, the activities of the Network were manifestly "socially harmful." The party's top mandarins concluded, "The extremists in Solidarność do not conceal that they aim to take over power first, in the immediate future, in the enterprises, and immediately after that in the whole country."[26]

But if a word like "extremist" was supposed to describe advocates of self-management, that epithet had to be applied to a remarkable range of Solidarność activists and advisors: so-called "radicals" like Kuroń and Frasyniuk, so-called "moderates" like Geremek and Wałęsa, orthodox believers in a free market like economist Stefan Kurowski, and socialist advocates of flexible decentralized planning like economist Ryszard Bugaj. Yet in specific ways, each of these participants had reservations about various components of the Network program. These doubts extended beyond people like Rulewski to include old-movement activists like Andrzej Gwiazda.

What the KKP debate brought into view was the central strategic problem facing the popular movement. Solidarność had to decide on its long-term political purpose before it could place a particular plan of self-management into workable context. In programmatic fine print, for example, one version of self-management rationalized the national economy through simple recourse to market mechanisms, a response that (on paper at least) insured structural unemployment. But so overwhelmed was Solidarność's leadership by its strategic problems with the state that such details of implementation did not become the focus of the discussion in the KKP.

Clearly, the existence of two "worker perspectives"—a rank-and-file union view on one side of the table and a rank-and-file "self-management" view on the other side—generated an expansive conceptual universe that illuminated hazards as well as possibilities. It dawned on some that if self-management bodies came to dominate the economy, they might dominate the work force as well as the party. Rulewski was of the opinion that a triumphant scheme of self-management would "move the union to the margins of social life" and "push forward a new group" of technical experts who "will soon be at odds with both the union and management." Under actually existing conditions in Poland, the party could step in as arbitrator and, in Rulewski's words, "really be the boss."[27]

Geremek focused on one element in Rulewski's analysis—the continuing geopolitical relevance of the state—in order to make a counterargument. Self-management was useful, said Geremek, precisely because it could serve as a lever to force other concessions from a recalcitrant party. He saw the issue as an "instrument of pressure . . . so that state institutions might be changed." Kuroń also had his eye on the state, but he favored self-management as a means of escaping agitation for free elections, which he felt would destroy the party and bring on the Russians. As for the party itself, Kuroń felt its inherited mechanisms of rule had simply "ceased to exist." The task for the movement was to find instruments of effective governance while maintaining the presence of a party shell as a shield against Soviet intervention.[28]

In its transforming scope, self-management so broadened the parameters of political debate that it opened up vast new areas for strategic analysis. Indeed, it became clear that the thrust of self-management created a terrain conducive to the broadest kinds of social and political thought, a democratic landscape that extended beyond the inherited confines of communist or capitalist practice. But the Network's most provocative cultural achievement was that its recruiting and educational campaign had brought these same expanded horizons into the day-to-day social world of the movement where thousands and, conceivably, millions of Poles could reevaluate the dimensions of democratic possibility. If one could somehow throw off the psychological fetters of socialization and deference, here, truly, was a space to think in. As Kuroń, for one, suggested, beyond the self-managing enterprise beckoned the self-managing Republic!

It is necessary to draw specific attention to a seminal fact about the politics of July 1981 in Poland. Outside observers were unable to follow the internal development of movement politics. To the party, the debate over the desired dimensions of self-management was "extreme"; to everyone else, it seemed merely fanciful. For both, the premises undergirding self-management were grounded in ideas that were too autonomous, egalitarian, and democratic for most observers to focus upon with serious attention. Faced with something beyond their own experience, observers seemed ready to grasp the first available expedient to hurry past the subject.

So it was that the July 24 meeting of the Solidarność national commission was largely unnoticed by journalists. A conference on self-management was not considered newsworthy; neither journalists nor their Western readers possessed the necessary political context. But subsequent actions by Solidarność on self-management in the ensuing months forced the world to pay at least some attention. As characterized by journalists in the autumn of 1981 and by subsequent academic observers, the subject of self-management was most generally considered in a context of "rising tensions" among movement leaders and palpable evidence of the emergence of "utopian" thinking within Solidarność. In the cramped world of the Cold War, any social ideas not in harmony with liberal capitalism or Leninist democratic centralism were considered to be, perforce, utopian. Self-management, thus, did not attract the interest of foreign observers habituated to the thought that their energies needed to be directed at matters of "high politics." The very subject of workers and engineers getting together to think about tangible ways to reorganize industrial relations was both too large a topic ("utopian") and too detailed ("mundane"). For either reason, it failed to pass the test of high politics. In its horizons, if not quite yet in its programmatic posture, Solidarność had gotten too democratic for the world to understand. In journalistic coverage at the time and in academic ruminations since, the commentary that emerged reveals rather more about the political paralysis of contemporary culture than it does about the forces at work in Solidarność in the summer of 1981.[29]

It is essential to be as precise as possible on this culturally central matter. The Network's creators did not live in some paradigm-shattered world beyond the inheritances of capitalist and Marxist-Leninist theories of governance. Most of them longed for a "Third Way" to such a world and, to a greater or lesser extent, were probing for ways to conceptualize its practical forms. But the tangible evidence they generated was incomplete and parts of it were internally contradictory. By mid-summer, the discussion and analysis had simply not proceeded long enough or far enough to bring all of these contradictions and their implications to the center of planning options. The significant achievement—the one that forms the outer limit of Solidarność's contribution to the world's democratic inheritance—is that the momentum unleashed by the movement finally broke through inherited constraints and generated an ongoing collective brooding over the possible forms of political and economic self-management.

In terms of human experience, the singular political achievement of Solidarność was that theories and practices of self-management were not being explored in some tiny kitchen conversation among isolated visionaries hopelessly remote from political power, but rather were taking place at the center of a huge popular movement. In the relationship of political theory to self-activity by living people, this achievement was almost unprecedented. Beyond this, it was an equally important fact that the exploration was not confined to the Network's leading circles or to the KKP itself; the basic subject matter of economic self-government had become an engaged topic in workplaces across all of Poland. Never in history had such far-ranging democratic premises been sanctioned for serious exploration within such a massive polity as Solidarność had constructed in Poland.

At a purely short-run political level, it may be recorded that Solidarność's presiding spokesmen formally pronounced "full support for the social movement in the cause of employees' self-management" and called on all levels of Solidarność's far-flung structure to provide "support and assistance" in creating shop-level self-management bodies. And despite their nervousness about the hazards to internal governance within the union represented by the Network's increasingly assertive spokesmen, the KKP for the first time mentioned the "network" by name. In the words of Henry Norr, one of the few scholars in the world who has researched and written about the subject with sustained attention:

> Among the intellectuals associated with the union, the self-management movement had come to seem a vehicle by which their various plans for Poland's future could be realized; for the economic reformers, the movement seemed to provide the political muscle that could force the state to act; for moderates like Geremek,

> it might be the lever that would push the authorities into an overall political settlement; for visionaries like Kuroń, self-management could serve as the model foundation for a new social system. As for the union's elected leaders, the shift had been most difficult for them, because they were most deeply attached to the old model of defensive trade unionism. But . . . with the economy deteriorating daily and the authorities still seemingly immobilized, the Solidarity leaders knew that they needed somehow to take the initiative or else risk the loss of their credibility and their hopes. . .[30]

Because the requirements of historical writing carry its devotees into the maelstrom of conflict, error, and bloodshed that seems to haunt the human experience, practitioners of the craft bear the occupational trait of caution. Historians become acutely aware of the element of contingency in human experience, including, when it happens, human achievement. They discover that most of the historical happenings that gain the reputation of "political breakthroughs" can be understood as very close things—as events that almost did not happen. Historians are forced, therefore, to come to terms with the fact that while individuals can sometimes come upon new ways to think, so that they travel along paths others regard as "deviant," it is extremely rare in history for large numbers of people to do so at the same moment of time. All the requirements of movement building that make collective action possible—the kinds of activities which overcome deference by stimulating individual self-respect and collective self-confidence—must, for conceptual breakthroughs to surface, be augmented by an environment that encourages rather than resists innovation. The rhythms of history indicate that this necessary environment can be encouraged by the prospect of utter disaster. It is the hint of doom that often encourages people to think.

But this also means that the conjunction of creativity with high politics is a moment of throbbing fragility. It is, of course, a circumstance that seems almost made to order for irony. So it was for Solidarność. For the earnest advocates of the Network, self-management did not lose its momentum at the base of Polish society because of the party's opposition (though it was implacable) or because of divisions within the popular movement (though they were real). Before anything else could happen to it, self-management essentially starved to death in the late summer and autumn of 1981 in Poland. The movement's shop-floor militants could not summon the energy for the patient development of self-management bodies because, in the words of one of them, they became preoccupied with a larger topic—"our biological existence and that of our families."

There was a moment of high promise before the slow, almost invisible dissipation of democratic energy. It came on the heels of the July KKP meeting, at a time when Network morale and confidence had reached an all-time high. In Wrocław, the regional Solidarność leadership committed itself to the Network's program as the "main task for the immediate future." The region's delegates returned from the Gdańsk KKP meeting to tell a mass meeting that the moment for total self-management organization at the grass roots had arrived. But the faces in the audience looked exhausted, the bodies tired. The question that had to be answered was, "How do we get something to eat?" The leadership proposed making plans for self-management, but a worker shouted back from the floor, "But I'm hungry!" Faced by men and women preoccupied with the most compelling human needs, the union spokesman, Jan Waskiewicz, soon reported in the Solidarność national newspaper that "I had to retreat."[31]

The irony was singular. By the end of July 1981, the Polish movement had used up much of the familiar democratic inheritance available as a historical legacy of the American, French, Russian, and Chinese revolutions and was trying to move beyond them into new social space of its own creation. To an extent never before attained in any country, large numbers of Poles, connected in a voluntary community of their own construction, were thinking and planning seriously about ways to erect that ultimate in democratic forms—a self-managing economy in a self-managing state. There had seemed to be visible energy for this task, but as July turned into August, there appeared a much more desperate energy in working-class Poland. Led by women textile workers, a spectacular three-day hunger march held the city of Łódź in thrall, even as transport workers achieved a massive tie-up in Warsaw on August 1 that effectively shut down the center of the nation's capital for fifty hours. The issue in both cases was food.[32]

In the first half of August, top-level negotiations between the government and Solidarność over the union's proposals for food distribution broke down amid a flood of deceit and arrogance by party officials that, even for Poland, set a new standard in effrontery. One of the party's counterdemands was that Solidarność "stop false messages about hunger in Poland."[33]

In a climate of growing popular anger and desperation, Network agitation on self-management suddenly seemed positively tame. The crisis in the economy altered the very nature of the internal dialogue in Solidarność and transformed ideas that had seemed innovative into something that appeared formal and bureaucratic. This did not mean, of course, that self-management abruptly disappeared as a goal; the issue, in fact, was to receive full attention at both sessions of the Solidarność National Congress in September-October. But it did mean that the gathering momentum coalescing around self-management as the ultimate expression of the popular movement lost its intensity as the food crisis deflected energy into other channels. Soon, from precincts beyond the Network would come new initiatives in the name of a new weapon, the "active strike." As a priority of Solidarność, self-management was to have much competition.

But while the movement responded to its mobilized base and became more internally cooperative in 1981, the party spent the first six months of 1981 deflecting similar impulses from its own rank and file. The party equivalent of the Network was the "horizontal movement." The horizontal movement sank its first roots within two months of the founding of Solidarność, when the party's foot-dragging began to stimulate the latent feelings of alienation that had historically coursed through the party's lower apparatus. Institutionally voiceless because of the rigid vertical organization of the party, workers in the Towimor Marine Engineering Works at Toruń reached out to fellow party members in the teachers' union at Toruń University. Led by a 32-year-old engineer named Zbigniew Iwanow, this move to establish lateral lines of communication among party organizations at the base of society was, of course, a fundamental breach of Leninist principles, one that threatened to put real meaning in the phrase "socialist renewal" then being bandied about by functionaries in Warsaw. Horizontal meetings soon followed between eight local party branches in Toruń. They sent signals up the line calling for secret elections at all levels of the party by January 15. Iwanow was expelled from the party and promptly reelected by plant workers, thereby becoming, in his words, "the only party First Secretary in Poland who is not a member of the party."

More important were the proposals for reform that emerged as the movement spread: complete separation of the party from the government (a sidestep around major elements of the nomenklatura), all political decisions to be made by elected party bodies rather than by individual functionaries, the appointment to high party posts to be made only through secret ballots of delegates elected from below. From Łodź horizontalists came a December demand for an immediate party congress dedicated to ending "totalitarianism" in the party and the reshaping of its structure by people "who can be integrated into the Solidarność movement." Clearly, the horizontalists were intent on democratizing the party.[34]

As the movement spread across Poland laterally, it also began to move upward from the base to envelop the party's middle apparatus. An enthusiast reported that "a considerable part of the apparat" had overcome initial hostility and had moved in "the direction not only of understanding the movement but of correctly appreciating it" as a means "to rejuvenate the party." In April, over seven hundred delegates from thirteen of Poland's provinces gathered in Toruń for a run-up to a party congress, which Kania, responding to this and other pressure, finally agreed to call for July.[35]

A remarkable feature of Polish society was revealed by the horizontal movement. Though demonstrably serious about reform and laced with young engineers, technical specialists, and a leavening of skilled mechanics, the horizontal movement had only insecure connections to the manual working class. And even those stalwarts of party reform whom the horizontalists could recruit from the working class were looked upon by rank-and-file workers as questionable participants in the movement to liberate Polish society. The continuing reality was that the Polish working class, led by younger workers in heavy industry, had its attention riveted upon Solidarność, not on the horizontal movement. This fact robbed the party reform movement of structural coherence. From the ranks of the party in the Lenin Shipyard came a ringing endorsement of the horizontal movement by Jan Labecki: "If we don't recognize [horizontalism], channel it, and use it for the good of the party, it can indeed, threaten us." But on the coast and in the rest of the country, rank-and-file workers were not intuitively drawn to functionaries who talked about "channeling" their democratic hopes. The party in the Lenin Shipyard unanimously elected the party reformer, Tadeusz Fiszbach, as its delegate to the party congress, but in most of Poland, there were few high silhouette officials in the party capable of giving serious expression to the cause of democratization. Beyond all these considerations, the very structure of the party placed immense organizational demands upon the proponents of horizontalism. As the Polish working class had discovered many times in the past, delegate elections proceeded through layers of apparatchiks. Victories at lower levels could, with proper institutional dexterity, be vitiated at the next.[36]

In the elections for delegates to the party congress, the party contrived to put up dependable functionaries with no previous track record that would leave them vulnerable to rank-and-file criticism. Provincial party secretaries met defeat in wholesale proportions, but in many cases, they were replaced by faceless—and trustworthy—functionaries. The reform of Leninist structures was an awesome political task. The extraordinary party congress in July became, the world discovered, a victory for Leninist orthodoxy. The party promptly illustrated the fact at the beginning of August in its high-handed negotiations with Solidarność over the food crisis.

But there was another dimension to this intraparty struggle. In profound ways, the horizontal movement was artfully undercut by the party's security police. Throughout the early months of 1981, as the party reeled from its smashing defeats of the autumn months and as the horizontal movement reactively spread through party ranks, the security service stepped up its harassment of Solidarność activists. These developments came in conjunction with still another emerging political force on the Polish scene—the movement for a free union for peasants. "Rural Solidarność," originally stimulated and later legitimized as an independent entity by the support it received from Solidarność, also drew the enthusiastic support of the Catholic episcopate. In March 1981, at Bydgoszcz, all these factors coalesced to produce a grotesque police provocation that turned into a national scandal and, very quickly, into a national crisis.

The "Bydgoszcz affair" began on March 16 with a sit-in by peasant activists at the headquarters of a party front group known as the United Peasants' Party. The action was given special meaning by the participation of the Solidarność regional chairman for Bydgoszcz, Jan Rulewski. On the nineteenth, as a group of about thirty-five Solidarność and peasant activists were working with some forty-five local municipal officials on a joint declaration after a day-long negotiation, some two hundred security police suddenly burst into the room and began arresting and beating people. Some twenty-seven activists were beaten, three of them, including Rulewski, rather badly.[37]

What the party immediately began downplaying as the "incident at Bydgoszcz" transparently was no incident. Rather, it constituted a frontal assault on the right of Solidarność to conduct its business with dignity or even with physical safety. A wave of outrage swept across Poland, as posters bearing photographs of bleeding union members began appearing in public places throughout the land. Led by Wałęsa, the KKP hurried to Bydgoszcz for an emergency meeting.

The police violence at Bydgoszcz came at a moment when the fabric of Polish society was already stretched taut. The party not only found itself under internal structural pressure from the horizontal movement, the Kania leadership had been exposed two weeks earlier to excrutiating pressure from the opposite direction. The top leaders of the Soviet Union had summoned Kania's inner circle to Moscow on March 4 to receive a sustained tongue-lashing for what the Soviets considered to be the party's unending series of "surrenders" of Leninist norms.[38]

But if party centrists had little room for maneuver, Solidarność had even less. Throughout Poland, at local, regional, and national levels, all the inherited folkways developed by the party over thirty-five years of imperial rule had produced a settled mode of arrogant and deceitful conduct that made daily life for Solidarność members exasperating and at times humiliating. The crudest kinds of examples existed in every region, indeed in virtually every locale. The "hospital troubles" at Łódź provided but one example of the process through which a minor matter of simple equity could escalate with breathtaking speed into a full-scale mobilization by an outraged citizenry. In February, employees of a hospital for policemen and their dependents discovered that the hospital director, a police colonel, had set aside for the exclusive use of management more than half of a consignment of ham dispatched as rations for all the hospital's employees. When five of the staff protested, they were promptly sacked. In the face of steadily mounting pressure from the Łódź Solidarność, the police colonel magisterially refused to relent. Only after a one-hour general strike of the entire working population of Łódź was reinstatement of the five secured. It may be taken as a symptom of the generalized sclerosis of the Polish party that a general strike was necessary to reinstate five workers whose only crime was to protest self-interested plunder in the routine distribution of food! Throughout the month of March, the security service engendered a series of such incidents across the nation. Some of these provocations, such as the sudden arrest of Jacek Kuroń and ostentatious surveillance of other prominent activists, had what can only be interpreted as intentionally high visibility. The Bydgoszcz beatings came little more than a week after the Łódź general strike and two weeks after the arrest of Kuroń. The point was not that the party was making "gains"—the hospital workers were reinstated and Kuroń was soon released—but that a steady signal was being sent to Solidarność that it existed only on sufferance, that it was, ultimately, illegitimate.[39]

The week of extreme political tension that subsequently surrounded the negotiations in Bydgoszcz and Warsaw have acquired a controversial presence in the chronicles of the Solidarność era. Transcripts of the KKP meetings in Bydgoszcz reveal a subterranean mistrust between Kuroń and Karol Modzelewski on the one hand and Lech Wałęsa on the other. The precise meaning of this division was obscured by the fact that both Wałęsa and most of the advisors held essentially the same strategic position on what should be done; namely, that the movement should avoid encouraging a precipitate national confrontation while at the same time moving with orderly and incremental precision to mobilize the entire national membership of Solidarność for a maximum demonstration of the union's determination to defend itself.[40]

It quickly became evident that a majority of the KKP delegates favored an immediate nationwide general strike to force a full-scale government investigation of the violence at Bydgoszcz. The KKP majority called for the prompt arrest of all policemen involved in the beatings and of all officials who ordered the action. Wałęsa, however, argued for an interim four-hour "warning strike" and, should that produce no response after three days, then a national general strike. When it became apparent this view was not prevailing in the KKP, Wałęsa angrily accused the leadership of not knowing the temper of "the lads in the factories." He shocked the KKP by walking out of the meeting and returning to his room, making it clear as he departed that he would resign if the KKP did not come to its senses. In discussions after Wałęsa left, the KKP decided that it was not really strategically possible to mobilize an effective general strike with Wałęsa occupying a public position of dissent. The next morning, under circumstances that were somewhat diminishing if not humiliating, a large majority of the KKP agreed to support Wałęsa's position. A presidium of ten was elected to negotiate with the government.[41]

It is necessary to pause at this juncture and note Wałęsa's leadership style. Even though mobilized in service of a strategically sophisticated analysis and in response to a democratic conception of the responsibilities of the leadership to the rank and file, Wałęsa's imperious behavior during the Bydgoszcz crisis preserved maneuvering room for Solidarność only at the cost of introducing authoritarian modes of conduct within the functioning structure of the movement. While Wałęsa could argue that he acted only in the name of preserving a democratic movement, this rationalization is always available as a justification for deviance from democratic norms. As such, it will not do. The outcry from Wałęsa's compatriots that ensued had a sobering and democratizing impact on his subsequent conduct. The Bydgoszcz crisis promptly unleashed a wide variety of forces both within Solidarność and within the party.

Not by any reasonable standard could the scenario that resulted from this imbroglio be considered "moderate." Such relative descriptive terms (a "moderate" Wałęsa versus a "radical" KKP) merely provided a way to misread the dynamics that were actually at work. At bottom, the controlling strategic judgment turned on an analysis that had to be based on conjecture. Was the police action at Bydgoszcz a conscious provocation by the party-state, or was it a tactical tightening of the screws on the leadership by disgruntled hard-liners in the Politburo and their all-too-willing allies in the security service? If the first contingency were the case, the ultimate weapon of the working class, the general strike, was the only possible way for the movement to resist public humiliation. But if the second possibility held, a warning strike preceding the unleashing of the ultimate weapon would create vital eleventh-hour time for party centrists to achieve a peaceful settlement with the union. Of course, no one in the KKP, including Wałęsa, could know the answer to this puzzle. But it was clear that an immediate general strike precluded the second option, while the reverse was not true. If the warning strike failed to produce results, the national general strike could still be employed. Indeed, the warning strike could serve as a much needed organizational dress rehearsal. If the long and embattled history of labor movements over the past two hundred years of industrialization has made one thing clear, it is that successfully mobilized general strikes do not materialize as easily as leaves falling from a tree.

The national warning strike of March 27, 1981, was, in any event, an awesomely successful event. For four hours, the country simply came to a stop. The strike was nothing less than a totally effective demonstration of national self-sovereignty by the people of Poland. In the words of one impressed observer, "For the party leadership, the most shattering feature of this national demonstration was the almost universal participation of Party members against the explicit orders of the Politburo." Over a million party members ignored party orders and participated in the strike against the party.[42]

The weekend negotiations that followed in Warsaw were conducted on the government side by Mieczysław Rakowski, the unctuous and patronizing editor of *Polityka*, the government's "insider" journal of politics. In an arrangement that Wałęsa himself preferred—particularly after the flare-up in Bydgoszcz—the negotiations were largely handled by small groups on each side. Such prominent Solidarność figures as Andrzej Gwiazda of Gdańsk and the regional chairman from Szczecin, Marian Jurczyk, spent much of their time in a downtown hotel room where they periodically received alarming reports of imminent government plans for a "state of war," the legal term for a military takeover. The Church also weighed in heavily, summoning Wałęsa to the quarters of the Primate of Poland. There, Stefan Cardinal Wyszyński made a personal plea to Wałęsa to avoid the general strike. Finally, both sides were reminded that the Russians were watching. The negotiations took place as Warsaw Pact armies maneuvered on Polish soil, an event lavishly covered on nightly television complete with footage of tanks and troop convoys rolling through the Polish countryside. The maneuvers, which had been scheduled to end earlier in the month, had been pointedly extended by the Soviets. It was a very ominous signal indeed.[43]

Because of the narrow maneuvering room available to both sides, the negotiations featuring Rakowski and Wałęsa were both tense and difficult. The initial basis of the talks was the government's official version of the violence, written under the direction of the minister of justice and summarized as the Bafia Report. This remarkable document acknowledged police responsibility for the beatings at Bydgoszcz in language that seemed drawn from some Polish theater of the absurd: "It is the opinion of the commission that medical expertise concerning the extent of the injuries does not provide grounds to assert that these injuries were self-inflicted." This non sequitur was followed by another: "The events in Bydgoszcz cannot, to any extent, adversely affect the general assessment of the conduct of the state organs with regard to Solidarność. This is most clearly testified to by the fact that ever since July 1980 the methods of settling disputes by political means has been used." A hundred events culminating in Bydgoszcz contradicted this fatuous assertion.[44]

Behind the scenes, as the whole world focused on Poland, the superpowers also weighed in. Washington announced that the Soviets might "undertake repressive action in Poland," warning against such a move, while Tass published an hysterical account of Solidarność militants blocking highways and occupying telephone exchanges. When Solidarność sympathizers among employees of the government television station saw to it that the nation's screens were filled with a written announcement, *solidarność-strajk*, Tass concluded, perhaps more in honest bewilderment than in conscious deception, that the movement's militants had "seized" a television transmitter. Meanwhile, the *People's Daily* in Peking cautioned Polish workers to be "prudent."[45]

In such a setting, the negotiators came to terms. The final agreement looked like what it actually was—a tortured compromise that, like its authors, had been put through a wringer. In no sense was it a totally honorable victory for Solidarność, but it also embodied yet another government retreat. The signatories agreed that the "use of the forces of order" in Bydgoszcz was "contrary" to "solving social conflicts through political means, above all through negotiations." The government also agreed, "in keeping with the demands of the Interfactory Strike Committee of Bydgoszcz," to order an official inquiry into the conduct of people "who have misused their prerogatives" in order "to place before a tribunal the persons responsible for the beatings. As soon as the investigation is closed, they will be tried according to law." More relevant was a structural change that did not relate directly to the beatings. The regime revealed it now accepted in principle the right of Polish peasants to form Rural Solidarność. But in the language of the agreement itself, there were no guarantees of anything concrete, other than the disclosure that the government would conduct an "inquiry." In any event, the Central Committee did move ahead the very next day to begin the process leading to the creation of a legal peasants' union.[46]

The crisis shook the internal structures of both Solidarność and the party. As Rakowski and Wałęsa labored through a seven-hour negotiating session on March 30, the Ninth Plenum of the Central Committee was the scene of angry recriminations. The dynamics of the controversy seemed to verify that the Bydgoszcz affair had its origins among the hard-liner–security-service faction. The leading conservatives, Olszowski and Grabski, came under withering assault and offered their resignations. For a moment, the entire Politburo considered the possibility of resigning. The pressure on the party—from the horizontal movement, from the Soviets, from Solidarność—had become unbearable. Events had, in fact, outrun the creative capacity of the party leadership. For a few weeks in April following the crisis, an increasing number of Central Committee members edged delicately closer to the horizontal movement. When Rakowski's *Polityka* provided public support for such moves, it appeared possible—just for a moment—that the party might try to save itself and Poland. But with the first sign of complications—renewed pressure from the Soviets—the party drew back into its ingrained posture of immobilized Leninism. The party trembled, but it remained static.[47]

If the party plenum seethed in discontent on the thirtieth, the Solidarność national commission underwent a similar trauma on the thirty-first. In a tumultuous KKP meeting in Gdańsk, Wałęsa and his advisors were excoriated for their undemocratic methods. An angry delegate from Łódź charged that the agreement was a "farce" that had been fixed ahead of time by government functionaries and Solidarność advisors. Wałęsa himself was personally attacked for his unilateral modes of procedure. Fueling these emotions were the deepest kinds of personal and political anxieties, some set in place by the Bydgoszcz crisis itself, others as old as the class divisions of Poland, still others as old as politics. The subterranean realities of social relations within a multiclass popular movement were exposed by the crisis.

The single action at the KKP meeting that most caught the attention of the world press was the dramatic resignation of Karol Modzelewski as the movement's national press spokesman. A historian trained as a medievalist, Modzelewski attacked Wałęsa by likening the structure of Solidarność to that of a feudal monarchy. Wałęsa was the "king," who governed with courtiers whom he himself selected. The worker prince and his advisors faced a "Parliament," (the powerless consulting commission) that they ignored when it so pleased them, as in the case at hand. Modzelewski said he could not go on as a spokesman for an imperial decision-making process that led to decisions with which he personally disagreed.[48]

Karol Modzelewski was an imposing and highly visible figure in the sprawling ranks of the Polish popular movement. He was the one Solidarność representative foreign reporters saw the most and felt they knew the best. (Wałęsa was hard to "know.") Modzelewski's credentials as a man of independent mind were absolutely impeccable. He had served a three-and-a-half-year sentence for his role in the joint-authorship, with Kuroń, of the famous "Open Letter to the Party" in 1964. Since the Polish August, he had been a reliable press spokesman, one reporters had grown to like and to trust. One of them later concluded that Modzelewski's analysis on March 31 "cut straight to the heart of the matter."[49]

But there was much more to the crisis in governance in Solidarność than could be explained by simple matters of structure—feudal or otherwise. Behind Modzelewski's expression of the anger of the intelligentsia was a seven-month struggle for influence within Solidarność. It was an effort animated by the most subtly ingrained presumptions as to what constituted the essence of politics. Matters of democratic governance provided an authentic point of tension, but the deepest gulf in Solidarność exposed by the controversy was the underlying cleavage between the Polish working class and the Polish intelligentsia. The controversy over the events in Bydgoszcz for the first time made this cleavage partly visible to foreign observers.

Modzelewski's resignation speech recalled his advocacy of a unified national structure at the original September 17 organizational meeting in Gdańsk. Because the views of Warsaw intellectuals that Modzelewski expressed had not then prevailed, the powerless "consulting commission" had been fashioned. The practical result was illustrated by Bydgoszcz: the KKP could appoint a ten-man negotiating team that Wałęsa could blandly ignore. Moreover, as the KOR milieu elected to see matters, Wałęsa picked advisors like Geremek and Mazowiecki who appeared to accommodate to his hip-pocket style of leadership. Lingering in the immediate background of Modzelewski's resignation on March 31 were the personal animosities that had surfaced a week earlier between Wałęsa on the one hand and Modzelewski and Kuroń on the other. And behind that moment lay many months of frustration on the part of KOR intellectuals at the independence, unpredictability, and, most of all, the power of Lech Wałęsa.

The written historical record of Solidarność contains two documents that bear on the efforts of KOR-connected intellectuals to enhance their role in Solidarność at the expense of Wałęsa. One is the complete transcript of the September 17 meeting, the other an article by Gdańsk writer Lech Bądkowski detailing the KOR-influenced efforts, four weeks later, to oust Wałęsa from his position as chairman of the Gdańsk MKZ. After these abortive maneuvers, KOR activists never really came to terms with their strategic marginality in the working-class movement. In these circles, a conventional wisdom materialized that gradually permeated the Warsaw intelligentsia as a whole. It took the form of a kind of parlor carping, a more or less permanent grumble of exasperation at the style of the working-class leader from Gdańsk.[50]

However, there were differences in the way various sectors of the intelligentsia distanced themselves from Wałęsa. Though the old *Robotnik* faction in KOR that was most familiar with the structure of authority in Solidarność knew that the Polish intelligentsia was not central to policy-making, less informed intellectuals (and less informed sectors of KOR itself) transformed their analysis in order to maintain their illusions. This bizarre achievement involved a basic redefinition both of the movement and of Poland's social classes. In pursuit of this cause, Kazimierz Wóycicki, an editor of a Warsaw intellectual journal and a prominent activist in the Clubs of the Catholic Intelligentsia, explained the sociological dynamics of the movement to foreign journalists: "Despite all the phraseology of a workers' movement, Solidarność was primarily a movement of the new middle class in Poland. It is this middle class which was most severely victimized by the regime." While conceding that workers in large enterprises rather quickly formed independent unions following the Polish August, Wóycicki emphasized that "enterprises which employed predominantly semiskilled and unskilled workers organized much later, several months later in fact . . . They lacked the skilled to do so. They needed outside assistance."

It was the old refrain: "more enlightened people" have to organize social movements. Similarly, Jan Józef Lipski, a literary critic and KOR activist, elaborated upon this redefinition. Writing in the American journal, the *Nation*, Lipski reported that "an entirely new and quite peculiar segment of the working class has come into being here. I refer to the workers of the 20 to 40 age group who have a secondary education, mental traits resembling those typical of the intelligentsia and quite developed needs and aspirations for high culture and high technology . . . In this respect, Lech Wałęsa happens to be rather atypical." Lipski summarized Solidarność as "primarily a coalition of that segment of workers and the intellectuals, each side understanding each other with amazing ease." One of the original founders of KOR, Antoni Macierewicz, also unwittingly betrayed his distance from the milieu of the workers by telling a Western journalist, "Most of us were surprised when workers at the very large factories gave up wage increases so that workers in the areas of culture and health care could win better wages." Macierewicz then went on to explain that Solidarność was "not a trade union." It was, he said, without further elaboration, "the nation organized."[51]

In short, a division of opinion materialized within the Polish intelligentsia. Among worker-oriented activists, led by an older generation that included Jacek Kuroń and Karol Modzelewski but extending to such younger oppositionists as Jan Lityński and Henryk Wujec, Solidarność possessed authentic working-class roots but suffered from the leadership of Wałęsa. It was Wałęsa's imperial tendencies and his independence, more than his cultural style, that worried them. A far broader sector, extending from the mainstream of KOR to the broad mainstream of the intelligentsia as a whole, seemed in psychological flight both from Wałęsa's workerish manner and from all other sectors of the working class that possessed similar cultural afflictions. In the redefinitions that accompanied their discovery of "an entirely new and quite peculiar segment of the working class," these deformations simply disappeared, to be replaced by a new worker-intelligentsia coalition that functioned "with amazing ease." In such ways, the capacity of cultural condescension to obscure reality was rather impressively demonstrated. Modzelewski knew better, of course. He had come to have, quite simply, a profound dislike for Lech Wałęsa.

As complex as all this was, there was still more to the embattled journey of the Polish intelligentsia through the opportunities and hazards of the Solidarność era. Indeed, a truly towering irony invested the Bydgoszcz rupture with a special poignancy. It grew out of the life of Solidarność itself and turned on little understood circumstances surrounding the internal dialogue that flowered at the very base of Solidarność.

As a direct function of their prior voicelessness, unempowered people bring to social movements deep personal longings coupled with uncertainty about the public conduct necessary to express those longings. This circumstance scarcely constitutes a historical secret—inasmuch as it is an organic feature of all modern societies. From the perspective of high culture, the general assumption is that politically inert people—the historic word is "rabble"—are inherently anarchic and fully capable of moving in some unpredictable, spasmodic swoop from total passivity to violent kinds of action. Indeed, much of the public rationalization of hierarchy rests upon the presumed need to guard social order against precisely this sort of unthinking "movement." The fear, however, is inexact and misplaced. The only historical confirmation to this class-based assumption is found in what can be called "shadow movements"—social formations that contain little or no internal dialogue. Such entities are so constructed that self-appointed spokesmen do most or all of the talking. Random social collections of this kind can appear in any society fairly quickly, often on the spur of the moment, as it were. In their most extreme and irrational forms, such collectivities (they are not, as shall become evident, "social movements") have two pronounced tendencies: they can kill (in America, the historically relevant word is "lynch") and they generally exist in public only briefly before they dissolve.[52]

The essential political distinction between "spontaneous" conduct and democratic conduct is most vividly evident in this context. Nothing enduringly democratic is ever spontaneous. Democracy is a dialogue, and to involve numbers of people, democratic forms must be fashioned that facilitate and protect such conversation. This is why those movements of democratic aspiration that have appeared throughout history have been—without exception—built by conscious action; they are never "spontaneous." That word has acquired wide application because it offers an explanation of historical causation that otherwise appears (because it was not inquired into) inexplicable. Observers use the word "spontaneous" to describe something whose origins they have not researched. In any event, because so much cultural confusion surrounds this rather fundamental political dynamic, many kinds of hierarchies that have materialized ostensibly to protect "order" against "anarchy" actually have the effect of inhibiting society in its further pursuit of democratic development.

But though movements, to be democratic, require internal democratic forms and thus cannot be understood as spontaneous happenings, the people in them necessarily harbor considerable potential for free thought. Indeed, to the extent movements encourage heretofore powerless people to overcome inherited forms of deference, movements foster a social life of expanding variety. People can effectively "come alive" in the Polish Augusts of history. That fact simply verifies the underlying value and meaning of the entire collective effort; the content of daily life is enriched. The practical effect of this energetic variety is that democratic movements can rarely achieve with ease what people are conditioned to regard as programmatic "unity." As an engaged conversation among many disparate people, a democratic movement can really be nothing other than a lively expression of multiple opinion—a sanctioned argument, as it were. Under the conditions of vitality that voluntary associations have or aspire to have, anything that might seriously be called "unity" can therefore only be a product of visible or masked coercion. Such a circumstance, of course, describes a social relation in which enforced silence replaces self-activity. Most political "unity" in history is achieved at precisely this cost, whether the social relationship in question is a marriage of two persons or a national war effort of millions.

But, it is fair to ask, can such endless democratic arguments ever acquire some sort of balanced maturity so that a social movement becomes capable of purposeful majoritarian action? The historical evidence indicates that this can occur, but only through the most candid interior dialogue between the relative few who channel policy and the relative many who expect to exercise some input into how that policy is projected. Classically in history, this dialogue is initiated from below in the form of autonomous actions at the base that threaten the stability of previous policy. For sixteen months, this internal dynamic constituted the essential democratic rhythm of Solidarność.

Thanks to the consistent duplicity and recalcitrance of the ruling apparatus in Poland, autonomous local actions became a constant of social life in the era of Solidarność. Indeed, in the face of police and party provocations that were designed to humiliate the entire citizenry, cooperative local assentions of dignity were absolutely essential to the continued social health of the movement itself. A central fact of life in Solidarność—its verifying democratic presence—

reposed in the following dynamic: widespread self-activity kept local, regional, and even national Solidarność spokesmen in earnest and often frenetic contact with the movement's rank and file. It brought leaders running to the base. The popular term that came into use to describe this phenomenon was "fire fighting." It connoted an active dialogue that actually was more democratic and carried greater long-term democratic implications than most other kinds of dialogue that managed to develop in Solidarność Poland (the activities of the Network also had this dual quality, see pp. 283–91).

At first inspection, such a characterization seems impossible. After the Polish August, Solidarność quickly developed a far-flung structure of local and regional citizen institutions that reached into every sphere of Polish life—save, of course, the ranks of the party nomenklatura itself. To administer this vast network of ten million people, over forty thousand men and women obtained leave from their jobs in the industrial enterprises and offices of the country and were paid full-time salaries out of union dues to serve the movement. Under Solidarność's bylaws, dues money was not only collected locally but funds also remained under local control so that the servants of Solidarność were structurally encouraged toward loyalty to the movement's grass roots. As a capstone of this self-organized institutional panoply, the movement launched a veritable fleet of local and regional newspapers, journals, and newsletters. In April 1981, after much government stalling, the well-written national weekly, *Tygodnik Solidarność*, became the movement's journalistic flagship. In facilitating the free expression of opinion, all of these arrangements contributed to what can be understood in the "movement culture" of Solidarność.[53]

But the ultimate social meaning of Solidarność was the flowering of free expression itself. And in political terms, the most tangible form of this "ultimate" expression did not come in the specific words that filled public discourse or imparted subtle meaning to behind-the-scenes argument; it came, rather, in the visible public acts that emerged from these discussions. These acts verified the democratic substance of Solidarność to a degree nothing else could. In mobilizing a local strike of one's co-workers to reclaim public space invaded by aggressive local party apparatchiks, the rank and file of Solidarność asserted their personal stake in the new Poland that the movement was seeking to construct.

The occasions that produced "fire fighting" can thus be understood collectively as a crowning moment of democratic self-activity in Solidarność Poland and one of the movements most intensive schoolrooms of self-education. It is not to be supposed for one moment that these discussions at the grass roots were institutionally unbalanced or mere exercises into which outside "firemen" stepped to dampen discontent or douse inopportune flare-ups. This happened, and more than once, but never without a complementary result—the sending from below of a sharp signal both to the local party and to the Solidarność leadership that violence was being visited upon the base of the movement by the ruling apparatus.

Local tactical results aside, the long-term meaning of these grass-roots colloquys lay in the reciprocal education they afforded. The movement's leaders learned starkly specific details about the status and morale of the citizenry while at the same time the people of the ranks participated—many of them for the first time in their lives—in a sophisticated political analysis of the possible consequences of their own actions.[54]

It is essentially through this means that movements "teach" democratic forms to people who have never been permitted to experience them. The teaching performed by the movement's spokesmen has a defensive purpose. If the instruction does not prove convincing, the local strike inevitably continued; if the lesson prevails, the fire goes out, at least temporarily. In either case, the historical causality is clear: local actions which recruit the movement's leadership create the process through which politics is discussed democratically at the very base of society.

For the men and women who gave public voice to Solidarność, "fire fighting" was both essential and educational, but it was also very draining. From a grass-roots perspective, recurring party outrages simply demanded a clear response (the "hospital troubles" at Łódź serving merely as one example of many hundreds of such outrages). The very self-respect of a local union and of each of its members seemed to hinge quite nakedly on the need to generate a firm reply to the abusing power. Any regional or national Solidarność leader who came to town to quash a local mobilization in the name of some broad but transitory "national policy" ran a very real risk of appearing to be callous and authoritarian or of appearing to trample upon the dignity of the very audience that was listening. From time to time, fire fighting was tactically imperative to protect Solidarność's national strategy; at all times it was reciprocally educational. But it also threatened the long-term political standing within Solidarność of everyone who practiced it. The job itself carried the risk of making leaders sound excessively fearful or weak.

The ranks of firemen, however, were not thin. Every Solidarność regional chairman was forced at one time or another into intense negotiations with self-mobilized local constituencies. When a local movement was literally "moving," democratic assertion was at its most intense, collective self-confidence at its most expressive, and individual self-respect most exposed to confirmation or rebuke. At moments of high conflict with the local party apparatus, a decision to stand down—no matter for what lofty reason—seemed to the rank and file in the trenches like abject surrender. Firemen thus had to care enough about Solidarność to risk the destruction of their own popularity. Finally, even with the best of will, they had to have genuine credentials to entertain any chance at all of prevailing. The very frequency of these encounters in Solidarność Poland testified both to the multiplicity of party crimes and the energy at the base of the democratic movement. Be that as it may, fire fighting demanded a high order of skill, an intimate knowledge of the world of work, a subtle ability to dissect and explain the strategic nuances of high politics, and, perhaps above all, unassailable personal credibility.

It is not surprising, then, that the most accomplished fireman in Poland was Lech Wałęsa. Everything in his personality and political outlook contributed to his effectiveness. Moreover, his loyalty to the very idea of the movement was unsurpassed in the nation. As Wałęsa's colloquial and staccato style illustrated in vividly concrete detail, fire fighting was not a job for the Polish intelligentsia. It not only required an easy familiarity with the working class milieu, it demanded the ability to be extemporaneously persuasive in an emotionally overheated environment. It was important neither to mince words nor to appear indecisive in manner. Workers who had willingly placed themselves at risk had an effortless ability to detect the slightest sign of weak resolve in some visiting oracle of caution. Such misgivings were extended to all outsiders who were suspected of not grasping the seriousness of the party's most recent local crimes. But the most emphatic skepticism was reserved for well-educated speakers. They necessarily carried the burdens of ingrained working-class distrust of intellectuals.[55]

Under the circumstances, it is a singularly poignant irony that, aside from Wałęsa, perhaps the most dependable and successful fireman available for service in moments of extreme confrontation in Poland was Karol Modzelewski's longtime associate, Jacek Kuroń.

Kuroń was *sui generis* among Polish intellectuals. It was not that he had a uniquely driving clarity of vision; many people throughout the country possessed a demonstrated ability to analyze with sophistication the complex historical and social forces that shaped the modern Polish reality. What distinguished Kuroń was that he lived by his ideas. He not only invoked them, judiciously endeavoring to select the proper circumstance, but he injected himself as effectively as he could into the maw of public life. His political life gave the word "self-activity" a vivid personal meaning in Poland.

In the sixteen years since he and Modzelewski had written the "Open Letter to the Party," Kuroń had acquired a subtle experiential knowledge of two contrasting human traits he believed to be politically counterproductive—fear, which was immobilizing; and unthinking "militancy," which sometimes appeared among people who had conquered fear. Kuroń's political goals were sufficiently large for him to be considered quite radical, but he routinely measured his dreams in a context of hard reality. His political confinement—Poland's confinement—conjoined with his expansive vision to provide a singularly unique tension. His political theorizing in the 1970s and in the era of Solidarność was imaginative and its trajectory not easily foreseeable. He was instinctively egalitarian and thus did not artificially isolate himself from workers. He also possessed an internalized self-confidence and a driving personal style that imparted a certain power to his ideas. He was a political strategist of a high order but he was not a philosopher; he was, rather, an intellectual engagé.[56]

The obvious ironies were therefore built into the very structure of his previous experience. As a man who had spent much of his adult life trying to start measured fires in Poland and who had pondered the ways these disparate flames, once ignited, might be fanned into a coalescing brilliance, Kuroń spent much of the Polish August in prison as Baltic workers set a blaze beyond anyone's imagination. Thereafter, he tried along with everyone else to fashion an efficient, modern combustion chamber that could drive Poland into democratic space without overheating the surprising new engine called Solidarność.

Kuroń lived the cause of the movement with unflagging intensity. He was asked to put out a number of fires, increasingly so as 1981 moved toward its fateful climax. He succeeded on occasions when local working-class leaders of Solidarność could not. He also stood in for Wałęsa when Solidarność's chairman was preoccupied fighting fires elsewhere. And in especially tense circumstances, Kuroń and Wałęsa combined their talents and fought the same fire together. Finally, Kuroń also helped put out some fires that were endangering Wałęsa.

Whatever Kuroń's role in the early KOR maneuvers against Wałęsa, and it seems to have been a bit more than marginal, the experience was fatally alienating for them both. Wałęsa did not trust Kuroń or Modzelewski, and he had seen enough of Konrad Bieliński in the closing hours of the Polish August to be wary of all the younger generation of activists whom Kuroń and his comrades in KOR had recruited. Nevertheless, Kuroń and Wałęsa shared so many strategic conceptions that they moved in harmony more often than not. But Kuroń persuaded himself that Wałęsa was too strongly Church influenced and thus prone to a fatally immobilizing posture; Wałęsa felt Kuroń was capable of artificially pushing him from the left, even when Kuroń agreed with his course, because such a stance undercut Wałęsa's political posture in a movement that had immense momentum. It is historically irrelevant whether either man had a factual basis for his fear; the operative political fact was that they did not trust each other. One commonality should, however, be specified. A very large number of Poles in every corner of the country gave all they could to the cause of Solidarność; it is hard to imagine any who managed to find ways to give more of themselves then Jacek Kuroń and Lech Wałęsa. After Bydgoszcz, Karol Modzelewski left, unable any longer to contain his frustration with Wałęsa. Jacek Kuroń stayed.[57]

The departure of Modzelewski was atypical for Solidarność. Compared with other major social movements that have appeared since industrialization, the continuity in leadership that Solidarność maintained at all levels over its legal lifetime was unusual. In contrast, the party went through convulsion after convulsion, from the Politburo down to the most remote party grouplet. With the measured perspective that time affords, it is now possible to formulate the broad outlines of the culminating causality that brought Poland to martial law and a State of War in December 1981.

If Poland were to extricate itself from the economic disintegration that thirty-five years of command politics had induced, the party had to meet certain minimum requirements. How much maneuvering room the party was ultimately able to create for itself with respect to the watchful Soviets depended largely upon its own actions. In the first six months following the settlement in Gdańsk, the party's tepid and equivocal responses added up to its first major blunder. Its endless deceptions—in recognizing Solidarność, in dealing with the issue of work-free Saturdays, in recognizing autonomous organizations for students and peasants, or in the simple implementation of any of the Gdańsk Accords—had the effect of precipitating a relentless litany of confrontations between the party and society. Party leaders might have persuaded themselves that they were engaging in an adroit delaying action and that each protracted tactical retreat somehow contributed to an eventual strategic triumph, but the practical effect was disastrous to the party on four separate fronts. Each retreat further alarmed the Soviets; each deception further demoralized the party rank and file; each confrontation lowered the self-esteem of the party leadership while increasing the prestige of Solidarność; and finally, the party's overall loss of any vestige of international respect infuriated the security services and the military apparatus as well as the party's orthodox hard-liners.[58]

By March 1981, the horizontal movement had spread through the party's lower ranks, the security services had begun acting like an independent state with a private agenda of semiofficial terrorism, the working class was losing its last remaining hope that the party could be negotiated with at all, and the Russians were in a state of high dudgeon. The imperial summons to Moscow early in March and the grim lecturing that followed seemed to complete the demoralization of Kania's inner circle. They returned home to find the hard-liners in an aggressive mood and the security services completely out of control. When the inevitable "incident" materialized at Bydgoszcz, Kania discovered that he was virtually locked into a static posture. It was not just that all of the Solidarność regional organizations seemed simultaneously to rise up as one indignant mass; the Central Committee also found itself inundated with angry telegrams from party first secretaries throughout the country. The beating of Solidarność activists at Bydgoszcz was such a grotesque violation of civility that it outraged part of the party's middle apparatus and almost all of its rank and file. But while Kania's ruling clique came under fire from below, it also found itself attacked by party hard-liners in the Central Committee. The subsequent bid on March 30 of Stefan Olszowski to oust Kania as a prelude to assuming power seems to have driven Kania, finally, into a self-protective response. Olszowski and Grabski were stared down, isolated, and defeated.[59]

It seems evident now that the ensuing weeks in April represented the party's one opportunity in 1981 to save itself. With his conservative opposition temporarily, if insecurely under control, Kania cautiously explored the prospects for genuine party reform through an accommodation with the horizontal movement. Rakowski was dispatched to beat the drums for reform in *Polityka*, and high party functionaries were sent to inspect and mingle with the huge April congress of the horizontalists at Toruń.[60]

Viewed strategically, the situation was the most promising that any satellite party had encountered during the entire postwar era. Leaving the Russians aside for the moment—they were always *there*, as a heavy but imponderable factor whatever one did—the party possessed an opening both to its own rank and file and to the entire surrounding society. This circumstance was unprecedented. Certainly, Nagy in 1956 in Hungary and Dubćek in 1968 in Czechoslovakia had never had the prospective help of such elaborately self-organized social formations as was represented by Solidarność and the horizontal movement. The working classes had mobilized themselves in Hungary and Czechoslovakia *after* the Russian tanks had rolled. In Poland, in 1981, all that work had already been done.

And then, sometime in late April or early May, the party stiffened. The interior dynamics are still not known. For whatever reasons, the party laboriously mobilized its vertical machinery to crush the horizontal movement. A form of planned noncooperation once again came to define the party's posture toward Solidarność. The July party congress was stage-managed for orthodoxy as Fiszbach on the left and Olszowski on the right were sent packing. Ironically, Olszowski's policies were to be carried out, but the task would be handled by the center. Leninist norms had been defended. The price was the final destruction of the party's own continuing relevance to anything approaching effective governance in Poland. With the nation organized by Solidarność, Kania's men left themselves no political opening to society. The party, thus, effectively removed itself from any possibility of resolving the socioeconomic crisis. Poles would eat less, and less . . . and less.[61]

Historical causation was left to others. The great autumn congresses of Solidarność attracted the sustained attention of reporters from all over the world. What could not be easily seen by the press was the slow ebbing away of the vital energies of the movement. To all appearances, self-management had become the breathtakingly appealing centerpiece of the Solidarność program. As everyone could see, the Network's chairman, Jerzy Milewski, was now installed as the national secretary of Solidarność. But those close to the movement's rank and file knew that the movement itself was in crisis. By September, Milewski had confirmed what his colleagues in Wrocław had found in late July: the mass base of Solidarność was worn out. "The membership was frustrated," Milewski later conceded. Hunger was the problem. They were spending "more and more time in lines." Organizing mass-based institutions of self-management across the nation required a poised purposefulness that—at the bottom, where people lived—the movement was in danger of losing.[62]

But the formal gathering of Solidarność on the Baltic coast in 1981 retained an international political significance. The first annual assemblage of the organized nation had been carefully planned for two separate conclaves, one in September—to consider a long maturing "draft program," followed by a break to allow delegates to go home and confer with their constituencies and to receive instructions—and the culminating congress in October. The September convention debated the intricacies of self-management, refined the Network's draft legislation, and adopted it for submission to the Polish Sejm for enactment. Much hope was invested in the proposed legislation as self-management had become a focus of the movement's identity and an anchor of its forthcoming new Action Program for the whole of society.

A moment of drama intruded between the two congresses. Negotiations with the Sejm on self-management were predictably tortuous, but Wałęsa, playing a major role as usual, effected a compromise. Though the revised document on self-management disappointed the Network, it sealed into place basic structural changes in the management of Poland's economy. But after the compromise was announced, the government promptly rewrote it and then resubmitted it to the Sejm for formal ratification. Surprisingly, that body rebelled—for the first time in its thirty-five-year history. The Sejm discarded the government edition and, after making a few cosmetic changes in the language of its original agreement with Solidarność, passed the compromise it had reached with the movement. This bizarre turn of events siphoned a bit of pressure away from Wałęsa, who had come under angry criticism from Network militants. Jacek Kuroń siphoned away some more by defending Wałęsa's actions.[63]

The principal achievement of the second congress was finalizing the movement's Action Program. It was assembled through intense debate extending back to the first working draft published six months earlier. With such continuous input from below, any program that emerged had the possibility of being utopian, contradictory, or otherwise unworkable. Such was not the case. Though its aims were sufficiently grand to be engraved on buildings—"justice, democracy, truth, legality, human dignity, freedom of opinion, and a reconstructed republic"—the document was practical as well as adventurous. It drew its coherence from a unifying egalitarianism that reflected the shape of the movement itself. "What united us is a revolt against injustice, abuses of power and monopolization of the right to speak and act in the name of the nation." As the chairman of the Action Program Commission, Bronisław Geremek, explained, the announced aims "were close to the daily life of working people" and were grounded in the idea of "people making a difference by their own activity."

Organized in eight chapters and thirty-seven theses, the Action Program had as its economic base the "socialized enterprise." Decentralization replaced the "command system" of "the Party and the bureaucratic apparatus," and flexibility in imports and exports would be achieved by connecting with the world market on a self-sustaining basis. While national authority could weigh in on matters of "prices, taxes, interest rates on credits [and] currency exchange rates," the new economic system would be rationalized by the profitability of local enterprises—though organically constrained in this intent by the imperative of full employment. Some inherited realities were honored, many inherited customs were not. The delegates asserted their acceptance of existing "international alliances"; on the other hand, the system of "private medicine" for the party apparatus had to go. No less than nine of the theses concerned the particulars of self-management and ranged far beyond industrial enterprises to include food distribution, social control of the media, and self-management of scientific institutes. In its conceptual emphasis on "universal public initiative," the Action Program of Solidarność embraced a structural empowerment of the citizenry on a scale that has not been achieved in any society. Indeed, under existing constraints embedded in communist and capitalist structures of governance, it is difficult to imagine that Solidarność's program could win majority approval in any existing parliament in the world. In its democratic thrust, the program was, as the saying goes, "ahead of its time."[64]

In one other relevant piece of business, the October convention reelected Wałęsa as chairman, with 55 percent of the vote, over three opponents. Szczecin's well-known regional chairman, Marian Jurczyk, ran second; Wałęsa's old comrade and bitter rival from Gdańsk, Andrzej Gwiazda, ran third; and Bydgoszcz's Jan Rulewski, recovered from his terrible beating, ran fourth. The campaign against Wałęsa, echoed as a choice between "dictatorship" and "democracy," was badly overblown, as its proponents conceded afterward. Wałęsa, worn out by interminable negotiations with the government and incessant duties as a fireman, did not help his cause with his campaign speech. He brought many gifts to the movement, and he had learned many new things in a tumultuous year, but like Solidarność itself, he showed the strain. He was tired and irritable and both symptoms were visible. In the new national structure of governance, the power of the purse was kept firmly at the grass roots. Income from membership assessments was divided 75 percent to local Solidarność unions in the factories, 22 percent to the regions, and 3 percent to the new National Coordinating Commission. The latter body was no longer a "consulting commission," but a voting body that possessed a structure ensuring that regional "barons," themselves fully accountable to their constituencies, would also possess the authority to check "King John" Wałęsa.[65]

In a worsening economic climate, Solidarność thus remained democratic at the base while becoming more democratic at the top at the very moment when the party's calculated politics of destabilization were driving increasing numbers of people to desperation. The last seventy days of legal Solidarność were frenzied as the movement fought for its poise in the face of unending party duplicity. The increasing chaos of daily living produced a crescendo of October strikes as Poles everywhere writhed in frustration and dismay. On October 23, the newly installed KKP, protesting that government stonewalling was turning "the name Solidarność" into "an empty term," called a one-hour national general strike for October 28. The intent was as much to bring order to the movement's multiple expressions of disappointment as it was to pressure the government. The minute the KKP meeting was over, Wałęsa hurried away to Tarnobrzeg to try to douse a blazing local general strike. The following morning, General Jaruzelski, in his fifth day as party first secretary after replacing Kania, dispatched military units to the countryside to "improve food supplies" and "bring order." Significantly, the term of service for army conscripts was extended for two months. If the announcement was an attempt at intimidation, it failed. The October 28 general strike was a total success and thus a clear reminder of who possessed popular legitimacy in Poland.[66]

The government had disdained contact with Solidarność's leadership since the food demonstrations of August, and the movement urgently pressed proposals for a new accord based on austerity and partnership. The Jaruzelski regime sent a sharp signal. It agreed to a meeting, and when it came on November 17, the government evaded all the movement's proposals. The status quo prevailed as the Polish ship continued to sink.[67]

The movement was at an absolute impasse. Given the balance of military forces on one side and the organized nation on the other, the movement's strategy—proceeding on the assumption that the only way out was for the party to make some sort of deal with society—seemed unarguable. Only such an eventuality could accomplish two necessary things at once: saving the party and thus keeping the Soviets out; and creating the political base for the necessarily stark economic solution—national austerity made real by broadened popular participation in decision making. But this strategic stance of Solidarność now began to come under heightened assault from within by two different segments. One was KPN and its newly organized off-shoot, the Club for the Service of Independence, which called for free elections as a means of creating some sort of popular hand-hold on power. The groups' chief spokesman in Solidarność was Bydgoszcz's Jan Rulewski. Not renowned for his patience, Rulewski attacked Wałęsa relentlessly. The other dissident faction, announcing itself on October 17, was the Founding Committee for a National Federation of Self-Governing Bodies. Carving out ideological ground to the left of the Network, the federation, centered in Łódź and Lublin, called for "active strikes" in which workers would continue on the job and distribute their products directly to the population. That this spacious dream was able to acquire an organizational form testified to the growing sense of desperation pervading society.[68]

Government passivity on policy coupled with strident media attacks on Solidarność had begun to have an effect. Within the movement, poise—which had been maintained so well for so long—began to crack. By the middle of November, over a quarter million Poles were out on strike, and the political tension in the country had begun to acquire an overlay of impending doom. Police had begun using tear gas to control movement demonstrations. A particularly volatile local controversy at Zielona Góra heightened emotions still further. Wałęsa and Kuroń joined forces, one final time, as firemen in Zielona Góra.[69]

Upon arriving back in Warsaw, Kuroń had one more move to make. He tired to defuse the situation with a risky maneuver—the formation on November 22 of Clubs of the Self-Governing Republic to siphon political energy away from the movement and leave the Solidarność national commission with a bit of maneuvering room in its struggle with the government. However, at the inaugural gathering of Kuroń's group, the security police swooped in and arrested him. Ten days later at an emergency meeting of the national commission at Radom, Wałęsa could hold back the tide no longer. Amid recurring talk about "confrontation," the KKP insisted it would "not accept price increases without economic reform." And—on a tape the government ominously played throughout the ensuing week—Wałęsa's voice could be heard wearily accepting the fact that "confrontation is inevitable . . . and it will take place." When the KKP met again on December 11 in Gdańsk, Wałęsa made clear he had been quoted out of context and reiterated, "I declare with my full authority we are for agreement and we do not want confrontation. The national Agreement must become a reality."

The consumer breakdown was not just in food, however. With winter coming, there was not enough heat, not enough winter clothing, not even enough shoes. The shortages could no longer be ascribed to simple mismanagement, even by an apparatus as mature in the art of error as the Polish party. The heightened anxieties of early winter and the economic chaos that fueled them were the external trappings of a military coup that had long been planned. In an atmosphere of palpable crisis, the tone of the KKP deliberations had a quality of abiding frustration appropriate to a land run by functionaries who had long since proved they had no ideas the nation wanted to hear. Since October, the party had possessed a general as its first secretary. In the early hours of the morning on December 13, after the Solidarność national commission had gone to bed, the general called for his army and put society under military occupation.[70]

For the first time in a part of the world functioning under Leninist norms, the leading role of the party effectively passed from the party apparatus to the army. Poland began to be operated as an old-fashioned military dictatorship.

8 ▪ The Party Declares a "State of War"

How many new churches is Wałęsa's head worth?

Polish worker

THE PRIMATE OF Poland, Józef Cardinal Glemp, was awakened in the early morning hours of December 13 and told by a messenger from the Central Committee of the party that a "State of War" had been declared as a means of avoiding Soviet invasion. The next morning, for the first time since the ill-fated speech by Cardinal Wyszyński at the height of the Gdańsk confrontation sixteen months before, the leading voice of Polish Catholicism offered a public sermon that constituted a major political intervention. Glemp, who had become Primate upon Wyszyński's death six months earlier, told Poland's Catholic faithful the next morning that the Church "received with pain the severance of dialogue . . ." He then went on: "There is nothing of greater value than human life. That is why I, myself, will call for reason even if that means I become the target of insults. I shall plead, even if I have to plead on my knees; do not start a fight of Pole against Pole." Like Wyszyński's Jasna Góra sermon sixteen months before, Glemp's homily received immediate and wide distribution in the party-controlled media.

According to the English observer Timothy Garton Ash, "The sermon seems to have played a part in reducing the immediate (passive) resistance to martial law." Ash added, "The Primate's words were bitterly resented by many Christian Poles who were, at that moment, preparing to risk their own lives for what they considered greater values. A week later the episcopate would come out with a much clearer defense of Solidarity; but that week was decisive."[1]

The Primate, of course, was a highly visible component of the Church's presence in Poland. But he was not the only component. In the bitter years of military rule that followed the December 1981 coup, many priests and bishops did what they could to restore the morale of the nation. None were more active or successful in this endeavor than a young Warsaw priest named Jerzy Popiełuszko. "Father Jerzy" became the energizing force at St. Stanisław Kostka Church in Warsaw where, beginning in January 1982, a "Mass for the Motherland" became a celebrated monthly occurrence. As Popiełuszko saw the situation in Poland, "The duty of the priest is to be with the people when they need him most, when they are wronged, degraded and maltreated."[2]

The "Mass for the Motherland" represented the Church politicized. The interior aisle at St. Stanisław Kostka, lined with balconies on both sides, supported two rows of flags—the red and white flag of Poland alternating with the blue and white flag of the Church. At the entrance was a poster proclaiming, "Mass for the Motherland and Those Who Have Suffered for the Motherland." In the center of the poster was a photograph of marching Polish workers holding aloft a banner of Solidarność.

Popiełuszko's Masses became a political event of the first magnitude in occupied Poland. On the last Sunday evening of each month, his church in Warsaw would be packed more than an hour before the scheduled beginning of the Mass, and by the time services began, the large expanse surrounding the church would be jammed with thousands more who listened over loudspeakers jutting from the four corners of the building. The service was a medley of music, poetry, and liturgy that lasted two hours. A succession of individual voices, male and female, could be heard over the loudspeakers in services that were deeply communitarian. They read religious poems, described as "very political," and political poems that were "very religious." The principal distinction between the two was that in the ones dubbed "political" the word "Solidarność" could be heard. A number of Polish anthems and religious hymns were also structured into the mass, many with new words at key junctures. The line, "Lord, keep our country free," returned, under martial law, to the old nineteenth-century version: "Lord, return our homeland to us free." These lines were punctuated by ten thousand hands shooting into the air, fingers pointing in a "V," and arms vigorously waving in rhythmic beat to the words. The effect was striking. The national anthem contained a line from eighteenth-century history imploring Polish troops to "March, march Dąbrowski from the Italian mainland to Poland." Under martial law, the words became, "Lead us, Wałęsa, from the sea coast to Silesia." The new final line was, "Solidarność will rise again and be victorious." In the aftermath of the military takeover, the "Mass for the Motherland" was a passionate political event in the Poland of the 1980s.[3]

In this manner, the democratic movement began to sustain itself in the era of martial law by mobilizing its own and the nation's symbols and recreating itself in myth. The instrument of this metamorphosis was—once again—sacred politics. When Danuta Wałęsa gave birth to her seventh child in January 1982 at a time when her husband remained in police custody, no less than fifty thousand people attended the "christening." This obvious political ceremony made young Maria Wiktoria Wałęsa the best-known infant in People's Poland. Under martial law, Jerzy Popiełuszko, a parish priest, gradually became a symbol of sacred politics. He not only toured the country soliciting funds for food and medical supplies for the families of thousands of so imprisoned Solidarność activists but also organized a letter writing campaign to the world. Food, medical supplies, and money poured into St. Stanisław Kostka from Western Europe and the United States and were judiciously distributed throughout the country. Popiełuszko also consciously tried to rally his parishioners to combat their own fears. The harsh new judicial codes and stringent censorship regulations imposed under martial law, augmented as they were by thousands of arrests, created a whole new climate in Poland. The conquest of fear was something that could be achieved, said Popiełuszko, "only if we accept suffering in the name of a greater value. If the truth becomes for us a value worth suffering for, then we shall overcome fear." As the months of martial law turned into years, Father Jerzy increased the pace of his ministry. With the "nation hanging on the cross with Christ," he said, "We are called to the Truth." Predictably, the Warsaw priest became the special object of police hostility. His phone was tapped, his apartment was ransacked, and his car was vandalized. A group of Huta Warszawa steelworkers (whose priest he was) acted as bodyguards as he moved about under constant police surveillance. He was publicly denounced by party functionaries for conducting "rallies hostile to the Polish state." When police provocateurs were sent to Sunday services at St. Stanisław's to interrupt Mass and raise political issues that would confirm the illegality of the proceedings, Popiełuszko calmed his irate congregation and told he police agents, "Brothers, you who were ordered to come here by others, if you want to serve the truth and regain your self-respect, let the people go in peace." The Christmas Mass at St. Stanisław Kostka in 1983 was attended by fifteen thousand people.[4]

The contrast between Glemp and Popiełuszko, between cardinal and parish priest, not only evoked comment throughout Poland and among the foreign press, it generated severe tensions within the Church itself. As early as December 1982, a group of thirty priests of the Archdiocese of Warsaw arranged a meeting with Glemp that turned into a remarkable confrontation over the Church's apparent acquiescence in totalitarian government. The Primate was informed that the Polish people felt that the Church had abandoned them and that he was making politics together with General Jaruzelski. It was put forward that priests who themselves "had not suffered" were being prevented by Church policy from supporting people who had. When one priest suggested that the Primate himself head a procession of the clergy against the militia, as did Pope Leo the Great against the Huns, "because then we should know what it means to be beaten up," the assemblage broke into applause. There was more. Another priest said the population felt the Church had embarked on a policy of collaboration in order to bring about a visit of the Pope. Such would be like a "visit to a great internment camp" in which the Church "took upon itself the task of keeping order." The only argument the regime understood, he said, was pressure. Soon the Church might find itself under attack from the regime. It would then have to try to survive with diminished support from a nation whose resistance it had helped to break. This accusation, too, was reinforced by an outburst of applause.

Surprised if not stunned by the tone of these remarks, Glemp nevertheless refused to back away. It was not the Church's task, Poland's Primate said, to create a "neo-Solidarność." It was not the Church's mission to change the system but to carry out its tasks under any system. The effort of the Solidarność underground was a struggle without a program, he said. It was blackmail for anyone to imply that the Church had an obligation to encourage people to heroics. Glemp concluded on a strategic note: for the Church to be with the nation, it must "penetrate the national institutions."[5]

If the priesthood felt that forthright expressions from below might alter the Primate's response, they were profoundly disappointed. As the shadow of military rule lengthened, Glemp held to his accommodating course. The government demonstrated its capacity for accommodation too—if not to Solidarność, then to the Church. On one occasion, Jaruzelski and Glemp huddled in conference for five hours; soon thereafter, on a trip to Gdańsk, the Primate was able to give Wałęsa fifteen minutes. The reception accorded Glemp by the coastal population was gracefully described as "cool." Finally, in 1984, a tense meeting of the episcopate brought together seven hundred priests who again confronted the cardinal with a concerted protest over the passivity of his ministry. The Primate steadfastly refused to be drawn into a discussion on the issue; his only response was to call to his priests to pray with him. Later in the year, on a visit to Brazil, Glemp was quoted as criticizing those sectors of the priesthood who worked closely with the Solidarność underground, explaining that he himself had chosen "a different, more difficult and more just path—the true pastoral path." Back in Poland, embittered partisans of the democratic movement began referring to the Primate as "Comrade Glemp." The Church, meanwhile, began to receive surprising benefits from the new regime. Though supplies of everything remained at critically low levels in Poland and the nation's housing crisis significantly worsened, the Church was able to receive permission to build new chapels and, impressively, the material to build them.[6]

The more Glemp accommodated, the more risks Popiełuszko took. It was as if the young priest had determined single-handedly to atone for the conduct of the hierarchy. From the pulpit of St. Stanisław, he cried out, "We are called to the Truth; let us not sell our ideals by selling our brothers." In July 1984, Popiełuszko was indicted for a variety of activities against the state. Most prominent was the charge that he harbored a vast cache of arms and ammunition in his apartment—an assortment that had been planted and then found by security agents while they held the priest under interrogation in a Warsaw police station. Though Popiełuszko was included under a general amnesty in August, he disappeared on the night of October 19, 1984, while returning to Warsaw from a special workers' Mass in Bydgoszcz. For days, he could not be found, and the whole nation went into a state of shock. On October 25, Warsaw radio announced that five persons had been placed under investigation for possible involvement in Popiełuszko's disappearance and the interior minister appealed for public help in the search. Huge crowds gathered daily for Masses at St. Stanisław and waited outside for news. On October 30, just as evening Mass was ending, word came that Father Popiełuszko was dead. The British journalist Neal Ascherson, who was in the church at the time, reported: "Nobody who heard it will ever forget the awful howl of agony that rose from the thousands waiting in the Church of St. Stanisław Kostka when a priest announced that Father Popiełuszko's body had been found." The cry "went on for many minutes until it was joined by the tolling of the bells."

It was learned in due course that the priest's automobile had been intercepted on the highway by agents of the security service, and he had been repeatedly beaten before his bound and weighted body was thrown into icy water above the Vistula dam at Włocławek. The country rocked in indignation and agony, and tens of thousands lined up to file past the coffin as it lay in the sanctuary at St. Stanisław Kostka. Delegations from every industrial center in Poland arrived in Warsaw, bearing Solidarność banners, which were held high as workers marched to the church. The farewell speeches following Mass seemed to draw from the deepest wellsprings of the Polish national experience. Wałęsa said, "The entire life of this good and courageous man, this extraordinary worker priest, pastor and leader of the national cause, bears witness to the unity of church and nation. We say goodbye to you with dignity and in the hope of peace and social justice for our country. Solidarność lives on, Father Jerzy, because you gave your life for us." A steelworker from Huta Warszawa cried out in grief stricken voice, "Jurek, our friend, you are still with us . . . Can you hear the bells tolling for freedom? Can you hear the hearts praying?" The funeral was attended by 350,000 people.[7]

The meaning of the Polish Catholic Church in the era of Solidarność ultimately did not turn on the cautious conduct of Cardinal Glemp or Stefan Wyszyński before him or even upon the selflessness and heroism of Jerzy Popiełuszko. The Church remained germane to the nation's social and political life because it existed, because it was *there*, because, in the deepest sense, it offered the people of Poland the prospect of another way to be. In its message, its symbols, and its rituals, the Church embodied the national past and the continuance of national aspiration.

There is no Catholic Church in the world quite like the Polish Church, a religious institution so self-consciously the repository of a specific nationhood and one that links its saints so intimately to the nation's struggle for independence. Nowhere else in the universe of Catholicism is the Church's musical liturgy so steeped in national history and national longing. If the Polish intelligentsia has historically looked upon itself as the mind and voice of the nation, the Church just as relentlessly has regarded itself as the nation's soul and spirit.

These religious and institutional elements conjoin as a very mixed blessing politically. Historically, for young priests who were trained and grew to maturity within such an institutionally self-righteous and nationalistic environment, the pastoral experience by no means imparted a profound sense of obligation to one's parishioners at the risk of the welfare of the Church itself. Rather, the classic effect of such a heritage has historically been to suggest to the priesthood that the well-being of the Catholic faithful was ultimately a function of the well-being of the Church.

For many generations, this insular view was controlling within the Polish episcopate. In the years before World War II, the Church demonstrated its willingness to run great risks with huge numbers of its powerless parishioners in order to avoid offending the tiny number who had political power. The Church looked the other way when striking miners were being shot in 1931, and the episcopate antagonized the peasantry with its consistent support of large landholders, most blatantly during the repression of the great peasant strike of 1937. After the war, the Church labored long and patiently for some sort of defensible accommodation with the communist regime. When an effective rapprochement was achieved with Gomułka in 1956, Cardinal Wyszyński became quite forthright in his support of Gomułka's regime. That message was sent West in 1956, and the effort to maintain a working relationship so largely defined the Primate's policy over the next twenty years that he was able to assure President Carter in 1978 that Gomułka's successor, Edward Gierek, "had the interests of Poland at heart."[8]

But as Wyszyński understood matters, all was done in the interest of the Church, not of communism. The confidence undergirding this view was a direct function of the profoundly proprietary sense that Wyszyński had of the Church's relationship to the nation: "In Poland the hierarchy has always had the nation's interest close to its heart . . . No one is more of the people and for the people than we priests, and that in no political or class sense. We are the emanation of the nation itself."[9]

The striking thing about the cardinal's religious nationalism was that it ultimately insulated the institution he headed against erosion at the hands of an authoritarian secular state. In historical terms, Wyszyński protected an anachronistic and medieval institution against incursions by the state in the very decades it was shaking off its provincial and anti-Semitic past and slowly but inexorably assuming a broadly humanistic posture. The postwar symbol of the old ways was August Cardinal Hlond, for whom the word "Christian" meant Catholic and who nursed a deep-seated religious prejudice against Jews. Wyszyński became Primate upon Hlond's death in 1948, just before the onset of Bolesław Bierut's militarized six-year plan heralded the arrival of Stalinism in Poland. In the early 1950s, Wyszyński accepted house arrest and internal exile rather than compromise what he considered the minimum institutional guarantees the Church needed for long-term survival. In the view of the French legal scholar Jacques Ellul, Wyszyński's strategic posture allowed Catholicism to survive in Poland while Protestantism was, with full assistance from its officials, being subsumed under the party-state in Hungary and Czechoslovakia. Under the leadership of J. L. Hromadka in Czechoslovakia and Bishop A. Bereczki in Hungary, Protestanism was, in Ellul's phrase, "given over in its totality" to the communist regime in a way that effectively destroyed the internal life of the religious institutions in both countries. In 1956 in Hungary and 1968 in Czechoslovakia, the Reformed Church "counted for nothing." Had they followed Wyszyński's course and conducted "a search for freedom within communism" while preserving "untrammelled Biblical teaching," said Ellul, the Hungarian and Czech Churches could have been "of great service in the search for possible solutions in 1956 and 1968."[10]

A certain strategic logic adheres to this analysis, though it is severely compromised on a practical level by the fact that Wyszyński in 1956 and 1970 and Glemp after 1981 did not actively employ their status of "counting for something" to defend Polish workers. Rather, they accommodated to government repression. But it is unarguable that Wyszyński defended his Church better than Hromadka and Bereczki defended theirs.

Meanwhile, the institution Wyszyński defended changed internally. As Daniel Singer has pointed out, the postwar Polish Church, deprived of its prewar allies, gradually altered its orientation from the rights of property to the rights of man. The sign of things to come was signaled in 1965 by an extraordinary letter from the Polish bishops to their West German Catholic counterparts. Recalling all the historic enmity between the two countries that culminated in the slaughter of six million Poles in World War II and the eviction of millions of Germans from their homeland at the end of the war, the Polish bishops said, "We forgive you and ask for forgiveness." The initiative was well in advance of Polish public opinion and made the Church vulnerable to a furious propaganda assault from Gomułka, who promptly seized the high ground of nationalism in an effort to prop up his sagging popularity through anti-German tirades. As the Church maneuvered to make peace with Germans, so it did with Jews. After the Polish government went on its insane anti-Semitic binge in 1968, a number of Jews purged from the party found employment at the Catholic University of Lublin. The Church's final move to transcend its preindustrial heritage came in a gesture to Polish workers. After standing to the side when Gomułka crushed the worker councils in 1956–57, Wyszyński, thirteen years later, began to take the Church on its first halting steps toward the terrain of the Polish working class. The Primate led the episcopate in expressing sympathy for the victims of the coastal massacre of 1970 and repeated this stand following the police violence at Radom and Ursus in 1976. Wyszyński even went so far as to say that "It is the clergy's duty to defend the workers' interests against hasty and ill-conceived government measures."[11]

But the aspect of politics the Catholic hierarchy understood best was the role that history and culture played in preserving the integrity of the nation. The episcopate went to herculean lengths to defend the Catholic University of Lublin—the only such Catholic institution of higher learning in the Eastern bloc; under constant harassment from the state, the university was sustained by collections taken up from the faithful five times a year in every parish in Poland. Then in the late 1970s, the Church's concept of knowledge at long last breached the bonds of strictly parochial considerations. In 1978, when the Flying University began to give lectures on the national past, Wyszyński publicly defended the professors and protested their harassment by police.[12]

In these ways, the Church demonstrated its determination to broaden its understanding of Poland's rich cultural inheritance. The gains were real. A learned Polish sociologist, who himself seldom went to church, explained to an American journalist why he insisted that his six-year-old son attend Mass: "The Church has preserved Polish history and culture. If my son did not receive exposure to the Church, there isn't a single Polish poem he'd ever understand."[13]

Gradually, this view of the Church began to extend beyond the mainstream of the intelligentsia to include the activists of the democratic opposition.

After the episcopate spoke up for imprisoned members of KOR in 1977, Jacek Kuroń described the Church in Poland as "genuinely independent because it represents a mass social movement while churches in other countries of the socialist bloc do not." The historically relevant question, of course, concerned the extent to which the Church chose to assert the independence it had so arduously won.[14]

The single event that most altered the shape of the internal environment of the postwar Polish Catholic Church was the arrival in national life of the independent trade union Solidarność. The change was visible in the manner individual Polish priests and bishops acted. Father Jankowski was a very different person in his daily rounds in St. Brigida Parish in Gdańsk in July 1981 than he had been as the obedient son of Rome in July 1980. Indeed, he was sufficiently altered by his intimate association with the coastal movement that, on occasion, he invited the controversial Popiełuszko to conduct Mass at St. Brigida's. "Father Jerzy," of course, was the outstanding example of the potential transformation within the Church generated by the Solidarność experience. But there were many more. Among them were men like Jan Zieja, one of the founders of KOR, Mieczysław Nowak of Ursus, Bishop Ignacy Tokarczuk of Przemyśl, Stanisław Małkowski of Warsaw and Józef Tischner of Kraków. None could idly withstand the moral imperatives that Solidarność raised and which police reprisals dramatized in ethically transcendent ways. In the era of martial law, parish priests and bishops throughout Poland found ways to look aside—sometimes for days on end—while the equipment used to print Church bulletins was employed to print the newspapers and journals of the Solidarność underground. The same churchmen did more than look the other way; they helped locate and pay for paper and ink and raise maintenance income for movement printers and activists.[15]

The Church, then, was an integral part of the hope of the nation. Whether a believer or not, every person in Poland could without the slightest twinge of unease view a personal trip to Mass as an authentic political act that helped to maintain a certain distance from the state. As an agency of emotional self-preservation, the Church played a political role of a very high order. It would be quite inappropriate, therefore, to underestimate the psychologically supportive role of Polish Catholicism. The Church preserved—defensively preserved, symbolically preserved—the spirit and, thus, provided a certain plateau beneath which the population could not be driven by the tyrannies of the party or the party's police.

Yet, that which constituted the Church's gift to the nation also defined the limits of the gift. The Church could console and thus help maintain morale for the struggle ahead; it could provide a bedrock of ethical belief; but it could not provide the explicit strategic or tactical ideas that might move Poland toward a freer social life. Those ideas had to come from elsewhere. In short, the Church in Poland, when it chose to be, was reactively relevant. But it was not a source of causation. It played no active role in the origins and development of the democratic movement. And however much improved the morale of Poles temporarily appeared to be after the visit of the Polish Pope in 1979, any "transformation of consciousness" (that might in some mystical way have been imagined from afar) had little to do with the mobilization during the Polish August of occupation strikes and interfactory strike committees along with the Baltic coast. Nor was the Church relevant to (or even aware of) the intensive internal working-class dynamics which over thirty-five years eventually produced both the demand for free trade unions and the tactical knowledge that led to the attainment of such a goal. Even by secular criteria, "Father Jerzy" was a courageous man and a democratic saint. But the Church was never more than a prestigious and emotionally helpful supporter of political initiatives that had their origin elsewhere. To speculate otherwise is to misunderstand the dynamics that produced Solidarność.[16]

Postwar evidence sustaining this basic conclusion about the politics of the Church was abundant in the years prior to the Polish August—the years of Hlond and Wyszyński—during the sixteen months of democratic promise that followed—the time of Wyszyński and Glemp—and during the long night of military rule that extended to 1989. In the era of Cardinal Glemp, when Church support of Solidarność became too publicly visible, as in Father Nowak's sermons in Ursus, such activity was brought to a halt by the Primate himself. Glemp decided to "promote" Nowak to a remote parish in the provinces. The messenger from the Warsaw episcopate who delivered this news to Ursus was booed by the congregation, and they subsequently initiated a hunger strike to protest the Primate's actions.[17]

In the 1980s, the Catholic Church became a diminished force for civility in Poland, even as Cardinal Glemp himself became an object of ridicule and bitterness. The word "Glempic" crept into the vernacular. It meant "to say nothing at length and in soothing terms." August Hlond had been an authentic expression of the Church of his era; Stefan Wyszyński moved ahead of his Church, consciously leading it until in the goodness of time it caught up with and passed him in the final year of his life. But Józef Glemp generated events in a new way; the Church prospered as the nation declined. The Primate allowed a great distance to develop between himself and his priests, even as he labored to narrow the distance between the Church and the state. The cool reaction of the Primate on the night of December 13, 1981, suggested a style of conduct that was to prove enduring.

Cardinal Glemp was not the only Pole to be awakened in the early morning hours of December 13. In Warsaw, activist circles among the intelligentsia also got the news from knowledgeable acquaintances in or near the party apparatus. The word quickly spread as people rushed to warn those who were the most obvious targets for arrest. But there was an odd short circuit in this cautionary network. Among the high-silhouette activists who needed to be warned were a number of workers prominent in the leadership of Solidarność. The difficulty was that, with the phones cut off, many intellectuals did not know how to reach them. They did not know where the workers lived. Intellectuals active in the movement had routinely met with their worker counterparts in Solidarność offices or, on special occasions, at the apartments of intellectuals. On the night of December 13, many in the Warsaw intelligentsia discovered that in all the months of freedom that dated back to the Polish August, they had not once set foot in the house of a worker.

The fact was symptomatic of a yawning cultural fissure in the body politic of Solidarność. Even though caught in the social vortex set in motion by the coastal working class sixteen months before, even though scores of local, regional, and national Solidarność spokesmen had emerged from the ranks of working-class Poland, and finally, even though every Pole's individual freedom was tied inextricably to the future of the independent trade union, intellectuals in the capital still instinctively thought of workers as "others."

Perhaps the best way to approach this seminal cultural fact is through the career of a Polish intellectual who eventually found a way, temporarily at least, to transcend the attitudes and conventions of his class. His name was Adam Michnik, and his history has no precise counterpart within the Polish intelligentsia.

Born in 1946, Michnik was the son of a party apparatchik. He learned the communist liturgy, but very early in life began to test the communist church. The discussion club he organized at the age of fifteen—"Seekers of Contradictions"—found enough of that very commodity to put the party under enduring scrutiny. One result was that the stunning "Open Letter to the Party" by Kuroń and Modzelewski seemed to 18-year-old Michnik in 1964 to be a forthright interpretation of reality. The young student was arrested for distributing the suppressed document and served a two-month prison sentence. This sobering experience can in retrospect be seen as having played a critical role in what might be called the "preventative education of Adam Michnik." The police brutality that accompanied the student protest of 1968 did not have the traumatizing effect on him that it clearly had on so many young Polish intellectuals of his generation who were more politically innocent.[18]

Though Michnik played a central role in the "March events," as the student protests of 1968 were called, and was again sent to prison for his efforts, his prior run-ins with authority insulated him against the kinds of interpretive errors so many of his fellow students subsequently made with respect to workers; that is, Michnik did not think the thugs the party trucked on campus to beat up students were a representative sample of the Polish working class. Many in the "class of '68" in Poland thought precisely that.[19]

In the 1970s, Michnik gradually developed a wholly original analysis of Polish society and politics. In two essays published in 1976 and 1977—"The New Evolutionism" and *The Church and the Left: A Dialogue*—Michnik explored a new strategic approach to social change and the tactical adjustments that the Polish intelligentsia needed to accept in order to make such a strategy practical. In "The New Evolutionism," Michnik wrote that "the lesson of Czechoslovakia is that change is possible and that it has limits. Czechoslovakia is an example of the fragility of totalitarian stability, and also of the desperation and ruthlessness of an empire under threat." Michnik ranged over the entire intellectual experience of postwar Poland, touching on the rise of Marxist revisionism in the 1950s and its collapse under the jackboots of the Czechoslovakia invasion in 1968; the limits of neopositivism; and the various currents that jostled against one another in the stream of Polish Catholicism. Significantly, he singled out the Catholic hierarchy for praise, applauding the fact that the old jeremiads against "godless ones" had given way to statements underscoring human rights. "Most important," said Michnik, the Church had begun "defending the civil liberties of the working people and particularly their right to strike and to form independent labor unions." He also offered a sober assessment of Soviet priorities and capabilities as a way of emphasizing the need for the Polish opposition to employ a gradualist strategy rather than cave in to apocalyptic longings for national independence.

Michnik devoted only two paragraphs in "The New Evolutionism" to the Polish working class. Though his analysis did not by any means reflect an informed knowledge of what had happened in working-class Poland during the postwar era, he nevertheless demonstrated a political sensibility that was well in advance of the opposition intelligentsia as a whole. "The new evolutionism," he wrote, "is based on faith in the power of the working class which, with a steady and unyielding stand, has on several occasions forced the government to make spectacular concessions." While conceding that developments were "difficult to foresee," he offered the thought, novel among the intelligentsia, that "there is no question that the power elite fears this social group most." He added, pointedly, "Pressure from the working class is a necessary condition for the evolution of public life toward a democracy."

However, not in this or any of his other writings did Michnik reveal he had a grasp of the role played by the Cegielski-led working class in the 1956 revolution or of the signal strategic imperatives those facts contained. And while he could speak in general terms of praise for "the new stage of worker consciousness" that seemed to have appeared on the Baltic coast in 1970, Michnik, in keeping with the rest of the intelligentsia, did not possess a sufficient understanding of the dynamics of shop-floor organizing to appreciate the strategic importance of the emergence of self-organized occupation strikes and, most important, the formula of interfactory strike committees as pivotal instruments of working-class autonomy. "It is hard to tell when and how other, more permanent institutions representing the interests of workers will be created and what form they will have," he confessed. "Will they be workers' committees following from the Spanish model or individual labor unions or mutual aid societies?"

Michnik concluded his two paragraphs on the working class with an exhortatory observation that inadvertently revealed the continuing cultural confusion persisting even within politically sophisticated sectors of the Polish intelligentsia: "But when such institutions emerge, the vision of a new evolutionism will become more than just a creation of a mind in search of hope." Every rhythmic nuance of this sentence established it as the concluding peroration of his ground-breaking article. Unfortunately, despite what his capacity for social analysis was telling him, Michnik went on—and on—to ruminate about what the Church needed to do and what intellectuals needed to do, concluding some pages later with a quote from the philosopher Leszek Kołakowski. The world Adam Michnik lived in—the world of the intelligentsia—remained, in

1976, the determining factor of where he searched for new routes to social change.[20]

While in the West in 1977, Michnik put the finishing touches on a book fated to become instantly controversial among opposition circles in Poland—*The Church and the Left: A Dialogue*. In his determination to shatter the instinctive anticlericalism that dominated "the secular left," Michnik went to great lengths to establish a working basis for a coalition of intellectuals and churchmen. He capped his argument with a striking paragraph:

> The children of the church, who had grown up and departed from her, in the hour of danger returned to their mother. And although in the course of their long alienation they have changed a great deal, though they look different and speak a different language, at the decisive moment mother and children recognized each other. Reason, law, civilization, humanism—whatever they are called—have sought and found at their source a new meaning and new strength. This source is Jesus Christ.[21]

It is not to be supposed that Michnik's cautionary warnings against excessive zeal for independence or his generous overtures to the episcopate won him instant acclaim, either within the revisionist Marxist milieu of which he was a one-time member or among the more nationalistic circles that hovered around the edges of the Warsaw intelligentsia. Indeed, his instinctive employment of the term "secular left" created some misunderstanding even within those circles inhabiting the Clubs of the Catholic Intelligentsia which might otherwise have looked with favor upon Michnik's eclectic and gradualist approach.

But his writings constituted a breakthrough nonetheless, for they had the great virtue of realpolitik. Michnik spoke to the political situation as it existed in Poland at that moment. He proposed modes of coalition building that were instantly employable, and he encased his ideas within an overall strategic analysis that made immediate sense. It is probably not too much to say that Michnik altered the trajectory of intellectual analysis in Warsaw circles and facilitated the emergence of a climate wherein KOR in the late 1970s could raise massive funds for its proliferating oppositional activities.

But in terms of building a movement, there is a fundamental difference between clearing away intellectual underbrush so that existing groups might coalesce and charting a course of action that would recruit significant numbers of people to those groups. The first had little meaning unless the second was achieved. And to this task, building the mass base of a movement, Michnik had little or nothing to contribute. Quite simply, the subject was outside of his experience.

Thus, the unanticipated and exhilarating emergence of Solidarność from the coastal shipyards in August 1980 revealed conclusively that Michnik's strategic vision did not further expand from 1977 to 1980. During the life of Solidarność and for several years thereafter, Michnik repeated the deeply self-serving KOR explanation about the origins of the independent trade union. Immediately following the creation of Solidarność, he wrote that the independent movement was a product of the joint efforts of the free trade unions of the coast and the editors of *Robotnik*. "They worked out and popularized the idea of worker self-organization and demands, and it is to them to a large extent that we owe the implementation of worker demands and the peaceful progress of the strike."[22]

After watching the internal dynamics of Solidarność for a year, Michnik began to become both abstract and remote in his assessment of causation: "The Church's opposition to atheistic policies, the villages' resistance to collectivization, the intelligentsia's defiance of censorship—all made up the 'Polish syndrome' that bore fruit in the form of the August strikes and Solidarność." It is perhaps just as well that abstractions such as "the Polish syndrome" do not seem to call for empirical support, for (to deal with only one of his three examples) Michnik would have been hard put to demonstrate how "villages' resistance to collectivization" bore on the dynamics of the August strikes on the coast. In any event, when Michnik did get specific, he was wide of the mark: "The actions of the intellectual groups that organized aid to the participants in the June 1976 strike played a special role. It was then that a common denominator for the activities of different social groups, especially the intelligentsia and the workers, was successfully created." The "common denominator" for the Polish August was, of course, the interfactory strike committee and the central demand for an independent trade union, a structural device and a strategic objective that were developed by coastal workers in 1970 independently of "intellectual groups."[23]

In a public speech in Warsaw which was one of the high-water marks of intellectual life in the capital in the autumn of 1980, Michnik ranged broadly over postwar Polish history. In many ways, his address was suffused with erudition. But he again misread the dynamics of the Polish October of 1956—presenting an analysis that contained no reference to the prolonged shop-floor struggles across Poland or the crushing of worker councils under the party's adroit "experimental program." He also misread the politics of 1968 and 1970, seeming to bless his own student generation of 1968 with the causative role in the Baltic rising of 1970. He accomplished this dismaying feat in one sentence: "1970 was obviously the consequence of 1968." No one in his Warsaw audience questioned either interpretation.[24]

Subsequently, in the Solidarność era, Michnik distributed his loyalties with respect to the policy disputes within the movement according to the criteria articulated by the *Robotnik* circle around Kuroń—criteria, it might be added, that gradually became the conventional Warsaw wisdom: "It is the young workers in the large factories that are the most radical," he judged. The members of the Gdańsk founding committee were "by their very nature more moderate, more susceptible to the arguments of their fellow negotiators from the government and—what is more important—to the voice of the Church, which has been toning things down."[25]

This judgment, offered in August 1981, showed a surprising distance from the interior dynamics of the movement. In a climate of agitation on self-management and proliferating hunger demonstrations and strikes, evidence of the "toned down" nature of public life was hard to find. But as the crisis accelerated in the autumn of 1981, Michnik began to discover the implications of this underlying reality. He thereupon embarked upon an evolution that over a period of four years was to bring him to a completely different understanding of the operative dynamics of the Polish movement. Though this intellectual passage was quite gradual, Michnik took station, once again, in advance of the intellectual mainstream. It is instructive to trace this evolution.

A year of steadfast non-cooperation by the ruling party apparatus had so exacerbated the economic crisis by midsummer 1981 that the resulting food shortages energized an impoverished but highly organized society to an unending series of eruptions at the very base of social life. The episodic surfacing of mindless police brutality in the midst of this larger crisis had the effect of testing to the outer limits the movement's ability to adhere to its nonviolent principles. One result was that virtually every thinking activist in the movement was called to the occupation of fire fighting. On such occasions, a number of people, Adam Michnik included, received a first-hand introduction to the environment that Lech Wałęsa lived in all the time.

At the town of Otwock, Michnik found himself in the midst of an enraged crowd bent on doing bodily harm to a policeman who had committed conspicuous acts of brutality. Michnik managed to acquire credibility with the crowd by announcing himself as "an antisocialist element." His ensuing pleas succeeded in calming people down and possibly saving the policeman's life. A lesson was implicit in these dynamics and, after martial law, Michnik indicated he had begun to internalize at least part of it: rhetorical "militancy" did not necessarily constitute a strength to be favorably contrasted with (weak-minded and possibly church-influenced) "moderation." Under certain conditions—such as those that were increasingly frequent in the Poland of 1981—"militance" might not only be short-sighted in a practical sense, but could lead to political positions that were counterproductive if not inherently reactionary in their effort. In such instances, prudence might simply be a sign of strategic sophistication rather than one of muddle-headed caution.

In point of fact, an oversupply of fiery rhetoric—from some movement activists as well as from party hard-liners—punctuated the national dialogue in early winter 1981. In the immediate aftermath of the military coup in December, the thousands of Solidarność activists who found themselves in prison possessed a new perspective from which to view the erstwhile policy struggles between so-called moderates and so-called radicals in the movement. From Białołęka prison in 1982, Michnik began to place less emphasis upon the merit of "young radicals from the large factories" in favor of a newly emerging sense of the strategic struggle the movement had endured. He assessed the politics of Solidarność in the final months before the State of War as an argument between those for whom strikes had become counterproductive and those who sought a national general strike to settle matters once and for all. Michnik commented: "It is difficult to know which group was more numerous, but it is certain that the more radical group could be heard more loudly. It was they—

mostly young workers from large factories—who urged radical solutions on Solidarność's leadership, and these were increasingly difficult to block (even though both Wałęsa and Kuroń tried to do so)."[26]

A subtle irony hovered over this shift in portraying the "young radicals in the large factories." Before the State of War, they had been pictured favorably as a healthy counterpoint to leaders like Wałęsa who were susceptible to cooptation by government negotiators and influential priests bent on toning things down; after martial law, they retrospectively appeared talking so "loudly" that their strategic drive toward confrontation had been "difficult to block." The irony was not that Michnik was in the process of changing his own strategic assessment—which, as shall become evident, he was not—but that he was in the process of altering the culturally narrow and thus politically partisan context in which his views were put forward. In the very first weeks after the unexpected appearance of Solidarność in 1980, Michnik, in keeping with everything he had written in the 1970s, reminded one and all of "the political boundaries that must not be crossed," geopolitical boundaries that had "properly been respected in the Gdańsk Accords" and which were rooted in the realities of "Budapest burning and Soviet tanks in the streets of Prague." A year later, this strategic consideration remained at the center of Michnik's analysis: "The task that faces the Polish nation is to work out a plan for a realistic system of political democracy, even while consciously restraining ourselves in order not to impinge on the state interests of our powerful neighbor." For Michnik, the conclusion was crystal clear: "I therefore declare myself for a compromise solution. Such is the demand of the moment. No reasonable person can promote a general confrontation today."[27]

These words described, of course, the strategic politics of Lech Wałęsa before, during, and after the appearance of Solidarność. Michnik's assessment emphatically did not applaud the politics of self-styled militants who regarded themselves as free of "toning down" influences. Indeed, if Wałęsa, in his multiple experiences as a fireman, had proved he was the most articulate extemporaneous advocate of strategic poise in the movement, Michnik was the movement's most articulate advocate of the same politics in carefully composed literary arguments. Michnik demonstrated this talent and this conviction once again in a general assessment of Solidarność written from Białołęka prison in 1982:

> The mighty and spontaneous social movement . . . did not possess a clear vision of specific goals or a well-defined concept of coexistence with the communist regime. It allowed itself to be provoked into fights over minor issues . . . Solidarność knew how to strike but not how to be patient; it knew how to attack but not how to retreat; it had general ideas but not a program for short-term actions. It was a colossus with legs of steel and hands of clay: it was powerful among factory crews but powerless at the negotiating table. Across the table sat a partner who could not be truthful, run an economy, or keep its word—who could not but do but one thing: break up social solidarity. This partner had mastered that art in thirty-seven years of rule . . . the Polish communist system was a colossus with legs of clay and hands of steel.[28]

One can only guess at what Lech Wałęsa must have thought as he read such words, not only in the first weeks after the Polish August, but in the crisis-filled days a year later and during the long darkness of martial law that followed. At the very time Michnik called for strategic poise in September and October 1980, his colleagues in the democratic opposition had launched a sustained struggle to diminish Wałęsa's authority within the new national structure of Solidarność and even within the regional structure in Gdańsk. Throughout 1981, as Wałęsa maneuvered with skill but also with increasing weariness against a simple-minded confrontational "fundamentalism" within the ranks of the movement, Michnik eloquently called upon the movement to exercise prudence. But he also joined in conventional KOR criticism of Wałęsa for his "moderation."[29]

The awkward truth about the strategic political stance of members of the KOR milieu in the era of Solidarność was that though they deplored the unthinking impetuosity of so-called "fundamentalists" who sought a confrontation with the state, their constant carping about Wałęsa's alleged moderation directly fostered a kind of rhetorical militancy that encouraged the ranks of the fundamentalists to grow. The conflagrations at the base of Polish society were first and foremost a response to the arrogant politics of the party, but KOR's constant second-guessing—by Modzelewski and others—had the effect of helping to ignite some of the very fires which Wałęsa, Kuroń, and many others were forced to try to extinguish. After the fact in 1982, Michnik belatedly began to grasp this dynamic at the heart of internal movement politics.

There were ironies everywhere. But surely the most striking concerned the respective relationships of Michnik and Wałęsa to the Catholic Church. It was a constituency whose strategic importance both men fully recognized. From 1976 onwards, Michnik pushed himself to extreme lengths to put the best possible face on the actions of the episcopate. Even when a kind of naked self-aggrandizement could be read into the Church's actions (as when high Catholic functionaries attacked Kuroń as an alleged "extremist" bent on misleading the movement), Michnik wrote, perhaps more in hope than in deep conviction, that the Church's pronouncement was caused by a certain "imprecision" in the episcopate's "diagnosis of the present situation." And lest this gentle admonition be taken as too harsh, he added: "But nothing can change the fact that the Catholic Church is a great asset for the Poles." Indeed, Michnik went on to pen a summary assessment of the Church that went beyond anything Wałęsa himself was ever quoted as saying: "And not only because churches serve as headquarters for committees aiding victims of repression, or because chaplains speak up on behalf of the wronged and the persecuted; not only because church buildings ring with the words of a free Polish literature and the sounds of Polish music, or because their walls are adorned with the works of Polish painters; and not only because the Church has become an asylum for an independent Polish culture. The Church is the most important institution in Poland because it teaches all of us that we may bow only before God."[30]

A great political truth about the dynamics of contemporary Western culture is here bared to public view. In their assessment of Poland's international position and its domestic social structure, Adam Michnik and Lech Wałęsa from 1976 onwards consistently occupied the same terrain. The understanding that both men possessed of the power of a Brezhnev-led Soviet Union insulated them against romantic, KPN-like fantasies about immediate "independence"; both understood the strategic relevance of the Catholic Church to the construction of a democratic movement in Poland. Yet, so powerful were the cultural assumptions that coursed through the Polish intelligentsia that Wałęsa was consistently attacked for espousing views that Michnik also held; and Michnik participated in the attack! What is evident here, then, is not a matter of fundamental differences in strategic considerations; there were none. Rather, what surfaces with powerful clarity is the impact of social class upon political perspective. Michnik's gestures toward the episcopate could be "understood" by intellectuals because Michnik was "one of us." Allowances could be made, Michnik was serving a higher strategic purpose. The same could not be imagined in Wałęsa's case, of course, because, as a worker, his capacity for independent thought was presumed to be circumscribed. Therefore, any association on his part with the Church (or with Church-connected intellectuals like, for example, Tadeusz Mazowiecki) constituted prima facie evidence of his "cooptation" by an episcopate determined to "tone things down." Two judgments may be offered about this habit of social analysis: the judgment was politically primitive—that is, ungrounded in realpolitik—and it was a naked, though unconscious, expression of class prejudice.[31]

Here was the Achilles heel of the Polish intelligentsia. It was one that Bronisław Geremek had forcibly brought home to him in the Lenin Shipyard in Gdańsk in the Polish August and one he actively transcended throughout the life of Solidarność. From his intimate vantage point in the Lenin Shipyard, with the courier war swirling about him, with the MKS daily enlarging in affiliated factories and taking on strength and resolve from the dynamics of the struggle itself, with an experienced Strike Presidium demonstrating an intimate knowledge of both the party police and the party's rhetorical tactics, and finally, with the constant attention being paid to the worker rank and file in nightly "vespers" by the chairman of the Strike Presidium, amidst all this evidence, Bronisław Geremek achieved the cultural leap necessary to grasp the interior dynamics of the workers' movement. With the same relative speed, though not quite with the intuitive strategic grasp that Geremek demonstrated, Tadeusz Mazowiecki made a similar leap. From their favored vantage point at the center of the strike scene itself, they saw the movement in ways imprisoned KOR activists could not. Geremek and Mazowiecki had no subsequent transformation to undergo because they accomplished their own private transition between August 23 and August 31, 1980 (see Chapter 1, pp. 29–35 and Chapter 5, pp. 218–22).

Adam Michnik needed four years to make the same journey. There were signs that in 1982 he had begun to reassess both Wałęsa and the explicit political meaning contained in the euphemism "moderate." The final leg of the journey came in 1984–85 after Michnik's release from prison.

As steadfast as he had remained throughout thirty-two months of imprisonment from 1982 to 1984, Michnik was unprepared for the depth of the popular resistance he found upon his return to civil society. It "exceeded not just my expectations but even my dreams," he wrote happily. Everywhere a thoughtful and vibrant underground press flourished along with seminars in private homes that surpassed in scope and frequency anything the flying University had ever achieved. He found "the people of Solidarność were wise, determined, ready for a long struggle." While Michnik also saw "sadness, exhaustion, ugly shrewdness, revolting mendacity, crafty operators pretending to be heroes, and Judases wearing the cloaks of political realists," these excrescenses were restricted to the fringes of society. Michnik found much that cast the politics of Lech Wałęsa in a new light and caused him to embark on a redefinition of political categories.

Michnik retrospectively found the source of discord inside Solidarność determined by the extent of the social crisis itself: "Acute tensions give birth to strong temptations. Deep humiliation may spawn proposals for extremely radical solutions. Lack of easy answers and clear prospects is conducive of demagogic bidding contests . . . We all saw this explosive mixture developing during the last three months of 1981 . . . The anarchization of daily life, consciously pursued by the authorities as they prepared for the military coup, made people susceptible to even the worst nonsense." Here, Michnik drew a bead on some very prominent people: "Those were golden times for certain suspect characters out to make a career." Under such conditions, old terms of political description lost meaning: "When I remember those weeks, in which everything was heading for the worst, when I try to reconstruct the atmosphere of illusion and the conflicts within Solidarność, it seems to me that the crucial problem did not concern the division between the 'radicals' and the 'moderates' or the dispute between Wałęsa and Gwiazda . . . Did this division correspond to the traditional division between the right and the left? I believed so then. Today, I think otherwise." Michnik then provided an analysis to undergird this shift. "The 'fundamentalists' say: no compromise. Talking about compromise, dialogue, or understanding demobilizes public opinion, pulls the wool over the eyes of the public, spreads illusion. Wałęsa's declarations about readiness for dialogue were often severely criticized from this point of view. I do not share the fundamentalist point of view."[32]

This strategic conception produced new tactical appraisals. "When Wałęsa declares the need for compromise, he unmasks the intention of the authorities; when the same thing is being said by a neorealist who avoids mentioning the word *Solidarność* like the plague, he sends word to the authorities of his own readiness to take part in murdering our union . . The TKK [the Solidarność underground coordinating committee] and Wałęsa are doing everything in their power to make dialogue possible."[33]

No more than other Warsaw intellectuals did Adam Michnik comprehend the long evolution of the Polish working class from 1945 to 1980, and he continued consistently to misinterpret the politics of 1956, 1968, 1970, 1976, and 1980 as a result; but he no longer saw a strategically crucial (though unfortunately low-consciousness) abstract entity called "the working class." Rather, he saw a complex and variegated social formation that contained vital forces with which he himself wanted to ally. Like Geremek and Mazowiecki before him, he saw a movement in which he might be able to play a supportive role, not one crying for his hands-on leadership. He was a notable addition to the interior of the movement, for he brought a keen long-range sense of possibility, and he was a man of moral courage who additionally had genuine literary gifts.

The night of December 13, 1981, also signaled the beginnings of a fundamental alteration in the life of Lech Wałęsa. Nothing he had experienced in his career of activism—not the lonely years of insurgency in the 1970s, not the tense days of the Polish August, not the subsequent war to deflect the constant grandstand quarterbacking of KOR enthusiasts, not even the exhausting rounds of fire fighting throughout 1981—could quite prepare him for the new tests to which he was subjected during martial law.

If the party in thirty-five years had not learned to function with a minimum of civil competence, it had emphatically learned the art of intimidation and the darker modes of animal cunning that accompanied that governing style. From the party's viewpoint, the politics surrounding Wałęsa's incarceration turned on the need to discredit him as a symbol of the movement's independence. Anything that would alienate him from other militants or from the population as a whole would prove useful. The tactics employed were threats, "reasonable" propositions, appeals to his sense of loyalty to the nation or to the cause of nonviolence, or veiled hints about future benefits or difficulties for his wife and children—or combinations of them all. The jailers of the military occupation judged Wałęsa to be a gregarious type who thrived on the give and take of political repartee. They isolated him in a house outside Warsaw for two months of intense interrogations (an effort that produced nothing useful for the regime) and then dispatched him to a room thirteen feet square in a remote hunting lodge at Arlamow near the Soviet border. Outside his windows, he could see nothing but forest; inside, he had only his guards for company. They kept a camera on his room, monitoring his every movement. The pressure was as intense as possible; he had neither privacy nor serious conversation.

He did, however, receive numerous political propositions, some from representatives of Jaruzelski (such as Mieczysław Rakowski, whom Wałęsa chased out of his room) and others dispatched by Cardinal Glemp. Notable in the latter category was Glemp's deeply reactionary personal representative, Father Alojzy Orszulik. Wałęsa turned everybody down, saying he would break his silence only when martial law was ended and all the nation's political prisoners were released. But in November 1982, Wałęsa wrote a letter to General Jaruzelski proposing the reopening of a national dialogue. The letter itself was unexceptional, a repeat of statements he had made throughout the era of legal Solidarność. Oddly, however, he signed it "Corporal Wałęsa." Small as it was, this slip was the best thing the strategists of the party had been able to extract from him in eleven months of cajolery and pressure, so they decided to try to capitalize upon it. Wałęsa was released from prison—even as thousands of other internees were retained. The effort failed. A huge sign, "Welcome Home, Leszek" greeted him outside his apartment, along with thousands of well-wishers. Wałęsa remained, as he had been since the Polish August, a symbol of independent Poland.[34]

He soon demonstrated that he retained his old skills from the surveillance days of the 1970s. Somehow, he shook off his round-the-clock comrades from the security service and pulled off a daring meeting with the Solidarność underground. He announced the details the next day—to the consternation of the Interior Ministry. Patiently, he set about the business of rebuilding all the old ties to the left, middle, and right in the working class, in the intelligentsia, and in the episcopate. He observed (with what response no one could say) the continuing though unsuccessful role of Father Orszulik as an attempted intermediary between the military regime and selected high-profile prisoners. This extended to the celebrated group known as the "Solidarność Eleven," comprised of an array of regional Solidarność leaders plus Kuroń, Michnik, Lipski, Wujec, and Lityński of the old KOR opposition. Michnik spoke for them all when he wrote: "Since I have not authorized anyone to act as mediator in my dealings with my jailers, since as a prisoner I do not feel entitled to participate in any negotiations, since my liberty cannot be the object of any deals, since I want to be put on trial in order to prove my innocence, I have declined to take part in these talks."[35]

A penchant for aggressive blundering became quite prominent in the pastorate of Cardinal Glemp, particularly his continued use of Orszulik. During martial law, the Solidarność underground announced a nationwide general strike for November 10, 1982. It turned out to be a disappointing failure. The untimely announcement on November 8 that Jaruzelski and the Vatican had reached an agreement for a papal visit to Poland did not help the cause. It was hard to boycott a regime with which the Church was making public deals. The papal visit itself—the following June—attracted international press attention. There was, however, another wrinkle to the story when the Pope returned to Rome. An editorial in the Vatican newspaper, *L'Osservatore Romano*, suggested that in the name of "national reconciliation" Lech Wałęsa should step aside from Polish politics! Again came an indignant public outcry; again, a flurry of denials and tactical retreats, including the dismissal of the paper's editor who wrote the editorial. But careful observers noted the similarity between the general tone of the editorial and the Pope's public statements during his visit to Poland. From the most embittered precincts of Solidarność came the muttered question, "How many new churches is Wałęsa's head worth?"[36]

Such a judgment, of course, focused solely on the immediate political effects of the odd diplomacy displayed by both the Polish Pope and the Polish Primate. The Church's long-term meaning as an independent institution surviving in a police state continued as a fact of Polish national life—whatever its leaders did. But the movement was forced to learn once again—as it had on August 26, 1980, when Cardinal Wyszyński offered his ill-timed sermon at Jasna Góra—that the Polish Catholic Church was a most uncertain ally, particularly in moments of high crisis. Under martial law, as during the Polish August itself, the movement had to rely on its own efforts. It was in this sobering sense that Poles recalled Wałęsa's words at the emotional funeral for Father Popiełuszko. With Poland's Primate standing beside him, Wałęsa had praised Popiełuszko as a "worker priest . . . and leader of the national cause." A parish priest, not the Primate, was "leader of the cause."

In 1985, the military regime tried another tactic to divide the leadership of the movement and, in the process, lessen Wałęsa's standing. The Solidarność Eleven had been amnestied just before their scheduled trial in 1984, and the following February, Wałęsa arranged a unique conclave to bring together four strands of the movement. To meet with him in Gdańsk, Wałęsa invited Władysław Frasyniuk, regional Solidarność chairman from Wrocław and noted "radical"; Bogdan Lis, one of the few party members to play an early and prominent role in the birth of the Baltic movement; and Adam Michnik from KOR. Among many topics, they discussed the possibility of calling a fifteen-minute warning strike—an illegal activity under new Polish law. Unfortunately for all concerned, the security services found a way to monitor their conversation, and everyone was arrested. But Wałęsa, significantly, was released. The other three were promptly indicted. The subsequent trial, aptly described by Michnik as "a classic example of police banditry," was notable for its interesting array of constraints. No foreign journalists or observers were allowed; the accused were permitted to give no testimony, even in answer to questions, beyond the words "yes" and "no."

Nevertheless, some high politics took place under the noses of the regime. When Wałęsa was brought in to testify, Michnik asked to be allowed to make a statement in his presence. This, of course, was denied, but when Wałęsa was being escorted from the courtroom, Michnik shouted after him, "Don't worry Lech, Solidarność will win in the end." Under these rather dramatic circumstances, the rapprochement between Wałęsa and a prominent member of KOR was finally achieved—formally and publicly. Frasyniuk was sentenced to three and a half years, Michnik to three, and Lis to two and a half years. Wałęsa went free. Yet against all odds, the experience brought the four men closer together.[37]

Meanwhile, Mikhail Gorbachev came to power in the Soviet Union, and this fact gradually and increasingly began to alter the strategic dynamics governing the functioning of political power in Poland. Within the more flexible circles of the party, some kind of accord between the government and society now seemed possible in ways that went beyond the narrow constraints of the Brezhnev era. In 1986, the government declared a limited amnesty, releasing Michnik, Lis, and Frasyniuk, and soon followed with a general amnesty accompanied by projections of impending economic and political reforms. The right of prominent Solidarność leaders to meet publicly was announced by the government, tested by the opposition, and verified.

These developments forced upon Solidarność an agonizing reexamination of its purposes and internal organization, one that lasted from 1986 to 1988 in a climate of continued economic disarray presided over by a virtually helpless party apparatus. The party's tentative introduction of selected market mechanisms succeeded principally in demonstrating to the population that members in good standing in the nomenklatura could enrich themselves through forms of state-chartered monopolies in a manner that had more in common with the early days of post-feudal mercantilism than with the structural invigoration of a modern industrial society. Living standards continued to decline even as trumpeted "rounds" of "reform" limped along aimlessly.

Meanwhile, within the ranks of the opposition, personal animosities and unresolved tactical and ideological differences that had been discernible but not controlling in the fifteen months of the movement's legality all resurfaced with enhanced intensity. Though disagreements, major and minor, had as many dimensions as might be expected in a society subjected to the searing social strains that characterized the period from 1980 to 1988, these tensions contained class and ideological components that can be summarized with a certain measure of economy.

As an organized expression of civil society, Solidarność had comprised a coalition of white-collar and blue-collar elements There were, unsurprisingly, divergent ideological tendencies within each sector, previously obscured by the vaguely social democratic thrust natural in a movement grounded in large-scale industrial organization. In 1980–81, this experiential reality armed trade unionists generally and Lech Wałęsa specifically with the organizational influence to withstand repeated efforts by intellectuals to supervise the movement. This interior struggle, alluded to in Chapter 7 for 1980–81, continued for the first three years of martial law. By 1984–85, it had subsided into a kind of resigned acceptance by Warsaw intellectuals to "put up with" Wałęsa, at least for the foreseeable future.

However, it is necessary to sketch in the broad outlines of this evolution, for it provides essential background for the transformed politics of 1989–90. In the original euphoric aftermath of the Polish August, KOR intellectuals had been a bit overconfident in their ability to seize cultural authority within what at that time remained organizationally a coastal movement (see Chapter 7).

Their abortive efforts on the coast in the first week of September 1980 and, more tellingly, in the September 17 meeting that shaped the national structure of Solidarność, convinced activist intellectuals that they would have to work through other working-class spokesmen if Wałęsa were to be successfully reined in. Initially, Andrzej Gwiazda, the engineer activist from Elmor in Gdańsk, seemed a likely prospect for this role. "Insider" analyses by KOR intellectuals in Warsaw—disseminated through Western correspondents to the world—painted a generalized portrait of a foot-dragging, "moderate" Wałęsa overly responsive to the Church, arrayed against more resolute forces implicitly (and, in some versions, explicitly) associated with Gwiazda. The interpretation necessarily had to be generalized and became increasingly so during 1981, because KOR intellectuals found themselves strategically in agreement with Wałęsa's approach far more frequently than they did with Gwiazda's. The latter's response increasingly struck close observers as characterized more by implacable obstinacy than by any discernible strategic pattern.[38]

In an effort to explain this awkward contradiction, two sets of descriptive terminologies appeared in interpretations of Solidarność. The tension within

Solidarność in 1981 was routinely portrayed as a contest between "moderates" associated with Wałęsa and "radicals" associated with Gwiazda, but also as one between provincial "fundamentalists" who wanted instant structural change and sophisticated "pragmatists" who understood the geopolitical realities embedded in Soviet military power. These descriptive formulations had the logical merit of explaining why KOR intellectuals, as thoughtful practitioners of realpolitik, so often found themselves in strategic agreement with Wałęsa, while the same terminology preserved room for (nonspecific) criticisms of Wałęsa's "moderation."

The strategic trajectory of this interpretation was clearly visible in the writings of Adam Michnik, but it also served as a conceptual thread in Lipski's definitive history of KOR, first published in 1983, as well as in such weather vanes of Warsaw intellectual opinion as *Kultura* in Paris, the *Uncensored Polish News Bulletin* based in London, and, more indirectly, the periodic advisories of Radio Free Europe. These descriptive formulations were also a centerpiece of the vast tide of articles and books on Solidarność published around the world.

On the surface, it would appear that such a harmonious thread of description could not be simultaneously employed by Western conservatives, socialists and latter-day Euro-communists, and by journalists and scholars, Polish and non-Polish alike. The explanation, though rather obvious, was one that was not readily sanctioned in most sectors of elite culture. For Western conservatives and liberals, as for members of the *Szlachta* intelligentsia in Poland, the necessity for leadership by the intelligentsia could be taken as a cultural given. If nothing else, the idiosyncratic manner and complex but often blunt speaking style of Wałęsa seemed to verify the continuing relevance of this ongoing cultural supposition.

Similar dynamics informed the analytical tendencies of participants in the Western anti-capitalist theoretical tradition. For socialists of various persuasions, nothing was more prosaic and effortlessly understandable than the need for intellectuals, particularly social-democratic intellectuals of the KOR persuasion, to function as the strategic brains of the popular movement. Though most such Westerners, after 1956 in Poland and Hungary and 1968 in Czechoslovakia, had come to have the most profound reservations about Leninist governance, so that the very meaning of the word "socialism" became increasingly problematic in meaning when employed by Western radicals, their political thought processes remained unconsciously Leninist in unreflected upon ways: *of course*, an intellectual vanguard had to lead a "mass movement."

The hierarchical components embedded in these traditions—conservative and radical, "capitalist" and "socialist"—thus contained an instructive symmetry that coalesced to inform the global interpretation of the interior of the Polish movement. This phenomenon easily incorporated various religious as well as ideological traditions. More than any of the participants seemed to sense, or could find the energy to consider, members of the *Szlachta* intelligentsia in Warsaw and abroad, overwhelmingly Catholic, shared many cultural affinities with radical agnostics in London and Paris, with Jewish liberals in New York, or with middle-class Protestant intellectuals worldwide. All had conceptually hierarchical assumptions about "civil society" that they unconsciously imposed upon any analysis of Solidarność. Thus, the continuing independence of worker leadership in Solidarność, verified by the high political visibility of Lech Wałęsa, constituted an analytical challenge that could not be coherently addressed within prevailing categories of analysis. Accordingly, most Poland watchers avoided this interpretive task altogether. This expedient was further encouraged by the fact that not much research has been devoted to the interior of the movement in the first place.

After the midnight sweep of Jaruzelski's military and police forces served to verify the common sense of Wałęsa's "moderation" and KOR's "pragmatism," the utility of Andrzej Gwiazda as a symbolic figure faded rapidly. As Solidarność's activists viewed matters from prison in the early months of 1982, Gwiazda's "resolute" strategy of late 1981 seemed even more strategically ill-conceived than it appeared to thoughtful Poles at the time.

But so ingrained were class-influenced cultural assumptions about politics that the drive of intellectuals to gain control of the popular movement continued through the early years of martial law. After Gwiazda's self-destruction, the new nominee for trade-union leadership was Zbigniew Bujak, head of Solidarność at the Ursus tractor factory and later of the Warsaw region. Bujak was a much more serviceable choice. Originally encouraged to political activism by KOR's Jan Lityński and Adam Michnik (KOR generated far more contacts in Warsaw than it did in Radom or anywhere else), Bujak seemed both more thoughtful and more malleable than the unpredictable Wałęsa. Bujak ably served the popular movement in 1980–81, steadily gaining stature in the Solidarność National Coordinating Commission and (even more relevant in the climate of martial law), he was the most prominent trade-union activist to escape the sweep of the security services on the night of December 12–13. If he could somehow be sustained amidst police surveillance, prolonged service by Bujak as the most prominent figure in the Solidarność underground might well equip him for a role in national leadership when the dark times of repression finally ended.

Once again, the international information network radiating from Warsaw intellectual circles put a visible "spin" on Polish politics that emphasized Bujak's actual and symbolic role. The most authoritative and frequently recurring source on martial law in Poland was the well-written *Uncensored Polish News Bulletin* from London, which provided a brilliant penetration of the government's blanket of disinformation and gave full attention to Bujak's activities. While Bujak was described in straightforward terms as a consistent symbol of continuing popular resistance, which, of course, he was, Wałęsa was handled much more distantly and elliptically.

Unfortunately for this scenario, social possibilities in martial-law Poland appeared very different from an embattled perspective underground than from the artificial security of prison. The latter sometimes encouraged romantic dreams of insurrection. To the surprise of many, both in Poland and in the West, Jacek Kuroń in February 1982 wrote from prison an open letter calling for the formation of a highly centralized Solidarność underground to provide coherent leadership for what he foresaw as a forthcoming "mass insurrection." To Kuroń, a popular rising seemed inevitable given the persisting authoritarianism of the regime. Entitling his missive "A Way Out of the Impasse," Kuroń departed from his own well-argued theories about social openness as a necessary component of a democratically functioning civil society: "we must organize ourselves—different from before August 1980—organize ourselves around a main center and display absolute discipline before it."[39]

From the underground, Bujak penned an open letter in rebuttal, one that not only drew on the lived experience of Solidarność as open debate rather than "discipline," but also reflected a much more accurate reading of social possibility under conditions of military repression. Thus, as it developed, Kuroń, not Bujak, was out of step with Wałęsa. Three months after the Kuroń-Bujak exchange, a defiant but peaceful demonstration in Warsaw resounded with cries of "We want Lech, not Wojciech." Simply enough, Wałęsa continued as the movement's symbol. In August 1983, Bujak offered his support to the call by the Gdańsk Solidarność underground for a one-week work slowdown if the regime failed to open negotiations with Wałęsa. In such a climate, KOR intellectuals gradually came to terms with the necessity of living with Wałęsa. Michnik penned his self-effacing reappraisals from Białołęka prison and, upon his release, publicly associated himself with Wałęsa. The inappropriate terminological distinctions between "moderate," "radical," and "pragmatic" were quietly laid away—to reappear in a new form in 1990.

In the era of martial law, two developments—one structural and one personal—seemed to have clear significance, at least in the short run. There was also a third development, far less tangible in its political implications, that conceivably carried more lasting meaning. Structurally, the most important result of the 1980s was the abject inability of the government to obliterate Solidarność as an ongoing political reality. Though the movement suffered through abortive general strikes, the periodic arrests of prominent underground leaders (eventually including Bujak), the murder of Father Popiełuszko, the beatings and deaths of relatives of Solidarność regional leaders, and the calculated arrogance and disinformation of the party's propaganda apparatus, the idea of independent civil society could not be expunged in Poland. The fact was a remarkable testament to the strength of Polish culture and the enduring capacities of the Polish people. The joy that Adam Michnik expressed at discovering this fact upon his release from prison was but a particularly well-expressed reaction shared in by Solidarność activists generally. Psychologically, if not in all other ways, Solidarność defeated the Polish party in the martial-law years.

A second phenomenon was the enduring political artfulness of the electromotor mechanic from Gdańsk. More than any other Solidarność figure, Wałęsa was singled out for a full treatment of government threats, bribes, innuendoes, deceptions, and a calculated "wearing-down" process that included the public pretense that he no longer "existed" politically. Criticisms of his actions, from any quarter of foreign and domestic opinion, were also seized upon and adroitly disseminated. All to no avail. Even foreign Poland watchers thoroughly influenced by the KOR information network were forced to concede, as one of them grudgingly confided to the author, "The man is unbelievably sure-footed. He takes chances, he makes a minor mistake or two, but he just hasn't made a single major one. It is amazing." It was also clear, however, that Wałęsa, despite some progress in this area, had by no means brought fully under personal control his capacity for self-celebration. It was a trait he shared with many of the world's politicians; the candidness of his style, however, was, to say the least, unique.

Less discernible than the movement's persistence or Wałęsa's continuing political relevance was the status of Solidarność's underlying capacity for democratic cohesion. The long years of economic privation and political humiliation were a terrible strain on popular morale. The fact was totally understandable. Indeed, the wonder was that Poles stood up as well as they did. Nevertheless, successful popular democratic politics necessarily requires—to remain democratic—enduring popular patience. It also requires generosity toward others—of the kind skilled industrial workers had demonstrated in 1980–81 in forgoing pay raises for themselves to ensure increases for less-skilled workers living on the edge of survival. The long agony of martial law put almost unbearable pressure upon this social ethos—upon the very idea of egalitarian generosity.

In such a manner, the long era of party authoritarianism wore itself out in 1988. Poland, tired as its people were, stood on the threshhold of great events.

9 ▪ The Re-emergence of Civil Society

It is important to distinguish the various ways in which mass parties were created. In the case of socialist parties, which gave the original impetus, the party was very largely an extention of an existing mass movement into the sphere of electoral politics whereas conservative and liberal parties, which already had a strong representation in parliament and government, created their mass organization from above, under the control of parliamentary leaders.

Tom Bottomore, *Political Sociology*

Mazowiecki chose not to put the fear of God into the two million nomenklatura of the old regime, but to incorporate them into his government. By doing this, he won their loyalty and kept the government from paralysis, but he never explained to the public what he was doing.

Martin Król, editor in chief, *Res Publica*

THE APPEARANCE IN 1989 of widespread democratic institutional forms in the Soviet sphere emerged out of a central precondition: the structural breakdown of the Leninist system of production. That these institutional forms first appeared where they did—in a shipyard on the Baltic coast of Poland in 1980—was the product of thirty-five years of effort that collectively produced for workers the specific instruments of social self-organization: the occupation strike capped by an interfactory strike committee possessing the strategic goal of achieving protected public space independent of the ruling party.

Among the resigned peoples of the East European nations, battered as they were by failed efforts at reform in 1953, 1956, 1968, and 1970 in East Germany, Hungary, Czechoslovakia, and Poland, the emergence of Solidarność quickened the hopes of millions of people. Though they could not know precisely how Solidarność came into being and thus could not emulate it, the very existence of a popularly based political institution outside party control gave heart to the weary in many lands. People everywhere watched for fifteen months until the Polish party, under pressure from the Brezhnev Politburo in Moscow, declared martial law and put the country under military rule. For a time, hopes sagged and resignation returned to Eastern Europe. But too much had happened in Poland—specifically, too much lived democratic experience—for social life to return to the status quo ante. Solidarność continued—in a thousand underground publications, in nationwide networks of cooperating activists, and in new civic habits tested in struggle. With the party bereft of any capacity for serious structural innovation, the economic crisis deepened with each passing year.

But the crisis in production was not national, it pervaded the Eastern bloc and was nowhere more pronounced than in the Soviet Union itself. Specific economic data do not exist—to this author's satisfaction at least—to date with precision the time period when the technological and human gears to the Leninist engine began to grind down. The material evidence is abundant that economic growth was real in the immediate postwar years and for a time thereafter, but signs of social strain and production malfunction were visible by the 1960s, took the form of an ignored crisis in the Brezhnev decade of the 1970s, and had become the object of covert rethinking and tentative replanning by the Soviet economic intelligentsia by the time martial law was declared in Poland in 1981. As the Soviet economy acquired a posture of stasis through the first half of the 1980s, market-influenced ideas increasingly became a part of planning scenarios in Soviet research institutes. The fact testified to the changing climate of which Mikhail Gorbachev emerged. By 1986–87, it had become clear to ruling elites in the satellite parties that Gorbachev was unlikely to countenance Brezhnev-like crackdowns as had occurred in Poland at the beginning of the decade because the Soviet leader was bent on transcending "Leninist norms," not defending them. By 1988, the signs were unmistakable that the Soviet leader was actively encouraging new departures in the relationship of the party to society.

But in only one country in the Soviet sphere did "society" have an organized presence—in Poland, where Solidarność had proved over seven years of martial law that it could not be broken by the police truncheon. In the spring of 1989, the Polish party, humiliated by the Solidarność experience, held in contempt by the army officers who had been forced to rescue it, and shorn of anything resembling programmatic cohesion, sat on the sidelines as General Jaruzelski sanctioned a high-level national "Round Table" to restructure the government and invited Solidarność in from the cold to bring legitimacy to the initiative. By early April, forces associated with Solidarność, "legal" once again, were deeply involved in election negotiations.

These developments promptly moved Hungarians in escalating sequence from private wish to cooperative thought and then to public action. In an effort to stay abreast of steadily rising political assertion by the population, the reforming Hungarian party took down its border fences in May; Solidarność won smashing electoral victories in June; and in July, East Germans began flooding through Poland to Hungary and then across the border to the West. In August, the candidate of Solidarność became premier of Poland; in September, sectors of the East German intelligentsia fashioned for themselves an independent political presence called "New Forum"; and in November, the Berlin Wall came down. At the end of the month, a democratic movement took life in Czechoslovakia, and at year's end, Romania exploded.

By mid-1990, a great deal of historical evidence had materialized from a number of societies in different stages of evolution from a Leninist form to something else. For millions of people spread over a vast territory, long suppressed hopes had given birth to new ways of thinking about social possibility. Indeed, in no other brief period in human history had such a wide variety of comparative evidence been generated through which to interpret the social meaning of the word "consciousness" and to relate that meaning to tangible political acts. In ways that heretofore had been impossible, numbers of people in all walks of life throughout the Soviet sphere became actively engaged in the political life of their region of the world.

At first glance, these tumultuous events seemed to contain a certain holistic unity—the appearance of "freedom" out of the wreckage of Leninist governance. Actually, however, wide gaps persisted in the social formations that had suddenly appeared in various East European societies. A sober investigation of the interior of these sundry civic expressions revealed conclusively that they had attained only a very early stage of movement building, scarcely reaching the level that Poles had begun to transcend in the 1950s at Cegielski. Through 1989 and the early months of 1990, self-organization remained at a very primitive stage of development restricted to the very apex of society where such grouplets as "New Forum" and "Civic Forum" emerged in East Germany and Czechoslovakia. Nothing approaching anything that might be called organized "civil society" yet existed, leaving the tiny grouplets that had emerged vulnerable to all manner of external events or internal anxieties.

Civil society did not exist because the popular presence in these countries was structurally formless; no interior lines of democratic communication were in place because there were no grass-roots organizations to which such lines could be connected. People wanted democracy but had not organized themselves to produce popular democratic forms.

It is precisely in this area of political thought that modern democratic theory is so weak and thus provides such thin guidance. To desperately aspiring populations possessing no structures of self-organization through which their aspirations could be expressed, it was not at all clear what people should do besides "oppose the party." Specifically, it was not clear what concrete political acts people needed to perform to create such popular institutions. The Solidarność achievement in Poland was not understood and thus could not be reproduced.

It was 1956 in Hungary all over again—with the decisive difference that the threat of Soviet invasion no longer existed. Thus, whatever new forms emerged could limp along until overwhelmed by the inevitable social crisis. What held the new East European societies together in 1990—what connected Czech and Slovak workers to Vaclav Havel, for example—was a single negative emotion and a single positive one: anger at the party and deep respect for Havel personally for his courageous opposition activities over the years. While these emotions might provide temporary cohesion, in the long run people had to eat. The transformation from the command economy to some form of market economy would inevitably produce fissures that the narrow forms of existing social dialogue could not possibly bridge. Civil society was an idea, actually not much more than an emotion; it did not, in fact, structurally exist in Eastern Europe—outside Poland.

Exposed here was the specific political anomaly that blanketed Eastern

Europe. The transforming democratic dynamics which characterized the 1980–81 period in Poland have been duplicated nowhere else in the Soviet sphere. Solidarność constructed an organized civil society on a scale of democratic participation rare in history. The infrastructure of the Polish movement, with thousands of elected local, regional, and national officials, imparted an added measure of structural integrity to the multiple forms of local democratic activity that emerged.

But even in Poland, where democratic political prospects were most firmly grounded, the events of 1989–90 illustrated the extreme hazards facing the emergence of workable democratic forms in Leninist states. As Gorbachev himself has been forced to learn in the Soviet Union, a party apparatus, left in place, possesses an intricate capacity to obstruct structural change. In Poland, the tepid pace of party-generated "reform" from 1986 to 1988 verified the final collapse of its martial-law and post–martial-law politics. The convincing verdict on these reforms came not from Gorbachev but from two rounds of industrial strikes once again centered in the Lenin Shipyard, that shook Poland in May and August 1988. These events proved—one last time—that the party could not govern without some sort of agreement with society. The Jaruzelski regime finally accepted this truth during the second Gdańsk strike and opened discussions with Solidarność. In symbolically appropriate Polish fashion, the exact date of the original signing of the Gdańsk Accords (August 31) was selected as the occasion in 1988 when an appropriate representative of the jailers—the interior minister, General Czesław Kiszczak—and an appropriate representative of the jailees—Wałęsa—met in formal negotiations. But the ensuing seven-month route to a political breakthrough revealed rather starkly the strategic alterations that time, widespread suffering, Gorbachevian politics, and martial law had inflicted upon the social fabric of both Solidarność and the Polish party.

An analysis of these interior dynamics sets out clearly and ominously the circumference within which conflict will characterize social life in Poland in the 1990s. The rhythm of Polish politics subsequent to the "Round Table" of 1989 provides instructive additional insight into the kinds of social evidence that intrudes most tellingly into human consciousness so that it shapes the dimensions and tone of public life. In essence, the Solidarność experience continued in 1990, as in 1980, as an extraordinary laboratory of large-scale popular politics in which democratic forms materialize and alter under great pressures.

Though the Kiszczak–Wałęsa negotiation in August 1988 was designed to lead to broader discussions, such did not immediately occur. Neither side was ready. Indeed, so deep were worker grievances that Wałęsa had a difficult time persuading angry young shipyard workers to suspend their strike so that the negotiation with Kiszczak could even begin. In a clear signal that another generation had arrived in the Polish working class, workers accused Wałęsa of being "soft" on the government and having, in fact, "sold out." Deeply affronted, Wałęsa pointed to his record as a strike leader in 1970 and 1980 and complained, "These young strikers are impatient with experience. They have no respect for us veterans."[1]

The workers were not alone. For different reasons, of course, party hard-liners were also unenthusiastic about any new opening to civil society. The behind-the-scenes struggle within the government became publicly visible in the divergent interpretations of competing party organs. It was not until December 1988, when the Warsaw party daily, *Trybuna Ludu*, gave prominent display to Solidarność organizing initiatives, that evidence became conclusive that the party was, in fact, finally altering its historic defense of its structural prerogatives.

But the form of these Solidarność initiatives also signaled a significant change within the popular movement. Wałęsa issued invitations to 128 oppositionists and on December 18 convened the gathering in Warsaw. The group announced it had formed itself into a "Solidarność Citizens Committee" ready to serve as a negotiating body with the regime. Though the party press demonstrated through its coverage the extent of its approval, protests were immediate from a variety of sectors of Polish society. The structure of the citizens committee revealed that the long, largely unseen battle between intellectuals and workers for control of Solidarność, one in which workers had prevailed throughout the era of the union's legal existence, had now taken a sharp turn. The citizens committee was overwhelmingly dominated by intellectuals in general and by Warsaw intellectuals in particular. Though the list of invitees carried Wałęsa's personal stamp of approval, the specific shape of the gathering had been determined by Henryk Wujec, a longtime KOR activist and a physicist with wide connections at Warsaw University. The Polish Academy of Sciences was well represented while the committee was quite thin in workers, farmers, women, and Poles of any description from outside the capital city. The prototypical participant in the citizens committee was a male, Warsaw-based professor. The justly famed broad base of Solidarność had shrunk to an elite stratum. Wałęsa's old "fundamentalist" rivals were predictably up in arms, as were KOR's critics in the middle class, centered in nationalist groups like KPN. But disappointment extended far beyond such circles to include veteran movement activists who felt slighted and whole cities and even regions of Poland that were transparently underrepresented.

Under such diverse pressures, Wujec was forced to expand his list. Eventually, over a hundred additional people were added to speak for civil society at the forthcoming "Round Table." But the initially skewered shape of the representation was not materially altered. Of the final membership of 232, no less than 195 were identifiable as members of the intelligentsia while only 26 had structural ties to the working class and 11 to agricultural interests, broadly defined. Twenty-six were women, most of whom were also of the intelligentsia. There was some overlap; of the eleven representatives of agricultural interests, four had college degrees as agricultural engineers, agricultural economists, or agricultural instructors. Similarly, some representatives from such places as Silesia and the Lenin and Szczecin shipyards were Solidarność attorneys or members of the technical intelligentsia specializing in mining and shipbuilding. While there were no fishermen, an economist attached to a coastal fishing institute was named to the Round Table. In keeping with the strategic posture of Cardinal Glemp, the Church was not directly represented through the episcopate; its interests, however, were protected by prominent Catholic laymen affiliated with Catholic publications and organizations.[2]

In such fashion, the working-class sinew of Solidarność disappeared, a feat that the combined propaganda of the party and the realities of martial law had failed to accomplish.

The obvious question, of course, concerned the role of Lech Wałęsa in all this. Clearly, the hero of the Polish August had not suddenly and publicly decided to transform his self-conceptualization of his own class identity or his sense of obligation as the functioning symbol of a movement grounded in and created by working-class assertion. What, then, had changed?

The answer is as complex as Polish culture itself. A bit more specifically, the answer is as intricate as the internal politics of Polish trade unionism, as tortured as the geopolitical fragility of Poland's proximity to the Soviet Union and Germany, as haunted as the *Szlachta*'s aggrandizing self-perception, as stressed as the political history of Solidarność's existence as the organizational embodiment of all civil society, and finally, as person-specific as the character of Lech Wałęsa himself.

Part of the explanation was fairly straightforward and turned on political analysis available to everyone in Poland. Looking back over the history of Solidarność, Wałęsa had every reason to feel vindicated—as many intellectuals were freely willing to concede—in his strategic conduct of high movement politics with the party, the army, the Church, the intelligentsia, the forces of Rural Solidarność, and the broad reaches of the Polish working class. Throughout all the crises of 1980–81, his twin guiding principles had been, first, the geopolitical imperative of negotiation rather than ultimate confrontation and, second (as a tactical component of the first), a persistent resolve to keep Solidarność as internally strong as possible by treating no movement opponent as a permanent enemy unless events forced him to acknowledge such permanence. Such worker spokesmen as Gwiazda, Jurczyk, Bujak, and Frasyniuk had all opposed Wałęsa on more than one occasion, and all possessed some degree of personal ambition of a kind organic to politics itself. But the latter two remained flexible and relational while the first two nursed angers apparently too deep to permit them to do so. Bujak and Frasyniuk passed Wałęsa's personal muster test; Gwiazda and Jurczyk did not.

Similarly, Wałęsa had found the strident nationalism of KPN partisans to be the organic fuel of fundamentalism; such a concretized political stance effectively destroyed a great deal of the tactical maneuvering room available to a movement desperately searching for operating space. Participants in the nationalist right wing, however, "egalitarian" or "socialist" a few of them professed to be, had a track record of shooting themselves in the foot. The forthcoming struggle with the party could not bear the added burden of all those who, in Jan Józef Lipski's phrase, instinctively rummaged through "the antique shop of Polish patriotic phraseology." As with Gwiazda and Jurczyk, so with KPN and the more outspoken zealots of Young Poland.

But this was as far as "straightforward political analysis" could carry any case for Wałęsa's defense. None of these stipulations could be broadly applied to the majority of the members of the Solidarność National Coordinating Commission, duly elected in the October congress of 1981. In this group—far more appropriately than for a unilaterally constructed "citizens committee"—reposed the representative democratic core of the Solidarność leadership. But despite initial pleas, soon escalating to demands, that he convene the Solidarność National Coordinating Commission, Wałęsa did not do so.

The explanation seems to lie in two absolutely paradoxical facts—Wałęsa's sense of his own commanding position as a spokesman for Polish workers, and his experiential awareness of the long-term vulnerability of any person saddled with precisely that political responsibility.

Both were grounded in recent history. During the fifteen months of legal Solidarność and in the seven years of martial law that followed, events had proved Wałęsa's Cassandra-like critics to be wrong. In 1980–81, there had never been a sound *democratic* strategy other than Wałęsa's policy of dignified but persistent negotiation. The only alternative, armed struggle, failed all tests of common sense. Similarly, during martial law, events had verified the logic of the movement's posture of maintaining a willingness to negotiate a new "national accord" while continuing to practice nonviolent resistance to the authoritarian state.

As for his relations with Polish intellectuals during the same time span, virtually every sector of the intelligentsia had fulsomely demonstrated its willingness to offer persistent advice, and almost every sector had at one time or another found something in Wałęsa's leadership to criticize. But there were decisive gradations of behavior. The original shipyard advisors—Mazowiecki, Geremek, Wielowieski, Kuczyński, Cywiński and Kowalik—had been steadfast in support, even when offering advice not taken. And even the most proprietary members of KOR ultimately came to terms with Wałęsa's continuing utility to the cause of Polish democracy. Moreover, there was too much intellectual and political talent and too much international influence in shaping attitudes about the meaning of events in Poland in the ranks of KOR and its associates for that sector to be safely ignored. Though the bonding between KOR and Wałęsa was in no sense intimate, a *modus vivendi* had been achieved on terms that, in the end, no longer involved an unending series of debilitating innuendos. Henryk Wujec, or someone like him similarly well connected to the nation's intelligentsia, could therefore draw up the list for the citizens committee. Recent history supported this conclusion also.

But the most telling events had come in 1988, which underscored Wałęsa's continuing political skill, but also the ephemerality of public reputations in contemporary Poland. Two occurrences in particular stood out. On the plus side was Wałęsa's artful triumph in a nationally televised debate in November with the overconfident head of the official trade unions, Politburo member Alfred Miodowicz. Wałęsa's public humiliation of the cliché-quoting party apparatchik seemed to have given the government a much-needed final push into serious preparations for the Round Table. But this memory had to be balanced against the constant surfacing of working-class anxiety about declining living standards, an emotion highly visible in the hostile worker reaction to Wałęsa's return-to-work pleas in August. The economy was in shambles and social volatility promised to persist for the foreseeable future.

The combination of these events seemed to suggest that while Wałęsa's personal popularity remained high, it was subject to instant collapse in specific situations. On the bread-and-butter issues that increasingly weighed on everyone's mind in Poland, with no quick remedies remotely possible, it seemed easier in the short run for Wałęsa to parley with intellectuals than with workers.

Such dynamics were mutually reinforcing both for intellectuals and for the electrician from Gdańsk. From Wałęsa's standpoint, it was clear that individual intellectuals could no longer threaten his power base within the working class itself. He had won that struggle. And for the intellectuals, it was a positive relief to leave the handling of worker discontent to Wałęsa. Since the Polish August, activist intellectuals had been seared by those flames more than once; "fire fighting" could be happily consigned to Lecho, and good luck to him! This stance was fine with Wałęsa; his own bargaining position with respect to the Polish intelligentsia would be stronger if everyone perceived that he alone among leading activists could sell the new accord—whatever it might be—to the working population. Though little of this reasoning became a topic for street-corner conversations in Warsaw in the spring of 1989, the structure of the Round Table inexorably proceeded from the realpolitik it contained.

There was, of course, one other matter of realpolitik that this solution blandly ignored: it was not remotely democratic. The function of the citizens committee effectively severed the internal lines of communication between the movement's elected leadership and its rank and file, lines that ran through the Solidarność National Coordinating Commission. As constructed by the autumn Solidarność National Congress of 1981, the national commission itself was structurally democratic, being composed of representatives of Poland's regions, democratically sanctioned by delegates elected by the local units in each region. In contrast, the citizens committee had been created out of thin air by a self-appointed elite.

This comparison underscores a somber development in 1989 in Poland. Collectively, the architects of the citizens committee suffered from a particular sociopolitical disease, rampant in societies everywhere, that is the organic fuel of hierarchy: a well-developed capacity for condescension. The architects of the citizens committee "knew better" what was needed for Poland, they told themselves. The Solidarność National Coordinating Commission had too many divergent working-class voices, too many rivalries, to function effectively in a necessarily tense and subtle negotiation with the party. The activist intellectuals who would dominate the citizens committee could do it better. Within the socially narrow confines in which this argument was made, it had great appeal. The intellectuals would tidy up the democratic loose ends later, they told themselves.

Historically, such is the all-weather rationale for would-be democrats on those occasions when they succumb to the lure of proffered rewards that come from behaving like mandarins. Had the citizens committee been created out of a political vacuum, as Civic Forum subsequently was in Czechoslovakia, it would have constituted a halting but understandable first step toward the fashioning of democratic forms. But as an instrument calculated to bypass a huge and experientially tested popular structure capped by a democratically constructed national commission, it was politically indefensible. The entire exercise leading to the formation of the citizens committee, one that extended over a seven-week period, was a systemic departure from the Solidarność heritage. Indeed, it was a betrayal of the movement's basic democratic tenets. In effectively bypassing an arduously constructed democratic polity, the elite committee represented the final triumph of the often attempted, always thwarted bid of Warsaw intellectuals for control of the popular movement. As such, it provided impressive evidence of the continuing cultural vitality of the class presumptions embedded in the *Szlachta* heritage.

This could not be said for Wałęsa, of course. His problem was his own empowered sense of self. In participating in a process that cut off the popularly constructed connecting links to his basic constituency in the Polish industrial working class, Wałęsa doubtlessly was relying on his personal ability to appeal over the heads of other officials or other institutions directly to the Polish population. As would soon become clear, there were certain burdens attached to this strategy, if it was, in fact, a conscious strategy. One deals here in the netherworld of human motive, a territory historians can never really know, regardless of the evidence they unearth. It is quite possible that Wałęsa, remembering the strident voices on the national commission calling for "confrontation" with the party in November–December 1981, had simply lost confidence in that body. If so, that is something democrats can never do—and remain democratic. One lives within the polity, or one does not live democratically at all. It is a political principle that dates at least from Socrates.

The ultimate social price inherent in these malfunctions was not immediately evident. On the contrary, things began to move with accelerating speed. The Round Table began on February 6, 1989, and ended in an agreement on April 5 for June elections. Under the terms of the accord, the party would have 65 percent of the seats reserved for its representatives in the old Sejm, but a new Senate would be created as a second chapter, and all of its one hundred seats would be contested in open elections.

As formulated, the agreement was a milestone in the political evolution of Leninist societies, and institutionalized formal participation of nonparty society in pluralistic politics at the level of the state. This achievement levered democratic forces in Hungary into motion in May, and the Hungarian border fences to the West soon came down. Following Solidarność's smashing election victories in June, East Germans began pouring through Hungary to the West in July. Energized by the breakthroughs in Poland, the East European revolutions were under way.

It is helpful at this juncture to draw a clear distinction between the creation and experiential testing of democratic forms on the one hand and the achievement of highly visible political openings on the other. In midsummer 1989, the societies of Eastern Europe began opening windows in the suffocating barracks of Leninism. Every one of these achievements, whether it was with a pickax on the Berlin wall, with a Prague street dance celebrating the election of an artist-statesman to national leadership, or through the evolving self-destruction of the Hungarian party apparatus, could be treasured as an authentically significant political event. But while these exhilarating moments stood as hopeful portents of possibilities to come, none of them in themselves touched upon the creation of self-generated democratic institutions of popular politics. Not only did all such labor still lie in the future, but the particular manner in which it was performed would shape in decisive ways the kind of political and economic society that would eventually emerge in each East European nation. The political openings offered the promise of new realms of liberty; but only the creation of authentic popular democratic structures could ensure that all sectors of society would have a voice in helping determine the specific directions of national policy.

For most of the decade of the 1980s, Solidarność Poland stood as a high-profile example that afforded edifying instruction for every aspiring democrat in the Soviet sphere. Not only did the Gdańsk Accords crack state censorship and open up social space in which individual Poles could find their subjective voices, but the specific organizational form that Solidarność created to occupy this social space encouraged far-ranging democratic communication and orderly speculation within society as a whole. It was a dialogue that ultimately produced the Action Program of 1981. "Freedom" and "democracy" can be mutually supporting, but they are words that do not describe the same thing. The more widespread the scope of democratic forms, the greater the freedom for democratic thought and action. The most important achievement of Solidarność in its fifteen months of legal existence was that it offered the world pioneering evidence in not one but both of these realms.

Viewed from this perspective, Solidarność generated three telling periods of political innovation. The first, of course, was in the Polish August, culminating in the creation of independent social space occupied by the new trade union. The second period extended from September to December 1980, during which time some ten million Poles partook of the personal democratic experience of joining and acting within that public sphere. The idea of free space, thus, was *experienced*. From each individual Pole, this experiential rock was a most essential and personally verifying cornerstone of subsequent democratic activity. It was an especially necessary social experience in a society whose people had been militarily prodded into forms of systematic deference.

In this second period from September to December 1980, Solidarność verified itself as an institution that afforded the Polish people protected social space. The protection was not only explicitly physical, it was psychological. One could now experience and experiment in creating ever-expanding democratic structures.

The third period of Solidarność, lasting through 1981 to martial law, saw the creation and employment of the vast democratic infrastructure of the movement, staffed by forty thousand people who were paid out of local dues and responsible to the people who paid them. Out of this self-generated world came the Network and its evolving plan for self-management, the Action Program, and the hundreds of local assertions through which self-organized communities endeavored to teach democratic conduct to each other and, collectively, to an authoritarian party apparatus.

It is only with these sequentially related democratic evolutions in mind that the surprising tone of Warsaw intellectual society in the summer of 1989 can be made comprehensible. What became visible was a new social fact that dramatically underscored the varieties of historical memory that can materialize in modern societies. Simply stated, very few activist intellectuals in Warsaw elected to understand the Solidarność experience in the context here outlined. Far from internalizing the 1980–81 era as a period of remarkable democratic achievement expressing itself in broad-based popular politics, or martial law as a time when that idea managed to survive systematic government repression, the underlying drives of Warsaw intellectuals were expressed both publicly and privately in ready acceptance, and even relief, that such a period was over. An entirely new trajectory of interpretation appeared, grounded both in observable political developments and in a certain partisan reading of those developments.

The cornerstone of the emerging philosophy was a demonstrable social fact—namely, what appeared to be the relatively slow pace of recruiting by the revived Solidarność. As of mid-1989, the relegalized independent trade union had two million participants while once there had been ten million. When foreign visitors asked individual intellectuals for an overview of the current situation, this circumstance was inevitably focused upon. It provided the first plateau of a three-tiered rationale. Since Solidarność "no longer had the appeal" it once had, the new focus was on organizing local "citizens committees" across Poland to work with the new national group that staffed the Round Table. The summary conclusion of this reasoning was that the political situation in Poland had "changed." The confidence and élan with which this litany appeared left no doubt that the change was intended to be understood as for the better.[3]

Interestingly, and despite this interpretation, Solidarność membership was once again concentrated in the industrial working class, particularly in the largest factories that possessed a commanding position in the national productive system. This fact imparted to Solidarność a structural basis that corresponded to its original thrust. Unlike 1980, however, this hard core had not yet expanded in 1989. Among the absent were precisely those white-collar social formations that Warsaw intellectuals now announced they wished to recruit to "citizens committees." In this manner, overt class politics became visible in 1989 with an interest group struggle as to who would be most likely to pay the necessary price for economic reform—Poland's white-collar class or its working class.

Considered in the absence of rhetorical pretense by one group of Polish partisans or another, there was nothing particularly surprising in these interpretive predilections. They merely verified that the class contradictions always present in Solidarność were now organizationally visible in two separate institutional formulations, a tiny but politically influential citizens committee and a large-scale organization of two million members, largely industrial workers.

What was new, however, was the public posture of the intellectuals who dominated the first group. They presented themselves as speaking for the whole of civil society, while at the same time they endeavored to emphasize the diminished political relevance of the organized sector of civil society—Solidarność itself. Also instructive was the clear absence of any sense of loss felt by intellectuals at what was characterized as the (significantly) slow organizational rebirth of Solidarność as a large-scale popular institution.[4] The underlying political assumption, never stated but quite implicit, was that while workers were worthy allies in a struggle against the state, they might well act as an encumbrance in the new climate of receding party authority.

It is necessary to be as explicit as possible as to the social and political nuances embedded in this strategically significant evolution at the end of the decade. Regardless of the specific organizational forms Poles created in the transition from a command economy to some sort of market alternative, class conflict would be an absolutely inevitable ingredient. Since Poland was not exempt from the social dynamics of industrialization prevailing elsewhere in the world, there was nothing new in this prospect. However, what was new in the emerging class politics of 1989 were the outlines of future social tension that these organizational forms foreshadowed. For the first time since the Polish August, favorable characterizations of Wałęsa by Warsaw intellectuals were not highly qualified with a variety of carping criticisms. His "artistry" and his "genius" in explaining complex matters of high politics to ordinary Poles, particularly at moments of acute tension, were now touted without qualification. That is, this talent was touted as a legitimating extension of the citizens committee itself or, more specifically, by midsummer 1989 as a legitimating extension of the intelligentsia-dominated parliamentary party.

The opposition deputies constituted a structural evolution of the citizens committee whom the Polish electorate had installed, in the name of Solidarność, in the Polish Sejm. Stories of Wałęsa's persuasive skills at worker gatherings now circulated in Warsaw intellectual circles with the same approval that had characterized similar appraisals from workers and members of the provincial intelligentsia in the tension-filled days of 1980–81. Wałęsa was now seen as being on the right track or, in some interpretations, as having made possible the emergence of a "sophisticated" or "workable" framework for oppositional activities.[5]

The flaw in this rosy portrait, however, was the obvious fact that Lech Wałęsa had not suddenly transformed himself into a card-carrying member of the Warsaw intelligentsia. The 1989 interpretation by intellectuals seemed to be grounded in the illusion that the erstwhile shipyard electrician did not know where his political power base lay. If Wałęsa were to be happily assigned the task of "taking care of the workers," he transparently could do so only by maintaining his own credibility as a worker spokesman. Given the pervasive anxieties of Polish workers about food costs and basic job security, this was more than an awesome task for one person; it was the greatest social threat to any post-Jaruzelski government. In the not-so-long run, this obvious reality was much more likely to lever Wałęsa into opposition to the Warsaw strategists than into a highly risky and ultimately self-destructive expenditure of his personal credit in their defense. The only strategic policy that could transcend this built-in contradiction was a miracle economic reform that simultaneously supplanted the command economy with market mechanisms that somehow did not produce mass unemployment and a systematic lowering of workers' already precarious standards of living. Unfortunately, if the outline of such a program existed, it was not making the rounds of the Warsaw intelligentsia in 1989.

In the context of working-class participation in the national politics of transition from Leninist governance, the ease and alacrity with which Warsaw intellectuals expressed their willingness to "leave the workers to Lech" was politically shortsighted. It also invited structural schism within civil society of a potentially volatile kind. Inferences that Solidarność had lost its appeal suggested that the Polish working class had suddenly disappeared as a political force. The contrary fact was that workers in Poland were more highly politicized than any working class in Europe.

In 1989, the political ingredients of precisely such schism were visible all over Poland, like dry kindling awaiting a timely economic match. The bypassing of the Solidarność National Coordinating Commission not only produced public outcries, but also pumped organizational energy into a dissident faction called the "Working Group of Solidarność," which had been originally created by the old fundamentalist faction within the trade-union leadership. Wałęsa's three rivals in the 1981 union election, Andrzej Gwiazda of Gdańsk, Marian Jurczyk of Szczecin, and Jan Rulewski of Bydgoszcz, were the most visible symbols of this grouplet in 1989, but its potential appeal was visible in the range of political tendencies with which it appeared tactically harmonious. This included not only a quasi-syndicalist strain, most discernible in the textile center of Łódź, but also old nationalist elements in the middle class associated with KPN. Temporarily at least, dissent could draw from both the right and the left. The Working Group of Solidarność was so affronted by the political maneuvering that produced the Round Table that it called for a national boycott of the June elections that the Round Table negotiations had produced. This was not a tactically compelling stance. It was hard for Poles to resist the opportunity to vote against the Jaruzelski government, even if in many working-class districts the vote necessarily had to be expressed in favor of a member of the intelligentsia rather than for a locally prominent trade-union leader. Additionally, the Working Group had no program other than a fiercely stated resolve to "defend the working class."

In any event, economic anxiety was real throughout the country. The volatility of the general social situation was underscored by the clumsy evolution of economic reform. Any sweeping alteration of the economy would necessarily be dislocating for vast numbers of people. At the very best, structural alterations might survive politically if they were painful but rapid or, alternately, if they were somehow slow and cushioned. Unfortunately, what Poland got in 1989 was the worst of both worlds—slow reform that was painful.

The party's ingrained habits of myopia temporarily came to the rescue. Though Jaruzelski's coterie had taken the necessary precautions in the Round

Table to ensure that the party would remain in parliamentary control whatever happened in the elections, it also announced as part of its rhetoric of "socialist renewal" that it would pay "due attention" to the outcome. But though the rejection of party candidates in contested election districts was breathtakingly uniform (Solidarność won 261 of 262 contested seats, and the other was captured by a free-enterprise businessman), Jaruzelski did not, in fact, "draw conclusions." Instead, he employed once again the time-tested party tactic of reshuffling government positions among the same people: Jaruzelski himself would move from party first secretary to president; Prime Minister Mieczysław Rakowski would become first secretary; and Interior Minister Czesław Kiszczak would become prime minister. It was *perestroika*—Polish style.

It did not sell. The regime's puppets in the satellite "Peasant" and "Democratic" parties, whose votes, after the party's crushing electoral defeats, were now mandatory for a simple government majority in the Sejm, decided to be wholly owned subsidiaries no longer. Enough of them voted against Kiszczak that his candidacy—the policeman as prime minister—failed. The party suddenly had a new crisis on its hands.

For a few days during this impasse, rumors circulated about a possible "grand coalition" of the party and Solidarność. But Wałęsa, working alone, had other ideas—a "little coalition" composed of Solidarność and the satellite parties. Wałęsa had played an active role in mobilizing the opposition to Kiszczak, and he went beyond this to work out Peasant and Democratic party support for the old Gdańsk advisor Tadeusz Mazowiecki as prime minister.

The announcement in the second week of August 1989 of this political coup not only stunned the party (and much of the world) but also caught the old KOR activists unprepared. It was not that they expected one of their own to become prime minister. Though they had staged a comeback, as the role of Wujec in the citizens committee verified, they no longer nursed illusions that one of their number could play a personally decisive role at the very apex of the government. But, had they a choice, they would have preferred Mazowiecki's close associate and fellow Gdańsk advisor Bronisław Geremek. Though he was not a part of the KOR group, Geremek also was not intimately connected to the Catholic hierarchy, as Mazowiecki was. Beyond this, the entire politics of the breakthrough had been yet one more act of solo entrepreneurship by Wałęsa. It stood as one more bit of proof that he could not be corralled. This discovery, or rediscovery as the case may be, was dismaying to intellectuals, even if the specific news about the shift in power was undeniably encouraging.[6]

The Polish nation now embarked on a course that would force it to explore and endure some of the more ironic rhythms of history. All across Eastern Europe, extending to the Soviet Union itself, aspiring reformers watched Poland with hope, trepidation, and, increasingly, genuine exhultation. The Round Table agreement in April not only energized activist sectors of the citizenry in Hungary but additionally emboldened reform elements in the Hungarian party. The stunning election victories by Solidarność in June galvanized individual East Germans to undertake arduous journeys through Hungary to the West and activated East German party reformers as well. There were many invigorating details in the stories emanating from Poland. During the Round Table, policemen sat down with scores of people whom they had sent to prison. The extraordinary vignettes that emerged from the ensuing "table talk" spread through Poland like wildfire—but also made their way across East European borders. Then in August, the chief jailer, General Jaruzelski himself, formally announced the ascendancy to prime minister of ex-prisoner Tadeusz Mazowiecki. In Czechoslovakia, the courageous circle of dissidents rallied by Vaclav Havel took due note. In East Germany, a hopeful band of democrats organized "New Forum" as a public expression of independent social aspiration. In Hungary, an overtly anti-Leninist faction within the party pushed itself, and was pushed by nonparty elements, into increasing confrontation with party conservatives. Gorbachev not only tolerated these departures but actively encouraged more than a few of them. Other Soviet politicians, most prominently Boris Yeltsin, drew even more ambitious conclusions. In all of these transformations, the causative role of Polish Solidarność was unmistakable.

The irony was the price the Poles had to pay for being first. The revolution in Poland, like most revolutions in history, had its own built-in constraints. In structural terms, some of them were rather severe. While the nonparty coalition possessed, at least for the moment, a parliamentary majority as well as its own prime minister, the old regime remained in control of the army, the security services, and the foreign ministry. Moreover, the far-flung party apparatus remained ensconced in ministries whose tentacles reached into every hamlet in the country. While some sort of "free market" initiatives were now clearly sanctioned, whatever specific new forms or proposed new forms materialized would to some undetermined extent have to percolate through the inherited bureaucracy. Though the dynamics were different, and relative power distinctly different, Mazowiecki thus faced many of the administrative hazards that were plaguing Gorbachev. As Mazowiecki gradually formed his governing team and as these newcomers labored to fashion related policies to transform a nation, it began to occur to some in Warsaw that change might come more slowly to Poland than to surrounding nations that had gotten a much later start.

To be sure, some changes gradually became real in 1990, though they were not quickly transformative. The złoty was stabilized against foreign currencies, and other financial preconditions were achieved for Poland's entrance into the world market. Goods became plentiful in stores, but only because prices had quickly become so high that many items were beyond the reach of the average family. Unemployment inexorably began to climb, to 200,000 and then to 500,000. Businessmen from Western Europe, Japan, and the United States visited in increasing numbers but frequently departed without being able to formalize new production or trade agreements. All quickly learned that Poland's industrial enterprises were not only huge but also inefficiently organized and technically obsolescent. Meanwhile, as the structure of party leadership that had prevailed for years and even decades fell apart and out of power in East Germany, Hungary, and Czechoslovakia, General Jaruzelski remained securely in his post in Poland, backed by a loyal army and a ubiquitous security apparatus. Given Solidarność's role in freeing Eastern Europe, it was all rather exasperating.

When the political realities were conjoined to the economic ones, the Mazowiecki regime did not seem to produce early results that were significantly different from those characteristic of the Jaruzelski era. General economic conditions continued to worsen, while at the political level, party apparatchiks were seen to be busily at work in the new age of reform.

But in a great many ways, the new government did achieve a revolution in public conduct. Throughout the 1980s, Solidarność had generated a remarkably sustained effort to produce public stenographic reports of its internal debates, even of contentious underground debates. The ministers of the new government followed in this tradition of openness. None had a more awkward task or measured up to it with more unrelenting candor that the new minister of labor, Jacek Kuroń. In his regular reports on national television, Kuroń spoke with the same combination of thoughtfulness and gruff bluntness that had characterized his relationship with the police during his years of oppositional activity, including his long years in prison. It is doubtful if a labor minister existed anywhere in the world who matched Kuroń in the somber clarity with which he spelled out his own understanding of the available options facing the nation.

Prime Minister Mazowiecki was responsible for all the social implications of high politics. While he performed in a manner that won the respect of most of the population—his personal integrity being his strongest quality—he faced an acute social problem that positively screamed for attention. A normative standard of equity pervaded the nation. No less reluctantly than anyone else did Poles anticipate the austerity that was built into the economic crisis. But if suffering were necessary, the general feeling was that the pain should be fairly distributed. While the inevitable class tensions were quite apparent, nothing was more galling than the continued public visibility of the party apparatus.

As Martin Król, the editor in chief of *Res Publica*, put it: "Mazowiecki chose not to put the fear of God into the one million nomenklatura of the old regime, but to incorporate them into his government. By doing this, he won their loyalty and kept the government from paralysis, but he never explained to the public what he was doing."[7]

This judgment was offered in midsummer 1990, after a fateful break between Wałęsa and Mazowiecki. But the seeds had been planted and had taken root much earlier—before Mazowiecki's accession to office. Indeed, they predated the appearance of Solidarność itself.

Poland is in class terms a deeply divided nation. Indeed, the tension between the intelligentsia and the working class, both broadly defined, has been evident throughout the era of Solidarność. While many specialists on Poland have taken due notice of the unifying functions of the Catholic Church, of the national history, and of Polish culture itself, there are limits to the social range over which these blankets may be spread. The specific approach of Warsaw intellectuals to the politics of the Round Table revealed both the depth of their participation in this organic class division and the cultural narrowness that seemed to blind them to the predictable consequences of such naked class aggrandizement.

As Król suggested, Mazowiecki might have done more to communicate to the nation the reasoning behind his accommodations to the party apparatus, but alleviation of the underlying social tensions facing his regime, or any Polish regime in the foreseeable future, was beyond the reach of ministerial verbal display. Especially given the structural depth of the economic crisis, Poland's political leadership clearly needed to have the benefit of the kind of dialogue with Polish workers that only a thickly textured internal communications network could provide. This extended beyond politically relevant individuals like Wałęsa and politically relevant structures of leadership such as the Solidarność National Coordinating Commission to include the entire panoply of regional, city, and enterprise representatives who provided, in 1980–81, the sinews of communication within the popular movement. The cause of democracy in Poland required the structural participation of the Polish democratic movement.

The currency for any such transaction would of necessity consist of some variety of cross-class equity in distributing the pain of economic austerity. And the politics necessary to make that currency negotiable would, also necessarily, involve endless conversation, debate, and contention between the political leadership and the organized social formation called Solidarność. Since the polity in Poland included an organized civil society, with all the (properly) democractic imperatives signaled by that structural fact, it is hard to see how the country possessed any other political route out of its economic dilemma. When the activist intelligentsia in Poland hid itself from this imperative, it lived in a dream world of *Szlachtan* dimension. The manageable social fissures that became publicly visible in 1988, widened a bit in the contentions over the Round Table in 1989 and widened a bit further in midsummer 1990, are destined to be viewed in retrospect as mere prologue. In terms of realpolitik, the workers of Solidarność have not suddenly disappeared—as the decade of the 1990s will surely verify.

What conclusions, then, can be drawn from the extraordinary time that will henceforth always live in Polish history as the decade of Solidarność? What summary interpretations can be offered concerning the emergence of the world's first majoritarian democratic social movement? What insights into "political consciousness" are embedded in the Solidarność experience?

The summary interpretive task is not an easy one—for a number of cultural and ideological reasons. It is perhaps useful to begin with a political commonplace, but to do so in order to emphasize a deceptive underlying rhythm it contains. All societies are governed by elites. In one sense, this reality is a tautology; even when nonelites gain power, the governing presumptions they attempt to codify into custom have the purpose of providing ethical protection for themselves as a privileged stratum. Indeed, what is generally understood to be culture itself is essentially an amalgam of ideas, values, customs, and enforcing mechanisms that serve to protect inherited forms of governance and the systems of privileges and rewards they enshrine. In this sense, the history of political thought itself can be understood as an elite cultural phenomenon, for it constitutes a body of philosophy that privileges the creation of social forms calculated to yield some manner of orderly continuity to social governance. And historically, the price of order has been the maintenance of hierarchy.

But the idea of democracy—the idea itself—is too open-ended and inclusive to permit its rigid confinement within a philosophical acceptance of permanent hierarchy. Elite forms distort social relations fundamentally because they coerce people into silence. The emotional price each generation of the living pays for its mannered participation in inherited rules of order not only ensures the reproduction of intense democratic yearnings in each new generation but provides ongoing vitality to philosophical speculation itself. In every society and in every generation, the idea of democracy is philosophically radical. What is called "tradition" helps provide political continuity and at least the pretense of social cohesion. But this result is achieved only at the cost of maintaining received habits of coercion. As Marx put it in one of his most quoted sentences, "the tradition of all the dead generations weighs like a nightmare on the brain of the living." Since social knowledge is experiential, incipient democrats routinely and unwittingly attempt to advance their cause by acting out acquired habits of hierarchical conduct.[8] They protest against hierarchy, but do so in hierarchical ways. In this sense, all advocates of what is called "change"—be they "reformers" or "revolutionaries"—live in imminent danger of reproducing a number of the hierarchical forms they purport to oppose, and in fact do oppose.

Nevertheless, it is in this terrain between idea and action that political ideology finds its home by providing an explanatory language of description that endeavors to connect the two. Conservative terminology sketches out a philosophical defense of hierarchy. The descriptive language of the radical tradition provides a rationale for any (allegedly temporary) hierarchical conduct that surfaces in movements dedicated to democratic objectives. In this manner, the ideologies of capitalism and socialism, and the languages of explanation they have generated, define the hierarchical societies that comprise the modern world.

To transcend these received traditions, a democratic philosophy necessarily must attempt to diminish the scope of prevailing contradictions between ideas and action by postulating democratic modes of individual and collective activity that can be performed in pursuit of democratic forms. It is necessary to say, however, that one does not find nearly as much helpful historical evidence to assist in this pursuit as is generally supposed to exist. Though prolonged study of social movements does indeed uncover moments of democratic activity, such moments have rarely endured for long, as evidenced by the actions of individual participants and by "movements" as a whole. For one thing, personal animosities within movements encourage the licensing of individual hierarchical behavior designed to outmaneuver rivals. For another, the dynamics of political struggle, especially struggle against powerful adversaries habituated to the use of a number of modes of cultural and physical repression, make the maintenance of communitarian behavior extremely difficult to sustain over time. Abused people can simply get too angry to behave democractically.

But the sustained study of social movements does uncover one recurring rhythm. The necessary predicate exists in the relationship of established systems of governance to the earlier revolutions that first brought them to power: the structural components of all existing regimes have their origins in the internal social relations and theories of politics at work in the insurgent movement that originally installed that system of governance. This relationship holds true whether "the revolution" happened last year or last century. This dynamic produces the following projection: it is unreasonable to expect any regime in power to behave in ways that are more democratic than were visible in the internal social relations within the movement that brought that regime to power. In twenty-five years of study of social movements, I have encountered no historical exceptions to this casual relationship. Rather, the historical evidence is compelling that revolutionary movements, once in power (whether for ten years or for two hundred), fashion modes of governance that over time become more hierarchical and less internally democratic. Only the appearance of another insurgent movement seems to offer the prospect of altering this pervasive historical tendency.[9]

In the modern world, sanctioned political terminology does not, of course, describe any regime's political evolution in so sober a manner as I have here suggested—that is, capitalist terminology does not so characterize capitalist evolution and socialist terminology does not so describe socialist evolution. Rather, the visible trajectory of governance is interpreted as "progress." The centralization of decision making in ever fewer hands—the evolution of elite forms—is described in terms of efficiency, not in terms of heightened centralization. However characterized, the evolution toward hierarchy has been a constant of modern life ever since industrialization began to alter mankind's social relations in transcendent ways some eight generations ago. While variously decentralized monarchies were gradually replaced by republics, the republics have gradually and steadily grown more centralized and their governing elites more remote from popular influence.

Because of its unprecedented democratic scope, Solidarność in Poland generated a veritable cornucopia of evidence about the opportunities and hazards of democratic politics, about the internal strains afflicting social movements, about the utility of a number of received democratic principles in providing a guide to transcending these strains, and about the descriptive precision of contemporary languages of politics to uncover some or all of these dynamics. Irrespective of what happens in Poland in the 1990s, Solidarność as it evolved in the 1980s will merit (and, one is convinced, will eventually receive) the most prolonged study.

The principal difficulty one encounters when confronting this Polish evidence, however, lies in the fact that it is so rich in both large-scale democratic conduct and intricate democratic tension that it outpaces normal standards of evaluation. Modern criteria for assessing society are, in democratic terms, very timid. The fact is essentially a function of lived experience. Since the presence of hierarchy is so ubiquitous in the modern world that only the uninformed or politically innocent can ignore it, it may be said that people are perceived to have become culturally sophisticated to the extent they have learned to live with this unpleasant truism.

No summary interpretations about the meaning of Solidarność can be offered with much confidence unless this modern habit of urbane resignation is clearly specified and consciously transcended.

What has become obscured is the actual price of modern "sophistication." To be politically graceful, one has to pass over the awkward social fact that standards of democratic evaluation throughout the West have been radically adjusted downward in the course of accommodation to the persistence of hierarchy and elite privilege in modern life. A stark cultural convention gradually became politically operative, though it was by no means polite or convivial to acknowledge its presence: public policy has been increasingly responsive to the asserted needs of an ever smaller number of people who have their hands on the most influential levers of concentrated economic power. In the modern world, a number of democratic forms remain highly visible—political parties, elected officials, a press free of overt censorship, as well as competing interest groups that seem to indicate the continuance of "pluralism."

But, in fact, the relevant political influence of civil society in all Western industrial nations has been systematically and progressively reduced over time. Indeed, in terms of decison-making influence on high national economic policy, civil society has virtually no political impact at all since it is populated only by a retreating labor movement and small sects who give voice to widespread but inchoate anxieties about accelerating environmental and economic strains or to long-unresolved cultural disparities relating to race and gender. Beyond this, civil society has over time become stripped of citizen participants to the point that it constitutes a social space that is virtually unoccupied. In such a culture, any reference to Jefferson's ancient conceptions about local-level "elementary republics" as the functioning base of democratic forms could scarcely be made by modern advocates of democracy because the references themselves turned on civic connections possessing such fragile contemporary context that they could only be understood as "quaint." This cultural folkway lays bare the operative dynamics of modern sophistication: contemporary standards of democratic evaluation are relatively low because, it seems, a "sober" reading of history reduces people to a reluctant (but presumably informed) acceptance of sharp limits upon their own public expectations. A cultural imperative, therefore, appears. It is not "wise"—that is, not sophisticated—to teach the young to aspire too grandly for too much democracy. The observable justifications of this organic rationalization are everywhere to be found in modern life. The citizenry is instructed, and in due course internalizes the thought, that democratic aspirations, unless quickly tempered, can easily be "unreasonable." Most such aspirations can be understood as "unrealistic."

Nothing illustrates the pervasive power of this modern cultural assumption more than the way Western observers of Poland, many of them quite self-consciously "progressive" in their democratic hopes, reacted to the discovery of

Solidarność's "Network" in 1981. So narrow has the range of contemporary imagination become in the West that the expansive aspirations of the Network could not be grasped. Though the political discourse in which Thomas Jefferson participated in the eighteenth century could have easily assimilated the self-management ideas of Network activists, these republican concepts, arrayed in Polish dress, simply struck sophisticated and therefore deferential twentieth-century reporters and scholars as "romantic." Sobering and even diminishing as this cultural evolution is, it pencils in the limits of contemporary political conversation. In urbane circles, serious democratic ideas are "utopian."

It is essential to specify, however, that this assumption, while inherently relative in its presumed or intended meaning, rapidly becomes absolute in its functioning cultural application. In daily life, it becomes less a warning about the dangers of excess than a convenient justification of prevailing habits of hierarchy. Most of all, in attaining the cultural authority it now possesses, this received tradition of political resignation has the practical effect of confining human imagination and truncating the range of aspiration itself. The end result is the sanctioning of standards of cultural evaluation that are so rationalized, so "reasonable," that they war against the very idea of egalitarian social relations. Thus, to a "reasonable" Westerner as well as to a "reasonable" party member in Poland, the aspirations of the Network were more than excessive; they were an offense to one's own politics because they challenged one's own resignation.

What the experience in Solidarność offers the world is compelling evidence that human capabilities in the sphere of democratic imagination and democratic performance can, on occasion, outpace sanctioned standards of "reasonable" expectation. Solidarność provides historically compelling evidence that people can perform more democratically than they have been told to expect they can—and more, accordingly, than they themselves believed they could. And they have done so over interestingly extended periods of time. At first blush, it might be expected that the evidence of democratic striving Solidarność presents might be seized by people across a weary world as a restorative and encouraging set of historical actions offering the possibility of satisfying emulation. That is to say, Solidarność might be expected to serve as a highly relevant laboratory of experience to be carefully studied with the purpose of extracting selected elements of useful social and political procedure.

Unfortunately, it is at precisely this juncture that received standards of evaluation intrude to set priorities on the "evidence" that should be studied. These standards, remote, culture-bound, and often theoretically abstract as they are, focus attention on sanctioned groups of historical actors (see Introduction). The same received standards consign people occupying the remainder of society to a descriptive wilderness where their actions can be vaguely suggested through abstract terminology rather than specifically reconstructed through sustained research. Ingrained habits of high culture speed this malfunction along its way. For example, it is much more culturally homogeneous—easier—for university-educated scholars and journalists to study the written words of university-educated Warsaw intellectuals than to unearth the complex actions over time of less accessible members of society who do not write about their political ideas because they are not trained to do so, but who are, in unobserved ways, acting out those ideas in social life. Easily enough, articles and books aimed at analyzing the historical causation underlying Solidarność focus on Warsaw intellectuals rather than on the Baltic sources from which the movement materialized.

In addition to these general strategic tendencies of research, certain enforcing dynamics of hierarchy have the tactical effect of shunting observers toward selected kinds of evidence and away from other kinds. The politics of protest in a police state—the activity that produced Solidarność—spews forth all sorts of variously relevant "evidences of insurgency." A sense of high drama, which in itself seems to impart an aura of vitality and centrality, surrounds revolutionary manifestos that can be unearthed after the fact and analyzed for their role in precipitating subsequent revolutionary events. In Poland, anyone searching for what is taken to be this kind of activating causal evidence may quickly be shown such literary efforts as the 1964 "Letter of the 34," the 1977 "Declaration of the Democratic Movement," or the 1979 "Charter of Workers' Rights." Whether published underground for dissemination to accessible Poles or abroad for the edification of world opinion, such political literature possesses, both for contemporary journalists and subsequent scholars, an obvious attraction. The documents themselves are grounded in familiar ideas clothed in a familiar idiom and, thus, can be rather effortlessly visualized as "social evidence." If no other kind of evidence is pursued, this material can routinely graduate to a comparatively high plateau of presumed relevance. The significant social fact here is that literary evidence is very easy to come by, a pleasant circumstance that serves to more than justify the brief research time invested in locating it. The manifesto writers, articulate and well connected as they are, prove simple to locate, and their documents are readily accessible to copying machines. These tactical conveyances rather effortlessly transform isolated pieces of raw evidence into fundamental strategic conclusions because, as a simple problem in literary organization, one can easily create from such free-floating evidence a seemingly logical narrative that "solves" the problem of causation: the manifestos "caused" the movement. What one would suppose to be the necessary connecting link between idea and action is not only evidentially missing, its nonappearance is not perceived as being germane because an abstract explanation will be sufficient to offer to a university-trained audience. Thus, a manifesto can be projected, without corroborating evidence, as having "raised the consciousness" of whoever it was who might later be seen carrying out the acts that comprised the revolution. It is precisely this fundamental causal leap that characterizes the interpretation of Solidarność propagated by specialists on Poland in all societies around the world.

For such historical explanations to be persuasive at the intellectual level, only one other requirement needs to be in place. Both author and audience need to participate fully in received cultural assumptions as to the presumed relationship of ideas about political protest and the tangible social acts necessary to movement building. Caught in the grip of the modern cultural habit of "viewing from afar," this conjunction of belief between author and audience is routinely present. In ways that journalists and scholars of Poland seem uniformly to avoid reflecting upon, there is a breathtaking simplicity to the causal connections unconsciously intuited to exist between the presence of an idea in one person's mind and the distant political acts by other people that seem to conform to the idea.

The literature on Solidarność repeatedly illustrates this simplicity. To employ only one of many examples, the English observer Stewart Steven explained that "Jacek Kuroń gave crash courses in politics and philosophy" (to Polish activists or would-be activists situated in unspecified locations), and as a direct result, "the consciousness" of something called "the workers" became "well-prepared." No matter that Kuroń did not give his "crash courses" to workers on the Baltic coast where the movement generated itself. No matter that he possessed no knowledge, experiential or theoretical, of the organization and political function of an "interfactory strike committee" as the critical instrument of the coastal movement. No matter that he advocated various forms of local-level worker councils, including infiltration of the official unions, to a national working class that over a thirty-year period had tested plant-sized councils in every conceivable form, within and outside of official unions, found them wanting, and had, accordingly, gone another route.[10]

The underlying reality, then, was elementary though culturally hard to credit. The structure of protest that produced Solidarność was grounded in organizational experiences the intellectuals of KOR had not had and, therefore, could not immediately grasp.

Nevertheless, the explanations provided Western journalists and scholars by Warsaw intellectuals as to how KOR engineered the dynamics of the Polish August have, in one incarnation or another, found their way into the literature of Solidarność so that how people throughout the world understand the origins of Solidarność is fundamentally out of touch with what actually happened.[11]

The claims for the intellectuals of the democratic opposition, and subsidiary supporting claims for Radio Free Europe, cannot survive sustained historical scrutiny—though they do have the appearance of political logic to the extent these claims reflect a widespread misapprehension shared by intellectuals the world over. This failure of perception concerns a central component of coherent popular politics and of the process of movement building that necessarily shapes such organized assertion—namely, the absolute centrality of thick lines of internal communication to the construction of large-scale democratic movements.

If history afforded more examples of such self-organized social formations, this relationship would be much more widely understood, but the task of political construction itself is so difficult that the appearance of instructive examples such as Solidarność are relatively rare. Confusion among intellectuals throughout the world is therefore endemic. This is a remarkable fact since the subject of "revolution"—how to have one and how to govern so as to prevent one—is such a central preoccupation in world politics. Since some kind of "movement" is organic not only to "revolution" but to the very idea of sustained popular political assertion itself, it is apparent that the area of confusion occurs at a very basic level of social and political understanding.

Indeed, the speculative problem about the components of revolution is so basic that it not only puzzled Marx throughout his life but eventually defeated him. Since his writings, accordingly, left a decisive political lacuna surrounding the question of how to make the Marxist revolution, the task was left to others. The solution formulated by Lenin affords a comparative framework for analyzing the Solidarność achievement. Four historical observations may be made about Lenin's resolution of the organizing problem: (1) the concept of the revolutionary vanguard party as the agent of the revolution proved politically workable—that is, the revolution succeeded in czarist Russia; (2) the internal structure of the resulting revolutionary movement was conceptually undemocratic; (3) the structure of the society that emerged after the revolution was necessarily structurally undemocratic; and (4) over time, this structure generated such massive social tension that it eventually ground itself to pieces.

One more dimension must be added to the revolutionary problematic. To answer the question "What are the actual internal components of revolutionary movements?" it is essential to inquire additionally into the components of *democratic* revolutionary movements. This is the elementary and potentially transformative historical question that can be brought to the historical phenomenon called Solidarność.

The challenge that faced the organizers of Solidarność incorporated precisely the political burden that history imposes on all people who feel they live in undemocratic societies. This burden is concentrated in a single, and awesome, problematic: how to build a structure of popular assertion that is strong enough to contest the existing power of the received culture while at the same time creating within that structure forms of social relations that are democratic. If the second cannot be done, the first is meaningless for, as I have suggested, the forms of social relations reproduced by the movement in power can only be the ones that were generated during the time of insurgency that brought it to power. The issue, then, can be simply framed by juxtaposing the two words necessary to describe both the process and the result—"organizing" and "democracy." Is it possible to organize democracy? What are the necessary sequential structural components of this process? In human terms, what is the structural relationship between the organizers and the people they organize? How is a structure of democratic accountability achieved?

Movements that postpone answering these questions "until later"—until "after we win" or "after the revolution"—doom themselves to undemocratic consequences since the structural forms through which these questions will be answered have already been put in place during the time the movement formed itself in insurgency. The limits of democratic practice in power are foreshadowed by the extent of democratic practice in insurgency. This generalization applies to all revolutions.[12]

The striking achievement of Solidarność is that it generated not only a mass of evidence that helps to illuminate these long misunderstood dynamics but also a considerable amount of social evidence useful to democratic aspirants far beyond the boundaries of Poland. The relationship of hierarchical and democratic forms is, of course, a dialectical one; the internal life of Solidarność is therefore a story of the tension between these two tendencies of social relations. The structural forms of assertion developed by the organizers of Solidarność in the course of building their movement against the Leninist state visibly defined the respective democratic and hierarchical components of the Polish movement.

Unfortunately, in the modern world, the status of democratic theory is so fragile and insecure, so underdeveloped, that contemporary observers possess very inexact and, indeed, murky criteria of evaluation to apply to the internal life of the Polish movement. The results are that the movement has been misread and that its more important democratic achievements have been passed over.

Though the fact is not widely understood, democratic movements can neither come into existence nor remain democratic in the absence of effective lines of internal communication. The first reason is defensive. The administrators of the received culture against which the movement must labor to build itself will, to say the least, not look with favor upon the effort. Indeed, upon discovery of the existence of the insurgency, the forces of orthodoxy will move sooner or later (sooner in Leninist states) to define the movement's participants in terms that bring immense cultural discredit upon them. Though rank-and-file members of social movements routinely regard themselves as advocates of social justice and democracy, the received culture defines them as highly unsanctioned renegades. Since established regimes everywhere have available elaborate instruments of mass communication, including police truncheons, through which to disseminate this message of social decertification, the movement will quickly discover that its potential recruits have suddenly become effectively immobilized. They shy away from performing the acts necessary to associate themselves with the movement—because of fear, because they have internalized the assessment propagated by the received culture, or because of some enervating combination of the two.

A movement, therefore, cannot passively allow the disparaging social and political communiqués emanating from the received culture to define its participants. The people of the movement must provide—constantly provide—a counterdefinition of reality in order to defend their morale, internal cohesion, and potential for movement growth. Given the power of received cultures, embracing economic and policing sanctions as well as social and cultural authority, an incipient democratic movement can never have "enough" internal mechanisms of communication to feel assured that its many adherents and potential adherents will not suddenly lose heart. The counteractions available to authority to encourage a resurgence of social fear and a loss of movement confidence are, of course, as numerous as the imagination of those in authority affords. The pervasiveness of this modern institutional reality helps make more understandable why so few large-scale democratic movements have appeared in history. Despite the depth of social anxiety historically visible across the globe over the past two hundred years of the industrial age, the organizing imperatives have simply been so huge that very few movements have been able to sustain themselves long enough to become politically significant.

But in addition to the obvious defensive need to protect the movement's élan by defending its sense of self, self-organized social formations also require interior lines of communication for an offensive purpose—to ensure that when the movement asserts itself against recalcitrant power, it does so with a structural coherence born of prior conversation among its diverse constituency. There must be a means for the movement's most experienced activists to convey their experiential knowledge to its least experienced rank-and-file member. On June 28, 1956, it was precisely because the Cegielski movement, which had developed elaborate internal communications in one enterprise, had not connected itself to the many hundreds of enterprises and thousands of workers who joined them in the streets of Poznań that the self-educated work force at Cegielski became the hostage and ultimately the victim of unrestrained rampages by sympathetic strangers who were not, in any structural sense, part of the movement. It was a lesson about the severe limits of plant-size independent councils that workers would learn at different times and at different speeds all over Poland in the years after 1956.

Finally, beyond reasons of defensive and offensive strategy, internal communications are necessary as a structural verification that the movement is what it says it is—a collective assertion for greater democracy in which the lines of conversation run both ways, not only downward from leaders to rank and file but also upward as a means of holding leaders accountable.

It is with these internal imperatives in mind that one can turn to a final evaluation of the dynamics that produced Solidarność. Since the Polish movement emerged from a strike of shipyard workers in Gdańsk, students of Poland have concluded that their task was to explain how workers managed to do this. Works by Timothy Garton Ash, Jerzy Holzer, Krzysztof Pomian, and Jan Józef Lipski established the general trajectory within which hundreds of books and articles on Solidarność have appeared.[13] With varying emphases, this literature collectively reflects perspectives actively propagated by what Lipski calls "the KOR milieu." A central KOR claim, implicit in many works, takes the following explicit form: "Jacek Kuroń directed the strike by long distance through Borusewicz on the coast." The role of KOR takes a slightly more elaborate form in a more complex version: "Information would come out of the shipyard to Kuroń, he'd tell the foreign press so it could get to Radio Free Europe, and they would broadcast it back to Poland. It was a perfect loop, with Kuroń as its nerve center."

None of this causality from afar can stand scrutiny, however moderated or qualified. The reasons turn on the realities of the internal mechanisms of communication within the Baltic movement. After the strike began on August 14, the authorities the next morning cut off all telephone and telex communications between Gdańsk and the rest of Poland. Kuroń did not learn of the formation of the Interfactory Strike Committee and the promulgation of the first Gdańsk demand, both developments of the sixteenth, until late in the evening of August 17 when Konrad Bieliński cut short his holiday and returned to

Warsaw to tell him. More important, the MKS structure was part of the history of coastal organizing and constituted an organizational strategy that contradicted Kuroń's strategy. Kuroń thought the first Gdańsk demand was unsound and actively worked to induce the strike presidium (of which Borusewicz was not a member) to back off from it. Moreover, Kuroń did not grasp the enormous organizational potential of the interfactory strike structure and was still trying on August 18 to ponder the implications of the twenty-one demands when the stunning 156-factory mobilization occurred.

Because he was so remote from the interior dynamics of the strike—as an ongoing practical development, but also as a theoretical construct pointing to the organization of civil society—Kuroń still did not understand the balance of forces at work on the coast after August 18. Though KOR publicly announced support for shipyard workers, Kuroń continued to try to induce them to retreat from the first demand. Since the interfactory organizing structure did not harmonize with Kuroń's previous theoretical speculations, he could not grasp the centrality of the workers' "preconditions," which focused on the courier war and the telephone blocade. The MKS had a complex message to deliver to Poland's worker activists: instructions on why and how factory work forces everywhere needed to meet, declare support of the MKS, and elect delegates to travel to Gdańsk to affiliate with it and add their weight to its negotiating posture with the government. Not being privy to the detailed listing by worker negotiators of the specific couriers who were being arrested and beaten in police stations, Kuroń thought MKS identification of "helpers" was a reference to him and other members of the Warsaw opposition.[14]

Essentially, Kuroń, in prison after August 20, remained unaware of the interior dynamics of the Baltic mobilization throughout the Polish August. This fact reflected his prior failure to understand the Baltic mobilization of 1970–71, the demands of December 1970 for independent trade unions, the demands of January 1971 for independent trade unions, and the pioneering development of the concept of an interfactory trade-union structure.

The understanding of political possibility that Jacek Kuroń brought to the Polish August, while much narrower than the strategic vision animating experienced coastal activists, was also experientially grounded. In 1976–77, Kuroń observed at very close range the power of the Leninist state to crush the efforts of the newly formed group called KOR to assist the worker victims of police repression following the risings at Radom and Ursus. KOR's efforts to generate free trade unions in the two cities ended in complete shambles. One union with one member appeared at Radom; nothing emerged at Ursus. What Kuroń saw was a state policing power so massive and one capable of generating a social fear so pervasive that the idea of independent trade unions seemed politically unworkable as an organizing strategy. Even before serious organizing efforts could get started, KOR activists and worker activists alike found themselves in prison. Kuroń and Lityński thereupon put aside the idea of free trade unions and thereafter opposed such initiatives when they were generated from below by workers in the provinces. As Kuroń's writings prior to 1980 and his responses during the Polish August confirmed, his strategic vision after Radom and Ursus was focused upon "reform groups within and close to the party" who, if the Gdańsk workers backed off in 1980, "would now have a better opportunity to make their voices heard." In the words of the 1979 Charter of Workers' Rights, free trade unions should be formed only "wherever there are strong groups of workers able to defend their representatives." Actually, the evidence seems conclusive that the 1977 repression had convinced Kuroń that prospects of achieving "strong groups of workers able to defend their representatives" was beyond reach. He therefore could not grasp the coastal MKS structure as a vehicle for such an achievement.[15]

While this miscalculation reflects a prior strategic conclusion, that conclusion had originally been based, in 1977, on an analysis of an organizational form. It did not occur to Kuroń that it was the particular form employed at Radom that failed, not necessarily the idea of free trade unions itself. Encased in a more elaborately protective organizational form, it might work. But it seems almost as if the moment he read the first Gdańsk demand in his Warsaw apartment, his brain clicked off. Kuroń never grasped the organizational structure surrounding the enormous mobilization of August 18. Outside the coast itself, no one else in Poland (or abroad) understood the MKS structure either. In the aggregate, the organizational meaning of the Interfactory Strike Committee constituted an experiential reality that had to be seen to be understood—as the academic advisors around Geremek and Mazowiecki saw and understood. Even this was by no means an automatic or effortless discovery. The strike structure had been institutionally consolidated for four days when, on the night of August 22, the academic advisors first arrived in the shipyard, but it was not until August 25 that the advisory group was finally persuaded by the unanimous resolve of the Strike Presidium to put aside the ameliorative Variant "B" as a strategic goal.[16]

Precisely the same dynamics applied to workers who lived and labored outside the experiential circumference of the coastal mobilization. Unless a given local work force had through its own efforts already become highly politicized, the mere knowledge that the Lenin Shipyard was striking did not in itself instruct people in what to do in order to position themselves to have delegates on the Interfactory Strike Committee—an unknown structure whose form and meaning could not be easily visualized. Large-scale popular mobilizations simply do not occur in such an effortless manner. One had to "see" the transformed power relationships between workers and the state physically verified by the scene in the shipyard. This sight, when first grasped, caused even veteran militants from central and southern Poland to break down and cry with joy (and anger at the censorship). A transformed Silesian worker said on August 22: "I was sent here by my factory to find out what is really going on. Someone [that is, Radio Free Europe] tells us that fifteen factories are on strike, then the [government] radio news says a similar number have gone back to work. So I am here to observe at first hand what is happening on the coast." The impact of this emissary's shipyard learning experience upon his political consciousness was contained in his final sentence, publicly broadcast over the shipyard loudspeaker: "I believe Silesia will join in" (see Chapter 1).

In short, if one could not "see" the strike, one needed to be told about it in verifying detail and told as well how to take the steps necessary to affiliate with the MKS. This was the job that couriers had to fulfill because this kind of hard "movement knowledge" was beyond the ken of remote sympathizers in far-off places, whether they were KOR, Radio Free Europe, or the Polish Pope.

All of these matters bear on the conceptual meaning of "consciousness" as a category of historical and theoretical analysis. Kuroń could not "direct" the Strike Presidium through Borusewicz because, throughout the Polish August, the strike leaders on the coast knew more than Kuroń did. Kuroń, like the Primate and like the academic advisors initially, thought the MKS was "going too far." In contrast, the Strike Presidium was convinced that, organizationally, things had gone almost exactly far enough. But it was precisely these things—these organizational things—that Kuroń did not grasp. He was not alone.

The second level of confusion concerns the spread of the strike and the role of Radio Free Europe or Kuroń or both in this achievement. What specific information did Kuroń or anyone else outside the coast impart to foreign press correspondents so that they could inform Radio Free Europe? What did Radio Free Europe broadcast that made a "loop" informationally germane—details of how work forces in factories across Poland could meet in support of the Gdańsk MKS and elect delegates to affiliate with it?[17]

The consolidation of the strike on August 18 through the 156-factory mobilization of that day was an achievement of the internal mechanisms of communications within the tri-cities and the immediately surrounding coastal region. The organizational feat was grounded in all the associations of coastal activists that dated back to 1970. The details of this mobilization (specified in Chapter 4), are details that Kuroń, the foreign press, and Radio Free Europe did not know before the fact on August 17 or after it on August 19. These details were the experiential product of ten years of coastal activism.

The specifics help clarify social relationships that are organic to political questions concerning which people, and which social groups, can play a coherent role in the initial organizing dynamics that attend the creation of large-scale democratic social formations. For example, Szczecin activists learned on August 14 that a "strike" had occurred in the Lenin Shipyard. They got this word before the phone lines to Gdańsk were shut down, and the same highly limited information was broadcast by Radio Free Europe, which was energetically pursuing its policy of publicizing the existence of strikes wherever they occurred in Poland. But Szczecin learned that something special was happening in Gdańsk on the night of August 18 when full information on the scope and structure of the MKS mobilization was delivered by a courier from Gdańsk.[18] This kind of thickly detailed information was no more available to Radio Free Europe on August 18 than it is now in the literature on Solidarność.

Had Jacek Kuroń or anyone else in the Warsaw intellectual milieu understood these organizing components of the Polish August, they would not have possessed such a fragile sense of the balance of forces on the coast; that is, Kuroń would not have taken the determined steps that he did in an effort to persuade the Gdańsk Strike Presidium to retreat from the first demand. In short,

Kuroń was not in "the loop." Radio Free Europe was not either. There was no "loop." Rather—simply enough—an organizing center existed in the shipyard from which organizing instructions went out to the needy across Poland. This imperative defined the courier war and the struggle over the preconditions during the tense opening negotiating session of August 23. The worker message to Jagielski and the party was emphatic: stop arresting and beating our couriers and immediately lift the telephone blockade.[19]

Though one observer of Poland has characterized the workers' posture during the August 23 negotiations as a kind of intimidated if not awe-struck "silence" in the presence of Jagielski's evasions,[20] the workers themselves understood that the struggle over the preconditions concerned the organizing lifeblood of the movement—its internal communications. The imagined role of KOR and foreign broadcasting was a conjecture from afar by observers who did not research the interior of the movement.

These distant interpretations have gained currency over the intervening years because of a general unfamiliarity on the part of journalists and scholars with the enormous hazards to large-scale popular mobilization in any society, particularly in police states. The perils the Baltic movement was attempting to overcome were the very ones that have wrecked social movements in many countries throughout the nineteenth and twentieth centuries. During the Polish August, the strategic structure of the Interfactory Strike Committee and the tactical role of a shipyard loudspeaker system ensured that two vital groups of workers would be fully informed at all times as to the manner in which the movement's goals, shared by both groups, were being faithfully put forward and defended by their own Strike Presidium. The two groups were the shipyard workers, whose occupation strike protected the worker presidium from arrest, and the "plenum" of the MKS which housed the hundreds of incoming delegates from the affiliated factories across Poland. This was the organic structure of democratic assertion that both provided internal communications and projected the first demand for an independent trade union.

The reason that organized civil society is rarely in place, even in Western industrialized countries whose political structures and customs guarantee certain minimum levels of civil liberties, is that such elaborate mechanisms of democratic organization and communication as were assembled on the Baltic coast of Poland in 1980 are extremely difficult to fashion. The greatest single outside aid to the coastal mobilization came neither from the opposition in Poland nor from the opposition outside Poland. It came from the Polish party. The party's oft-proclaimed and thoroughly visible "leading role in society" guaranteed its public transparency as the visible social cause of the population's misery. No delegates to the MKS, no shipyard worker in the occupation strike, possessed any doubt as to who the oppressor was. Such transparency, when it can be made clear by activists to their potential recruits, can be a major boon to social movements; that is, it helps make movements happen. Indeed, if enough infrastructure is in place to "hold" recruits who appear, such transparency can help make movements "large-scale." But it is important to specify that transparency alone is not enough. Workers in Lublin, Łódź, Warsaw,

Kraków, throughout Silesia, and elsewhere also knew "who their oppressors were." This fact alone did not translate into an organized movement because it did not address the organizing problem. Movement building turns on a compelling recruiting message and a coherent structure to receive and hold the recruits; both are products of existing internal communications.[21]

Clearly, broad gaps in strategic reasoning separated the Strike Presidium from most of the rest of Poland's activists, except, of course, nonworker activists on the coast who were generally able to keep up with the pace of events. But the roots of KOR's unresponsiveness were partly an understandable outgrowth of some strategic choices the Warsaw oppositionists had made during the first year of KOR's existence. It is relevant to recall that when KOR activists emerged from prison in 1977 and reconstituted their group out of the wreckage of the Radom and Ursus initiatives, Kuroń and Lityński not only abandoned the tactical goal of trying to instigate free trade unions, they, together with their associates, changed their name from the "Committee for the Defense of Workers" (KOR) to the "Committee for Social Self Defense" (KSS). The new focus was not so much a retreat as a recognition that the political and cultural task was larger than originally feared and that the democratic opposition needed to concentrate its initial energies on demonstrating an ability to create an alternative public space in a police state. This was an essential goal, but it also introduced a subtle conceptual problem. The struggle to create a viable civil society, even in microcosm, necessarily threw KOR into a dialectical relationship with the state. This, in turn, caused Kuroń to become even more deeply preoccupied in analyzing the party and the party's power.

Polish workers were also trying to create an independent civil society, of course, though they did not write about it in articles for the émigré journals. The difference was that the focus of the workers remained, after 1977 as before, on themselves—on the huge potential constituency represented by the Polish working class itself. Experientially, this had led them, over decades of struggle, to the development on the coast of interfactory structures of assertion. These efforts produced a specific kind of interior organizational knowledge that, in Poland, only coastal workers possessed.

Thus, KOR's central miscalculation during the Polish August cannot be understood solely as a result of a lack of theoretical imagination. An unseen experiential gap existed. Kuroń did not know the organizational details of the enormous mobilization of August 18, and the evidence seems compelling that even had he somehow heard about them, he did not possess an experiential basis for understanding the meaning of what he heard. Outside the coast itself, no one, in Poland or abroad, understood these complex social imperatives.

A final component of the misplaced causality undergirding worldwide interpretations of Solidarność concerns Jacek Kuroń's role in propagating the myths surrounding him. He has played very little role. As he told a Western journalist in 1982, with respect to the demand for free trade unions, "I thought it was impossible, it was impossible, and I still think it was impossible." However, the power of ingrained cultural presumptions is such that the role of KOR and its Kuroń-centered "loops" to Borusewicz or to Radio Free Europe march on through the literature despite Kuroń's disclaimers. Kuroń was wrong on August 25, 1980. He has said so publicly. The many other participants who were similarly wrong have not been so self-effacing. For this reason, among many others, Jacek Kuroń will probably wear better as a historical figure than a number of his colleagues and rivals.

The causal claims for KOR, Radio Free Europe, and the Pope have materialized in the literature on Solidarność essentially because they seemed to fill an apparent yawning evidential gap concerning the origins of Solidarność. If the elaborate organizational structures necessary to overcome the impediments to large-scale movement building were properly understood, the claims made from afar in behalf of all these distant actors, Polish and foreign, would not have been offered at all.[22]

The intellectuals of KOR were unable to divine the relative strength of the workers' position because the organizing politics that culminated in the Baltic mobilization constituted Act I of a long drama that was many years in the making and which they had not seen. From beginning to end, the shipyard confrontation was a play within a play unfolding before the entire world. In terms of political consciousness and historical causality, the distinguishing feature of the Polish August was that the most vital and germane evidence existed absolutely at center stage of the drama with the principal actors bathed in kleig lights. Though everything seemed to be visible, the historical literature on Solidarność reveals conclusively that very little was audible. When the actors talked, no one seemed to hear the words. It was as if a brilliant Polish-speaking cast played before a world audience that admired the gestures, the costumes, and the kleig lights, but unfortunately could not understand a word of Polish. The operating presumption, deeply lodged in the thought processes of the observers, was that the important lines in the play were necessarily being spoken off-stage, in the language of the audience. Though it was pleasant to enjoy the stage scenery and the gesturing worker actors, the essential task was to filter out the noise in order to hear a sophisticated Greek chorus of intellectuals and churchmen chanting somewhere off stage. In this manner, the world's scholars have unintentionally constructed a fantasy play called "The Origins of Solidarność" that is a parody of the authentic drama.

Proof of this is as stark as it is overwhelming. Evidence from center stage on the first day of negotiations, August 23, 1980 (it can be visualized as Act II of the drama), vividly illustrated the experientially based consciousness of Baltic workers; similarly, the melodic lines of the Greek chorus on August 26 (a bit later on in Act II) spell out the truncated understandings of opposition intellectuals and Catholic prelates.

To understand the consuming cultural momentum that diverted the attention of the members of the audience to the wings of the stage, it is necessary to grasp the controlling presumption they brought with them. Stated simply, this presumption was that the workers of the Lenin Shipyard should not be seen as reasoning individuals who shared with one another a common occupational and social experience in Leninist Poland but should be regarded, in terms of originality and purpose, as role players—mere supporting actors who bore an abstract label, "the workers." The absolutely striking ingredient in the coastal mobilization is that the evidence of precisely what worker members of the Strike Presidium knew that no one else in the intelligentsia or the episcopate understood was tangibly visible in the public transcript of the Gdańsk negotiations. If Act I of the Polish drama is understood as the long thirty-five-year working-class ordeal of assertion and suffering that culminated in the interfactory organizing structure that forced the government to the negotiating table on August 23, 1980, Act II was the negotiating process itself. This drama that was to culminate in the birth of a national civil society in Leninist Poland opened with an intense debate over "preconditions." The strange dialogue between impassioned workers and an elusive government labor negotiator named Mieczysław Jagielski possesses a vivid clarity with thirty-five years of coastal history in mind. To complete that history, it is necessary to insert a brief biographical description of one of the speakers at center stage—Florian Wiśniewski of the Strike Presidium.

Wiśniewski was a product of the worker mobilization at Elektromontaż in 1979, of the repressions that ensued, (including the murder of Tadeusz Szczepański), and of the counterorganizing and counterrepressions that followed. At that factory in 1979, Wiśniewski had watched what he called the "amazing facility" of the shop-floor organizer, Lech Wałęsa, and had himself played an active role alongside Wałęsa in organizing the 168 workers who had signed the original founding document of the plant's independent shop council. He had watched these workers expose themselves to repression by presenting a united demand for the reinstatement of the independent union's president, Wałęsa, and he had also watched them withstand the interrogations and detentions that followed. Wiśniewski knew that far more than 168 Elektromontaż workers had attended the emotional funeral of Szczepański in May 1980; finally, he knew that virtually the entire 500-person Elektromontaż work force had been "ready" when the Lenin Shipyard raised the flag of revolt on August 14, 1980.

In short, Florian Wiśniewski understood in very precise terms the specific process through which people remote from power acquired the particular kind of knowledge conducive to autonomous action—knowledge that enabled people to act in the presence of power. This is what Florian Wiśniewski said to Poland's deputy premier: "I consider a man of great standing and broad outlook capable of taking great steps. This means releasing a communiqué in the usual fashion, through the national press and television. Making our demands known in this way will restore our reputation after all the slanders that were heaped upon us."

The topic was the censorship, and the new balance of forces created by the Interfactory Strike Committee bore directly on the topic. "We won't tolerate prevarications. Truth cannot be concealed. The work force will make sure of that. The great barrier here is censorship. It is enough to call something political and it won't be published. But these are not political matters." The reality that needed to be discussed—the experiential reality—was the enforcing mechanism of the censorship. Wiśniewski went on:

> I have worked in trade unions for many years and know that the problem is really serious. At the moment, it looks difficult from the legal angle because the Labour code is presently so formulated that anyone who goes on strike can get the sack. People come to us from southern Poland—one brought a paper to show that he lost his job for going on strike. That would be impossible in the present atmosphere on the coast. Things have gone further here . . . We are fully representative. Prime Minister, this is not the first time such events have taken place on the coast . . . In the past, the strikers tried hard to make contact, often reaching out directly to the government, but there was no response. Instead, lies were circulated about us and various incidents staged.

Wiśniewski thus explored the censorship, deception about the censorship, and brute repression of those who violated the censorship. But his most strategically vital statement was that the government's habit of firing workers for striking "would be impossible in the present atmosphere on the coast. Things have gone further here."[23]

In these words, which came so easily to him as he confronted Jagielski, Wiśniewski made publicly visible the experiential development of the Baltic movement. A new state of citizen-party dialogue was at hand because traditional power relationships had been altered through the workers' own organizing achievement. Over the next three days, it became clear—again, publicly—that none of the other sectors of Polish society, whether in observable opposition to the regime or not, had gained the level of awareness that Florian Wiśniewski so effortlessly articulated. Bronisław Geremek engagingly explained why the academic advisors had been so stunned by the first Gdańsk demand: "what could we know? We were Warsaw intellectuals, living close to state power. The shipyard was a revelation. No police! And the way the workers acted—as if there were no police!"[24] What Geremek and Mazowiecki, in common with other thoughtful Poles, had brought with them to the shipyard was a profound awareness of one ingredient of Polish society—the power of the state. Within forty-eight hours, they had seen and reflected upon the meaning of a second ingredient that they had encountered for the first time in their lives—the social formation represented by the organized workers of the coast. By August 24, both men had accepted the fact that the first Gdańsk demand was an understandable, if bold, objective, one worth supporting in whatever way they could. When the full advisory committee recruited by Geremek arrived in the shipyard on the twenty-fourth, the "grave doubts" about the demand for a free trade union once again dominated all intellectual discussion. Though Geremek and Mazowiecki no longer fully shared this view, they acquiesced in a decision to see how the Strike Presidium would react to a second conception of worker organization, one that had the merit of being both "strong" and stopping short of being "free and independent of the party." The unanimous rejection by the Strike Presidium of the advisors' proposed Variant "B" completed what might be called (to paraphrase the terminology of Jan Józef Lipski) "the preparation of the consciousness of the advisors for the strikes."

On the same day that the workers turned aside this proposal, the strategic advice of KOR was given to the Swedish correspondent in Warsaw for dissemination to the shipyard workers. The KOR call for a retreat from the first Gdańsk demand had no more structural consequences than did Variant "B." The strategic reasoning, however, was different in its direct focus on party reform. In this manner, Kuroń repeated the strategic course that the "October Left" had followed in 1956—a focus upon reform-minded intellectuals in and near the party rather than upon workers.[25]

As the KOR advisory went out over Swedish radio, the Primate of Poland, Stefan Cardinal Wyszyński, intervened in the crisis with a public statement released to the press from Jasna Góra:

> I think that sometimes one should refrain from too many demands and claims, just so that peace and order may prevail in Poland; this is hard, especially as demands can be justified and they usually are sound; but there never is a situation where it would be possible to fulfill all the demands at once, right away, today. Their implementation must be gradual, and that is why we must talk. We shall first satisfy those demands that have priority, and then others. That is the law of everyday life.

It would be quite inappropriate to assess in traditional ideological terminology these distinctions between Baltic workers and everyone else in Poland. It was not that Geremek and Mazowiecki were "more radical" than KOR or that Florian Wiśniewski, Andrzej Gwiazda, Anna Walentynowicz, or Lech Wałęsa were more radical than KOR, the advisors, or the Primate. Rather, the coastal activists had lived through a different experience than had the Polish intelligentsia or the episcopate, possessed a different, experientially based understanding of what was possible in organizing against the state, and, above all, possessed a different appraisal of the transforming meaning of the coastal mobilization in terms of the balance of forces at work in Polish society at that moment. Therefore, as rational activists rather than as "more radical" activists, the workers understood that the union free of the party was much more within reach than outsiders understood it to be. Social knowledge is experiential; they had the experience; the Warsaw intelligentsia had not.

The decisive difference was the fundamental distinction in perception that came to hover over the Polish August. While the advisors initially, and KOR and the Church hierarchy throughout, felt the workers were "going too far" and were thus precipitating their own arrest and repression, Florian Wiśniewski and the entire Strike Presidium understood that it was precisely the fact that "things have gone further here" that made traditional government repression "impossible." Under the conditions created by the Interfactory Strike Committee, no one worker could now be arrested; all fifteen thousand in the shipyard would have to be arrested. Should that happen, hundreds of affiliated factory work forces would also have to be dealt with by the authorities. The existence and meaning of an Interfactory Strike Committee with over four hundred affiliated factories on August 26 was not only an unprecedented organizational achievement, but also a political fact that could be easily grasped only by the people who had created the fact.

This last reality, so difficult for the world's intellectuals to accept[26]—and psychologically almost impossible for Polish intellectuals to accept—can be easily verified by a comparative analysis of KOR's most advanced statement on worker organizing, the 1979 charter, and KOR's subsequent analysis at the height of the Baltic crisis. The charter's highly qualified recommendation was that the formation of independent unions should proceed only "wherever there are strong groups of workers able to defend their representatives." This conclusion, reflecting Kuroń's experientially based discovery at Radom, had the appearance, at least, of being conditionally subject to change. It is to be supposed that a coordinated coastwide general strike of over 500,000 workers would fulfill the requirements for "strong groups of workers able to defend their representatives." However, Kuroń did not possess a strategic grasp of what an interfactory strike committee was or the organized power it represented. The immediate evidence, of course, is contained rather starkly in the August 25 KOR advisory to the workers to retreat from their first demand.

But evidence of KOR's basic strategic preoccupation goes quite beyond this one fatal tactical miscalculation. A complete accounting of all KOR's expenditures from 1976 to 1980, meticulously listed in Lipski's institutional history, verifies the group's intense concentration on intellectuals, on intellectual journals, on intellectual organizing materials, and on the publication of novels, poems, literary criticism, and scholarly works in politics and the humanities. The financial investment in the worker-oriented newspaper, *Robotnik*, was but a tiny fraction of KOR's expenditures. Similarly, all of the writings of Adam Michnik published in these years and extending to his post-Solidarność analysis written while in prison reveal the same almost instinctive preoccupation with the intelligentsia—a preoccupation that caused his continuous and systemic misreading of the organizational dynamics that produced the Polish August.[27]

KOR's subsequent extreme claims about its own causal role detract from the courageous and imaginative role its activists played in creating a tenacious democratic opposition in Poland. It may be fairly assumed that KOR's partisans dared to make such claims because they were so distant from shop-floor organizing that they had no idea how hollow their assessments actually were. They advanced their assertions for perfectly understandable political reasons—namely, to buttress their own credentials for leadership roles in Solidarność.

In any case, the anxious KOR message got through to the coast where it fit easily within a cacophony of advisories representing a broad consensus that the Strike Presidium was "going too far." Foreign journalists inadvertantly joined in this consensus and, indeed, added substantially to it by incorporating its components into the questions they put to shipyard workers: "Are you afraid the Russians will come?'" "Do you think Polish soldiers will fire on Polish workers?" Almost all foreign journalists arrived after the strike had been consolidated on August 18, and it required only a few days of this sort of questioning to make the militants of the Strike Presidium rather testy—since the implicit substance of such inquiries was that coastal activists were unsophisticated and had gotten themselves into water over their heads. In such a context, the KOR advisory, like the Primate's, had no chance of influencing policy. The workers knew of their own organizational achievement, even if far-off observers did not.[28]

Since the literature on Solidarność generally reproduces Polish workers essentially as a faceless abstraction—"the workers"—undoubtedly the single most neglected area of research on the Polish movement concerns political distinctions within the working population. What can be said about Baltic workers generally? About the shipyard workers of Gdańsk? About the militants of the Strike Presidium? Or about the organizing role of Lech Wałęsa? What does the organizing experience of ordinary Polish citizens have to teach the world about the creation of democratic forms in stratified modern societies?

The answer is clear, though it is neither uniform in application nor simple in meaning. No single "level" of consciousness existed among the Polish intelligentsia generally or even among the various self-selected vanguard strands that included KOR, ROPCiO, and KPN, and the same variety existed among Polish workers. In every sector of the Polish population, extending to the peasantry and to the hierarchy of the Church, there persisted a received tradition of opposition. Polish nationalism was a central component of this heritage, for traditions of opposition and of insurrectionary activity had materialized out of the centuries-long tragedy of foreign occupations of the homeland. Anti-Russian sentiment was an activated component of a more generalized Polish anxiety about Germans and Austrians and, indeed (among a lesser number of Poles), about Jews, Lithuanians, Ukrainians, and Czechs as well. But "the antique shop of Polish patriotic phraseology" provided something less than a dependable path to political clarity. One of KOR's many valuable contributions to the nation's contemporary political culture was its sustained attempt to induce Poles to focus upon society as a democratic opportunity rather than upon the nation as a sacred and conclusive symbol.

Nationalism and religion and religious nationalism worked their will on Poles in every stratum of society, including workers. But for the working class in Leninist Poland, the most compelling drive was for hour-by-hour and day-by-day relief from the terrible confinement of life lived under the thumb of the party. The party hovered over shop-floor production with an omnipotent presence and obsessive tenacity that, in psychological terms, could not be matched in the offices that housed the white-collar populace. In the course of two generations of Leninist production methods, all Poles gradually developed and polished elaborate mechanisms of escape and evasion from the party, whether in terms of self, group, or job task. Every occupation group, even individual crafts and individual office cliques, experimented in its own methods of circumventing the party's insatiably consuming sense of its own "leading role." Large or systemic victories were impossible, but small ones, either on the edge or somewhere beyond party doctrines, were possible and were endlessly sought.

But no group in the country had more subtle memories of institutional possibility than the Polish working class. This was the great, silent, and unacknowledged truth about Leninist Poland. The heaviest kind of ironies surround this persisting social ingredient about life in the Poland of Gomułka, of Bierut, of Gomułka again, and finally of Gierek. The precipitating irony was that the working-class memory itself was dangerous. Even the most "conscious" shop-floor militant who carried within his or her psyche this cultural knowledge of working-class tradition knew, as an ingredient of the memory itself, that shop-floor militance was personally dangerous. This reality was not immediately evident in the very first postwar days when the new ruling party spoke publicly in the name of "the workers." So shop-floor militants drew on prewar traditions, defended themselves and other workers, and, in so doing, exposed themselves. In the fierce struggle of 1945–47, the natural shop-floor advocates, particularly in heavy industry, were killed, exiled, blacklisted or otherwise beaten into submission. The memory, therefore, went underground, where it was passed on—very judiciously—to selected members of the younger generation. It was passed on to those who "proved themselves" in one way or another that they were "strong workers" rather than "weak careerists." In terms of content, the memory contained two fundamental ingredients. The first was a way to be—a functioning participant of a genuinely independent trade union; the second was the weapon of struggle—the occupation strike.

In postwar Poland this heritage was conveyed to young workers by the "old militants" of the 1950s. One of the recipients of this instruction was Henryk Lenarciak of the Lenin Shipyard. Now an "old one" himself, Lenarciak can look back on a life in which he served on the shipyard strike committee of 1970, as a shop-floor representative in the "reformed" official union of the early 1970s, and as the honored comrade chosen by shipyard workers in 1980 as the chairman of the committee to oversee the building of the now famous Gdańsk monument. In many ways his story, like Florian Wiśniewski's or Lech Wałęsa's, explains the interior dynamics that produced the Polish August.

Henryk Lenarciak is a quiet and modest man. He is also cautious and honest, and these last two qualities stamp him as an enduring militant—a man who knows and can convey the heritage because he is cautious enough to have survived to convey it and honest enough to understand the need to convey it. It was Lenarciak who in the late 1960s passed the thoughts of the "old ones" to the young militant named Lech Wałęsa. It is an awesome indication of the terror under Stalinism in the 1950s that it is more revealing to discover not how shop-floor traditions were passed to Wałęsa but how they were passed to Lenarciak. The simple fact was that, because the knowledge itself was dangerous, Lenarciak did not receive it when he first went to work in the shipyard in 1952. He had to earn it—and it took years. It was only after he had proved to the satisfaction of the "old ones," during the struggle for shipyard democracy in 1956–57, that he was told about how the enterprise ought to be organized, how work should be organized, and how the workers themselves had to do the organizing. Around a kitchen table, on the Baltic coast, in the 1950s, he was told these things. In 1980, foreign correspondents covering the great shipyard strike found among the workers a certain respect for a brigade leader named Mośiński and of another old-timer named Nowicki. But when old men like Lenarciak speak of old men like Nowicki, the respect has an intimate quality to it—of comrades in an old tradition.

Traditions so carefully preserved, so circumspectly passed on to the young, quite obviously can never constitute anything that might remotely be regarded as "mass consciousness." There was, in fact, nothing "massive" about it at all. But to the historian, the relevant point is not initially numbers but rather the quality that is being conveyed. To qualify as "strong" and "honest," one did not have to swagger, to drink more ale than other workers, to have the ability to make a compelling speech, or to have "charisma." Henryk Lenarciak possesses none of these characteristics. Rather, to be "strong" meant that one did not routinely acquiesce to the party's power on the shop floor; to be "honest" meant that one could be counted on not to betray other workers, not because of fear or for personal gain or for any other common reason.

On this basis, the Baltic movement built itself, in Gdynia and Elbląg and Szczecin as well as in Gdańsk. The heritage, however, was not equal to the full dimension of the task workers faced. If the idea of an independent trade union was a cornerstone of the faith as well as a necessary goal, the occupation strike was clearly not in itself an adequate instrument through which to realize the goal. Though the occupation strike was the most potentially powerful and certainly the most dramatic weapon available to a highly organized local union and, thus, constituted the ultimate assertion of a plant-sized worker council, it was a prewar weapon that had applicability only under capitalism. Such a strike in one plant potentially could bring one boss to the bargaining table. But how to strike against the state? That was the problem posed for workers by the existence of a Leninist party. No matter how well organized or how brilliantly it conducted industrial struggle, a plant-sized worker council was simply too small in scope to contest a boss who controlled the whole nation.

In the parlance of historical description of social conflict, "militant" is a word applied—rather indiscriminately— to "old ones," down to and including young ones like Wałęsa. As such, the word tends to convey—though it in no way explains—the idea that a person of "high consciousness" is being characterized. "Militants" are most easily recognizable after the fact, when they have appeared heroic in a losing struggle or when they have somehow emerged successfully from a campaign of insurgent politics. The Polish experience can be used to improve upon this sort of vague allusion. The 1971 worker leader in Szczecin, Edmund Bałuka, was a militant but not a person of "high consciousness." He was too angry (see Chapter 3, pp. 126–28). Wałęsa was a militant and also angry, but he additionally understood that the only political alternative to armed triumph over one's adversaries was to negotiate with them. While Wałęsa was—and remains through midsummer 1990—always willing to negotiate, he spends most of his organizational energy trying to strengthen the political base from which he negotiates. He is a tireless organizer because he is angry. The Lenarciak of 1970 and 1980 was also a militant, and as an honest man, he possessed a full measure of indignation toward the Leninist party that controlled his life. But he lacked the confidence in his own capacity as a spokesman, did not feel at ease in negotiation, and thus was content to express himself through his loyalty to those around him. Bałuka, Wałęsa, and Lenarciak all possessed a consciousness that was, in Lipski's phrase, "prepared for the strikes," but only Wałęsa additionally possessed the strategic sense of how to go about the necessarily communitarian business of maximizing the chance to win the strike.

It is necessary to emphasize, however, that existence of this quality does not necessarily foretell the presence or endurance of such traits of character or range of competence as capacity for growth, integrity, or commitment to a set of political values. The Solidarność experience brought into view thousands of militants, in the working class, in the intelligentsia, and even in the Church, but all were socialized in discretely different milieus, all were variously subjected to the strains of personal ambition, and all possessed different capacities to act with poise under pressure. "Militancy" is a word that refers to the capacity to act; it does not foreshadow the quality of action. All of which is to say that "militants" can shepherd a society toward glory or toward ruin.

During the thirty-five years of Leninist organization that preceded the Polish August of 1980, literally thousands of Polish workers at different times became sufficiently militant that they experimented in how to act and what to do. They did so in 1945–47, in 1956–57, and in 1970–71. I have focused only upon some of these experiences; research of greater comprehensiveness necessarily awaits the day students of Poland can gain access to the copious police files that are—other than oral research—the principal repository of social evidence of Polish "militance." Even without access to these files, however, it is possible to trace the main outlines of the human dynamics that culminated in the Baltic movement of 1980.

While experiential knowledge transparently extends to knowledge acquired through reading (including the perusal of manifestos and other calls for social struggle), the essential knowledge that must be acquired by people structurally oppressed by hierarchy reposes at a very practical level. It consists of learning what to do to construct civil society—to build a movement.

"Education" began very early in Leninist Poland when the independent shop councils created at the end of the war were crushed by the party police. It happened again in June 1956 when the Cegielski-based mobilization of virtually the entire working population in the industrial city of Poznań ended in disaster. Stanisław Matyja's militants learned they had not experimented enough and thus learned enough to fashion an interfactory structure of purposeful cooperation and assertion.

Over the next 289 months, an unending series of organizing failures on Poland's shop floors taught an array of lessons. After the Polish October of 1956, the autonomous shop council at Cegielski was outnumbered, outflanked, and defeated by the party-organized councils of the "experimental program." The party's victory was codified in the Council on Workers Self-Management, formalized in 1958. In the 1960s, at individual enterprises across Poland, the continuing dream of worker independence from the censorship of party-controlled trade unions was boxed in and destroyed by the party's vertical system of industrial organization. Plant-sized worker councils did not work.

Nevertheless, in defeat, organizing experiences were accumulated and received traditions tested and enlarged upon. For every lesson, a fearful price was usually paid by the leading worker activists. But though knowledge and organizing talent was lost with the repression of the Stanisław Matyjas and Lechosław Goździks of working class Poland, group experience could not be repressed for it became part of the memory of the entire working class.

The idea of interfactory cooperation was born on the Baltic coast in 1970, amid the terrifying carnage of the December massacre. But the interfactory committee at Gdynia was arrested within twenty-four hours of its formation and the interfactory structure in Szczecin, though possessing the nominal allegiance of ninety-four enterprises, was effectively a wholly owned subsidiary of the Warski and Gryfia shipyards. Following the repression, as the party gave the appearance of accommodating to worker assertion, such activists as Edmund Bałuka in Szczecin and Lech Wałęsa in Gdańsk, like Stanisław Matyja before them, were encouraged to experiment in "reform of the official trade unions." The utter hopelessness of this tactic was brought home to Baltic activists in 1971–73, as Bałuka was forcibly removed from the official unions and Wałęsa resigned in disgust.

Thus, "interfactory cooperation" was in 1970 merely an idea, one almost wholly devoid of experiential meaning. For this idea of 1970 to move to action in 1980, interfactory cooperation had to be made experientially real. In the maritime provinces of Poland, this process—wholly "off stage" as politics is understood when viewed from afar—unfolded.

The most active coastal militants learned to talk to workers by being among them; they learned to talk to the party by negotiating with it and discovering thereby a new context for the received forms of elite deception; and they learned to talk to the police in detention cells where accumulated stylistic knowledge had the practical effect of minimizing as much as possible the enervating force of state-empowered retaliation. It is not only understatement but also misleading to describe this process simply as one in which a person's "consciousness" is "raised." For those who experience these dynamics over a sustained period of time, the accumulated impact of this sequential process has the effect of generating a completely different understanding of human possibility. The experience not only heightens one's skepticism about official piety but also provides critical new insights into the absurdity of much of everyday life.

This describes what is normally perceived as a "political" consequence; namely, a cultural attitude that is understood as a sign of ideological maturation. But in the more deeply political sense through which people learned autonomous conduct (and thus internalize through their own actions the experiential prerequisites necessary to participate as self-conscious actors in an authentic "civil society"), what was really absorbed on the Baltic coast in the 1970s was an extremely subtle level of knowledge about the working class itself. These were the men and women who had heard Edward Gierek's anguished pleas in 1971, who had watched his subsequent betrayals, who had watched "reform through the official unions" until they were sick of it, who headed occupation strikes in their own enterprises in 1976, who had met people like Aleksander Hall and his minions from Young Poland and Tadeusz Szczudłowski and his band of white-collar reformers from ROPCiO, who knew Bogdan Borusewicz and had heard, through him, of people like Jacek Kuroń in Warsaw. But they were also the people who knew most intimately the ways of life and styles of politics that prevailed in two Polish social formations: the party and the working class.

In this manner, the eighteen-person Strike Presidium of the Gdańsk MKS was also born in the years of the 1970s—years when Henryka Krzywonos's buses periodically came to a stop in effective demonstration strikes, teaching Krzywonos herself about her own bus and tram drivers, about the ways of the party, and about the militants from the other enterprises in the tri-cities—people like Anna Walentynowicz, Lech Sobieszek, Stefan Lewandowksi, Florian Wiśniewski, Joanna Duda Gwiazda, Andrzej Gwiazda, Lech Wałęsa, Bogdan Lis, and Krzywonos's own colleague in the transport workers, Zdzisław Kobyliński. Of all the coastal militants, they were among the most experienced. Under great pressures for seventeen days of intense national crisis, the members of the Gdańsk Strike Presidium performed with steady poise that outdistanced that of the episcopate or that of KOR.

But this still does not quite explain the victory at the Lenin Shipyard. The essential "transformation of consciousness" that occurred in Poland between 1945 and 1980 occurred during the Polish August itself. The transformation occurred among those who participated actively in this event. These people were changed—and changed fundamentally in their conception of political possibility—by the concrete events in which they participated. One speaks here literally of thousands upon uncounted thousands of people who, in the words of one of them, "found my real self." The most articulate observers of this process were certain strike leaders and certain intellectuals who, coming to the shipyard as advisors, underwent the experience themselves. One of the most psychologically penetrating descriptions emerged from Waldemar Kuczyński, the economist who was invited to Gdańsk by Geremek and Mazowiecki after Wałęsa and the presidium of the Interfactory Strike Committee had sanctioned the team of advisors. In Kuczyński's description, the social product of thirty-five years of Leninist governance was an "internalized fear" carried by the entire population. Kuczyński spoke of the strike as "gigantic reduction glass" that burned away "fear," and said that during the strike, during the emotionally powerful scenes of collective action that comprised the strike, "you could feel this barrier of fear go down." Though the economist was commenting on himself rather than workers, the judgment, quite obviously, applied to every participant.

But a crucial distinction lurks within these social dynamics. Because social knowledge is experiential, this discovery of the means of conquering fear was something that Poles outside the Baltic region could internalize only by having a similar experience: it could not be learned simply by reading the text of the twenty-one Gdańsk demands or by listening to Radio Free Europe or to KOR intellectuals. For most Poles, the autumn months, beginning immediately in early September after the Gdańsk Accords were signed, became the "gigantic reduction glass" through which their consciousness was altered. This was the time when individual Poles "came alive." As the provincial school teacher said, "I was born on September 12, 1980, when we stood up to the party secretary in a meeting."

As an opposition activist on voluntary duty as an advisor in the shipyard during the final week of August 1980, Waldemar Kuczyński stood at a vital junction between private courage, which as an implacable oppositionist he had long possessed, and collective confidence, which came to him as a function of his experience in the shipyard. It is instructive to analyze this bridge of experience as it applied in different ways to individual activists and to the thousands who comprised the movement.

For most coastal workers in August, as was the case for most of the rest of the population over the next four months, the organizational dynamics that removed "fear" not only produced that emotional moment when people said they were "born" and "came alive"; it was also the moment that made the performance of an autonomous political act possible. This moment may be understood as Solidarność's gift to the people of Poland, but it was one they themselves had to earn by their own actions. Those who waited, who did not act, did not receive the gift; in subtle ways, they remained deferential to state authority. It was the energizing rhythm of organizing that pushed people through the "barrier of fear" and into public life. It was this specific experience that created "political consciousness."

But this sequence does not describe the dynamics of autonomy that functioned in people like Kuczyński or in the worker activists on the coast who had for years labored for independence from the party. The difference was that the Kuczyńskis of Poland—the tiny rank and file of the democratic opposition that filled the narrow ranks of KOR, ROPCiO, and KPN— did not need to come to the Lenin Shipyard to acquire the necessary psychological ingredients for autonomous action. To whatever degree they felt "fear," they had found a means to act anyway—as Kuczyński himself had acted in his oppositional life in the years before 1980. The same was true of Gwiazda, Borusewicz, Walentynowicz, Wałęsa, and hundreds of other worker militants who from Szczecin to Elbląg had long been active on the coast. Fearful or not, they had acted, and they, too, did not require the Polish August to move them to public expressions of opposition.

All societies at all times have such people, such "activists" and "militants," but the historical record is quite clear that this circumstance does not translate into "movements." The prevailing assumption, worldwide in its range, is that "activists" have to sharpen the awareness of aggrieved others and that this feat is achieved, if it is achieved at all, most generally through literary exhortation that demystifies for them the real causes of their grievances.

Unfortunately, the thought that KOR or anyone else can in such a manner "prepare the consciousness of the workers for the strikes" is grounded in the intellectual illusion that insurgency fails to occur in society because people do not understand that they are oppressed and therefore require some authority to instruct them in this regard. The problem for opposition organizers does not turn on this circumstance at all. Aggrieved people know they have grievances;

their problem is that they do not have a clear idea of what to do about their condition or are reluctant to try possible remedies for fear of a retaliation from authority that will leave them in an even poorer condition. While the failure of aggrieved people to protest can be judged (when viewed from afar) as a sign of "apathy," in reality it is simply a fairly coherent belief as to the predictable outcome embedded in real power imbalances. In essence, their caution is reasonable.

This is not to say that "workers" or "people," whether individually or in identifiable groups, intuitively know the precise form or scope of their grievances or possess an easily acquired understanding of how to proceed to analyze their ongoing social situation as a prerequisite to conceptualizing a possible social solution. Indeed, people may actively repress their grievances, feeling that any "real solution" is beyond reach, or they may attempt to numb themselves by escaping into one or another form of mysticism or nationalism, or they may search for people with an even lower status in society as a temporary scapegoat to increase their own sense of value. They may also seek solace in forms of determinism or fatalism or be persuaded, if all else fails, to place blame on themselves. Indeed, it is precisely this complexity that reveals the simplistic inadequacy of vague generalizations, whether from Durkheimian or behaviorist sources, about "consciousness raising" as a causal explanation of social movements.

The transcendent reality about the politics of social life is one that is so sobering, so disarming, that people almost instinctively shrink from its full contemplation; namely, that the route to the creation of a reasonably coherent social environment in which democratic initiatives can become seriously possible is a path strewn with a truly awesome variety of cultural and psychological impediments. Above all, existing social forms—which in the aggregate constitute the received culture of any society—are massively reinforced by the capacity of existing political and military power to protect itself.

"Fear," then, is not the only hazard to democratic self-activity in modern society. But it demonstrably is a prime component of social passivity and political resignation and, as such, must be confronted in some broadly effective manner if large-scale social recruitment and mobilization is to occur. Politically, the "barrier of fear" diminishes in the presence of tangible indications that power relationships are shifting, or offer the prospect of shifting. The necessary will for people to participate in a collective action of high confrontation is not a product of reading a manifesto; it is a response to an organizing achievement that is in the process of happening in one's presence. Isolated here is the "gigantic reduction glass" that the strike represented for Kuczyński.

Still another social illusion pervading intellectual circles worldwide concerns the supposition that something called (necessarily abstractly) a "climate of insurgency" can be brought into being by efforts of intellectuals. In reality, this causal relationship exists only in marginal ways, and then solely within the confines of sanctioned argumentation within the normal non-fear-laden procedures of politics; that is, marginal social adjustments or reforms can conceivably be encouraged through logical exhortation, but only within received cultural tradition. When the objective at hand is truly structural—to shatter the traditional scope of received authority—the "barrier of fear" immediately becomes the centrally operating political fact because real power is being fundamentally challenged and real power is known to be dangerous. For organizers, then, the problem (the depth of this problem is a central prop of received cultures worldwide) is to find a tactical way for people to perform an act of protest that offers at least minimum prospects either for a measure of success against a powerful establishment or a measure of survivability in the event of failure. The presence of this level of fear and anxiety is the central challenge facing all activists seeking serious structural change, and it is one that cannot be fundamentally addressed by manifestos, newspaper articles, scholarly analyses, foreign radio broadcasts, or religious homilies.

In understanding the rise of Solidarność, it is crucial, therefore, to draw a distinction between "militants," who can act despite their fears, and the rest of the population, who possess grievances but do not know how to give candid expression to them. It was precisely this long-term structural and experiential impediment to democratic activity that the great Baltic strike of 1980 successfully removed. The "barrier of fear" came down in stages, piece by piece, but as historical time is measured, it came down with great speed for the thousands of people who made the strike possible.

For Poland, the process, these "stages," can now be summarized with a fair amount of precision. Like all incipiently large social formations, the working population of the Lenin Shipyard contained its share of people who not only had endured all they could stand of the regime but were willing to act upon this settled conclusion. They were the "activists" or "militants" of the shipyard. People like Anna Walentynowicz and Henryk Lenarciak had learned over time who they were. Roughly two hundred in number,[29] they were the contact people in each shop section. It was in their lockers that leaflets were hidden, at their work sites that the word was spread, and their familiarity with the work force that provided the human connecting links of the strike action itself. On the morning of August 14, 1980, they were the ones who nodded and knew what to do when one of their most relentless associates, an electromotor mechanic of some local fame, spoke the words, "I declare an occupation strike." And two days later, on the Saturday when all hung in the balance after the premature settlement, they were the ones who understood that the outflow of workers from the shipyard had to be stopped. Indeed, they were the ones so enmeshed in the confrontation itself (because it expressed what they individually wanted to say to the regime) that they waited patiently while negotiations proceeded and had no intention of leaving until their own leadership signaled it was all right to leave.

As best as can be determined, the number of shipyard workers who stayed through the first decisive weekend, when all the informational and coercive power of the state was in motion to end the strike, was about six hundred.[30] Two days of strike action, a time of intimate cooperation and autonomous activity, had lowered the "barrier of fear" enough so that the ranks of "militants" had tripled. It may be observed that a rather stout measure of personal resolve was necessary to stay, alone and exposed, in the shipyard over the weekend when the social protection of most of one's workmates was no longer present. The admission ticket to this special community of militancy was not something one acquired effortlessly.

These, then, were the Poles who stood on the inside of Gate No. 2 on the fateful Monday dawn of August 18, 1980, when the electro-motor mechanic made his short talk to all the Poles who stood on the outside of the gate. The appeal itself, the specific one made by Wałęsa that morning, was more than an evocation of collegiality or comradeship or an appeal to something that might be called "collective consciousness." It was rooted, indeed anchored, in the most intimate understanding of what the precise political problem was that had at that instant to be overcome. Wałęsa did not talk about the specifics of the worker demands. Rather, he addressed two more pressing topics. First, he gave an explanation for the need to act—to stand by the other workers from the smaller enterprises in Gdańsk who had stood by them. The "idea" of an interfactory strike committee was thus put to the work force in clear programmatic terms. With the reason to act in place, Wałęsa then addressed the central issue. Because the stakes were high and because the medium contributes to the message, he spoke very quietly and very calmly: "Come with us, shipyard workers, there is nothing to be afraid of."

This moment occurred one week before Waldemar Kuczyński first set foot in the Lenin Shipyard, but it was the occasion when his "barrier of fear" was most strategically operable—when it determined whether the Polish August would, in fact, become a historical event that would add itself to the long calendar of Polish insurrectionary moments.

Abstractly, it can be said, with a certain incontestable historical precision, that the breakdown of Leninist production generated the alienated social relations that created the "militants" of the shipyard, that created the Interfactory Strike Committee, and that created the worker activist named Wałęsa with his effectively telling command of worker idiom. But this antiseptic specification of preconditions destroys the meaning of the human achievement that lay at the heart of the Polish August. For the same preconditions prevailed all over Poland, indeed all over the Eastern bloc.

The Polish August happened because people labored to make it happen, because people had developed a form of political communication that was both authentic in meaning and persuasive in mode of presentation. It happened, above all, because it was organized. It is germane to specify that the precise form of organization could not be predicted as a function of preconditions. It emerged, rather, as a function of human striving and out of the experience that thirty-five years of this striving generated. The admonitions of the "old ones" were important, but the final and conclusive pieces of the democratic edifice were put in place by young ones who had learned from their own experiences.

That Monday—August 18, 1980—the Baltic movement consolidated itself as an occupation strike of the entire shipyard work force and as an Interfactory Strike Committee that enrolled, on that day alone, 156 enterprises on the Baltic coast. The activities that flowed from this circumstance—worker security guards, new rules of social order and food logistics, new forms of expression that soon festooned Gate No. 2 with slogans and pictures, and above all, the couriers and loudspeaker system at the heart of the interior lines of democratic communication—all had the effect of building confidence and diminishing fear.

By the seventh day of the strike, two days before the academic advisors arrived and began to express their alarm over the sweep of the first Gdańsk demand, the strike leadership knew that a decisive change had already occurred in the shipyard's relationship to the ruling party. As Wałęsa endeavored to explain to an inquiring foreign journalist on August 20, the gains were already permanent because the fear was gone and "the shipyard workers are not the same people anymore." Thomas Jefferson had once said, "A great deal of knowledge about the revolution is not on paper, but only within ourselves." Jefferson said this more than a quarter of a century after the American Revolution. In specific historical detail, this particular kind of experiential knowledge about the American Revolution is still "not on paper."

The descriptive term, "the politics of protest," conceals the social reality it purports to describe, just as does the more organic term "consciousness." Underneath such phraseology lies that elusive area of human activity that embodies autonomous conduct. This mode of activity is deviant from received traditions of behavior which are routinely orchestrated and conveyed to all societies as "proper conduct" and, hence, as "normal conduct." Autonomous activity is, therefore, inherently experimental. The point of departure necessarily must be grounded in received tradition, a fact that makes the initiative all the more difficult. But it is facilitated by the awareness, partial at first, that elements of received tradition do not work fairly and have been self-servingly justified by sanctioned authority. This understanding is the starting point of democratic experimentation. The experiences generated by the act of "starting" have the effect of peeling away additional layers of the received culture through which established tradition is socially justified.

The recipients of this mode of education, those commonly described as "activists" and "militants," are placed under a new kind of social pressure by their experiences. It is a test that many do not pass for the relatively simple reason that the circumstances of their "education" are so brutal that they routinely generate an extremely high level of anger. Since humans do not like to be alone, the angry ones try to continue functioning in society only to discover that their anger routinely alienates their peers in the context of day-to-day life. They thereupon tend to gather together. In the metaphor of this book, they gather around kitchen tables to exchange their unsanctioned interpretations and to find consolation in their isolation.

A great deal of human creativity is stifled at this juncture. The occupants of the kitchen, desperate for social and political relevance, give themselves a name (there are many names but "radical" will suffice for the moment to characterize them all), and they verify through discussion and adjustment that their codified views have achieved an acceptably logical consensus in the kitchen. They then sally forth with the intention of "raising the consciousness" of a mystified and oppressed populace, attracting a mass base through conversion experiences and, in such a manner, initiating the revolution. In ways that are, to say the least, not broadly understood as part of the received tradition of "the politics of protest," the focus of the militants tends to center upon the logic of spoken and written exhortation. Since industrialization began to engulf the world some eight generations ago, an enormous amount of desperate energy has been channeled into this programmatic cul-de-sac. The stark historical evidence is overwhelming that this mode of "protest activity" or "revolutionary activity" leaves ordinary people (abstractly described and distantly understood as "the masses") insufficiently moved. Battered and defeated, the planners retreat once again to the kitchen, either dazed or embittered by what they take to be the "low consciousness" of the surrounding population. The rhetoric of protest is accelerated to new and often strident levels, an understandable development, perhaps, but also one which commonly increases the remoteness and isolation of the militants.[31]

Their seminal error lies in the exhortatory premise—that social knowledge is essentially purely intellectual and thus can be conveyed in the form of argumentatively creative advisories to the population. Unfortunately, social knowledge cannot be conveyed through mere reflection; what people, all people, necessarily require is an opportunity to participate in experience. Democratic conduct, like hierarchical conduct, is experientially learned and tested.

As the accumulated experience of thirty-five years of Polish insurgency revealed, "militants" acquired gradually and at high personal cost a singularly useful body of knowledge. Luckily, coastal activists did not try to convey this knowledge primarily by writing logical articles about the necessary interrelationship between occupation strikes, interfactory strike committees, and free social spaces called independent trade unions. Rather, they endeavored to create the social circumstances in which these relationships would become apparent to all who participated in the circumstances.[32]

Twelve days into the Polish August, the incoming academics of the advisory committee presented to the presidium of the Interfactory Strike Committee their considered opinion that the first Gdańsk demand was excessive. When they took their more modest proposal to the militants of the presidium and then to rank-and-file shipyard workers, the answer, on August 25, 1980, was everywhere the same: "no." It is transparent that had the same professors gone to the Baltic coast a month earlier and gathered a group of workers together for a café conversation about possible ways of improving life in Poland, Variant "B" would have appeared as a most appealing proposition. That is to say, the consciousness of workers in July was sufficient to have enabled them to appreciate Variant "B," but not sufficient for them to have understood the necessity for rejecting it. This additional knowledge was acquired by workers between August 14 and August 25 in the course of the extraordinary social experiences that comprised the strike environment itself. Everyone had become a Florian Wiśniewski. It was knowledge of this fact that Wałęsa tried to convey to a foreign journalist on August 20.

By isolating and specifying the sequences of this process, it is possible to conclude that by living through the stages of the Polish August, workers were able to produce the end result of the Polish August: Solidarność. But the knowledge necessary to fashion the stage setting whereon these experiences could materialize required thirty-five years of experimentation.

What summary judgment, then, can be made of the role of the Polish intelligentsia? Is the implication simply that the efforts of KOR and the entire democratic opposition in Poland were strategically irrelevant to the Polish August? Not at all. Indeed, a small but far-reaching irony reposes in the relationship of the body of evidence presented herein and any overall assessment of the oppositional role of KOR in pre-Solidarność Poland. Had the partisans of KOR advanced no claims of their own with respect to "raising the consciousness" of workers, and had scholars of Poland pursued more obvious sectors of society in quest of the origins of Solidarność, a foreign social historian would, on the evidence, have to agree wholeheartedly that KOR played a relevant role in helping to create the preconditions for the Polish August.

Jacek Kuroń did not teach Polish workers how to organize occupation strikes,[33] and neither did he nor any other KOR activist ever grasp the centrality to large-scale movement building and communications of the interfactory strike structure. Nor did opposition intellectuals in any organization propound the doctrine of free trade unions as the capstone demand of an interfactory mobilization articulated by a leadership protected from arrest by a huge occupation strike. KOR, ROPCiO, KPN, Young Poland, the Clubs of the Catholic Intelligentsia, DiP, the Flying University—none of these self-organized social formations played any sort of structural role in formulating either the organizational or the essential programmatic content of the Polish August.

Nevertheless, what KOR did achieve was absolutely vital. The activists of Poland's democratic opposition did what any self-organized group of intellectuals can do if they have the will and personal courage—they reached out in fraternal association to the indigenous activists of the social groups most in need in society. KOR performed these acts publicly and in a police state. In so doing, the men and women of KOR gave organizational birth to civil society by breathing experiential life into the very idea of civil society in Leninist Poland.

In this structural sense, the most germane compliment ever paid to KOR in the era of Solidarność came not from the ranks of the democratic opposition or from the surrounding intellectual milieu that watched the KOR activists with admiration, and from a distance, but rather emanated from a worker on the Baltic coast. What KOR brought to the world of the Baltic activist Lech Wałęsa was "my first taste of genuine human solidarity."

The small irony is, therefore, that KOR was a vital link to the Polish August, irrespective of the emptiness of its organizational claims. Because the exaggerated projections of KOR's partisans conjoined both with the methodological oversights of foreign scholars of Poland as well as with the class presumptions of the Polish intelligentsia, the mystique created around KOR had the effect of concealing the much more instructive worker dynamics that actually undergirded the remarkable mobilization of civil society which materialized on the Baltic coast. To make this larger reality visible, it has been necessary in this book to endeavor to peel away the encrustations of romantic mythology that have been layered over the Polish August in hundreds of articles and books. This done, what remains of KOR's contributions is every bit as relevant as that which surrounds the erroneous claim that KOR "prepared the consciousness of the workers for the strikes."

The Baltic coast was the experiential tinderbox of popular democracy in Poland. There, the flames gradually gathered force, blazed, subsided, and flickered again, as the party blundered from 1945 to 1980 toward its self-immobilizing dead end. KOR activists, by organizing themselves and sustaining themselves in public acts of democratic assertion in the national capital of a police state, did what people of education and privilege can do to keep alive the flame which flickers—which is always flickering—where social oppression weighs most heavily. KOR not only proclaimed the existence of democratic civil society in Poland, but its activists lived political lives that proved the proclamation to be experientially true. They suffered a great of physical abuse, confinement, emotional stress, and material deprivation as a result. But they pressed on with no certainty as to their fate.

In this manner, the people of KOR, through their actions rather than through their political ideas or their literary skill, gave hope to a nascent social dream. Where resignation prevailed, KOR brought hope. Where lonely oppositional activists struggled against their own private despair, KOR brought to them knowledge they were not alone; where blacklisted activists faced the immediate economic danger of being repressed into silence, KOR sustained the spirit by raising money for maintenance income.

Taking into full consideration the vitality and past experiential achievement of the Baltic movement, KOR's efforts played a discernible role in helping to tip the scales against the party of Lenin. In courage, imagination, and staying power, if not always in humility, the opposition intellectuals of KOR set a new standard of democratic striving by which the activists of the world may measure themselves. The result was a worker-structured civil society large enough to house all who wanted to participate. The national coalition that erected itself on this base gave Poland and the world the astonishing experience called Solidarność.

With these relationships in place, it becomes quite apparent that no such elaborate democratic evolution has occurred anywhere in Eastern Europe. To see the events of 1989–90 in Poland, Hungary, East Germany, and Czechoslovakia as an undifferentiated "triumph" of something approximating "freedom and democracy" is a political misjudgment of considerable dimension. Within four months of the Gdańsk agreement in 1980, Solidarność was a self-organized structure of almost ten million Poles in a decentralized institution that licensed widespread participation, conversation, thinking, and decision making at the local level. The vitality of the movement—the range of democratic activity it housed—was extraordinary, perhaps even historically unprecedented, in its range of social, artistic, intellectual, political, and, in speculative terms, even economic expression. As a democratic social formation, there is, at this writing, nothing remotely comparable to it anywhere in Eastern Europe.

The skeletal political parties emerging across the Eastern bloc are quite traditionally hierarchical both in conception and in internal organization. None has attempted to set about the business of creating a democratic agenda by first creating a popular base through which incipiently democratic citizens can participate in the creation of the agenda. On the contrary, the parties are being built essentially from the top down with the resulting party program being offered to people as an accomplished fact to which they, acting solely as passive voters, can endeavor to find a coherent way to respond.

All of the trivializing instruments of electoral propaganda that have been so destructively polished in the United States in recent years by media consultants and pollsters are now being systematically exported to Eastern Europe in a manner that—though encountering resistance—nevertheless further attenuates the democratic relevance of the political environment that is in a state of emergence. This does not mean that self-generated institutions of popular civic initiative might not be erected in Eastern Europe. I only stress that they have not yet been successfully launched.

The idea of free expression, or social openness in the Russian word, *glasnost*, is the necessary precondition for restructuring, or *perestroika*. In the rhythms of human experience, restructuring, once achieved after a certain amount of social wrenching, becomes a kind of inherited normality that over time acquires a visible rigidity corresponding to the political needs of empowered elites. The resulting social conformity that eventually emerges is not called "rigidity," of course; it is called "socialism" or "freedom" or "democracy" or something similar that is symbolically charged and socially inexact. Free expression, or *glasnost*, may continue a nominal existence within this new received culture, but the unseen social constraints of habit limit its range and vitality. As this condition becomes visible—over decades or over centuries—the need for another effort at restructuring becomes evident.

The rigidities of Leninist production methods strangled Leninism, giving rise to Solidarność in Poland in 1980, to the emergence of Mikhail Gorbachev in the Soviet Union in 1985, and to the social loosening of Eastern Europe in 1989–90. The first of these was by far the most difficult to achieve politically, for it necessarily had to come entirely from below. In terms of the democratic experience embedded within its construction, Solidarność clearly represents the most striking of these experiments in *perestroika*, though Gorbachev's initiatives contain larger geopolitical implications in the short run.

But the long-term implications of Solidarność are not yet fully understood. The idea of a self-managing economy in a self-managing republic, explored briefly but with intensity during the summer of 1981, will not go away—not in the spacious time frame to which historians are habituated. For this reason, the ultimate geopolitical consequences of the Polish achievement, for the West as well as for the East, are not yet evident. For the moment, the states of Eastern Europe seem to have charted for themselves a rather unimaginative and doubtless costly path toward unprotected integration into a crisis-ridden world market afflicted with its own structural rigidities. In this global sense, the contemporary political adjustments in Eastern Europe in 1989–90, while liberating in many ways for East Europeans, do not now seem to offer serious prospects for innovative structural initiatives that can substantially enrich the world's heritage of democratic experience.

The springtime of Solidarność in 1980–81 is another matter altogether. It will be awhile, it seems, before it is clear what the Poles accomplished, what they tried to accomplish, and what, with a grossly inadequate economic infrastructure, they are still trying to accomplish.

But there is a deeper truth to Poland. For a brief instant of historical time—fifteen months—Poland was the most democratic society in the world. Solidarność was more than just a name affixed to a movement; it was an experience that literally millions of people lived with immense energy. In so doing, the Polish people recovered their own subjectivity and lived public lives marked with social and political candor. Despite the political fact that the policing power of the authoritarian state was still intact in Poland, despite the corresponding political fact that Americans in the same time period were "free," Poland in 1980–81 was easily the more democratic society. The citizens of the United States enjoyed private liberty but possessed such a truncated public space that they had no basis for feeling they were significantly participating in the public politics of a commonwealth.

There was no functioning "civil society" in the United States, nor in the other advanced industrial societies of the West—not in Germany or France or England or Sweden. Though the future in Poland was very uncertain in 1980–81, the people of Solidarność *counted* politically—to themselves individually, to each other, and, conceivably, possibly even to the state. That the latter did not happen does not diminish the extraordinary social content of what did happen.

In centuries past, Poland's people have surged with insurrectionary impulses fueled by explosive hungers. Their apocalyptic moments have almost always ended badly, providing the raw material for romantic poets and a tenacious national despair forever in search of something realistically believable. The Polish Church has long thrived on this public yearning. Its nuns, priests, and bishops have thrown themselves earnestly into the ritualized duties of consolation. Despite the shared urges that parishioners and clergy alike have brought to this tribal rendezvous, enduring consolation has rarely been an authentic product of their formal encounters. Most Poles know this, though as a private truth unsuitable for public discussion. And the most sensitive nuns and priests know it, too. They have accordingly been drawn into areas beyond religion—into politics and social welfare and conspiracy, so that the Church becomes progressively less Catholic and more Polish. It settles in as the continuing national home until that elusive day, so long postponed, when Poles can feel they finally have a real measure of control over their own fate. The Polish Church has remained unique in world Catholicism because that day has been so long in coming. It seems now that the Church can return itself to God.

For the people of the nation, the reappearance of civil liberties in 1989, accompanied this time by the knowledge that non-Party Poles now possessed access to state power, effectively brought into the glare of public view the class divisions that had persisted in a more or less manageable state throughout the era of legal Solidarność. The *Szlachta* intelligentsia's proprietary sense of itself (see pp. 188–90, 342–50), given elaborated historical form in the KOR myth, and Lech Wałęsa's own proprietary sense of himself as a symbol of the working-class origins of Solidarność (p. 346), conjoined to produce all the necessary raw materials for a political conflagration. As one foreign observer, David Ost, carefully noted, the politics of 1989 suggested that the individual preoccupations of Geremek, Mazowiecki, and Wałęsa tended to reflect the respective sites of their influence—Geremek in the parliament, Mazowiecki as the head of the state bureaucracy, and Wałęsa in the working class. The divisions implicit in this dynamic became visible for all to see in 1990—with an emotional intensity that surprised and angered them all. Inevitably, the idea of civil society, given such powerful expression by Solidarność, was weakened by the petulance and vituperation that accompanied the 1990 presidential election. But in structural terms, the political divisions were predictable manifestations of the cultural disjunctures long present in Polish society, fissures that, it is now clear, were not effectively transcended by the Solidarność experience.

The fact illustrated once again that the communitarian outlook encouraged by the development of a "movement culture" is a very fragile social form, indeed—one that is most vulnerable at the precise moment social formation moves from large-scale self-organization to electoral politics. Class and cultural memories plus the added strains generated by the lure of pesonal power and influence—in this case by Tadeusz Mazowiecki, Lech Wałęsa, Adam Michnik, and others—inevitably weaken the democratic cohesion that movements endeavor to bring into being. But this fact did not diminish the long-term historical importance of the Solidarność achievement in setting in motion the dynamics that liberated not only Eastern Europe but also the Soviet Union itself from the organically authoritarian social relations inherent in the Leninist conception of a modern "progressive" society. In Poland, the public sphere at long last belongs to the Poles. It is terrain conquered with suffering, contention, imagination, and sustained courage. In accordance with their differing memories of the national history and of the Solidarność era particularly, the people of Poland are free now to contend with one another in the course of creating for themselves a post-Leninist society.

One historical verdict, in any case, is now clear: the Polish August of 1980 is the most significant date on the long and anguished calendar of Polish national aspiration. As a laboratory of democratic striving, it stands as a vast civic experiment whose lessons are not confined within Polish boundaries. Indeed, it seems complacent to suppose that the newly empowered citizens of Eastern Europe can study with profit the Polish experience. The larger point is that aspiring democrats everywhere may do so as well.

Notes

Introduction

1. David Montgomery and Herbert Gutman have played a pioneering role in instructing American historians in how easily workers can get lost from view in studies that purport to be about them. For example, in writing about the dynamic interplay in America between the National Civic Federation and the Socialist party in the Progressive Era, Montgomery was still forced to record, as late as 1979, that while "the literature is abundant . . . attention to the workplace and community of the (socialist) movement is still rare." David Montgomery, *Workers Control in America: Studies in the History of Work, Technology and Labor Struggles* (London, 1979), 185. See also Herbert Gutman, *Work, Culture and Society in Industrializing America: Essays in Working Class and Social History* (New York, 1976).

2. Jan Józef Lipski, *KOR: A History of the Workers Defense Committee* (Berkeley, 1955), 424.

3. Early examples: Ignacio Ramonet, "La Stratégie des intellectuels: Vers la solidarité," *Le Monde Diplomatique* (October 1980); Stan Persky, *At the Lenin Shipyard* (Vancouver, 1981); Krzysztof Pomian, *Pologne: Défi à l'impossible*, (Paris, 1982).

4. Richard Hofstadter, *The Age of Reform* (New York, 1955).

5. Timothy Garton Ash, *The Polish Revolution: Solidarity* (London, 1983), 29–30; Neal Ascherson, *The Polish August* (New York, 1982), 77.

6. Jay M. Shafritz and J. Steven Ott, *Classics of Organization Theory*, 2nd revised edition (Chicago, 1987), summarize large elements of contemporary sociological theory. For a review of popular initiatives, see William A. Gamson, *The Strategy of Social Protest* (Chicago, 1975).

7. The specific historical contours of this reality have now been rather well documented for early industrial America, though not always with the implicit social dynamics consciously spelled out by the authors involved. In addition to the work of Gutman and Montgomery, see Daniel Walkowitz, "Working-Class Women in the Gilded Age: Factory, Community and Family Life among Cohoes, New York Cotton Workers," *Journal of Social History*, 5 (Summer 1972): 464–90; Stanley Aronowitz, *False Promises: The Shaping of American Working-Class Consciousness* (New York, 1973); Katherine Stone, "The Origins of Job Structures in the Steel Industry," *Review of Radical Political Economy*, 6 (1974): 115–80; Barbara Garson, *All the Livelong Day: The Meaning and Demeaning of Routine Work* (Garden City, N.Y., 1975); Dodee Fennell, "Beneath the Surface: The Life of a Factory," *Radical America*, 10 (Sept.–Oct. 1976); Alan Dawley, *Class and Community: The Industrial Revolution in Lynn* (Cambridge, Mass., 1976); David F. Noble, *America by Design: Science, Technology and the Rise of Corporate Capitalism* (New York, 1977); Sean Wilentz, *Chants Democratic: New York City and the Rise of the American Working Class* (New York, 1984); Christine Stansell, *City of Women: Sex and Class in New York, 1789–1860* (Chicago, 1987). A pioneering work in this area was C. D. H. Cole, *Workshop Organization* (Oxford, 1923). A work of singular influence, however, was E. P. Thompson, "Time, Work-Discipline and Industrial Capitalism," *Past and Present*, 38 (Dec. 1967): 56–97.

There is no counterpart in Polish sociological and historical literature to this body of scholarship. The methodological approaches that have matured in the course of producing this literature are also not a part of any of the research designs of the works that have appeared on Solidarność; that is, the studies of Solidarność produced by Poland watchers, both in Poland and in the West, may without exception be classified as a view from afar (see "Ways of Seeing: A Critical Essay on Authorities," 443–60).

8. The conservative historical tradition is, of course, not grounded in a social vision that turns on the creation of an activated and large-scale civil society but rather is focused on forms of social organization that serve to minimize and otherwise control such "elementary republics" (see Chapters 7 and 9). The great bulk of radical historical literature has similarly been focused elsewhere than on civil society. This is partly a function of Leninist condescension to the very idea of civil society. In the academy, this predisposition has rather easily extended itself to radical theorists who consider themselves non-Leninists or anti-Leninists. In the late twentieth century, political theory is still struggling to extricate itself from this self-generated cul-de-sac, one that, in complex ways, sanctions hierarchy both in politics and in social relations.

Gorbachevian politics clearly encompasses a sliding trajectory from Leninist to conservative principles of social control. Gorbachev's "man in Moscow," Mayor Popov viewed the politics of the Russian people ("the masses") with the same alarm and disdain that routinely characterizes the approach of American conservatives and neoconservatives and the same anxiety and condescension easily detectable in the political exhortation of American liberals and neo-liberals. One would suppose that the pervasiveness of social loneliness in advanced industrial societies will sooner or later undermine the cultural confidence that continues to sustain these traditions. In democratic terms, it is difficult to see what, at this late day, they rest upon. Only by maintaining extremely low standards of political measurement can contemporary political forms be defended as "democratic."

9. G. Konrad and Ivan Szelenyi, *The Intellectuals on the Road to Class Power* (New York, 1979). The idea of the political centrality of intellectuals, while latent in the minds of many at an earlier stage, did not acquire hegemonic scope until the twentieth century. True, as a way of thinking about social life and politics, one stream of hierarchical thought, Social Darwinism, began to acquire a certain amount of cultural authority in the nineteenth century, but it was always contested both politically and culturally by republican traditions that stemmed from the Enlightenment and that received enormous legitimization from English political thought and from the American and French revolutions of the eighteenth century. The republican motto, "Equal rights for all and special privileges for none," provided the centerpiece of a producer ideology that in America extended from early nineteenth-century workingmen's associations through the late nineteenth century when the phrase was emblazoned on the flags of organized groups of farmers.

But republican ideology and particularly its more muscular form—artisanal republicanism—began to come under immense pressure in the late nineteenth and early twentieth centuries from two directions. The rise of industrial capitalism set in motion a politics and a cultural environment that sanctioned anti-egalitarian theorizing that extended beyond Social Darwinism to embrace somewhat softer modes of rationalizing laissez-faire economics. As these latter conservative and liberal approaches drew from both republican and social Darwinian traditions, egalitarian approaches gradually became attenuated in the political environment of the twentieth century. The domestication of national labor movements under capitalism in the West over the past four decades further weakened the political base of egalitarian doctrines. Under these circumstances, popular or citizen politics became increasingly difficult to sustain. Civil society began to disappear both as a social presence and as a site of political thought.

Meanwhile, intellectual opponents of capitalism also diminished the role they were willing to accord popular constituencies in the new societies they hoped to bring into being. The most influential assessment, of course, was V. I. Lenin's *What Is To Be Done*, published in 1902. In asserting the need for a vanguard party of revolutionaries to transcend what he perceived as the crippling "trade-union mentality" and "economism" of workers, Lenin inserted an authoritarian component into the very structure of his politics. Though softened in some respects in his 1915 treatise, *State and Revolution*, Lenin's 1902 formulation outlined the coercive configurations that in fact came to describe all communist parties that subsequently came to power.

By the last quarter of the twentieth century, a kind of worldwide paradigm of hierarchy could be seen to have settled into place at a theoretical level. While remaining easily within the circumference of the paradigm, capitalists and marxists could comfortably criticize the structural costs of each other's hierarchy—pivoting, for example, on the presence of immense social stratification and permanent unemployment in the West and the destruction of civil liberties in the East. To proponents of each approach, egalitarian traditions seemed "romantic" or "unscientific." Such democratic conceptions were commonly regarded, at the level of high culture, as either "quaint" or "innocent." Everyone had become conceptually hierarchical as the Cold War engendered combat between capitalist mandarins and Marxist-Leninists. What was demonstrably missing from both conceptions of social organization was any sustained theory of politics encompassing democratic social relations through the whole realm of social life. In intellectual circles, condescension toward the population, sometimes described as "the masses," assisted in obscuring the extent to which twentieth-century political horizons had become decisively narrowed. Had such narrow conceptions prevailed in the Polish working class, Solidarność simply could not have happened.

But hierarchical habits of thought die hard. Though Baltic workers created Solidarność, over the specifically rendered advice of the democratic opposition, the workers' own academic advisors, and the highest authorities of the Catholic Church, credit for the energizing role in the creation would go to precisely these other onlookers. It is the great irony of the Polish democratic experience.

1. An Event of Unknown Origins

1. The most detailed accounts in English of the evolution of the strike in the Gdańsk shipyard are Stan Persky, *At the Lenin Shipyard*, and Jean-Yves Potel, *The Promise of Solidarity* (New York, 1982).

2. Stanisław Starski, *Class Struggle in Classless Poland* (Boston, 1982) 72; *Kultura* (Sept. 14, 1980); Paris interview no. 14 (Stawek, Gdańsk), Feb. 16, 1983.

3. *Głos Wybrzeża*, Aug. 20, 1980.

4. The letter of the Secretariat of the party Central Committee, dated Aug. 19, fell into the hands of strike sympathizers and was published in its entirety in *Solidarność Strike Bulletin* no. 2, Aug. 24, 1980, *Labour Focus on Eastern Europe*, vol. 4, nos. 1–3, published separately as *The Polish August: Documents from the Beginnings of the Polish Workers' Rebellion*, edited by Oliver McDonald (San Francisco, 1981), 36–37, hereafter cited as *Documents, Labour Focus*.

5. *The Birth of Solidarity, the Gdańsk Negotiations*, as tape recorded and transcribed by Mirosław Chojecki and translated by A. Kemp-Welch (New York, 1983), 22.

hereafter cited as *Gdańsk Transcripts*. The role of Pyka is effectively analyzed in Starski, *Class Struggle in Classless Poland*, 69–73.

6. Persky, *At the Lenin Shipyard*, 94.
7. Persky, *At the Lenin Shipyard*, 106; Polish interview no. 3, Apr. 24, 1983.
8. Kemp-Welch, *Gdańsk Transcripts*, 37.
9. Ibid., 37–38.
10. Ibid., 39.
11. Starski, *Class Struggle in Classless Poland*, 80.
12. Kemp-Welch, *Gdańsk Transcripts*, 39–49.
13. Persky, *At the Lenin Shipyard*, 3–24, 58–80.
14. The courier war on the Baltic coast during the Polish August provides the first empirical example of the kinds of causal discontinuities that—as a result of what I have termed "research from afar"—suffuse the scholarship on the origins of Solidarność (see Introduction). In its logistical and organizational components and in its fundamental function as an ingredient of large-scale popular mobilization, the courier war cannot be said to exist in the literature on Solidarność. As an exercise in the logic of research agendas, this circumstance flows from a failure to perceive the central role of internal lines of communication as a prerequisite to the creation of large-scale social movements. Thus, the struggle of the Gdańsk workers to communicate the organizational details and cooperative requirements of the Interfactory Strike Committee to other workers across northern and central Poland has not become a subject of research. The empirical specifics—the printing and distribution of informational leaflets and the reciprocal repression of couriers by the police—is simply passed over. Even when evidence of this struggle surfaces in public, as when the Strike Presidium, to Jagielski's distress, forced it upon the stage of the negotiations over the preconditions, the resulting social evidence is not perceived to bear on the central controversy the workers have with the state.

For example, the Polish sociologist, Jadwiga Staniszkis, can find no political context for the workers' argument with Jagielski over the police repression of couriers and the telephone blockade. She regards the strategic thrust of their active defense of the preconditions to be so tangential that she characterizes their conduct on August 23 as one of "silence." The only person Staniszkis perceives as addressing real issues in the negotiations of August 23 was Jagielski! See *Poland's Self-Limitiing Revolution* (Princeton, 1984). The English observer, Neal Ascherson, similarly failed to detect the courier war, though he was present in the shipyard when loudspeakers controlled by workers were announcing the license-plate numbers of the unmarked cars of the security police. Ascherson instead took note of the "passivity" of the shipyard workers. See *The Polish August*, 159–60. Other contributors to the literature on Solidarność either pass over the courier war without comment or merely note some element of the struggle as a discrete social detail. The structural role of couriers as an ingredient of movement building does not seem to be understood.

The reason is straightforward: as social causality is perceived from afar, the spread of the Baltic strike was an achievement of Radio Free Europe! The station routinely reported work stoppages across Poland in the summer of 1980, and this act—the act of reporting itself—has been taken as the essential component of causality in the coastal mobilization.

At this early stage of our inquiry, evidence from the interior of the movement is still too scanty to reveal the rather grotesque misreading of causality implicit in the "radio thesis." One cautionary comment does seem to have adequate context, even at this early stage. If mere knowledge of strikes in other locales was a fact sufficient to cause Polish workers to coalesce into a concerted movement, the obvious question is: Why did the other disputes—in textiles at Łódź, in steel at Huta Warszawa, and in scores of other enterprises across Poland—not produce the organizational components and programmatic demands achieved by the Interfactory Strike Committee of the Lenin Shipyard? Radio Free Europe announced those stoppages, too. As the interior dynamics of the coastal movement become steadily more visible, the marginality of Radio Free Europe will become quite apparent.

Contributing to this organic misreading of causality is a basic structural component of modern politics that is not widely understood—namely, the fact that large-scale social movements are extremely difficult for people anywhere to create. In all societies, cultural and political impediments to the creation of large-scale social movements are pervasive. Were this not the case, the number of such movements would be far greater than it is, especially given the extent of popular discontent across the globe. Intermittent news from distant radio stations, particularly news that is empty of crucially relevant organizational detail, cannot begin to transcend these impediments. The role of Radio Free Europe in the Polish August will be considered in due course.

15. Kemp-Welch, *Gdańsk Transcripts*, 49–52.

16. Ibid., 52–53.

17. Ibid., 53–54.

18. Ibid., 54.

19. Ibid., 56.

20. Starski, *Class Struggle in Classless Poland*, 81.

21. The demand continued "in accordance with convention no. 87 of the International Labor Organization concerning the right to form free trade unions, which was ratified by the government in Poland." *Solidarność Strike Bulletin* no. 1, Aug. 20, 1980, *Documents, Labour Focus*, 27.

22. Kemp-Welch, *Gdańsk Transcripts*, 40–42.

23. Lipski, *KOR*, 27; Paris interview no. 18 (Waldemar Kuczyński), Feb. 17,1983.

24. Lipski, *KOR*, 11, 164–66, 252, 407, 422.

25. Persky, *At the Lenin Shipyard*, 103; Kemp-Welch, *Gdańsk Transcripts*, 145; Polish interview no. 1. The citation "Polish interview no. 1" is a collective one, reflecting the author's conversations within the Warsaw community of foreign correspondents of American and English print and electronic media and conversations within the same milieu in Paris.

26. Kemp-Welch, *Gdańsk Transcripts*, 180–81.

27. I offer a summary comment on the penetration of the interior of social movements through oral investigatory techniques. This general overview of the interior of the Gdańsk movement is derived from émigré oral sources pursued by the author over the years of martial law, principally in Paris in 1982 and 1983 and in the United States from 1983 to 1986, and oral sources contacted in Poland in 1989. This brief summary is intentionally discursive in that it makes no attempt to assess the comparative value to the movement of different organizational roles or the individuals who became proficient in one or more of these roles. It is merely a summary glance at the interior of coastal insurgency, a subject that is pursued in greater depth in Chapters, 3, 4, and 6, with appropriate sources cited.

Chapter 2 provides another example of the penetration of a Polish social movement, at Poznań in 1956, with the aim of illuminating the plateaus of organization achieved and, equally relevant, the plateaus not achieved. I wish to call attention to the fact that the decisive evidential material for Chapter 2 is grounded in the work of a team of Polish social scientists working in Poznań in 1980–81: Jarosław Maciejewski and Zofia Trojanowicz (eds.), *Poznański Czerwiec* (Poznań, 1981). This work possesses two qualities that make it absolutely unique in Polish scholarship. It is a systematic study that goes beyond a mere *collection* of random interviews to produce a *written* interpretation, derived both from written as well as oral sources, and it attempts to probe and analyze at least some of the interior social dynamics that produced the Poznań movement. In short, the Poznań study stands as a pioneering work of Polish social history. In form, depth of research, and analytical intent, it has no Polish counterpart with respect to the Solidarność era. To be a bit more specific on this central and therefore inherently controversial topic, the studies of Solidarność by K. Pomian in Paris or J. Holzer in Warsaw do not employ methodological techniques that permit them to penetrate to the interior of the social movement they are attempting to analyze. As a direct result, the underlying causality animating the coastal movement is misread. Both authors would have been well served had they applied to the coastal movement the research techniques demonstrated by the Poznań team. The same injunction applies to the rest of the literature on Solidarność, with the exception, as noted in the Introduction, of the work of Roman Laba (see "Ways of Seeing: A Critical Essay on Authorities").

28. Historically, the appearance of popular politics in any society has almost invariably caught outsiders, including "opposition" or "revolutionary" outsiders, by surprise. The time period during which the intellectual community could appraise and respond to the first Gdańsk demand was quite short—from Aug. 18–19, when residents of Warsaw first began to hear about the text of the demands through nonparty sources, to Aug. 31, when the crisis ended. During this period, intellectuals in Warsaw were not being asked for interviews because their opinions were not relevant to the decision-making process in the shipyard and because—from Aug. 20—almost all of the prominent members of the democratic opposition, whose opinions would at least have been of general interest, were in jail. The view of some oppositionists, at least, were made known to Western journalists on Aug. 25 and were reported by Swedish sources (see Chapter 5). After Aug. 31, of course, the topic of one's prior opposition to the first Gdańsk demand was scarcely an appealing one for public dissemination. In real terms, the issue was settled conclusively on Aug. 25–26 when the Strike Presidium rejected Variant "B"' proposed by their academic advisors and committed the movement to the first demand irrevocably. In the absence of timely sources for this pivotal one-week period, I have constructed the trajectory leading to the intelligentsia's deep reservations about the first worker postulate by applying to the analysis those general attitudes, long in place and extensively discussed in Chapter 5, known to be representative of the milieu of the intelligentsia, as augmented by two published sources—the explanations by Tadeusz Kowalik of the team of experts at Gdańsk and by Jacek Kuroń of KOR. Kowalik's elliptical but nevertheless revealing account may be found in Kemp-Welch, *Gdańsk Transcripts* 145–60, and p. 420, n. 34). Kuroń's contrastingly generous and self-effacing retrospective is recorded in interviews with Stewart Steven and quoted in Steven's study, *The Poles*, 175–77. For Kuroń's role in the Polish democratic movement, see Chapters 5 and 9; for his role in Solidarność, see Chapter 7.

29. Denis MacShane, *Solidarity: Poland's Independent Trade Union* (Nottingham, 1981), 37.

30. Worker skepticism about the negotiating utility of intellectuals is an intermittent but endemic feature of the retrospective account of the Polish August published by Solidarność's weekly newspapers on the first anniversary of the strike. *Tygodnik Solidarność* (Aug. 14, 1981). See also note 33.

31. Polish Interview No. 2 (Gdańsk), Apr. 24, 1983; nos. 24 and 28 (Gdańsk)

August 8, 1989; American interview no. 8 (New York), June 10, 1984; Paris interview no. 4, Jan. 25, 1983; Kemp-Welch, *Gdańsk Transcripts*, 181. Mazowiecki could be described as a man of conscience: he most easily expressed his political views as a writer and through activist politics centered in a religious milieu. He had played an affective role in 1977 in a fast at St. Martin's Church in Warsaw in defense of KOR—at a time when that organization was more defenseless than it later became (Lipski, *KOR*, 161–65). The issue discussed here, however, concerns Mazowiecki's "feel" for working-class politics during the Polish August. Like virtually everyone else in Poland (including the security police), Mazowiecki would learn many things about the Polish working class in the sixteen months subsequent to Aug. 23. These elusive relationships are further explored in Chapters 3, 4, 5, and 6.

32. Polish interview no. 2; American interview no. 8; Paris interview no. 18 (Waldemar Kucyński), Feb. 17, 1983.

33. This summary of the gap between the two groups, as manifested from the perspective of workers, is common in the secondary literature on Solidarność, in which, generally, it appears as a fact that is present but not controlling. The interpretation that characterizes the paragraph herein and the two quotations from workers are derived from oral interviews of Solidarność activists conducted by the author in Paris in 1982–83 (esp. nos. 4 and 9), in the United States in 1984–85 (nos. 3 and 16, and in Poland (no. 2 in 1983 and nos. 24, 34, 35 and 38 in 1989). Corroboration, of course, is in the Gdańsk transcripts of the negotiations themselves (Kemp-Welch, *Gdańsk Transcripts*, 145–60). I offer this comment on the general subject of group "consciousness" with respect to intellectuals and workers. The understanding that workers regard intellectuals with considerable reserve is scarcely restricted to the literature on Solidarność. It is present in literally hundreds of other scholarly monographs on social disorder around the world, ranging from local labor disputes to national revolutions. One observes that on those comparatively rare occasions when their lives intersect with workers, individual intellectuals function as if this generalization does not apply to them; that is, they carry with them the same sorts of assumptions that some of the Warsaw intellectuals carried to Gdańsk. These assumptions are class-based and as such license an attitude that is proprietary toward workers. It inevitably leads to the kinds of tensions and disputes over policy that run like an open wound through the history of Solidarność. This unhappy circumstance is merely one more piece of evidence than when the internalized class values of intellectuals clash with empirical evidence, class values prevail as a guide to conduct. In drawing attention to this prevailing modern folkway, I am aware of no compelling evidence that would suggest the need to make a distinction between conservative and radical or bourgeois and marxist intellectuals. Indeed, the point, rather, is to emphasize the power of cultural assumptions to prevail over ideological ones in cases where they are, within a single individual, competing. "Left" intellectuals may be marginally more sensitive, but the differences do not seem to be significant. All exceptions are freely conceded. Bronisław Geremek is a striking illustration of such an exception, one that would not apply, until very recently if at all, to, for example, Adam Michnik. These relationships will be dealt with when, chronologically, they become politically relevant.

34. The working-class experiences that led to this conclusion are explored in Chapters 2 and 3.

35. Kemp-Welch, *Gdańsk Transcripts*, 36; American interview no. 16 (New York), July 11, 1985; Andrzej Drzycimski, "Growing," in *The Book of Lech Wałęsa* (New York, 1982), 74.

36. Kemp-Welch, *Gdańsk Transcripts*, 36, 143–46.

37. Ibid., 145.

38. Jadwiga Staniszkis, "The Evolution of Working-Class Protest in Poland," *Soviet Studies* (Apr. 1981), 213. Kemp-Welch, *Gdańsk Transcripts*, 147.

39. Kemp-Welch, *Gdańsk Transcripts*, 146.

40. Their condescension to workers is tellingly visible in the retrospectives of both Staniszkis and Kowalik. For Kowalik, see "Experts and the Working Group," in Kemp-Welch, *Gdańsk Transcripts*, 145–60. For Staniszkis, see Jadwiga Staniszkis, "On Some Contradictions of Socialist Society: The Case of Poland," *Soviet Studies* (Apr. 1979), 167–87; "The Evolution of Forms of Working-Class Protest in Poland: Sociological Reflections on the Gdańsk-Szczcerin Case, August 1980," *Soviet Studies* (Apr. 1981), 204–31; and *Poland's Self-Limiting Revolution.*

41. Coastal *Evening News* (Gdańsk), quoted in Persky, *At the Lenin Shipyard*, 74; *Los Angeles Times*, Aug. 27, 1980; Associated Press, Aug. 29, 1980; Reuters, Aug. 29, 1980; United Press International, Aug. 299, 1980; *New York Times*, Aug. 29, 1980; *Solidarność Strike Bulletin* no. 10, *Documents, Labour Focus*, 82; Radio Free Europe Research, "The Free Labor Union Movement in Poland: A Short Background," reprinted in Robinson (ed.), *August 1980, The Strikes in Poland* (Munich, 1980), 60, 71–72.

42. *Le Figaro* (Paris), Aug. 21, 1980; UPI (Gdańsk), Aug. 20, 1980, quoted in Robinson (ed.), *The Strikes in Poland*, 339.

2. The Workers Encounter a "Communications" Problem

1. See, for example, Alex Pravda, "The Workers," in Abraham Brumberg (ed.), *Poland: Genesis of a Revolution* (New York, 1983), 68–91. It is interesting to observe that though Pravda's attempt to use polling data to penetrate the Polish working class is unsuccessful on its own terms, he draws on a keen sense of inference to conclude that Solidarność was a product of worker experience rather than an outgrowth of guidance from intellectuals. This verdict is an extremely rare one among writers on Poland.

2. David S. Mason, "Membership of the Polish United Workers' Party," *Polish Review*, XVII, no. 3–4 (1982):138–39.

3. Poland was a crossroads of migration at the end of the war. In addition to the exodus of a quarter million Jews, some eight million Germans were sent off to the west; five million Ukranians in provinces that had formerly been Polish found themselves citizens of the Soviet Union; and approximately seven million Poles moved from the extreme east—newly annexed by the Soviet Union—with a substantial percentage settling in the extreme western part of the reconstituted country on land previously part of Germany. The definitive study of the Polish national experience is Norman Davies's *God's Playground: A History of Poland*, two volumes (London, 1981; New York, 1982). See also Hans Roos, *A History of Modern Poland*, (1964), 211–29; Władysław Majkowski, *People's Poland: Patterns of Social Inequality and Conflict* (Westport, Conn., 1985), 29–38. World War II losses per 1,000 population for Poland were 220: for Yugoslavia, 108; for France 15; for the United Kingdom, 8, for the United States, 1.4.

4. Andrzej Szczypiorski, *The Polish Ordeal: The View from Within* (London, 1982), 41.

5. Jean Malara and Lucienne Rey, *La Pologne: D'une occupation à l'autre, 1944–1952* (Paris, 1952), 131–34.

6. George Kolankiewicz, "The Polish Industrial Manual Working Class," in David Lane and George Kolankiewicz, *Social Groups in Polish Society* (New York, 1973), 90–

92; Malara and Rey, *La Pologne*, 131–34. Scholarly characterization of worker-party tension within the official unions are inherently in danger of conveying an emphasis, intended or otherwise, that is traceable to the kind of sources readily available. In the closed informational world of Leninist societies, the absence of competing sources enhances this danger. For example, the party explained its volatile relationship to workers in the immediate postwar years in the following elliptical manner: "In many factories, cooperation between management and the factory councils has had positive results. At the same time, management recruited primarily from specialists and functionaries of the old mould [who] are often incapable of resolving conflict, working as they do with old methods, disregarding the new attitudes of the working class, as well as ignoring the representatives of the working class in deciding upon fundamental problems." These "old methods"—the inherited traditions of the Polish working class in defense of itself—were perceived by the party as the "anarcho-syndicalist demands of the factory councils." See Kolankiewicz, "The Polish Industrial Manual Working Class," 91. In the era of Solidarność, this party description of shop-floor democracy would be heard again—directed against the proposals that emerged from Solidarność's Network. Curiously, though competing sources were readily available to scholars in the Solidarność era, the old habit of utilizing party sources could—again whether intended or not—affect the way this worker-party tension was described. See Chapter 7, note 29, pp. 425–26.

7. Jean-Yves Potel, *The Promise of Solidarity: Inside the Polish Workers' Struggle, 1980–1982* (New York, 1982), 3–5; Michael Checiński, *Poland: Communism, Nationalism, Anti-Semitism* (New York, 1982), 63. Malara and Rey, *La Pologne*, 133–34.

8. A particularly vivid account of the destruction of the Peasant party and the political rout of its chairman, Stanisław Mikołajczyk, may be found in Stanisław Mikołajczyk, *The Rape of Poland: Pattern of Soviet Aggression* (Westport, Conn., 1973).

9. Kolankiewicz, "The Polish Industrial Manual Working Class," 92–125; Malara and Rey, *La Pologne*, 134.

10. Potel, *Promise of Solidarity*, 4; Ascherson, *Polish August*, 58. The party characterized purged workers as those who "had accidentally found themselves in the party over the years" or who "were opposed to and in conflict with its policies." Noting the decline in worker membership in the party that followed upon the party's "verification" policy, Kolankiewicz comments that "this decrease did not augur well for the party, but by expelling some members, it was able to increase its authority over others." Kolankiewicz, "The Polish Industrial Manual Working Class," 117.

11. Czesław Miłosz, *The Captive Mind* (New York, 1953).

12. Richard Hiscocks. *Poland: Bridge for the Abyss?* (London, New York, Toronto, 1963), 97–136.

13. Chris Harman, *Bureaucracy and Revolution in Eastern Europe* (London, 1974), 57.

14. Checiński, *Poland*, 69–83; Szczypiorski, *Polish Ordeal*, 49–59; Lewis, *A Case History of Hope: The Story of Poland's Peaceful Revolution* (New York, 1958), hereafter cited as Lewis, *A Case History*. During the period from 1946 to 1950, over seven million peasants were moved from the countryside to employment in newly built factories in the towns. Kolankiewicz, "The Poland Industrial Manual Working Clas," 93.

15. Konrad Syrop, *Spring in October: The Story of the Poland Revolution* (London, 1957), 11–12.

16. The most authoritative investigation of this subject is David Montgomery, *Workers' Control in America* (New York, 1979).

17. Malara and Rey, *La Pologne*, 131–34; R. Hiscocks, *Poland*, 287.

18. Malara and Rey, *La Pologne*, 272–73.

19. Kazimierz, Grzybowski, "The Evolution of the Polish Labor Law, 1945–1955," *Studies of the Assn. of Polish Lawyers in Exile in the U.S.* (1956), 31–56.

20. The thirty-five-year effort of Polish workers to free themselves of the state-owned trade unions is most aptly comprehended as a struggle against the party's closure of social life both in the workplace and in the state. Such individual components of the struggle as the firings of worker activists, the intimidation of the rank and file, and the elaborate efforts to curb institutions of shop-floor democracy were all carried out by the party in the name of maintaining a voiceless work force. In the deepest meaning of the word, the issue was censorship. No one understood this relationship more deeply and more personally than the working-class activists themselves.

21. Karol, *Visa for Poland*, 133.

22. Władysław Majkowski, *People's Poland: Patterns of Social Inequality and Conflict*, (Westport, Conn., 1985), 52–53; Karol, *Visa for Poland*, 21. Hiscocks, *Poland*, 157.

23. Quoted in Lewis, *A Case History*, 148.

24. Thad P. Alton, *Polish Postwar Economy* (New York, 1955), and J. M. Montias, *Central Planning in Poland* (New Haven and London, 1962), provide the necessary overview. The figures quoted, based on later data, are from Majkowski, *People's Poland*, 51, 55–58; 163–65. See also Tadeuz N. Cieplak, *Poland Since 1956*, (New York, 1972), 287–91.

25. Potel, *Promise of Solidarity*, 11; Lewis, *A Case History*, 147–53.

26. Lucja Łukaszewicz, "The Background of the Strike at H. Cegielski Engineering Plant in Poznań in 1956," in Jarosława Maciejewski and Zofia Trojanowicz, *Poznański Czerwiec* (Poznań, 1981), 49; see also Kazimierz Grzybowski, "Social and Economic Roots of the Workers' Uprising in Poznań (June 28–30, 1956)," in *Highlights of Current Legislation and Activities in Mid-Europe*, vol. IV., no. 11 (Nov. 1956): 379–94.

27. See Chapter 1, pp. 15–21. My intent here is to suggest the utility of a focused comparison of the internal mechanisms of communication developed in 1956 at Poznań to those developed on the Baltic coast between December 1970 and August 1980, as a means of emphasizing the central political relevance of such self-organized internal machinery as a building block of democratic movements.

28. The most extensive account is Lewis, *A Case History*, 139–62, which is generally sympathetic, focused on the sensational aspects of the riot and preoccupied so exclusively on the international implications for Eastern bloc politics that the author passes over the political role of the Polish working class in that equation. Suffering the same preoccupation is Syrop, *Spring in October*, 48–53. Both accounts treat the matter of possible causes of the Poznań events in an extremely cursory fashion and, as a direct consequence, intuit pocketbook issues as solely determinant. S.L. Schneiderman, *The Warsaw Heresy* (New York, 1959), Hansjakob Stehle, *The Independent Satellite* (New York, 1965), and Hiscocks, *Poland* all refer to but do not analyze the Poznań upheaval. The impact of this literature, and of sundry shorter pieces by scholars and journalists dealing with the same material, can be measured by secondary accounts relying on such sources. See, for example, Joseph W. Żurawski, *Poland: The Captive Satellite* (Detroit, 1962), 64–70.

Later studies, such as Jakub Karpiński, *Countdown: The Polish Upheavals of 1956, 1968, 1970, 1976, 1980* (New York, 1982), and Raina, *Political Opposition in Poland, 1954–1977*, pass over the Poznań affair for a different reason—namely, the announced intent of both authors to emphasize the role of intellectuals rather than that of workers.

Despite the time period contained in his title, Raina deals with the 1956 crisis without comment about the Poznań revolt. Karpiński presents a one-page summary which he uses as further evidence of the regime's illegitimacy; that is, his focus is upon the state rather than upon Poznań workers.

The observers with the surest feel for the rhythms of working-class assertion in this period are Karol, Malara and Rey, Harman, Barton and Jamie Reynolds, "Communists, Socialists and Workers, Poland, 1944–48," *Soviet Studies*, 30 (1978). Karol, an independent socialist who had broken with the regime and had gone into exile, returned to Poland and observed the dynamics of the October Revolution at first hand. Noting that the regime had attempted to buy off disgruntled Silesian coal miners with substantial pay raises, Karol found the rank and file profoundly disappointed in Gomułka: "I realized at once that their disappointment did not arise from the question of wages." Karol, *Visa for Poland*, 200–201.

The party's official version of both the Poznań events and of the subsequent Polish October was reported in a special issue of *Nowe Drogi* (1983) and later published in the West: Adam Bromke, *Eastern Europe in the Aftermath of Solidarity* (New York, 1985), 120–75. In the ease with which the party passed over the role of the working class in this tumultuous year, the account fit rather well into the mainstream of Western literature on the subject. There was an interesting exception to this general trend—J. Chałasiński in *Przesztość i Przysztość Inteligencji Polskiej*. The workers demonstrated not only against the party, "but also against the isolation from the masses of the 'progressive' intelligentsia who wrote about and propagated socialism . . . and against its social mythology which screened the reality from view." ("The Polish Industrial Manual Working Class"), Kolankiewicz adds that for Chałasiński, "the nobility-intelligentsia, in the ghetto of their prewar isolation from the masses, did not differ significantly from the progressive intelligentsia." This is a seminal insight.

29. The literature of American labor history yields scores of examples. One that focused on a decisive moment is Sidney Fine, *Sit-Down! The General Motors Strike of 1936–37* (Ann Arbor, 1969).

30. Maciejewski and Trojanowicz, *Poznański Czerwiec*, esp. Stanisław Matyja, "Memories, We Acted Openly and Loud," 215–33.

31. Lukaszewicz, in *Poznański Czerwiec*, 48–51.

32. Ibid.

33. Paris interview no. 13, Feb. 10, 1983; Berlin interview no. 3, Apr. 28, 1983.

34. Matyja, in *Poznański Czerwiec*, 216

35. Matyja, in *Poznański Czerwiec*, 217–19.

36. The "Club of the Crooked Circle," also functioning at this time, received its originating organizational impetus from the security services. Witold Jedlicki, "The Crooked Circle Club," in Tadeusz Cieplak, *Poland Since 1956* (New York, 1972), 120–29.

37. Matyja, in *Poznański Czerwiec*, 218–19.

38. Quite obviously, these analytical and terminological formulations intrude into scholarly discussions ongoing at the moment in several spheres of sociological, philosophical, and historical speculation. The matter is intermittently addressed throughout this book, receives extended attention in chapter 9, and is summarized in the "Critical Essay on Authorities."

39. Matyja, in *Poznański Czerwiec*, 221. The focused inquiry into the interior of the Cegielski movement by social scientists at Poznań University has been augmented by a more conventional political history of the Poznań events published under the auspieces of Warsaw University: Witold Morawski (ed.), *Poznań Wydarzewa Czerwcowe.*

Essays in this collection are sequentially concerned with the origins of the Poznań crisis as perceived by high party sources in Warsaw, official actions by the party, Warsaw during the Poznań crisis, the subjective impressions of a Poznań writer, the politics generated at Poznań University by the June events, the street riot as viewed from a party building in downtown Poznań, and some thoughts about the strategic implications; that is, the book on the Poznań affair is not about the workers who are at its center. However, in one forty-two page essay by Andzej Choniawko ("Przebiej Wydarzen Czerwcowych w Poznaniu") four pages are devoted to events at Cegielski prior to the fateful June 26 meeting of Poznań delegates with Warsaw party officials. Since Choniawko's four-page account is grounded in party sources in the plant, the functioning dynamics do not turn on worker organizing but rather on the turmoil and confusion created among party plant committees by worker organizing (Choniawko, pp. 30–34).

In its approach both to the subject and to sources generating empirical evidence on the subject, the Warsaw essay collection by Witold Morawski is typical of the "view from afar" that dominates scholarship on Poland, both by Poles and by Western observers of Poland (see "Ways of Seeing: A Critical Essay on Authorities"). Thus, in its conceptual methodology, the literature on Poznań in 1956 presaged in precise ways the literature that subsequently emerged on Solidarność. The fact serves to emphasize the striking exception in Polish academic literature that is represented by *Poznański Czerwiec.* It is a landmark in Polish scholarship, a pioneering work of Polish social history that illuminates political causality through sustained inquiry into the social dynamics underlying political events.

40. Lewis, *A Case History*, 159–61; Matyja, *Poznański Czerwiec*, 221–22.

41. Matyja, in *Poznań Czerwiec*, 222.

42. Ibid.

43. Ibid., 223.

44. Lewis, *A Case History*, 139–42.

45. John Kulczycki, "The Beginnings of the Solidarity Movement in Poznań," *Polish Review*, XXVII, no. 3–4 (1982): 155; Lewis, *A Case History*, 139.

46. Majkowski, *People's Poland*, 80–81; Starski, *Class Struggle in Classless Poland*, 26; Lewis, *A Case History*, 141–42.

47. Lewis, *A Case History*, 141–42.

48. Majkowski, *People's Poland*, 85.

49. Raina, *Political Opposition in Poland*, 21–51; Lewis, *A Case History*, 187–92; adam Ważyk, "Poem for Adults," *Twentieth Century* (London, 1955).

50. Lewis, *A Case History*, 187–92; Harman, *Bureaucracy and Revolution in Eastern Europe*, 104–6; Karol, *Visa for Poland*, 134–40; Żurawski, *Poland, The Captive Satellite*, 80–82.

51. Berlin interview no. 3, Apr. 28, 1983; American interview no. 7, June 9, 1984. Lewis also inferentially notes the party's collapse at Cegielski.

52. Harman, *Bureaucracy and Revolution in Eastern Europe*, 113.

53. Jan Wszelaki, "The Workers," *Polish Review*, IV (Summer 1959) 73.

54. Lewis, *A Case History*, 160–61; American interview no. 7, June 9, 1984.

55. Karol, *Visa for Poland*, 12–13.

56. Stehle, *The Independent Satellite*, 32, 64; Karol, *Visa for Poland*, 142–45; Raina, *Political Opposition in Poland*, 52–57; Lewis, *A Case History*, 221–23.

57. Władysław Gomułka, address to the Eighth Plenary Session of the Central Committee of the Polish United Workers' Party, Oct. 20, 1956.

58. Lukaszewicz, in *Poznański Czerwiec*, 55.

59. Berlin interview no. 3, Apr. 28, 1983; Matyja, in *Poznański Czerwiec*, 220. J.J. Marie and B. Nagy, *Pologne-Hongrie* (Paris, 1966); Lewis, *A Case History*, 228.

60. Not much concrete historical evidence exists from which to draw upon at this base level of society, a circumstance about which democratic theorists are, perhaps, more precisely aware than other interested persons. In *On Revolution*, a justly acclaimed book, Hannah Arendt has systematically attempted to utilize such empirical evidence as is known to exist. The movement at Cegielski in Poland and the strategic national politics it helped to generate in 1956 add highly relevant detail to what is known about democratic self-activity at the base of stratified modern societies. It constitutes detail about which Arendt was unfamiliar when she undertook to interpret the Hungarian Revolution of 1956. See note 62.

61. More scholarly and journalistic attention has been focused upon the Polish October of 1956 than on any development in Poland between the communist seizure of power and the Polish August of 1980. This would include inquiries by Hiscocks, Bethell, Syrop, Schneiderman, Lewis, Stehle, Żurawski, Bromke, Malara and Rey, Karol, Harman, Barton, Reynolds, Karpiński and Rania, previously cited (summarized in note 28, p. 396), methodologically analyzed (note 39, p. 401–2), and theoretically considered (note 62 below).

As seen in note 28, J. Chałasiński has offered a general caveat based upon the distance between what he calls "the masses" and various strands of party-connected "progressive intellectuals" (who wrote about socialism in a way that concealed the Polish social reality from view) and the "nobility-intelligentsia" (who remained confined within "the ghetto of their prewar isolation from the masses"). Though it is necessary to note in passing that Chałasiński unwittingly betrays a rather profound distance between himself and the highly differentiated social formations he compresses into the term "the masses," his general commentary on the narrowness embedded in the political perspectives of intellectuals seems incontestable. Indeed, on the evidence generated by the authors cited above, this social narrowness may be seen to extend quite beyond Polish intellectuals of the right and left to include scholars writing about Poland from abroad. Only Malara and Rey, Harman, Barton, Reynolds, and Karol demonstrate any degree of ability to write about the Polish populace as an aggregate of people comprising highly differentiated social milieus rather than touching upon them solely through oblique and holistic references to "the workers," "the peasants," and, as industrialization proceeded, "worker-peasants." As I hope to demonstrate, the narrow trajectory of this social vision continued to dominate scholarship on Poland after 1956 so that when the Polish August occurred, these constraints proved decisive in shaping the way Solidarność was researched and interpreted.

62. With more sustained intensity than most theorists, Hannah Arendt has probed the historical experience of popular democratic striving as part of the empirical base of her theoretical speculations. In *On Revolution*, she devotes sustained attention to the instructive precedents established in the American and French revolutions, in the Paris Commune of 1871, in the Soviets of 1917, in the 1919 German Revolution, and, finally, in the Hungarian Revolution of 1956. Grounding herself in the research of Oska Anweiler (*Die Rätebewegung in Russland, 1905–1917*), Frank Jellinek (*The Paris Commune of 1871*), and Helmut Neubauer on the German Revolution, she unfortunately sees all of these formations of popularly based councils as "spontaneous" happenings. This is a familiar but unproductive global scholarly habit. It can almost be taken as a law of research that the word "spontaneous" is employed by observers to characterize insurgent mobilizations that they have not researched.

For 1956—and again drawing upon the work of Anweiler—Arendt concludes that "The Hungarian Revolution *from its very beginning* produced the council system anew in Budapest, from which it spread all over the country with incredible rapidity" (*On Revolution*, 266, emphasis added).

Two points need to be stressed at this juncture in the analysis of the evolution of Solidarność. The move toward worker councils in Hungary in no sense emanated "from the very beginning" of the Hungarian Revolution, and this fact points to the democratic weakness of the revolution, not to its democratic strength and legitimacy. On the other hand, in the Lenin Shipyard in 1980, Polish workers constructed mechanisms of internal communications that imparted to the coastal movement the enormous democratic political potential upon which it was subsequently able to capitalize. For reasons that illuminate the weakness of modern democratic theory, these achievements by coastal workers have not been understood and are, therefore, not an integral part of the literature on Solidarność. Indeed, many organizational components of the achievement have not been subjected to research, such components not having been grasped as essential ingredients of popular democratic forms.

For Solidarność, the view from afar has produced necessarily distant explanations of causality—such as the Pope's visit in 1979 and the supposed vital role of Radio Free Europe and of KOR during the Polish August. These estimates of causality will be explored in due course.

While instructive, the Hungarian Revolution of 1956 is not as structurally relevant as a demonstration of democratic striving and the forms of elite politics it summons forth as were the dynamics of worker organizing at Cegielski-Poznań and the complex politics of containment the Poznań mobilization imposed on the new Gomułka regime in 1956–57. In comparing the Hungarian rising of 1956 to the emergence of Solidarność in 1980, the Hungarian effort may be seen as the early stage of a democratic structure of achievement and Solidarność as a much later stage. Poznań stands between them. While this may seem to suggest that Arendt studied the wrong country in 1956, it seems relevant to point out in her defense that no empirical research on the interior of the Cegielski movement had been conducted at the time she wrote *On Revolution* in 1960–61. Given the corpus of Arendt's work, it may be safely inferred that she would have been as impressed by the research achievement of the Poznań University social scientists, and through them by the evidence of democratic activity on the part of Cegielski workers, as I have been. In terms of research that both penetrates to the base of voluntary social formations and explores the dynamics that subsequently unfold at that level and above it, it may be said that the works of Anweiler, Jellinek, and Neubauer on the Paris Commune and the Russian and German Revolutions of the twentieth century leave pivotal questions unexplored. The same may be said of the revolutions that preceded these events in eighteenth-century America and France. Research lacunae at the base of these revolutionary moments are not perceived to exist because the ideological traditions that provide the framework through which modern scholars formulate research questions have been conceptually narrower in scope than the social range of self-activity generated by people at the base of society during the course of these revolutionary periods. Since scholars cannot "find" what they do not look for, the experiential knowledge developed in such periods of energetic social activity is not passed on as part of the received heritage of historical evidence. Cegielski-Poznań is merely one more case in point. As will become clear, Solidarność is an even more profound example.

At the practical level of modern politics, this circumstance has a powerfully confining effect, for it instrumentally minimizes democratic aspiration and democratic achievement by past generations. This, in turn, justifies low expectations, particularly by sanctioned authorities but also by everyone else. Sustained over time, the very standards of democratic measurement are codified at a low level, a condition that encouraged condescension toward specific popular social formations and toward the idea of popular democracy itself. The cumulative force of all of these cultural considerations is to provide an impressive rationale for the continuance of authoritarian conduct by empowered elites. Over time, these customs of social relations inform and warp daily life in families, communities, schools, and workplaces—that is, far beyond the formal range of "politics." At issue here, then, are the very traditions governing how human possibility is thought about.

63. Checiński, *Poland: Communism, Nationalism, Anti-Semitism*, 123–25.

64. Kazimierz Grzybowski, "Workers' Self-Government in Poland: A Year Later," *Polish Review*, III, no. 1–2 (Winter-Spring, 1958): 138.

65. Lewis, *A Case History*, 140; Ascherson, *Polish August*, 234. *Tygodnik Solidarność*, no. 30 (Oct. 23,1981).

66. *Tygodnik Solidarność*, no. 30.

67. Harman, *Bureaucracy and Revolution in Eastern Europe*, 113, 117; Jacek Kuroń and Karol Modzelewski, *An Open Letter to the Party*, translated and published in English as *A Revolutionary Socialist Manifesto* (London, 1968), 45.

68. Harman, *Bureaucracy and Revolution in Eastern Europe*, 111; Bethell, *Gomułka*, 232.

69. American interview no. 9, June 11, 1984; Karol, *Visit for Poland*, 199; *Po prostu* (Jan. 6, 1957); Harman, *Bureaucracy and Revolution in Eastern Europe*, 116. In Szczecin, workers sacked the Soviet consulate after word reached the Baltic coast that Russian tanks had flooded into Budapest. "Hooliganism" was reported in Bydgoszcz and Wrocław. *Głos Pracy*, Dec. 18, 1956.

70. American interview no. 7, June 9, 1984; Karol, *Visa for Poland*, 199: At the time, Gomuła's speech stressed "parity with the Soviet Union," freedom for peasants, and for the Church and, within proper "socialist" limits, freedom of criticism—that is, even at the level of rhetoric Gomułka carefully excluded the working class. Gomułka's speech of Nov. 29, 1956, is quoted in Stehle, *The Independent Satellite*, 32.

71. *Po prostu* (Dec. 9, 1956), cited in Harman, *Bureaucracy and Revolution in Eastern Europe*, 117. Cf. Marie and Nagy, *Pologne-Hongrie*.

72. For the fate of Matyja, see pp. 99–101.

73. Raina, *Political Opposition in Poland*, 32–33. In November, Kołakowski wrote a satirical piece for *Po prostu* entitled "What is Socialism?" which focused on a number of antisocialist practices that every reader could recognize as standard procedures of the Polish party. Gomułka personally saw to it that the piece was not published and later grumbled: "when you read what socialism must not be, you find as well as correct ideas, a profound slander about the idea of socialism." Harman, *Bureaucracy and Revolution in Eastern Europe*, 114.

74. Kazimierz Grzybowski, "Workers' Self-Government in Poland—New Style," *Polish Review*, IV, no. 4 (Autumn 1959): 57–58.

75. This subject, including the efforts of party revisionists and of independent intellectuals in 1956–57 and the actions of opposition intellectuals in the period from 1977 to 1980 is explored in Chapters 4, 5, and 6 and is a component of the concluding interpretation of the Polish democratic movement from 1980 to 1990 in Chapters 7, 8, and 9.

76. Grzybowski, "Workers' Self-Government in Poland—A Year Later," 137–38; Grzybowski, "Workers' Self-Government in Poland—New Style," 57–59; Wszelaki, "The Workers," 73; Harman, *Bureaucracy and Revolution in Eastern Europe*, 114–16.

77. Paris interview no. 13, Feb. 10, 1983; Berlin interview no. 3, Apr. 28, 1983; American interview no. 7, June 9, 1984; Ascherson, *Polish August*, 80n; Matyja, in *Poznański Czerwiec*, 231.

78. Berlin interview no. 3, Apr. 28,1983.

79. Karol, *Visa for Poland*, 199–200.

80. Grzybowski, "Workers' Self-Government in Poland—New Style," 57; *Po prostu* (Jan. 6, 1957).

81. Karpiński, *Countdown*, 99, Raina, *Political Opposition in Poland*, 58–62; Karol, *Visa for Poland*, 235–37.

82. Matyja, in *Poznański Czerwiec*, 231–33.

3. The Movement Finds a Democratic Form

1. Melvyn Dubofsky, *John L. Lewis* (Urbana, Ill., 1986).

2. William L. Standard, *Merchant Seamen: A Short History of Their Struggles* (New York, 1947), 106–11, 135, 154, 182–86; Charles P. Larrowe, *Harry Bridges: The Rise and Fall of Radical Labor*, 2nd rev. ed. (Westport, Conn., 1972).

3. Vernon H. Jensen, *Strife on the Waterfront: The Port of New York State 1945* (Ithaca, 1974); Eric Taplin, *The Docker's Union* (New York, 1985).

4. Karol, *Visa for Poland*, 196–203, 241, 254. Though Karol reported on some experiments in worker councils, he focused essentially on the workers' relationship to the state rather than directing his inquiries toward worker self activity. See also George Sakwa, "The Polish October: A Re-Appraisal Through Historiography," *Polish Review* 23, 3 (1978):62–78.

5. Paris interview no. 1, Nov. 12,1982; American interview no. 13, May 20, 1985.

6. Paris Interview no. 1, Nov. 12, 1982; no. 15, Feb. 14,1983; American interview no. 21, Apr. 14, 1987.

7. Paris interview no. 13, Feb. 10, 1983.

8. Paris interview no. 6, Jan. 26, 1983.

9. *Jedność*, no. 8 (Oct. 20, 1980); W. J. Twierdochlebow, *Solidarność: A Biblio-Historiography of the Gdańsk Strike and Birth of Solidarity Movement* (Menlo Park, Calif. 1983).

10. Alton, *Polish Postwar Economy*, provides an overview. See also Cieplak, *Poland Since 1956*, 287–91; Raina, *Political Opposition in Poland*, 85; Majkowski, *People's Poland*, 51–58, 163–65.

11. Adam Ważyk, "Poem for Adults," *Twentieth Century* (London, 1955).

12. In the aftermath of the Polish October, Stanisław Starski concludes that intellectuals and workers acted separately "not because they wanted to limit themselves, but because the prevailing consciousness was still the monopoly of the ruling class and did not allow awareness of the basic dependence of everyone on a sole employer, the state." *Class Struggle in Classless Poland*, 29.

13. Polish interview no. 2, Apr. 24, 1983 nos. 38, 39, Aug. 8, 1989; Paris interview no. 9 Jan. 30, 1983; nos. 34, 35, 36, May 5, 1983; American interview no. 19 (Washington, D.C.), June 16, 1985.

14. American interview no. 19 (Washington D.C.), June 16, 1985.

15. Lawrence Weschler, *Solidarity: Poland in the Season of Its Passion* (New York, 1982), 44–45.

16. Wyndham Mortimer, *Organize! My Life as a Union Man* (Boston, 1971); Sydney Fine, *Sit-Down! The General Motors Strike of 1936–37* (Ann Arbor, 1969); Art Preis, *Labor's Giant Step: Twenty Years of the CIO* (New York, 1964).

17. Roman Laba, "The Roots of Solidarity" (unpublished Ph.D. dissertation, University of Wisconsin, 1989), 31–66. The evolution occurred with great speed: workers into the street on the fourteenth; occupation strike under the control of the party on the fifteenth; occupation strike under the control of workers on the sixteenth.

18. For a similar understanding by a trade unionist, see MacShane, *The Independent Trade Union Solidarity*, 14–16.

19. A historically elaborated view of this interpretation of class first appeared in E. P. Thompson, *The Making of the English Working Class* (New York, 1964).

20. Laba, "Roots of Solidarity," 32–35; Paris interview no. 1, Nov. 12, 1982; George Blazynski, *Flashpoint Poland* (New York, 1979), 8.

21. Though some youths who augmented the crowd seem to have come from a variety of workplaces, Laba cites evidence from an internal study (Plynność Kadr w Stoczni," Nov. 29, 1974) that a sizable number came from the occupational school of the Gdańsk Shipyard. A demographic breakdown of these students shows that 72 percent came from worker families, 15 percent from white collar families, and 13 percent from peasant families. There is evidence from worker sources, presented by Laba and Blazynski, that police set fire to kiosks and engaged in looting to provide a predicate for the government's interpretation (soon abandoned after the fall of Gomułka) that the anti-party activity of Dec. 14 came not from workers but from "hooligans" and "anti-social elements." Laba also provides evidence, however, of a party member who regarded with equanimity the burning of official government cars by rioters: "If you take into account that one person rides in an automobile and the other doesn't have bread to eat, it's not so strange that cars irritated people." Emil Białkowski, "Diary," K-2, Lenin Shipyard, quoted in Laba, "Roots of Solidarity," 39.

It seems reasonable, on the evidence, to conclude that not much that resembled a "spontaneous rising" occurred on Dec. 14. The crowd was orderly for a prolonged period, and when its objective of finding someone in authority with whom to talk was thwarted, it wandered in search of a strategy—to the radio station, to the college, and—at 7 p.m.—to a planned joint rally with students at Gdańsk Polytechnic. The students, under threats of expulsion not to participate, did not appear. It was only after all these somewhat *ad hoc* but orderly measures were thwarted and militia appeared that street fighting broke out in Gdańsk in the late evening of Dec. 14.

22. A. Ross Johnson, "The Polish Riots and Gomułka's Fall," *Problems of Communism* (July–Aug. 1971); Zbigniew Pelczynski, "The Downfall of Gomułka," *Canadian Slavonic Papers*, 15 nos. 1–2 (1972): 1–21.

23. Paris interview no. 15, Feb. 14, 1983; Polish interview no. 42, (Henry Lenarciak), Aug. 9, 1989.

24. Laba, "Roots of Solidarity," 65–66.

25. Paris interview no. 15, Feb. 14, 1983; *Głos Wybrzeża*, quoted in Blazynski, *Flashpoint Point*, 10. The "eyewitness" account is from Potel, *Promise of Solidarity*, 18.

26. R. Bajalski in *Polityka* (Belgrade), Feb. 23, 1971, cited in Majkowski, *People's Poland*, 82–83.

27. Laba, "Roots of Solidarity," 65–66.

28. Ibid.

29. Ibid.

30. I wish to acknowledge the generosity of Roman Laba for sharing with me information from the private diary of a Gdańsk shipyard official containing an hour-by-hour account of the futile negotiation between the Paris Commune Shipyard and Kliszko's "command post" in the fateful hours leading up to the tragedy of Dec. 16, 1970. Professor Laba acquired this document in the course of his work with the historical research team established by Solidarność following the Polish August.

31. Barbara Seidler, in *Życie Literackie* (Kraków), Feb. 21, 1971, quoted in Blazynski, *Flashpoint in Poland*, 16.

32. Blazynski, *Flashpoint Poland*, 16–17.

33. Weschler, *Solidarity*, contains both the original photograph and a still photograph from Wadja's *Man of Iron* (illustrations between pp. 47–48).

34. Laba, "Roots of Solidarity," 94–122. The first of the twenty-one demands in Szczecin called for the Central Trade Union Council to be replaced by "independent trade unions dependent on the working class." Laba, "Roots of Solidarity," 112.

35. Laba, "Roots of Solidarity," 113–25.

36. Michael Costello, "Political Prospects," *Survey* (Summer 1971), 53–73.

37. Laba, "Roots of Solidarity," 117, 125, 129. Some of these events are also traced in Blazynski, *Flashpoint Poland*, 18–20, 35–45.

38. Extensive excerpts from Gierek's extraordinary appearance before workers in Szczecin and Gdańsk have appeared in the West. *New Left Review*, no. 72 (Mar.–Apr. 1972); Blazynski, *Flashpoint Poland*, 45–48.

39. Bolesław Sulik in *La Pologne: Une Société en dissidence* (Paris, 1978); MacShane, *Independent Trade Union Solidarity*, 37–38.

40. Ascherson, *Polish August*, 244n.

41. Szczesiak, "Notes On Biography," *Book of Wałęsa*, 31–35.

42. American interview no. 22 (Durham, N.C.), Apr. 14, 1987; Szczesiak, "Notes on Biography," 33–34, 51. On the first occasion of monument politics—May 1971—shipyard workers carried a banner "demanding that their dead be honored." Szczesiak, "Notes on Biography," 51.

43. American interview no. 22 (Durham, N.C.), Apr. 24, 1987.

44. Paris interview no. 18 (Waldemar Kuczyński), Feb. 17, 1983; John Farrell, "Growth, Reform and Inflation," in Maurice D. Simon and Roger E. Kanet (eds.), *Background to Crisis: Policy and Politics in Gierek's Poland* (Boulder, 1981), 299–307.

45. MacShane, *The Independent Trade Union Solidarity*, 16.

46. Keith John Lepak, *Prelude to Solidarity: Poland and the Politics of the Gierek Regime* (New York, 1988), 86–87 Paris interview no. 18 (Waldemar Kuczyński), Feb. 17, 1983; Farrell, "Growth, Reform and Inflation," 304.

47. "Interview with Lech Wałęsa," in Majkowski, *People's Poland*, 173; Szczesiak, "Notes on Biography," 34–35; American interview no. 22, Apr. 14, 1987.

48. Szczesiak, "Notes on Biography," 35.

49. Ibid.

50. Ibid., 36. Wałęsa's view a decade later: "I broke my teeth."

51. Ibid.

52. Ibid., 37; Mary Craig, *Crystal Spirit* (New York, 1986) 140, 150.

53. Blazynski, *Flashpoint Poland*, 257–58, 266, 331.

54. Lipski, *KOR*, 33–36.

55. MacShane, *Independent Trade Union Solidarity*, 37; Blazynski, *Flashpoint Poland*, 257.

56. Conspiracy theories are a highly developed art form in Poland, and several

have been generated concerning factional maneuvering within the party in both 1968 and 1970. One such theory holds that many regional party secretaries anticipated a severe popular reaction to the 1970 price increase, but withheld telling Gomułka because they wanted to see him replaced by Gierek. This account fails to explain, of course, why pro-Gomułka regional functionaries also remained silent.

4. The Movement Experiments in Self-Activity

1. Craig, *Crystal Spirit*, 158–59.

2. Ryszard Kapuściński "Revolution," *New Yorker*, (Mar. 4, 1985), 79–96.

3. Paris interview no. 4. Jan. 25, 1983; American interview no. 8, June 10, 1984.

4. I am persuaded on the evidence generated by interviews with former residents of Gdańsk in the working class and in democratic opposition that the role of Young Poland and to a lesser extent ROPCiO has been somewhat underestimated. Paris interview no. 4, Jan. 15, 1983; no. 7, Jan. 28, 1983; no. 14, Feb. 13, 1983; American interview no. 4, July 20, 1983; no. 8, June 10, 1984: no. 14, May 21, 1985. See also Majkowski, *People's Poland*, 181–84. The role of President Carter's human-rights campaign in Polish politics is discussed in Garton Ash, *The Polish Revolution*, 19.

5. Szczesiak, "Notes on Biography," 44. Polish interview no. 2, Apr. 24, 1983.

6. American interview no. 4 (New York), July 20, 1983; Paris interview no. 2, Nov. 20, 1982.

7. Lipski, *KOR*, 79–95, 241.

8. Polish interview no. 20 (Krzysztof Wyszkowski), Aug. 4, 1989; L. Shatunov, *Kak Nachinalas Solidarnost* (London, 1981), 191; *Tygodnik Solidarnosść*, no. 2 (Apr. 10, 1981); Lipski, KOR, 242–45.

9. Polish interview no. 20 (Wyszkowski), Aug. 4, 1989; Wałęsa, *Way of Hope*, 96.

10. Lipski, *KOR*, 230, 492–96.

11. Wałęsa, *Un Chemin d'espoir*, 132; Wałęsa, *Way of Hope*, 97, 130.

12. Stan Persky and Henry Flam (eds.), *Solidarity Sourcebook* (Vancouver, 1982), 69.

13. American interview no. 8 (New York), June 10, 1984; Anna Walentynowicz, *Who's Who, What's What in Solidarnosść* (Gdańsk, 1981).

14. Potel, *Promise of Solidarity*, 28–31; Wałęsa, *Un Chemin d'espoir*, 140; Polish interview no. 2 (Gdańsk), Apr. 24, 1983; American interview no. 4 (New York), July 20, 1983.

15. MacShane, *The Independent Trade Union Solidarity*, 42.

16. Ibid., 38; Szczesiak, "Notes On Biography," 38–43; Majkowski, *People's Poland*, 172; Polish interview no. 2 (Gdańsk), Apr. 24, 1983. Though some authorities have reported that Wałęsa was completely supported by KOR funds, the coastal movement had other, local sources in addition to outside help from KOR. Dariusz Kobzdej, in Wałęsa, *Un Chemin d'espoir*, 153. Wałęsa also received help from ZREMB and Elektromontaż workers and from the Church. American interview no. 4, July 20, 1984.

17. Potel, *Promise of Solidarity*, 26–33; Majkowski, *People's Poland*, 173–75; Persky, *At the Lenin Shipyard*, 71–72; Wałęsa, *Un Chemin d'espoir*, 152; Szczesiak, "Notes on Biography," 45; Polish interview no. 2, Apr. 24, 1983; American interview no. 16 (New York), July 11, 1985.

18. Leszek Kaczyński, a movement lawyer affiliated with ROPCiO, reflected the tensions which operated at what might be called an endurable level in Gdańsk. For

example, he intuited Krzysztof Wyszkowski's motive in initiating the free trade union on the coast to Wyszkowski's desire to "head off" the growth of ROPCiO in Gdańsk. Kaczyński, in Wałęsa, *Un Chemin d'espoir*, 159. Kaczyński was critical of Borusewicz's organizing style ("He was not one to put himself in the forefront") and implied that Wyszkowski's vibrant anticlericalism alienated people.

This perhaps says more about criteria functioning in middle-class society in Gdańsk than it does about the shop-floor capabilities of worker activists. Within the working class in Gdańsk, the activists of the free trade unions—like activists in enterprises throughout the city—almost without exception enjoyed the quiet respect of their peers. They were an embattled group, and they needed each other. Thus, Lewandowski, Sobieszek, Szołoch, Krzywonos, and Kobyliński were looked upon as dependable advocates of the workers' cause, just as Wałęsa, Gwiazda, Borusewicz, Pieńkowska and Walentynowicz were. But within this well-respected group, Wałęsa was the best known. After the appearance of the *Coastal Worker* attracted the attention of foreign journalists, Borusewicz, responding to reporters' requests to meet a "real worker activist," took them to see Wałęsa in his tiny two-bedroom apartment in Stogi. See Alina Pieńkowska in Wałęsa, *Un Chemin d'espoir*, 161. Later rivalries have obscured these relationships which prevailed in the embattled 1970s.

19. Disagreements between Wałęsa and KOR, lesser ones between Wałęsa and Borusewicz, and greater ones between Wałęsa and Gwiazda constitute one of the elements of the workers' movement about which many participants know a great deal and are willing to discuss. Though this legacy of disputation sometimes surfaces in the literature of Solidarnosść with respect to the Bydgoszcz crisis or the Solidarnosść Congress of 1981, the less strained history in the years of organizing leading up to the Polish August is virtually nonexistent either in journalistic or scholarly accounts. This is apparently traceable to the fact that worker organizing in the 1970s does not play a central role in studies of Solidarnosść.

Lipski (*KOR*, 424) refers to Wałęsa as "endowed with great stubbornesss" and as one who "strives steadily toward his goal of free trade unions" but, other than linking the two judgments in the same sentence, does not comment further. For his part, Wałęsa did not, in his autobiography, explore movement tensions and rivalries beyond saying that the original free-trade-union group did not go about organizing in the proper manner.

20. I endeavor here to touch upon in a general way some of the interior dynamics of the organizing process—and the widely differing understandings about that process—existing within nascent social movements. Though specific evidence from the interior of the coastal movement bearing on this process will be presented throughout this book, the ways these prevailing assumptions played themselves out within the KOR milieu are explored in detail in Chapter 5 and reviewed in a summary analysis in Chapter 9.

21. Pieńkowska, quoted in Wałęsa, *Way of Hope*, 110–12. See also *Un Chemin d'espoir*, 148, 150, 160, 176.

22. Szczesiak, "Notes on Biography," 38–44; Paris interview no. 9, Jan. 30, 1983, and no. 15, Feb. 14, 1983; American interview no. 4, July 20, 1983, and no. 8, June 10, 1984.

23. Wiśniewski, quoted in Szczesiak, "Notes on Biography," 41; Lipski, KOR, 340; Wałęsa, *Un Chemin d'espoir*, 163–64, and *Way of Hope*, 101–3.

24. Szczesiak, "Notes on Biography," 41–42; Lipski, *KOR*, 340–41; American interview no. 4, July 20, 1983; no. 8, June 10, 1984; and no. 16, July 11, 1985; Wałęsa, *Un Chemin d'espoir*, 144.

25. American interview no. 4, July 20, 1983; Fac, in *Book of Wałęsa*, 54. At

various times, Wałęsa was represented in court by himself, Leszek Kaczyński of ROPCiO and Jacek Taylor of KOR. Wałęsa agreed to repair Taylor's care "whenever he needed it" in exchange for legal services. For views of Kaczyński and Taylor, see Szczesiak in *Un Chemin d'espoir*, 150, 159, 170–71.

26. American interview no. 4, July 20, 1983; Paris interview no. 2, Nov. 20, 1982; Mieczysław Wachowski, in Wałęsa, *Way of Hope*, 175.

27. Polish interview no. 27 (Henryk Lenarciak), Aug. 9, 1989.

28. Polish interview no. 2, Apr. 24, 1983; Paris interview no. 2, Nov. 20, 1982; American interview no. 16, July 11, 1985; Polish interview no. 27 (Lenarciak) Aug. 9, 1989. See also Jozef Drogan and a second, anonymous worker, in Wałęsa, *Un Chemin d'espoir*, 164–65, 167–70, 176–77.

29. Andrzej Gwiazda, in *Who's Who, What's What in Solidarność*; Paris interview no. 2, Nov. 20, 1982; Polish interview no. 2, Apr. 24, 1983.

30. Bogdan Borusewicz, in *Who's Who, What's What in Solidarność*; Paris interview no. 2, Nov. 20,1982.

31. Gwiazda, in Wałęsa, *Un Chemin d'espoir*, 149; Steven, *The Poles*, 223–34. Paris interview no. 2, Nov. 20, 1982.

32. William F. Robinson (ed.), *The Strikes in Poland* (Munich, 1980), 3–5.

33. Potel, *Promise of Solidarity*, 36–37; Paris interview no. 12, Feb. 8, 1983; Robinson, *Strikes in Poland*, 6–7.

34. Robinson, *Strikes in Poland*, 7–8.

35. Ibid., 8–9.

36. Joseph McKay, "The Polish Opposition," *Survey* (Autumn 1979), 10.

37. Ascherson, *The Polish August*, 145; "Secretariat of the Party Central Committee," reprinted in McDonald, *Labour Focus*, 36–37; Krzysztof Pomian, "Miracle in Poland," *Telos*, 47 (Spring 1981): 78–91.

38. Persky, *At the Lenin Shipyard*, 62.

39. Ibid.

40. *Tygodnik Solidarność* (Aug. 14, 1981), Krzysztof Wyśzkowski interview with Jerzy Borowczak; Persky, *At the Lenin Shipyard*, 21–22; American interview no. 8, June 10, 1984.

41. Persky, *At the Lenin Shipyard*, 22.

42. Timothy Garton Ash has called attention to the multiple versions of this impromptu speech that have appeared in Polish, French, and English. *The Polish Revolution*, 39, 72. The version here is drawn from Persky, *At the Lenin Shipyard*, 23, 45, and from Garton Ash, *The Polish Revolution*, 39.

43. Persky, *At the Lenin Shipyard*, 23.

44. American interview no. 4, July 20, 1983; Potel, *Promise of Solidarity*, 42–46; Persky, *At the Lenin Shipyard*, 59.

45. Persky, *At the Lenin Shipyard*, 44; Potel, *Promise of Solidarity*, 46.

46. Persky, *At the Lenin Shipyard*, 64–65.

47. Bogdan Szajkowski, *Next to God . . . Poland* (New York, 1983) contains the most detailed summary of the original demands of August 14 (p. 90).

48. Garton Ash, *The Polish Revolution*, 40; Potel, *Promise of Solidarity*, 48–50; Persky, *At the Lenin Shipyard*, 74–75.

49. Persky, *At the Lenin Shipyard*, 76.

50. Potel, *Promise of Solidarity*, 51; Fac, in *Book of Wałęsa*, 58; Garton Ash, *The Polish Revolution*, 41; Persky, *At the Lenin Shipyard*, 76–77; American interview no. 4, July 20, 1983.

51. American interview no. 4, July 20, 1983. This figure, widely used, may be

high. The number of workers who stayed the first night might have been as low as 300–400. Whatever the number, these workers were, in truth, the "militants." See Chapter 9, pp. 374–86, esp. 382–83.

52. Roman Laba, *The Roots of Solidarity* (Princeton, 1991), 159.

53. Ibid.

54. Garton Ash, *The Polish Revolution*, 44; Wałęsa, *Way of Hope*, 130; Potel, *Promise of Solidarity*, 51–53.

55. American interview, no. 4, July 20, 1983; Paris interview no. 2, Nov. 2, 1982.

56. Paris interview no. 2, Nov. 2, 1982.

57. Ibid.

58. Persky, *At the Lenin Shipyard*, 83–84; Garton Ash, *The Polish Revolution*, 45.

59. Paris interview no. 2, Nov. 20, 1982; American interview no. 4, July 20, 1983.

60. Garton Ash, *The Polish Revolution*, 45; Potel, *Promise of Solidarity*, 51; Polish interview no. 2, Apr. 24, 1983.

5. Tributaries

1. Garton Ash, *The Polish Revolution*, 20, 32.

2. W. J. Twierdochlebow, *Solidarnoćś: A Biblio-Historiography of the Gdańsk Strike and Birth of Solidarity Movement* (Menlo Park, Calif., 1983), 28.

3. Tadeusz Kowalik, "Experts and the Working Group," in Kemp-Welch, *Gdańsk Transcripts*, 143, 150.

4. Garton Ash, *The Polish Revolution*, 19–21, 30; Majkowski, *People's Poland*, 153.

5. Intellectuals are a minority in the broad definition of the intelligentsia favored in Eastern Europe. For a penetrating and judicious appraisal of the Polish intelligentsia, see Alexander Gella (ed.), *The Intelligentsia and the Intellectuals* (Beverly Hills, 1976). See also Leszek Kołakowski, "The Intelligentsia," in Brumberg (ed.), *Poland: Genesis of a Revolution*, 54–67; Jack Bielasiak, "Recruitment Policy: Elite Integration and Political Stability in People's Poland," in Maurice D. Simon and Roger E. Kanet (eds.), *Background to Crisis: Politics and Policy in Gierek's Poland* (Boulder, 1981), 95–134.

6. Kazimierz Brandys, *Warsaw Diary* (New York, 1983), 41–42.

7. Kołakowski's address is reprinted in Peter Raina, *Political Opposition in Poland, 1954–77* (London, 1978). Raina's study is essentially a compendium of documents by varied sectors of the opposition intelligentsia. As such, this 1977 work differs from Mr. Raina's later efforts, which are full of political interpretation that can perhaps most gently be described as idiosyncratic. In any case, the crisis of 1968, which spread immense discontent through intellectual and student circles, was wholly unnecessary, a product of utterly pointless censorship, factional maneuvering within the Central Committee, and latent anti-Semitism at the elite level of the party.

8. Maria Hirszowicz, "The Polish Intelligentsia in a Crisis-ridden Society," in Stanisław Gomułka and Antony Polonsky, *Polish Paradoxes* (London and New York, 1990), 139–59; Leszek Kołakowski, "The Intelligentsia," in Brumberg (ed.), *Poland: Genesis of a Revolution*, 54–67.

9. Daniel N. Nelson, "Subnational Policy in Poland: The Dilemma of Vertical Versus Horizontal Integration," 65–89, and Andrzej Korboński, "Victim or Villain" Polish Agriculture Since 1970," 271–98, both in Simon and Kanet (eds.), *Background*

to Crisis; Jack Bielasiak, "The Party: Permanent Crisis," in Brumberg (ed.), *Poland: Genesis of a Revolution*, 10–25.

10. Jeffrey C. Goldfarb, *On Cultural Freedom: An Exploration of Public Life in Poland and America* (Chicago, 1982), 66–101; American interview no. 9, June 11, 1984.

11. Brandys, *Warsaw Dairy*, 41, 42.

12. Alexander Gella, "An Introduction to the Sociology of the Intelligentsia," in Gella (ed.), *The Intelligentsia and the Intellectuals*, 9–34; Gella, "The Life and Death of the Old Polish Intelligentsia," *Slavic Studies*, 30 (1971): 1–27; Norman Davies, *Heart of Europe* (Oxford, 1984), 331.

13. So pronounced were these views that observers in the West and in Poland have had little difficulty in characterizing attitudes within the Polish intelligentsia. The contradictions that the rise of Solidarność introduced into this universe necessarily generated a certain disarray—again both in Poland and in the West. The chapter represents an effort to cope with this disarray.

14. Steven, *The Poles*, 216–21; Kazimierz Brandys, A Question of Reality (New York, 1980), 161.

15. Brandys, *Warsaw Diary*, 157.

16. Jan Józef Lipski, *KOR: A History of the Workers Defense Committee of Poland* (Berkeley, 1985), 79, 281–84, 575.

17. Lipski, *KOR*, 178–82.

18. Paris interview no. 26 (Mirosław Chojecki), Feb. 24, 1983; Lipski, *KOR*, 182.

19. Lipski, *KOR*, 73, 77, 161–65, 346; Adam Michnik, "The New Evolutionism," *Survey*, vol. 22, nos., 3–4, (1976).

20. Lipski, *KOR*, 87.

21. Oliver McDonald, *The Polish August* (San Francisco, 1981), 142–45; Lipski, *KOR*, 56, 85–86; Kuroń interview in MacShane, *Solidarity*, 41.

22. Lipski, *KOR*, 98–112, 165–68, 186–93.

23. Kuroń's lack of enthusiasm for the Katowice initiative is described in detail in Lipski, *KOR*, 240–42, 254. For KOR's reaction to the Gdańsk initiative as articulated by Lityński, see *Tygodnik Solidraność*, no. 2 (Apr. 10, 1981).

24. Adam Michnik, *Letters from Prison and Other Essays* (Berkeley, 1985), 135–48, esp. 144–45.

25. The Solidarity Archive at Harvard provides the most convenient access to this periodical available in America. Rather than an organizer's guide, *Robotnik* served principally as a vehicle for providing news of opposition activity, broadly defined. Stories that touched in concrete ways on shop-floor organizing were not a part of *Robotnik*'s fare. Such activity at that point of production was beyond the ken of the editors. Kuroń's and Modzelewski's manifesto was translated and published in English as *Revolutionary Marxist Students in Poland Speak Out: An Open Letter to Communist Party Members* (New York, 1969).

26. Staniszkis, *Poland's Self-Limiting Revolution*, 120–46; Majkowski, *People's Poland*, 153.

27. Borusewicz's role is traced in Chapter 4, pp. 170–73.

28. Lipski, *KOR*, 340–41; *Tygodnik Solidarność* (Aug. 14, 1981); Steven, *The Poles*, 223–25; Alina Pieńkowska, quoted in Wałęsa, *Way of Hope*, 110–11.

29. The depth of worker self-education on the Baltic coast throughout the months of struggle in Dec. 1970 and into the spring of 1971 is richly detailed in Roman Laba's forthcoming book, *The Roots of Solidarity*. See "Ways of Seeing: A Critical Essay on Authorities."

30. As detailed in Chapters 3 and 4.

31. The matter of the structural base of a democratically functioning "civil society," is discussed in terms of evidence generated in the Solidarność era in Chapter 7 and as a theoretical and cultural proposition in Chapter 9.

32. I have been able to identify, through oral investigation, organized worker assertion in the 1970s at some half-dozen enterprises in Gdynia (specifically including the Paris Commune Shipyard), Gdańsk (including the docks), and at both Elbląg and Szczecin. Systematic oral investigation of individual enterprises, with a particular view to isolating the structural and ideological evolution of worker self-activity over time, would, I am confident, expand this list manyfold. Now that the police grip on social information has been broken, the way seems clear for such studies in Poland.

33. The failure to interpret the dynamics of working-class assertion before, during, and after the "Polish October" of 1956 is endemic in the literature on Solidarność. This necessarily includes those portions of books on Solidarność that deal with the politics of 1956. See, for example, Neal Ascherson, *The Polish August,* in which the author asserts that worker councils in 1956 not only "demanded control of management and production, ignoring proposals for trade-union reform, but they also—the Warsaw car plant at Żerań especially—became both discussion clubs for anti-Stalinist revolution and potential bastions of armed force to fight for revolutionary gains" (p. 234). This judgment, especially if toned down and stripped of romantic hyperbole, would apply much more to the self-organized workers at Cegielski than to the somewhat mystified rank and filers of Żerań, who had been forced to try to assert themselves in political context shaped by the party plant committee and the somewhat beguiled Lechosław Goździk.

In this and in many similar scholarly and journalistic examples, the failure to perform research on worker self-activity is concealed within a term of description. Acts of popular assertion are characterized as "spontaneous." Thus in 1956, "The spontaneous development of worker councils was a phenomenon to which the new regime was bound to give its blessing." Richard Hiscocks, *Poland: Bridge for the Abyss?*, 245. In the literature of Solidarność, the failure to research the interior of the workers' movement has given rise to fundamental misinterpretations of the role of intellectuals and the episcopate in the causal dynamics of the Polish August, as interpreted by Holzer, Pomian, Garton Ash, and indeed, Poland watchers in the aggregate (cf. "Ways of Seeing: A Critical Essay on Authorities," pp. 443–60).

34. Polish interview no. 1, Apr. 23, 1983. As previously noted, this is a collective citation for foreign correspondents contacted by the author in Warsaw. The group includes journalists from Reuters, the BBC, the London *Daily Telegraph*, the *Los Angeles Times*, and the American Broadcasting System.

35. Robinson (ed.), *August 1980, The Strikes in Poland*, 49.

36. Oral evidence here is so rich that I believe it is capable of providing the evidential base for a monograph on August 17–18, 1980: "The Consolidation of the Polish August." Virtually every activist in the tri-cities has a story to tell of this passionate evening of government propaganda and worker organizing activity and of the dramatic consequences the following day in the Lenin Shipyard. The subtopics of the organizing effort include printing, leaflet-composition, leafleting, impromptu telephone committees, flag gathering, and, as one activist said to me in Paris, "very uncoordinated coordinating meetings."

Perhaps the most intricate subtopic, one touched upon intermittently throughout this book and summarized in the concluding chapter, concerns the dynamics of social "fear" and its transcendence through organizing and movement building. It is one of the truisms of the organizing process, particularly applicable to successful efforts at moments of high opportunity and genuine pressure, that the process itself appears inherently "disorganized," however much its underlying strategic thrust is experientially coherent. Voluntary social formations almost always come into being under crudely experimental conditions. Indeed, the fact verifies their democratic quality.

37. Paris interview no. 4, Jan. 25, 1983; Polish interview no. 2, Apr. 24, 1983; American interview no. 8, June 10, 1984, and no. 14, May 21, 1985; Polish interview no. 27 (Lenarciak), Aug. 9, 1989.

38. The courier war also merits a monograph. Subtopics include methods of concealing leaflets in trucks and automobiles; varieties of police interrogations; forms of brutality by the security services; the activities of couriers to report on the arrest of couriers; and inadvertent detentions of foreign journalists.

I cannot conclusively verify whether the first loudspeaker announcement about license-plate numbers of unmarked police vehicles occurred on Aug. 18 or 19. Sources interviewed in the West during martial law conflicted, and I was unable to resolve the matter on my final trip to Gdańsk in 1989. The more important point, of course, is that the courier war is a proper topic for systematic research on the Polish August, for it concerns internal communications, an essential component of large-scale movement building.

39. This explains why interfactory strike committees and postulates emphasizing free trade unions did not materialize elsewhere in Poland in range of Radio Free Europe. In Warsaw generally and at the Ursus tractor plant specifically, where KOR had claimed it had, through Bujak, "prepared the consciousness of the workers for the strikes," no MKS and no first demand appeared. This result was traceable not only to the fact that neither KOR nor Radio Free Europe delivered such a message; more organically, this kind of experiential knowledge was not part of the organizing heritage of worker activists at the Ursus plant. That is, even had an explicit organizing message been conveyed specifying how to form an MKS and explaining why free unions were necessary, no experiential context existed at Ursus that could have served as a basis for the sequential organizing steps that necessarily had to follow. Social knowledge is experiential; there are sharp limits to what can be learned from press releases and distant radio stations, particularly when central components of the organizing purpose and organizing structure are not included in the message.

40. Kuroń had one tested contact among the coastal militants—Bogdan Borusewicz. While he was not on the Strike Presidium, Borusewicz had impeccable movement credentials as one of the original members of the "Free Trade Unions of the Coast" formed in 1978. After Kuroń first read the twenty-one demands, he had endeavored to dispatch, through Bieliński, his reservations about the first demand to Borusewicz; apparently, to no effect. Why, Kuroń, after the twentieth, could not know since he was in prison. In the event, Borusewicz, like all the coastal militants, was *for* the first demand. (Chapter 9 contains a summary discussion of the KOR myth and some of its fanciful explanatory components, including "the long-distance line to Borusewicz," and the "loop through Radio Free Europe.")

41. *Svenska Dagbladet*, Aug. 25, 1980. The translation used here is drawn from Radio Free Europe's detailed summary on Aug. 26 of the report on KOR's position by correspondent Sheutz.

The subject of Radio Free Europe's role in the Polish August brings into review the relationship between the subtleties of simple cognition on the one hand and the structural requirements of movement building on the other. Foreign broadcasts were quite instructive to people in Warsaw and elsewhere in Poland who were not privy to the organizational dynamics that lay behind occupation strikes, interfactory mobilizations, and the twenty-one Gdańsk demands. During the August telephone and telex blockade, what those distant from the coast knew of the strikes they learned from Radio Free Europe and the BBC or heard by word-of-mouth from other Poles who had heard it from those sources. Such information was certainly exciting, thoroughly educational, and, therefore, seemingly very important. In the absence of any information from the interior of the movement as to the specific organizing process through which the Gdańsk strike committee recruited its proliferating affiliates, one could conclude—from afar—that foreign broadcasts were playing some sort of seminal causal role in the spread of the strike. In effect, what observers at the time and subsequent scholars have done is to project the informational role foreign broadcasts had for them personally onto the coastal workers. This is precisely the kind of causal misinterpretation that suffuses the literature on the origins of Solidarność. Sources include *The Polish Challenge* by Kevin Ruane, the BBC correspondent in Warsaw, and William F. Robinson (ed.), *August 1980, The Strikes in Poland.*

The specifics of the informational flow do not support substantive conclusions about a causal role of either Radio Free Europe or the BBC. The adamant stand on the preconditions that characterized the Strike Presidium's first meeting with Jagielski on Aug. 23 not only threw the party into a negotiating crisis which lasted for forty-eight hours but also spread deep anxiety through both the democratic opposition and the episcopate in Warsaw. Both KOR and the Primate broke under this strain. Radio Free Europe's analysis on Aug. 26 paralleled KOR's strategic conclusion that settled achievements had already been obtained short of independent unions: "The strikes have already achieved some substantial gains, the most significant of which is Gierek's promise of secret and democratic trade-union elections with an unlimited number of candidates proposed from the floor." Roman Stefanowski, "In the Wake of the Leadership Reshuffle, 26 Aug. 1980," in Robinson (ed.), *The Strikes in Poland*, 57.

This language precisely tracked the reasoning in the Aug. 25 KOR advisory to back off the first demand in order to consolidate the gains already made. Strategically, this response is implicit in Kuroń's original analysis of Aug. 17. In any case, these are matters of high strategy. In terms of movement building, neither Radio Free Europe nor Kuroń possessed the hard movement knowledge necessary to materially assist the MKS in its recruiting challenge (see Chapter 9).

KOR's analysis spread through the affiliated émigré network of intellectuals. In Paris, Krzytof Pomian, in an interview on Aug. 25 with *Le Monde* emphasized the need for a "compromise."

42. This perspective provides an interesting counterpoint to the government view, expressed by *Polityka*'s editor in chief, Mieczysław Rakowski, on August 23: "Representatives of anti-socialist forces are putting words in workers' mouths, words aiming at a permanent destabilization of the state" (*Polityka*, p. 11).

The underlying reality was far removed from these two views, offered, respectively, by the democratic opposition in Warsaw and government officials in the same city.

43. Paris interview no. 26 (Mirosław Chojecki), Feb. 24, 1983; Lipski, *KOR*, 200, 302.

44. Paris interview no. 26 (Mirosław Chojecki) Feb. 24, 1983; American interview no. 6 (Boston), Feb. 13, 1984; the *New York Times*, Aug. 18, 19, 1980.

45. Bernard Guetta, in *Le Monde*, quoted in Barbara Torunczyk, *Gdańsk 1980, Oczyma Świadkow* (London, 1980), 28; *Trybuna Ludu*, Aug. 21, 22, 26, 29, 1980; *O Estado de Sao Paolo*, Mar. 2, 1984, Foreign Broadcast Information Service, II, Mar. 7, 1984, cited in Roman Laba, "Worker Roots of Solidarity," *Problems of Communism*. The episcopate's attack on Kuroń was made by Father Olojzy Orszulik, deputy secretary

of the Episcopal Conference in an interview with Western correspondents in Dec. 1980. Unsurprisingly, this position, with which the new Polish Primate, Józef Cardinal Glemp, episodically associated himself, generated a flood of complaints, subsequent denials, and clarifications. Szajkowski, *Next to God . . . Poland*, 113–14.

46. American interview no. 6 (Boston), Feb. 13, 1984.

47. Starski, *Class Struggle in Classless Poland*, 66–70. Robinson (ed.), *The Strikes in Poland*, 13–14.

48. Lech Bądkowski, "The Man of What?" in *The Book of Wałęsa*, 110–11; Polish interview no. 2, Apr. 24, 1983.

49. Lipski, *KOR*, 421–22; Kemp-Welch, *Gdańsk Transcripts*, 180–81.

50. The "Memorandum of the 59" is reproduced in Raina, *Political Opposition in Poland*, 213.

51. The origin of the name "Solidarność" has been attributed to Bieliński (Lipski) and to Krzysztof Wyszkowski (Laba).

52. *Trybuna Ludu*, Aug. 24, 25, 26, 27, 1980.

53. Szajkowski, *Next to God . . . Poland*, 92.

54. Ibid., 95–96. The party, not content with the utility of facts, "improved" the Primate's address by selective editing. When the full text appeared, the Church was able to defend its good name by claiming the sermon had been doctored. While true, the full, uncensored version was damaging enough.

55. I am indebted to Tadeusz Szafar for this bit of lore from the Polish August. In his autobiography, Wałęsa treats the episode with remarkable generosity toward the Church (*Way of Hope*, 137).

56. The bishops—seeking to ground themselves as firmly as possible while taking a position independent of the Polish Primate—were quoting in this passage from the Second Vatican Council. Szajkowski, *Next to God . . . Poland*, 97.

6. The Movement Consolidates Democratic Space: The Birth of Solidarność

1. The broad scholarly account of Polish history that effectively integrates the role of the Church is Norman Davies's definitive study, *God's Playground: A History of Poland*. Works focusing on the post–World War II Church and emphasizing the period immediately prior to and extending beyond the Solidarność era include: Szajkowski, *Next to God . . . Poland*; and Tischner, *Spirit of Solidarity*.

2. This brief reference centers on one of the central themes of this book—the manner in which certain ingrained assumptions of high culture make it difficult for journalists, scholars, and other intellectuals to engage the complexity, diversity, and meaning of the social experiences of nonintellectuals. This entire matter (inherently more controversial among intellectuals than it is among workers) is discussed in some detail in Chapters 7, 8, and 9 as well as in the text that immediately follows in this chapter.

3. Paris interview no. 18, (Waldemar Kuczyński), Feb. 17, 1983; Garton Ash, *The Polish Revolution*, 73, 105. Book-length expressions of the frustrations of Polish intellectuals prior to and during the period of Solidarność include: Szczpiorski, *The Polish Ordeal: The View from Within* (London, 1982); Starski, *Class Struggle in Classless Poland*; Brandys, *Warsaw Diary*, and *A Question of Reality*; and the study compiled by the Experience and the Future discussion group, translated and published in English as *Poland Today: The State of the Republic*. In a much larger sense, of course, much of Polish literature can be seen in this light—as a cry for freedom written under the

pressures of alien occupations. Nevertheless, in the context of the worsening economic crisis of 1980, the specific professional frustrations of economic specialists such as Kowalik and Kuczyński acquired a singularly vivid quality. Waldemar Kuczyński, *Po Wielkim Skoku* (Warsaw, 1981).

4. The panel of advisors assembled by Mazowiecki and Geremek was shaped, understandably enough, without consultation with Wałęsa or anyone else on the Strike Presidium. The quite valid assumption among the intellectuals was that coastal workers did not have workable knowledge of the relative utility to their cause of individual members of the Warsaw intelligentsia. The professional qualifications of the assembled panel were rather broader than they needed to be—philosophy, history, economic history, sociology, economics, and, in the case of Mazowiecki himself, a practicing editor of an intellectual journal. Only one was a practicing lawyer, Leszek Kubicki of the Institute of Law at the Polish Academy of Sciences. Unfortunately, Kubicki went to Gdańsk under the assumption that he would be advising both sides, and when he learned he would be representing only workers, he withdrew. The fact emphasized the political nature of the entire undertaking—which was, in fact, quite delicate. As the statement they brought with them to Gdańsk indicated, Mazowiecki and Geremek saw their role as being one of mediators who took the workers seriously and, as such, could be counted on to respect the workers' cause. At the same time, the panel could not appear to be anything that could be described as "too radical." Thus, the Flying University was well represented, but not KOR, though the economist Kuczyński had ties to both. But Kubicki, who declined, and Staniszkis were more conscious of the party as an influence than the others. In general, the advisory group could be described as composed of independent-minded intellectuals. But with Kubicki's departure, it had no practicing lawyers. This circumstance proved rather awkward for members of the Strike Presidium to explain to their worker constituents in the MKS and the shipyard—a condition that accounted for the "selective way" the credentials of the panel were described. Kowalik's misreading of worker politics did not augur well for his ability to grasp the meaning of the negotiating strains soon to follow.

Kowalik's "Experts and the Working Group" forms a part of Kemp-Welch's English translation of the Gdańsk negotiations, *Gdańsk Transcripts*. The cited passage is from pp. 143–44.

5. Jadwiga Staniszkis, *Poland's Self-Limiting Revolution* (Princeton, 1984);

6. Robinson, (ed.), *The Strikes in Poland*, 18, 121. As a critic of Gierek's economic policies Olszowski could be interpreted, and often was, as a "focal point of reformist elements in the party." Ewa Celt and Roman Stefański, "Major Shake-up in Poland's Leadership," *The Strikes in Poland*, 121). See also Staniszkis, *Poland's Self-Limiting Revolution*: "The best guarantee for [the workers'] movement was to create strong links with the anti-bureaucratic movement in the Communist party (which at the same time was anti-clerical, ideologically oriented, and opposed to Gierek's "window-dressing" liberalization), represented by the faction headed by Olszowski and Grabski" (p. 49).

7. Persky, *At the Lenin Shipyard*, 115.

8. *Trybuna Ludu*, Aug. 24, 25, 26, 1980.

9. In the final pre-talk negotiations, the demands of the Interfactory Strike Committee, specifically addressed to Gdańsk prefect, Jerzy Kołodziejski, opened with these words: "We categorically demand that all repression of Interfactory Strike Committee helpers cease and that any further repression be prohibited. In the last few days, the militia and security service have repeatedly picked up, interrogated and detained people distributing Strike Committee publications . . ." Kemp-Welch, *Gdańsk Transcripts*, 37.

10. *Solidarność Strike Bulletin* no. 2, Aug. 24, 1980, attempted to dramatize the issue of the preconditions by describing the opening negotiating session the preceding day in the following terms: "As the authorities have not compiled with our precondition regarding the reestablishment of the telephone service, the meeting did not in practice proceed beyond the statement by Deputy Premier Jagielski." This, of course, was not true; but most of what the workers did say concerned the preconditions.

11. Kemp-Welch, *Gdańsk Transcripts*, 69, 71–72.

12. Ibid., 75.

13. Ibid., 77–79.

14. Ibid., 82.

15. Ibid., 83.

16. Ibid., 146–47; Persky, *At the Lenin Shipyard*, 123.

17. American interview no. 8 (New York), June 10, 1984.

18. Potel, *Promise of Solidarity*, 150–52; Paris interview no. 3, Jan. 4, 1983.

19. Staniszkis, *Poland's Self-Limiting Revolution*, 58–59.

20. Kowalik, in Kemp-Welch, *Gdańsk Transcripts*, 148–49.

21. Staniszkis, "On Some Contradictions of Socialist Society," *Soviet Studies* (Apr. 1978), 167–87; *Poland's Self-Limiting Revolution*, 51, 119, 122, 137, 313–14.

22. Staniszkis, *Poland's Self-Limiting Revolution*, 112, 131.

23. Staniszkis, *Poland's Self-Limiting Revolution*, 59; Staniszkis, "Evolution of Forms of Working-Class Protest," *Soviet Studies* (Apr. 1981), 204–31.

24. Staniszkis, *Poland's Self-Limiting Revolution*, 59–62.

25. The academic advisors—a bit stunned by the coldness of the party's attitude (in contrast to the conviviality of their first meeting) and the firmness of the party's pronouncements with respect to its "leading role"—did not anticipate the workers' relative indifference. Kowalik wrote: "I do recall that this cardinal point did not give rise to much interest at the plenum. Much greater interest was shown in lesser matters . . ." "Experts and the Working Group," in Kemp-Welch, *Gdańsk Transcripts*, 149.

26. The bulk of the text of Jagielski's television address is available in Raina, *Independent Social Movements in Poland*, 545–50.

27. Robinson (ed.), *The Strikes in Poland*, 18.

28. Cegielski waged a 24-hour warning strike on the twenty-ninth, announcing that "should the shipyard workers postulates not be met, a full-fledged strike would be proclaimed." Starski, *Class Struggle in Classless Poland*, 87. Also on the twenty-ninth, word was received in the shipyard that twenty thousand Silesian copper miners had struck—the "first significant action to occur close to the 81 coal mines centered near Katowice." The *New York Times*, Aug. 30, 1980.

29. Sanford, *Polish Communism in Crisis*, 52–58. Sanford, an engaged student of the party for many years, concluded: "The public, and apparently sincere, acceptance of the strikers' conditions was arguably the greatest humiliation which any ruling CP has ever had to accept." This view was undoubtedly dominant among the party's upper-level apparat. Coastal party leaders such as Fiszbach and Kołodziejski, of course, had already made it clear they saw agreement as an opportunity for party redemption, not "humiliation."

30. Paris interview no. 31, May 8, 1983; Kemp-Welch, *Gdańsk Transcripts*, 83; American interview no. 16, July 16, 1985.

31. Kowalik, "Experts and the Working Group," in Kemp-Welch, *Gdańsk Transcripts*, 149–50.

32. Paris interview no. 3, Jan. 4, 1983; American interview no. 8 (New York), June 10, 1984.

33. Kowalik, "Experts and the Working Group," in Kemp-Welch, *Gdańsk Transcripts*, 166–68.

34. Unfortunately, "satisfactorily" constitutes an understatement of the advisory group's sense of its own literary and political achievement, as interpreted by Kowalik. The paragraph "drawn up by our group of experts, headed by Mazowiecki," was viewed as containing a "symmetrical political formula" as follows: "The Interfactory Strike Committee would undertake to observe the Constitution, to accept the common ownership of the means of production as the basis of the system, and not to question the leading role of the party or its international alliances. There followed a series of undertakings on the government's side: to guarantee the complete independence and self-government of the new trade unions, to provide conditions for fulfillment of its basic functions and to introduce appropriate changes of legislation." Kowalik, in Kemp-Welch, *Gdańsk Transcripts*, 149. Such was the "important notion of the reciprocity on which the entire Agreement is based."

The formula of the Gdańsk agreement did not turn on fine phrasing. Rather, the regime faced a functioning coast-wide general strike and the increasing threat of a wider national general strike. In exchange for going back to work, the strikers sought independent trade unions. This was the "reciprocity" achieved by the agreement. Tadeusz Kowalik seems to have been guided by the presumption that the real test of power in Gdańsk was not between the worker representatives of a coastal general strike and a regime that urgently needed to get people back to work but rather consisted of a contest of intelligence between the advisors and the government. This is the essential theme of his restrospective account!

35. Kowalik, in Kemp-Welch, *Gdańsk Transcripts*, 156.

36. Ascherson, *The Polish August*, 164; Persky, *At the Lenin Shipyard*, 132; Kowalik, in Kemp-Welch, *Gdańsk Transcripts*, 156; Potel, *The Promise of Solidarity*, 166–67.

37. Kowalik, in Kemp-Welch, *Gdańsk Transcripts*, 155–56. Amid the intense pressure that surrounded the Gdańsk negotiations, Bieliński lost his poise because he was unable to differentiate between KOR's long-term goals, independence from one-party totalitarianism. and the short-term goal of an immediate democratization within the existing party system. Earlier, KOR's venerable apostle of democratic socialism, 91-year-old Edward Lipiński made the distinction that eluded Bieliński. In his critique of the moderate reforms proposed by the Experience and the Future group (DiP), Lipiński was sharply critical of the report's long-term strategic shortcomings while being more tolerant of its short-term utility. Potel, *Promise of Solidarity*, 123.

38. Wałęsa passes over the Bieliński intervention without comment in his autobiography. So does Lipski in his history of KOR. Kowalik, deeply offended by the KOR intrusion, presents a detailed account in Kemp-Welch, *Gdańsk Transcripts*, 155–57.

39. Ibid., 156–57.

40. Kowalik, in Kemp-Welch, *Gdańsk Transcripts*, 158–59; Persky, *At the Lenin Shipyard*, 131–32; Potel, *Promise of Solidarity*, 166–70.

41. Garton Ash, *The Polish Revolution*, 49–50.

42. Staniszkis, *Poland's Self-Limiting Revolution*, 121–22; Garton Ash, *The Polish Revolution*, 50.

43. Wałęsa, *Way of Hope*, 42–43.

44. What might be called the generalized "tendency of inquiry" informing the literature on Solidarność concerns the holistic terms of description used to characterize workers and to replace research on them. Various aspects of this problem are treated in this and the following chapter, in the conclusion, and in the "Critical Essay on Authorities."

45. Ascherson, *The Polish August*, 160–66; Potel, *Promise of Solidarity*, 171.

46. Bolesław Fac, "Lech Wałęsa—The Man Who Spoke Up," *Book of Wałęsa*, 65–66.

47. Garton Ash, *The Polish Revolution*, 65.

48. Bronisław Geremek, in *Who's Who, What's What in Solidarność.*

49. Some writers have elected to see the presidium's effort to restore the organizational integrity of the MKS by checking credentials as evidence of the disappearance of democratic poise or proof of Solidarność's undemocratic character. Ascherson, *The Polish August*, 164; Staniszkis, *Poland's Self-Limiting Revolution*, 50.

50. Potel, *Promise of Solidarity*, 171–72; Garton Ash, *The Polish Revolution*, 66.

51. Persky, *At the Lenin Shipyard*, 140.

52. Waldemar Kuczyński personally communicated the sense of embarrassment he and some of the other advisors felt at the exaggerated acclaim they received upon their return to Warsaw. Paris interview no. 18, (Waldemar Kuczyński), Feb. 17, 1983.

7. Solidarność Creates a Democratic Culture: Civil Society in Leninist Poland

1. Research on the empirical building blocks of popular democratic forms has been intense for the past thirty years both in England and in America. While research by social historians has produced a revolution in the methods of historical inquiry, it is gradually becoming clear that parallel work by political scientists contains the seeds of a potentially transformative theoretical understanding of the prerequisites for the achievement of an empowered and democratic civil society. Pioneering and influential works in both these areas are Sheldon Wolin, *Politics and Vision* (Boston, 1960), and E. P. Thompson, *The Making of the English Working Class* (London, 1963). Among other works of special significance are: Christopher Hill, *The World Turned Upside Down* (New York, 1972); Hannah Arendt, *On Revolution* (London, 1973); J. G. A. Pocock, *The Machiavellian Moment* (Princeton, 1975); David Montgomery, *Workers Control in America* (New York, 1979); John Schaar, *Legitimacy and the Modern State* (New Brunswick, 1981); Sean Wilentz, *Chants Democratic* (New York, 1984); James Scott, *Weapons of the Weak* (New Haven, 1985); Barrington Moore, *Authority and Inequality Under Capitalism and Socialism* (Oxford, 1987); Wolin, "What Revolutionary Action Means Today," *Democracy*, 2 (Fall 1982).

With the creation of Solidarność as a legal institution on Aug. 31, 1980, one emerges from the obscure dynamics of movement building to the more familiar world of high politics. This circumstance bears directly on the structure of the present chapter. The abundant scholarly and journalistic accounts of events in Poland over the fifteen months of legal Solidarność detail the activities of the Central Committee of the party, the upper reaches of the episcopate, and the Solidarność National Coordinating Committee as these three entities interacted at the level of state power. Tensions within each of these groups has also been a focus of attention, essentially embracing the following subjects: factional maneuvering within the Politburo and the Central Committee, both exacerbated by the emergence of the horizontal movement in the party's lower echelons; strains between Józef Cardinal Glemp and his priests; and struggles within Solidarność, described as being between "moderates" and "radicals" or sometimes between "pragmatists" and "fundamentalists." Hovering in the background in most accounts is the Soviet military threat and the activities, real and potential, of the Soviets' domestic proconsuls in Poland, the security services, and the Polish army.

The following two chapters touch on all these matters, though I have not been particularly interested in intruding into certain scholarly disputes about General Jaruzelski and the Soviets, First Secretary Kania and the Soviets, the strategic sagacity or myopia of the high episcopate, or personal rivalries within Solidarność in ways beyond their underlying structural significance.

Rather, the trajectory of this chapter flows from the following developments in Poland which seem to me to contain the most significant long-term meaning. For fifteen months in 1980–81, Poland possessed the most self-conscious and self-empowered civil society in the world. In a formal political sense, the vast social formation that was Solidarność could not, and knew it could not, formally take state power, for the collapse of party authority would manifestly bring Soviet troops to Warsaw. But this fact, strategically constraining as it was, did not prevent Solidarność from becoming the most expansive and enduring laboratory of organized popular assertion that has existed in any society at any time in the twentieth century. I am essentially concerned with the task of clarifying the structural components of this large-scale experiment in democratic forms and the complex social relations that subsequently materialized among millions of people. These altered relationships were experienced by the Polish people as class and cultural assertions; that is, Poles experienced politics as party members, as workers, as intellectuals, as peasants, and as Catholics. But social life was also experienced by people possessing certain controlling political visions—as believing Leninists, as state functionaries, as Polish nationalists, and, broadly speaking, as social democrats and as Christian democrats.

From these complex dynamics, it seems to me that certain social drives materialized that merit sustained attention and analysis: the emergence of individual and collective self-activity as a national phenomenon in the course of the organizational spread of Solidarność from the Baltic coast to the whole of Poland between September and December 1980; the emergence in 1981 of the movement for self-management under the "Network"; the intense internal dialogue between Solidarność's blue- and white-collar grass roots on the one hand and its local, regional, and national leadership—an institutionalized conversation that came to be known as "fire fighting"; and the "horizontal movement" engendered inside the party by the appearance of Solidarność. The structural confinement of the horizontal movement proved, one last time, the institutional invulnerability of Leninist forms to democratic transformation from below.

Four sites of strategic interest also materialized—in the party, in Solidarność, in the Church, and in the intelligentsia. In this chapter, the central focus is upon the struggle between the ruling party and the popular movement. In Chapter 8, the Church is investigated, and an analysis is made of the social and political interpretations of the foremost strategic thinker within the Polish intelligentsia who publicly put his ideas into a sustained body of written commentary: Adam Michnik. High politics is visible as the shifting terrain within which the social infrastructure of Solidarność endeavored to create itself as a nationwide democratic polity.

2. Ryszard Kukliński, "Wojna z narodem widziana od środka" ("War Against the Nation Seen from Within"), *Kultura* (Apr. 1987), 3–57.

3. Paris interview no. 18 (Waldemar Kuczyński), Feb. 18, 1983.

4. Paris interview no. 10, Jan. 30, 1983.

5. Wałęsa voiced this opinion on August 20. Four days later, he told BBC reporter Brian Walker that the regime would have to respond to the workers' demands because the workers "could strike for five years." Robinson (ed.), *The Strikes in Poland*, 386–87.

6. In theoretical terms, the rather tedious debate about "what Marx meant" and "what Lenin meant" has been complicated by the fact that Marx's original phrase "Sein bestimmt das Bewusstsein" is a play on words in German: "Sein" (being) and "Bewusstsein" (conscious-being) with the verb "bestimmt" (untranslatable in most languages)

most often rendered as "determines." The word "bestimmt" is probably most accurately translated with three English words—shapes, limits, and conditions. I am indebted to my colleague Dirk Philipsen for calling these relationships to my attention. A simplistic interpretation of this sentence would generate the following trajectory—being, or experience, determines one's consciousness; in societies industrializing under capitalism, one's relationship to the means of production determines one's political view; the proletariat is the class whose experience reveals the exploitive nature of capitalism; the proletariat is thus the revolutionary class, the "universal class" in whose name the capitalist mode of production is to be transcended. That is, a simplistic reading is mechanistic in the manner it informs the subject of production-consciousness-politics with what appears to be a scientific dynamic. But "conscious-being" is a human achievement, a product of reflection; it is not programmed or otherwise historically determined. Stripped of the suggested power inherent in a play on words, the phrase, "social knowledge is experiential" is transparently less mechanistic in appearance, in that any suggestion of a "scientific" quality is missing. Rather, the inherent subjectivity of the relationship of "being" to "consciousness" is emphasized.

In the Polish August, the power of the "vanguard" was present in the strategic analysis projected by the activists of KOR, by the academic authors of Variant "B" among the shipyard advisors, and by the Aug. 26 homily of the Polish Primate. In unanimously rejecting all three of these analyses, the members of the presidium of the Gdańsk Interfactory Strike Committee drew upon one ingredient—the strategic understanding that had grown over time out of their organizing experiences. These experiences were sufficiently powerful, sufficiently grounded in an understanding of real power relationships between the party and society, that they enabled the worker activists to overcome the accumulated wisdom of the entire Polish oppositional culture.

The inability of contemporary capitalist and Marxist political traditions to provide a conceptual framework of sufficient breadth to illuminate these relationships perhaps helps measure the limits of their democratic serviceability.

7. Paris interview no. 2, Nov. 20, 1982; no. 11, Feb. 5, 1983.

8. John J. Kulczycki, "The Beginnings of the Solidarity Movement in Poznań, 1980–1981," *Polish Review* 27, 3–4 (1982): 164.

9. This area of social activity—movement building—constitutes a central lacuna in political science and political theory. It is often subsumed under the term "spontaneous," a word which characterizes popular assertion without touching upon its origins and development. As a general principle of social investigation, the use of the descriptive term "spontaneous" may be understood as constituting prima-facie evidence that the journalist or scholar employing it has not performed research on the event being described. It is also germane to note that the kind of activity alluded to—self-activity by "ordinary people"—can be abstracted before the fact as "trade-union consciousness," "economism," or the product of "petit bourgeoise individualism" or of a "peasant mentality." Aside from the research laziness and intellectual pomposity signaled by such terminology, these descriptive euphemisms routinely conceal the social relations and the political consciousness their users seek to describe. This habit is absolutely organic to the way scholarship on Solidarność has been practiced by Poland watchers, in the West and in Poland.

As a category of social and political investigation, popular self-activity will disappear as a lacuna precisely to the extent it is subjected to sustained research. As stressed in the "Critical Essay on Authorities," this methodological addition to the research armory of modern scholarship proves generally instructive, but it is absolutely mandatory as a part of any research design concerned with clarifying those comparatively rare occasions in history when a large-scale popular movement attempts to shatter the existing political paradigm in any society.

In an interpretive sense, the task is essentially a research challenge—to place oneself inside the social formation in order to ascertain what is taking place. Descriptive formulations may then be risked. (For a specific example of this problem in the literature on the period of legal Solidarność, see note 29, on research methods applied to the "Network.")

Failure to take this precaution in describing popular insurgency is common among historians of politics, even more common among historians of ideas, and almost endemic among sociologists. There are consequences—of both an empirical and theoretical dimension. For the former, see "Critical Essay on Authorities"; for the latter, see Chapter 9.

10. Paris interview no. 3, Dec. 2, 1982; no. 12, Jan. 18, 1983.

11. Malara and Rey, *La Pologne*, 131–34, 275–76; Raina, *Political Opposition in Poland*, 74–76, 82–91, 95–100, 178–96; Maciejewski and Trojanowicz, *Poznański Czerwiec*, 215–33; Witold Jedlicki, "The Crooked Circle Club," in Tadeusz N. Cieplak (ed.), *Poland Since 1956* (New York, 1972), 27–28; Leopold Labedz, "The Polish Intellectual Climate," *Survey*, no. 35 (1961): 3–11.

12. Brandys, *Warsaw Diary*, 152.

13. That the Polish movement could mobilize society and yet not take state power created a historically unique situation, one that rendered as inappropriate a number of received categories of political description. For example, a Pole who did not understand the geopolitical situation would easily be drawn to the "radical" position of favoring "independence" and might appear more "militant" than a sophisticated insurgent. Thus the Polish Right, represented by KPN, and sectors of the British and French Left could criticize both KOR and the coastal movement for being weak and "moderate." This particular conjunction of terminology is further discussed in Chapter 8, note 31, in terms of its application to KOR.

14. Polish interview no. 20 (Krzysztof Wyszkowski), Aug. 4, 1989. The transcript of the Sept. 17 meeting has been published in *Krytyka*, no. 18 (1984). The translation I have followed here—from a recording by a union activist, was made by Roman Laba. It is hereafter cited as "Sept. 17 transcript." It differs in minor details from the *Krytyka* version. See also Laba, "The Roots of Solidarity, 223.

15. Kemp-Welch, *Gdańsk Transcripts*, 188, n. 8. Olszewski's important protective legal role—the one workers hoped to have performed for them when they agreed to a panel of academic advisors—is not focused upon in this account by Kowalik.

16. Sept. 17 transcript.

17. Ibid.

18. "It is certainly necessary to understand that natural human authority has been overwhelmed by the continued impact of the very forces, structures and intellectual and moral orientations that we identify with modernity." John H. Schaar, "Legitimacy in the Modern State," in William Connolly, *Legitimacy and the State* (New York, 1984), 124. In the contemporary world, further substantive democratic evolution is probably not possible until the intellectual and political prestige of modern "progress" is systematically and conclusively undermined as a determinate of social policy. Here, indeed, are the chains that bind us; contemporary "radical" and "conservative" traditionalists busily compete in providing the buttressing links. A powerful new treatment of this central modern condition has recently appeared: Christopher Lasch, *The True and Only Heaven: Progress and Its Critics* (New York and London, 1991).

19. The moral tension that Solidarność generated within the families of party

members was a topic that surfaced in a number of interviews, particularly the earlier ones recorded in 1982–83. An especially vivid account was given to the author in Paris in Feb. 1983 (Paris interview no. 15, Feb. 14, 1983). Party bureaucrats had inured themselves to public unpopularity and had polished the uses of cynicism in combating it, but it was difficult to respond to the questions raised by one's children through such a posture.

20. I had explored some ramifications of this problem in two essays, one as an evaluation of standard Marxist approaches ("The Cooperative Commonwealth and Other Abstractions: In Search of a Democratic Premise," *Marxist Perspectives*, no. 10 (1980)), and the second as an evaluation of standard liberal approaches ("Organizing Democracy: The Limits of Theory and Practice, *democracy* 1, 1 (1981)). The French sociologist, Alain Touraine, has approached the same conceptual challenge in the following language: "The idea of contradiction which had been imposed by capitalist exploitation, did not lead toward social movement but towards the party, and the party became a state . . . the implacable logic of which no country in the world was able to disentangle . . . From bourgeois violence we have passed on to state violence, as though a free space for social movements could never have existed." Touraine's understated conclusion: "The idea of a society enlivened by social conflict would seem to have had considerable difficulty in becoming entrenched." It may be observed that if the words "social life" are substituted for "social conflict," the focus on the constrained political and social dialogue so characteristic of modern society becomes even sharper. Alain Touraine, *The Voice and the Eye: An Analysis of Social Movements* (Cambridge, 1981), 4.

21. "The Movement for Self-Management," in Stan Persky and Henry Flam, *The Solidarity Sourcebook* (Vancouver, 1982), 177–79; Henry Norr, "Solidarity and Self-Management," *Poland Watch*, no. 7, pp. 102–4.

22. Paris interview no. 4, Jan. 25, 1983.

23. Norr, "Solidarity and Self-Management," 109.

24. Ibid.

25. Ibid., 110–11.

26. Quoted in Peter Raina, *Poland, 1981: Toward Social Renewal* (London, 1985), 298–99.

27. Norr, "Solidarity and Self-Management," 112–13.

28. Ibid., 105, 114–15.

29. For example, Ascherson, a socialist, passes over the subject in one sentence while Garton Ash, a liberal, treats it briefly and from a great psychological distance. Clearly reacting to what he regards as the extreme radicalism of the Network, Garton Ash explores the use of condescension: "It would be wrong to seek too sophisticated or Machiavellian explanations for the popularity of the self-government idea. As I heard someone say at the gate of the Lenin Shipyard in August 1980: 'If the whole economy was organized as well as this strike . . .' " (*The Polish Revolution*, 187–89; Ascherson, *The Polish August*, 272). The political meaning of Solidarność as a threat to the party is well understood; but the more subtle nuances of the Polish movement as a psychological threat to Westerners has not received comment. Indeed, within the truncated languages of traditional "liberalism" and "socialism," it cannot be treated coherently. Condescension thus becomes an urbane means of avoiding a professional task that one is not psychologically equipped to handle.

As an additional example, George Kolankiewicz elects to view the subject from afar, devoting more space to a historical review of the party's sophistic excursions into "self-management" than he does to the Network. Since Kolankiewicz engages the Network through a preponderance of party sources and other remote observers, it is perhaps not surprising that he finds the effort a "morass." The movement emerged as an "almost accidental formation"; there could be "little wonder" that the party opposed initiatives in this area. In this methodological approach, Kolankiewicz cannot penetrate to the interior of the social formation he is writing about, so that its own internal dynamics and its affect on Solidarność itself remain incomprehensible—because untreated. ("Employee Self-Management and Socialist Trade Unionism," in Woodall (ed.), *Policy and Politics in Contemporary Poland*, 129–47.) While the technique of endeavoring to "read through" party sources was a widely employed scholarly tool in the censored culture of the pre-Solidarność era, the unconscious application of this custom to the study of a social movement that somehow materialized in opposition to the party necessarily produces a static and distanced view. In effect, Poland watchers became prisoners of their sources at the apex. Unfortunately, since the movement for self-management in Poland was an innovative organizational and conceptual form that grew out of a larger and still embattled insurgency, party sources are a bit too far removed to be evidentially informed. Kolankiewicz's term "almost accidental" is a causal description closely akin to "spontaneous" (see p. 403, n. 62; p. 414, n. 33; and p. 423, n. 9 for related interpretive problems that flow from the use of such descriptive terminology).

This research habit, unfortunately, pervades the literature on Solidarność, helping to ensure that worker self-activity remains invisible. Given his international prestige and influence with Polish graduate students, Stefan Nowak of the University of Warsaw must, I think, accept some responsibility for this persisting condition in Polish sociological literature. I wish to thank Roman Laba for pointing out to me an observation by Jan Malanowski in *Polscy Robotnicy* (Warsaw, 1981) that is relevant in this context: "We did not create and are not creating any body of knowledge on the working class." The "blank" on sociological inquiry into social activity in Poland, alluded to at some length in the Introduction of this book, has been allowed to persist to a degree that is remarkable, given what has been going on in Poland since Aug. 14, 1980. A recent observation by an American sociologist, Arthur Stinchcombe of Northwestern, would seem to have an enhanced application to Poland watchers: "We must get social reality into the discipline." "General Discussion," in Siegwart Lindenberg, James S. Coleman, Stefan Nowak (eds.), *Approaches to Social Theory* (New York, 1986), 58.

30. Norr, unrivaled as a scholar of the movement for self-management, unfortunately here employs descriptive political terminology that, as we have seen, unravels when rubbed against the complexities of Solidarność. To the movement's more impatient activists, Kuroń was not "visionary"; he paid too much attention to geopolitical concerns—in the same way people like Geremek and Wałęsa did. To such activists, Kuroń, too, was a "moderate." If the term must be used, the "visionaries" of the movement were in the Network. Adam Michnik's earnest efforts to come to terms with the meaning of the Polish movement similarly became entangled in the descriptive vagueness of "moderation" and "radicalism." See pp. 324–27.

31. An activist in the Wrocław Network said: "I wanted it, other workers wanted it, we had good leaders who wanted it, but we had no immediate means . . . and we were getting *so* tired." American interview no. 7, June 9, 1984.

32. The American journalist Lawrence Weschler, who left Poland in May 1981, was struck by the fundamental difference in mood he found upon his return in September, a time of endless queueing for diminishing food supplies. "No one is laughing and the vitality has gone out of life in Poland this sorry autumn." *Solidarity*, 78.

33. Robert Eringer, *Strike for Freedom* (New York, 1982), 150; American interview no. 27, Aug. 20, 1986.

34. George Kolankiewicz, "The Politics of 'Socialist Renewal,' " in Woodall (ed.),

Policy and Politics in Contemporary Poland, 65–71; W. G. Hahn, *Democracy in a Communist Party* (New York, 1987), 31–32.

35. Paris interview no. 28, Mar. 5, 1983; Hahn, *Democracy in a Communist Party*, 32–35; Bielsiak, "The Party: Permanent Crisis," in Brumberg (ed), *Poland*, 24.

36. The party's performance at the July congress has been variously characterized. For W. G. Hahn, the party was prevented from having enough maneuvering room to accommodate to the horizontal movement because of ill-timed pressure from Solidarność. *Democracy in a Communist Party*, 127–60. For George Sanford, the party congress "confirmed the reform analysis of the causes of the crisis and put its imprimatur on the democratization process. The scale of the leadership renewal . . . was breathtaking." Kania comes through as a sophisticated leader. Sanford, "The Response of the Polish Communist Leadership to the Continuing Crisis" (to the Ninth Congress, July 1981); "Personnel and Policy Change," in Woodall, (ed), *Policy and Politics in Contemporary Poland*, 52. Jadwiga Staniszkis goes further: "The new PUWP elite consisted of ordinary people, uncorrupted, and elected through relatively democratic procedures." But when they found that society's anger was now directed at them personally, their despair mirrored that of Solidarność's increasingly anxious rank and file. "The mutual hostility between the two groups deepened, reinforced by the similarity of mentality and status of both." Staniszkis, *Poland's Self-Limiting Revolution*, 133–34. George Kolankiewicz is more complex. Democratic procedures are noted and attention is called to the increase in the number of workers over nonworkers elected to high party office. Nevertheless, he concludes: "Bald sociological statistics hide the other major unanticipated consequence of the democratic elections, namely the electoral annihilation of the 'horizontal' activists." Kolankiewicz, "The Politics of 'Socialist Renewal,' " in Woodall (ed.), *Policy and Politics in Contemporary Poland*, 70–71. Timothy Garton Ash ventures more in fewer words: "If 'democratic socialism' meant internal Party democracy, then the breakthrough had not come . . . If 'democratic socialism' meant a ruling Party which was answerable to the majority of the electorate—or at least responsive to their clearly expressed wishes—then the failure was complete." Garton Ash, *The Polish Revolution*, 181. This seems straightforward.

37. Szajkowski, *Next to God . . . Poland*, 117–20; Ascherson, *Polish August*, 264.

38. Kukliński, "Wojna z narodem widziana od środka," 34–35.

39. Garton Ash, *The Polish Revolution*, 149.

40. Transcript of meeting of Solidarność National Coordinating Commission in Bydgoszcz: "Shall We Call a General Strike," Persky and Flam (eds.), *Solidarity Sourcebook*, 161–66. American interview no. 10, June 12, 1984.

41. Joining Wałęsa in opposing an immediate national general strike were Karol Modzelewski, Bronisław Geremek, Tadeusz Mazowiecki, Jan Olszewski, and Władysław Silas-Nowicki. Transcript, *Solidarity Sourcebook*, 163–66. Andrzej Celiński, national secretary of Solidarność, provides this insight: "As soon as Wałęsa was nominated as one of the negotiators, Jacek Kuroń made a motion by which the negotiators would not have the power to sign any agreement that had not been ratified by the Commission. I interpreted this gesture of Jacek's as a clear stand in favor of the general strike—and that was exactly what it was. Wałęsa grumbled but had to accept the condition." Celiński, in Wałęsa, *Way of Hope*, 187–88.

42. Paris interview no. 27, Feb. 25, 1983.

43. Though ill, the Primate, Stefan Wyszyński, on Mar. 26, received Prime Minister Wojciech Jaruzelski for three and a half hours of wide-ranging talks that extended beyond Bydgoszcz to include the registration of Rural Solidarność. The episcopate then

"intensified its pressure on Wałęsa" on Mar. 27. Wałęsa and the entire Solidarność negotating committee were called to the cardinal's residence where the Primate's remarks were described as "unequivocal". He is quoted as saying, "I am not a melodramatic person, but I insist the situation is dangerous." Szajkowski, *Next to God . . . Poland*, 120–25.

Szajkowski provides background to an essential strategic fact to emerge from the Bydgoszcz crisis: though both the party and Solidarność were wounded in the affair, the Church succeeded in its aims concerning Rural Solidarność.

44. The Bafia Report is reprinted in Raina, *Poland, 1981*, 86–95.

45. Kukliński's *Kultura* account helps provide the necessary context for this crisis. See also Weschler, *The Passion of Poland*, 46, 140–41; Ascherson, *Polish August*, 264–65.

46. The text of the agreement is in Raina, *Poland, 1981*, 98–101. Wałęsa wrote: "I had the feeling of having defused an enormous charge of dynamite, of having staved off something that would have proved irreversible." *Way of Hope*, 160.

47. Hahn, *Democracy in a Communist Party*, contains the details of the Ninth Plenum of the Central Committee in Warsaw on Mar. 29–30, pp. 98–104.

48. In Persky and Flam (eds.), *Solidarity Sourcebook*, are excerpts from the transcript of the KKP meeting in Gdańsk on Mar. 31–Apr. 1, 1981, pp. 167–71, including the full text of Modzelewski's resignation speech, 168–71.

49. This was the view of Garton Ash, *The Polish Revolution*, 161.

50. Lech Bądkowski, the Gdańsk writer whose resolution of support of the Interfactory Strike Committee electrified the MKS on Aug. 21, 1980 (see Chap. 5, pp. 217–18) and who was subsequently elected to the MKS presidium, wrote a detailed and indignant account of attempts by "the supporters of Jacek Kuroń in the presidium of the Gdańsk MKS" to engineer a "coup d'état directed against Wałęsa or at least to accept their political supremacy." Bądkowski, in *The Book of Wałęsa*, 120–21. To my knowledge, Bądkowski's account of this abortive effort was not subsequently ever denied by the principals involved. A reference to the incident, and Anna Walentynowicz's role in it, is contained in Wałęsa, *Way of Hope*, 148–49, in a passage that serves to introduce the entire subject of the strategic debate within Solidarność, and the role of Kuroń, Modzelewski, and Gwiazda in that debate that dominated the Sept. 17, 1980, meeting. *Way of Hope*, 150–56.

This rupture came early—before Solidarność had factions that could materialize out of the struggle with the party. That is to say, the factional fight was not ideological but turned on preconceived rights to positions of prominence in the movement (see note 57).

51. Adrian Karatnycky, "Polish Conversations: Kazimiera Wóyciki," *Workers Under Communism*, no. 1 (Spring 1982): 12; Jan Józef Lipski, in the *Nation* (May 26, 1984), 637; Karatnycky, "Polish Conversations: Antoni Macierewicz," 11.

52. The phenomenon of "shadow movements" is discussed in Lawrence Goodwyn, *Democratic Promise: The Populist Moment in America* (New York, 1976), 388–401, 590–92; and *The Populist Moment* (New York, 1978), 128–29, 310–12. Frequently mistaken for social movements, shadow movements can be easily identified through research into the origins and development of the social formation that produced the movement; that is, shadow movements, like social movements, are difficult to detect accurately from "afar."

53. "Movement culture" is a descriptive term used in *Democratic Promise* to characterize the social relations that develop inside social movements that have the effect of encouraging people, through self-activity, to overcome acquired forms of deference.

Internal communication networks are one of the essential components necessary to the development of a "movement culture." *Democratic Promise*, 47–51, 540. The term is descriptive rather than analytical and cannot be safely carried too far. Individual behavioral changes dependent on collective social formations are not permanent. Modern historical evidence suggests that just as the "socialist new man" did not appear at any time after the revolution, the creation of social opportunities that encourage the development of democratic modes of social relations is necessarily a very long-term social and political process, one that is inherently contingent.

54. From a worker who activated himself during the "hospital troubles" at Łódź: "It was important. I remember it very well. It was the first time in my life that I really was able to say what I meant to the party . . . After that, I was really ready for the general strike over Bydgoszcz." In short, this moment was "important" not because of its strategic meaning for Poland, but for its personal meaning to one citizen. The worker was actually describing the moment of his entrance upon the stage of Polish politics—when the "barrier of fear" had been shattered (see Chapter 7, pp. 264–74, and Chapter 9, pp. 379–80). The worker became such an activist after this incident in Mar. 1981 that he was one of the "leaders" interned by the security police following the declaration of martial law. Paris interview no. 30, Mar. 9, 1983.

55. Attitudes of workers toward intellectuals in Poland are discussed in Chapter 1, p. 36 and note 50, p. 428, and those of the intelligentsia toward workers, Chapter 5, pp. 189–90.

56. Kuroń lived an engaged political life of unusually high intensity for a longer period (1964 to the present) than any other intellectual in Poland. In the era of Solidarność, a number of intellectuals became deeply engaged—Mazowiecki as MKS advisor and later as editor of *Tygodnik Solidarność*, the movement's national weekly; Modzelewski as press spokesman; Geremek as MKS advisor and chairman of the drafting committee for the Action Program of the movement; Jerzy Milewski, Edward Novak, and others of the Network and, to conclude with a well-known figure who is discussed at length in Chapter 8, Adam Michnik.

57. Krzysztof Wyszkowski provides explicit details of a meeting in the apartment of Mariusz Muskat in Gdańsk that lasted several days in the first week of Sept. 1980. The subject was the command staffing of the movement. Among those present, according to Wyszkowski, besides himself, were Kuroń, Helena Łuczywo, and Bogdan Borusewicz. Lech Wałęsa was not present. Though I have no particular reason to doubt this fascinating story of behind-the-scenes maneuvering by KOR, I have been unable to corroborate Wyszkowski's account. It is not a subject people want to talk about at length.

The Polish August found KOR's leading activists in prison in Warsaw, even as new worker leadership emerged, accompanied by the Mazowiecki-Geremek panel of advisors. As detailed in Chapter 7 herein, KOR was forced by these circumstances to back date its credentials through a publicity campaign to familiarize all of Poland, and particularly the newly influential leadership of coastal workers, with details of the organization's oppositional efforts over the years. The campaign was accompanied by several bold bids to assume a more active role in the Solidarność leadership.

Rumors concerning this incident, coupled with the maneuvering at the Sept. 17 national meeting of Solidarność activists and other similar leadership contentions created deep divisions within the movement in the first month of its life. Foreign press accounts (known to reflect the Western contacts of Michnik, Lipski, and Kuroń) of KOR's alleged olympian role—pulling the strings, as it were, of worker activists in the shipyard—further alienated the Mazowiecki-Geremek group, the Strike Presidium generally, and Wałęsa particularly. The government added even more to this tension by suddenly launching a press campaign against Kuroń, accusing him of trying to incite a "bloodbath" on the Baltic coast during the Polish August.

Isolated as he suddenly found himself from the protective cover of the movement, Kuroń faced the immediate prospect of going to prison at the very onset of the liberation of civil society in Poland. Understandably worried, Kuroń retired to his apartment and wrote a long, at times powerful, at times plaintive, "Open Letter to Shipyard Workers and All Coastal Workers." It was grounded in the (sound) strategic argument that the threat to his own freedom of action was also a threat to the movement's freedom. In his attempt to buttress this analysis, Kuroń included in his defense against the party's attack a statement that has subsequently been carried to absurd lengths by observers of Poland. Because these interpretive leaps have had such distorting consequences in assessing the origins and development of the popular movement in Poland, Kuroń's open letter merits a measure of attention.

After first thanking the workers for securing his freedom from prison on Sept. 1, Kuroń said: "At first sight, [the government attack] concerns Kuroń, but in fact the authorities want you to accept that they can decide who has and who has not the right to be active in trade unions. Polish television maintains that in an interview given to Swedish television, I invited workers to set fire to party committees. In fact, in July and August, warning the authorities against using force against the strikes, I reminded them of the party committees set on fire in December 1970 and June 1976. At the same time, I repeated: 'Do not burn down party committees, set up your own instead.' It is strange that Polish television should make me responsible for the blood of coastal workers."

Kuroń went on to explain why KOR was formed after the government began its repressions at Radom and Ursus in 1976: "We were ashamed that the intellectuals had been silent in 1970 and 1971 and we wanted to restore its good name." At the time of the strikes in the summer of 1980, KOR worked hard to spread information, and he wanted workers to know this. "We realized that the authorities can only use repression and suppress a strike when nobody knows about it. We knew that one plant's experiences can be of help to others. When on 18 August 1980 the region of the Baltic coast on strike was cut off from the rest of the country, we spread information about the strike, about the formation of the Interfactory Strike Committee and its twenty-one demands to dozens of big enterprises. On 20 August 1980 over a dozen or more active 'KOR' members and collaborators were arrested. It was not my first time in prison, this time a few days only, but altogether I have spent six years in prison. Now every additional day behind the bars is more difficult. It is obvious that I have very personal reasons for gratitude and I am turning to you now, when I am being slandered by every possible means in the press, radio, and television. I am approaching fifty, I have no position, no property, no titles, my only possession is my good name of which they are trying to deprive me. They resort to lies, forgery, and provocation, and they have at their disposal the most sophisticated technical means, a powerful propaganda machinery, and large sums of money for bribes. You remember well that when the Interfactory Strike Committee was formed it was said that it was the work of antisocialist elements who had found their way into the ranks of the strikers; in inner circles, it was said that 'KOR' was directing the coastal strike action.

"I do not know why the West German weekly *Der Spiegel* chose to publish this version of events . . . The authorities were also hoping that Lech Wałęsa—called a frontline lieutenant [of Kuroń]—would take offence. Ridiculous small-minded people. They do not know that only they want to become at all cost generals, secretaries, and ministers. We: Lech, Anna, Andrzej, Bogdan, myself, we are all frontline officers. We are and always want to be in the first line of the struggle for human rights, workers rights, national rights. As a matter of fact, they have somewhat promoted us, for an officer in the army gives orders, while we only can and want to listen to the people who are fighting together with us and for whom we are fighting, who believe us and who must control us.

"There are many difficult tasks in front of us . . . We cannot count on a good party secretary; we have to organize ourselves democratically and take the affairs of the country into our own hands . . . This is the meaning of the Gdańsk agreement, which is mainly your achievement. To fulfill this agreement, not to waste this great national opportunity, is our common task, the task of all Polish citizens who do not think only of their personal or group interest."

The sense of unjust isolation that pervades this letter was by no means restricted to Jacek Kuroń. Other prominent KOR activists, most notably Jan Józef Lipski, Adam Michnik, and Jan Lityński were also stunned to discover the privileged position now accorded to Mazowiecki and Geremek and even to certain lawyers who, though previously having defended KOR, acquired enhanced status not because of these actions but because of their service to the MKS in Gdańsk. In the eyes of the KOR militants, none of these other members of the Warsaw intellectual community had been remotely as active in the democratic opposition as they had been, none had paid as high a personal price, and none had specifically wrestled with the conceptual problems surrounding the subject of worker organizing in the way Michnik, Lityński, and, particularly, Kuroń, had. As Lipski's own account makes abundantly clear (*KOR*, p. 431), they felt it was outrageous that they had suffered four years of abuse, harassment and periodic imprisonment only to be cast to the sidelines at the very moment when an authentic civil society had come into being in Poland.

In 1980–81 and subsequently, each of these KOR activists, in their own way and in their own language, endeavored to tell Polish workers and the outside world what their own contributions had been. Nothing was more human than for them to have done this. Kuroń's quotation of his own statement ("Don't burn down party committees, set up your own") was not made in a context of aggressive self-aggrandizement but in defense of himself against pernicious and wholly unjust attacks on his politics by the party's propaganda apparatus at a moment when he felt extremely isolated and vulnerable.

But it is remarkable to track what subsequently was made of this one sentence. It soon became an integral, energizing feature of the seminal role of KOR in engineering the Polish August. Though I have found no concrete evidence that Kuroń himself zealously pushed this interpretation in his own analysis of KOR's role in Solidarność that was passed on to foreign journalists and scholars, there is absolutely no question that other very prominent KOR partisans did so freely and repeatedly. (Two other components in this elaborate explanation of causality concern Kuroń's "long-distance line to Borusewicz" and his "loop through Radio Free Europe," both of which are treated in the summary analysis of Solidarność in the concluding chapter, pp. 366–69. In any event, as an explanation of causality, Kuroń's one-liner, offered amidst the extreme pressures on him from the party in Sept. 1980, made it all the way into Wajda's famous motion picture of the Polish August, *Man of Iron*. Perhaps more important in terms of influence, it also became a central causal explanation in the most widely read book on Poland to be published in the West.

There are so many things wrong with this story—it is so patently ridiculous as causal evidence—that one hesitates to take time to deal with it, were it not for the fact it has been quoted to me in dead seriousness by half a dozen or so intellectuals—in Paris, Berlin, Boston, Washington, and New York, as well as Warsaw.

The purpose of the story, if it is not already self-evident to an American audience, is to suggest that it contains the wisdom that brought the Polish working class to the use of occupation strikes in 1980, following upon the conflagration at Radom in 1976, the coastal rioting in 1970, and the upheaval in the streets of Poznań in 1956. As Timothy Garton Ash summarized its centrality: "Kuroń's slogan, 'Don't burn down Party committees, found your own!' could hang as a motto over all the workers' protests of 1980." *The Polish Revolution*, 23.

Any scholar who had tracked coastal workers through the demand-gathering campaigns of 1970 and 1971, who understood that the recovery of worker-controlled occupation strikes was a result of organizing *after* the rioting in December 1970, and who understood that occupation strikes were widespread in Poland in 1976 and at many individual plants in other years between 1973 and 1979 would be aware that no "slogans" were necessary.

The story also illustrates, almost embarrassingly, how effortlessly unsubstantiated assertion by intellectuals passes into scholarly "fact" in Poland without the intervention of research. One wonders to whom Kuroń spoke those magic words—to Swedish television, to the party, to *Der Spiegel*, to the organizers of the Gdańsk MKS? Not only to whom, but when and in what context were these words spoken? The questions have not been asked because, as the literature of Solidarność copiously demonstrates, ideas move to action in Poland without the need for corroborating evidence about the self-activity inherent in action. Like Stewart Steven's stories about Kuroń's "crash courses" on philosophy and the censorship given in unknown places to "the workers," the tale of the "slogan" stands against certain known facts: (1) Kuroń proposed to worker activists that they infiltrate (cf. Lenin: "enterism") the official trade unions, not "build your own"; (2) Kuroń opposed independent union initiatives when they appeared in Katowice and Gdańsk in 1978; (3) though the subsequent continued existence of independent unions on the coast forced KOR to acknowledge their existence, KOR, *Robotnik*, and Kuroń continued to accept the idea only with the qualification that such unions be preceded by the organizing of "strong groups of workers able to defend their representatives"; (4) Kuroń opposed the demand for free trade unions when he first read it on the evening of Aug. 17, 1980, and draped a remarkably adroit veil over it in his lengthy interpretation given to the foreign press on Aug. 18 (his alternative emphasis on civil liberties and on lesser labor issues dominated the analysis broadcast by Radio Free Europe that day); (5) Kuroń continued to oppose the first demand throughout the Polish August (as he has himself freely acknowledged) and played an energizing role in the Aug. 25 KOR advisory to the Strike Presidium to back off from the demand; (6) if Warsaw intellectuals (Michnik, Kuroń, Staniszkis, Nowak, Holzer) or foreign Poland watchers (Kolankiewicz, Potel, Persky, Oliver McDonald) had kept abreast of working-class organizing in Poland during the thirty-five years of Leninist rule, no one would have *dared* to spin this yarn! and (7) the thought that a weapon as subtly dependent upon shop-floor organizing skills and experience over time necessary to meet the attendant logistical and organizational requirements could be instantly brought to social reality by the simple one-line pronouncement of a "slogan" by a distant intellectual betrays a profound innocence about the dynamics of large-scale movement building in any society and particularly in a police state. It is relevant to note that of the "dozens of big enterprises" to which Kuroń sent word "about the MKS and the twenty-one demands," none of them, (including KOR's carefully cultivated, Bujak-led platoon of activists at Ursus) proceeded to form MKS's of their own and move to a similar level of programmatic action. Experientially, they did not know how. Kuroń and KOR, experientially or theoretically, did not know how either. To suggest, as Garton Ash and so many others have done, that KOR taught Poland's workers "how to have an occupation strike, what to demand" simply bares the awesome experiential gap between intellectuals on the one hand and Baltic workers on the other in the area of movement building. It also overlooks the evidence of what KOR did do in the final two weeks of August.

In the event at hand, Kuroń did not go to prison in Sept. 1980. The party, dazed by the spread of Solidarność across all of Poland, had to seek more organically structural ways to combat the popular movement. While continued attacks on KOR and other "antisocialist elements" were necessary as an explanation of the awkward fact that the party of the working class was being challenged by a self-organized working class, the regime understood, even before the Polish August, that KOR was dangerous precisely because it was geopolitically sophisticated and therefore not "incendiary."

The struggle for control of Solidarność in the early fall of 1980 was not only intimately known by Wałęsa, Mazowiecki, Geremek, Bądkowski, Gwiazda, Borusewicz, Walentynowicz, and their sundry allies and opponents among the activists of of KOR, but it also added a specific political edge to the tensions that surrounded the Bydgoszcz crisis. Wałęsa was outraged at Kuroń for diminishing his maneuvering room during the crisis. For that matter, Wałęsa's highhandedness in the emergency meeting of the Solidarność leadership outraged Kuroń. (*Tygodnik Solidarność* tilted in Wałęsa's direction in its subdued and careful reporting of the meeting on Apr. 3, 1981.) The controlling irony was that, leaving all matters of personal conflict aside, Kuroń and Wałęsa were in basic agreement over a proper strategic response to the party's provocations that precipitated the Bydgoszcz affair. In terms of mutual trust, Wałęsa and Kuroń were now estranged. That was the heart of the matter. The roots of the problem dated to the events of Sept. 1980.

58. Kukliński's account documents this rhythm in some detail. *Kultura*, esp. pp. 32–37 for the Soviet response to Bydgoszcz and the party's internal strains attendant thereto.

59. Sanford, "Response of the Polish Communist Leadership," in Woodall (ed.), *Policy and Politics in Contemporary Poland*, 47–48.

60. Hahn, *Democracy in a Communist Party*, 98–103.

61. Kukliński in *Kultura*, 34–40; Andrew Michta, *Red Eagle: The Army in Polish Politics, 1944–1988* (Stanford, 1990), 163. Michta suggests that the army's headlong campaign against Solidarność in military publications had the effect of highlighting the party's weakness, further discrediting the Kania regime (pp. 89–90).

62. Karatnycky, "Polish Conversations: Jerzy Milewski," *Workers Under Communism*, 13–14.

63. Norr, *Solidarity and Self-Management*, 118–20; Lipski, *KOR*, 441–42; Weschler, *Solidarity*, 117.

64. Persky and Flam, (eds.), *Solidarity Sourcebook*, 205–25.

65. Paris interview no. 1, Nov. 12, 1982; Weschler, *Solidarity*, 100.

66. Michta, *Red Eagle*, 165–68.

67. There is a certain majesty in the capacity of high-level Polish apparatchiks to engage in public mendacity. In his 1985 book, Mieczysław Rakowski described November as "a month of intensive effort to reach a national accord." Rakowski, *Ein Schwieger Dialog*, quoted, without comment, in Hahn, *Democracy in a Communist Party*, 190.

68. For the rationale of the active strike, see Zbigniew Kowalewski, "Solidarity On the Eve," *Labour Focus on Eastern Europe*, vol. 5, nos. 1–2, (Spring 1982), reprinted in Persky and Flam (eds.), *Solidarity Sourcebook*, 230–40.

69. Michta, *Red Eagle*, 186–93.

70. Ost, *Solidarity*, 140–48, presents a thoughtful summary of the final weeks, informed by Kukliński's account.

8. The Party Declares a "State of War"

1. Garton Ash, *The Polish Revolution*, 269.

2. Grazyna Sikorska, *Martyr for the Truth: Jerzy Popiełuszko* (London, 1985).

3. This description is derived from the author's attendance at a "Mass for the Motherland" in 1983.

4. Craig, *Crystal Spirit*, 245, 250, 272–75.

5. "Primate Glemp Criticized by the Priests," *Survey* 26, 4, (1981): 200–203; Tadeusz Kamiński, "Poland's Catholic Church and Solidarity: A Parting of the Ways?," *Poland Watch*, no. 6, 1980. *Uncensored Polish News Bulletin*, no. 22, (Dec. 10, 1982). A comprehensive summary of this remarkable exchange is in *Zeszyty Historyczne*, no. 64, pp. 206–18.

6. Kamiński, "Solidarity," 79–81; Szajowski, *Next to God . . . Poland*, 194; Weschler, *Solidarity*, 191.

7. Kamiński, "Solidarity," 80–81; Ascherson, the *Observer*, Nov. 11, 1984; Wałęsa *Way of Hope*, 386.

8. A. Korbonski, *Politics of Socialist Agriculture in Poland* (New York, 1965), 207 and passim; Tamara Deutscher, "Poland—Hopes and Fears," the *New Left Review*, no. 125 (Jan.–Feb. 1981): 61–74; Harman, *Bureaucracy and Revolution in Eastern Europe*, 112; Anna Kaminska, "The Polish Pope and the Polish Church, *Survey* 24, 4 (1979): 204–22. Ironically, the rise of Stalinism in postwar Poland may have done much to overcome accumulated peasant anger at the Church for its prewar support of the rural gentry. In her study of peasant politics, Olga Narkiewicz writes: "A further gain for the anti-Communist forces was the increased influence which the Church now gained . . . Because politics of an independent kind were completely forbidden, the Church acquired far more influence in the region of religious practice than had been the case before." Olga A. Narkiewicz, *The Green Flag: Polish Peasant Politics, 1867–1970* (London, 1976), 259.

9. Stehle, *Poland, The Independent Satellite*, 60.

10. A precise and instructive history of postwar Polish anti-Semitism is contained in Michael Checiński, *Poland: Communism, Nationalism, Anti-Semitism* (New York, 1982); and, with greater economy, in Tadeusz Szafar, "Anti-Semitism: A Trusty Weapon," in Brumberg (ed.), *Poland: Genesis of a Revolution*, 109–22; see also Stewart, *The Poles*, 305–23; Jacques Ellul, "Lech Wałęsa and the Social Force of Christianity," *Kattallagete* (Summer, 1982), Supplement, p. 5.

11. Daniel Singer, *Road to Gdańsk* (New York, 1981), 191; Singer, "How Many Masses Is Poland Worth?," the *Nation* (Sept. 3–10, 1983), 173; Szajowski, *Next to God . . . Poland*, 21–22; Blazynski, *Flashpoint Poland*, 275. On Jan. 1, 1971, the episcopate asserted in a letter read in every parish in Poland that national life could not develop "in an atmosphere of intimidation." In 1976, the Church called upon the party "to respect civil rights and to conduct a true dialogue with society," quoted in Raina, *Political Opposition in Poland*, 407.

12. Vincent C. Chrypinski, "Church and Nationality in Postwar Poland," in Pedro Ramet (ed.), *Religion and Nationalism in Soviet and East European Politics* (Durham, 1984), 128–29.

13. Weschler, *Solidarity*, 21.

14. Jacek Kuroń, "Reflections on a Program of Action," *Polish Review*, 22 (1977): 62. But KOR was also "disappointed" in the Church's excessive caution. Lipski, *KOR*, 298.

15. American interview no. 14 (New York City), May 21, 1985; Tadeusz Walendowski, "The Polish Church Under Martial Law," *Poland Watch*, no. 1 (Fall 1982): 54–62; Dominik Morawski, "The Polish Church and the Government," *Survey* 26, 4 (1981): 193–97.

16. Szajowski, *Next to God . . . Poland*, 194.

17. Ibid.; Kamiński, "Solidarity," 82.

18. Lipski, *KOR*, 12–15; speech of Adam Michnik to the Warsaw Voidvodship Court, quoted in Raina, *Political Opposition in Poland*, 178–96.

19. Starski, *Class Struggle in Classless Poland*, 37.

20. Adam Michnik, "The New Evolutionism," *Survey* 22, 3–4, reprinted as "The New Evolutionism, 1976," in Adam Michnik, *Letters From Prison and Other Essays* (Berkeley, 1985), 135–48, esp. 144–45.

21. Adam Michnik, *Kościół, Lewica, Dialog (The Church and the Left: A Dialogue) (Paris, 1977)*, quoted in Szajkowski, *Next to God . . . Poland*, 47.

22. Michnik, "A Time of Hope, 1980," in *Letters From Prison*, 105.

23. Michnik, "A Year Has Passed, 1981," in *Letters From Prison*, 125.

24. Michnik, "What We Want To Do and What We Can Do," *Telos*, no. 47 (Spring 1981). In 1982, Michnik added to his analysis of 1956 that "one thing is certain: the Polish October was not repressed by force." "On Resistance," *Letters From Prison*, 48. It is possible to interpret Michnik's one-sentence causative connection of 1968 to 1970 as a reference to maneuverings by anti-Gomułka factions in the Central Committee, but the surrounding context seems to suggest his comments at that juncture in his talk were focused on society rather than the state.

25. Michnik, "A Year Has Passed, 1981," 126–27.

26. Michnik, "The Polish War: A Letter From Białołęka, 1982," *Letters From Prison*, 31.

27. Michnik, "A Time of Hope," 105; "A Year Has Passed," 130–31.

28. Michnik, "The Polish War," *Letters From Prison*, 29–30.

29. Michnik, "A Year Has Passed," 126–27.

30. Michnik, "Letter from Gdańsk Prison, 1985," *Letters From Prison*, 94.

31. For another view of these relationships, see Jerzy Holzer, "Solidarity's Adventure in Wonderland," in Stanisław Gomułka and Antony Polonsky (eds.), *Polish Paradoxes* (London, 1990), 97–115.

The politics of popular movements creates strange bedfellows among outside observers. KPN's fantasies about overthrowing the party were shared by sectors of the British and French Left, which joined the Polish Right in regarding both Wałęsa and KOR as "weak" in their opposition to the ruling regime. These criticisms at least had the merit of being a logical evolution of a certain strategic analysis (however geopolitically inappropriate the analysis itself). The striking ingredient in the criticism of Wałęsa by KOR was that members of that milieu generally shared his perspective. Stripped of verbiage, KOR's resentment of Wałęsa turned on class-based cultural nuances, not on strategic disagreements. Given how high the stakes were, this is a rather remarkable example of the power of class perspectives to overwhelm strategic reality.

32. Michnik, "Letter From the Gdańsk Prison, 1985," in *Letters From Prison*, 89–90, 95.

33. Ibid.

34. Craig, *Crystal Spirit*, 252.

35. Michnik, "About the Elections," in *Letters from Prison*, 73.

36. Lawrence Weschler, *The Passion of Poland* (New York, 1984), 189–94; American interview no. 7 (Washington, D.C.), Aug. 20, 1983.

37. Craig, *Crystal Spirit*, 291–95, provides an economical summary.

38. Gwiazda had an abiding, experientially based hostility to all things Soviet. As a boy, he had been deported with his family to Siberia. I thank Tadeusz Szafar for pointing out this relationship to me.

39. Kuroň and Bujak, in *Labour Focus on Eastern Europe*, 5:34 (Summer 1982).

9. The Re-emergence of Civil Society

1. Interview with Lech Wałęsa, *Konfrontacje* (Sept. 1988), quoted in David Ost, *Solidarity and the Politics of Anti-Politics* (Philadelphia, 1990), 184.

2. Individual biographies of the participants in the Round Table have been collected and published in book form: *Okrągły Stół Kto Jest Kim, Solidarność Opozycja* (Warsaw, 1989).

3. Interviews with Henryk Wujec, Warsaw, July 30, 1989; Stefan Bratkowski, Warsaw, Aug. 3, 1989; Jan Lityński, Warsaw, Aug. 6, 1989; Andrzej Celiński, Warsaw, Gdańsk, Aug. 9, 1989.

4. Wujec smilingly dismissed as mere carping by disappointed individuals the general complaint that both his original and final projections of Round Table membership were "too narrow." Bratkowski was quite forthright in his willingness to "leave the workers to Lech." However, the most noticeable thread in the general reaction was the clear absence of any sense of loss over the structural omission of the organized popular movement from the high politics of the Round Table and the subsequent parameters of public politics that were structured through the Round Table agreement in April.

5. "Artistry" was Lityński's word; "genius" was Bratkowski's. Wujec and Lityński were part of the Kuroń circle in KOR; Bratkowski had been head of the journalists' union before being expelled from the party, and Celiński was national secretary of Solidarność until ousted during the internal dispute generated by the Bydgoszcz crisis. All were prominent activists in 1980–81 and 1989–90.

6. Among those who had no inkling of what Wałęsa was doing behind the scenes to create a majority coalition against the government in Aug. 1989 was Janusz Onyszkiewicz, Solidarność's national press officer.

7. Martin Król, quoted in the *New York Times*, July 1990.

8. As Marx himself did, repeatedly, throughout his life.

9. I first suggested the possible utility of comparing the level and democratic quality of structural forms developed within insurgent movements to the forms those movements created once in power in "Organizing Democracy: The Limits of Theory and Practice," *democracy*, vol. 1, no. 1 (1981).

10. Stevens, *The Poles*.

11. There are instructive gradations among the exceptions to this rule. For some scholars of Poland who have not endeavored to trace the historical origins of the movement, the visible evidence of KOR's surprise at the first Gdańsk demand has proved sufficient to limit the degree to which KOR is seen as an organically causative force. An example of this scholarly ability to capitalize correctly on negative inferencing is

Jerzy Holzer, *Solidarność, 1980–81* (Paris, 1984). This, of course, leaves unresolved how the movement did occur. For a much stronger thread of causality, grounded in research rather than negative inferencing, see Roman Laba, *The Roots of Solidarity* (Princeton, 1991).

12. Goodwyn, "Organizing Democracy."

13. The first attempt at a comprehensive account to appear on Solidarność was by the British author and journalist, Timothy Garton Ash. In his well-written and thoughtfully argued analysis, *The Polish Revolution*, Garton Ash found the movement to be a product of "a great convergence" of the Church, the intelligentsia, and the workers. The 1979 visit of Polish Pope generated huge throngs in vast public spectacles where the Pope called for a "fruitful synthesis" between love of country and love of Christ. When the Pope announced that "the future of Poland will depend on how many people are mature enough to be nonconformists," Garton Ash said, "this language, this vision, came like revelation to countless Poles." The result was a "transformation of consciousness" that prepared the way for the Polish August. But this was a subsidiary influence. "The single most important initiative," Ash concluded, grew out of the Radom and Ursus trials in 1976, which persuaded a group of Warsaw intellectuals in September of that year to form KOR. This energetic oppositional effort culminated in a "Charter of Workers' Rights" in 1979, which spelled out the need for independent trade unions.

Garton Ash brought "the workers" into his portrait in his final interpretive summary of causation: "It is naturally difficult to separate the workers' own autonomous political learning process from the direct influence of KOR, but certainly these tiny free union cells, *Robotnik*'s translation of KOR's general strategy into specific tactics (how to organize an occupation strike, what to demand), and the nationwide opposition network played a major role in helping discontented workers to generalize their grievances, formulate remedies, and co-ordinate their activities." As a consequence, KOR succeeded in "raising the political consciousness of the proletariat in key industrial centers." *The Polish Revolution*, 24–29. By employing these perspectives encased in this literary form, the author sidestepped research into the self-activity of Polish workers, contenting himself with a narrative overview of less than two pages for the pre-1980 era.

While *The Polish Revolution* became the single most widely read book on Solidarność in the West, Jan Józef Lipski, Krzysztof Pomian and Jerzy Holzer brought Polish perspectives to bear. Lipski's *KOR*, the definitive account, credited that organization with "preparing the consciousness of the workers for the strikes," but other than a generalized reference to the effect that "this was largely of work of *Robotnik*," Lipski does not pursue the matter. In *Pologne: Défi à l' impossible?* Pomian also credited the Polish intelligentsia, and specifically KOR, as the energizing force of the movement. In *Solidarność 1980–81*, Holzer is more distant, not only from the intelligentsia as a causal force but also from workers. His survey of oppositional activities, perceived through a global geopolitical perspective that serves to obscure the structural relevance of all such activity, is descriptive rather than analytical. This methodological approach is in keeping with the author's announced intention to write about the Solidarność era rather than on "the social movement" that preceded the dates in his title. While noting the tactical utility of Michnik's concept of "the new evolutionism," Holzer does not perceive KOR as playing quite the seminal role that Garton Ash, Lipski, and Pomian claim it played (see note 11). What these four observers share in common is an absence of research on the pre-Solidarność organizing experiences of Polish workers. Specifically, Holzer's analysis of oppositional activity focuses on ideological content rather than on the relevance of such activity to organizational efforts. In common with the overwhelming bulk of the sociological literature on Poland, "movement building" is not a category of analysis for

Holzer and Pomian, both historians; for Lipski, a literary scholar; or for Garton Ash, a journalist. The static categories of analysis employed might be quite adequate for a narrow study at the apex of society (a monograph on the U.S. Supreme Court, for example, or the rules of procedure in the Polish Sejm), but as a strategy for discovering the dynamics of a social movement on the scale of Solidarność, it is a formula for failure. These matters are explored in greater detail in "Ways of Seeing: "A Critical Essay on Authorities.

14. This assumption is also common in the literature.

15. See Chapter 5, pp. 194–200.

16. Michnik also did not grasp the structural components of the Baltic mobilization. See Michnik, "KOR i Solidarność," *Zeszyty Historyczne*, no. 64 (1984). See also Chapter 8, pp. 321–29.

17. Radio Free Europe reported on the Lenin Shipyard strike like it did every other strike in Poland: it rendered status reports. Since the strike lasted so long, the Gdańsk area gradually (not before Aug. 18) received more attention, particularly after the government began talking to the MKS on Aug. 22. By that time, reporters from all over the world were in Gdańsk. The basic point is simpler however. It concerns the purported role of Radio Free Europe in "spreading" the strike, a causal conclusion that rests on the assumption that mere knowledge of a strike in Gdańsk was, willy-nilly, sufficient to induce whole enterprises in other parts of Poland to join the MKS. Aside from the rather compelling historical reality that large-scale social movements cannot be constructed in such an effortless manner, the elementary fact was that workers outside the maritime provinces did not know what an MKS was, how it functioned, or the organizational tasks necessary to join it. The confusion of the Silesian worker cited in the text is typical. It applied to all the enterprises in the Warsaw area, where no MKS appeared. A status report by radio could not begin to address these intricate programmatic and organizational imperatives. Movements cannot recruit with a wave of the hand; if they could, social life and governance around the world would be considerably more energized, contentious, and, perhaps, democratic than it is. See Robinson (ed.), *The Strikes in Poland*, for the information the station did and did not convey.

The structural fulcrum of the coastal mobilization and the centerpiece of its information network was the Interfactory Strike Committee. When coupled organizationally with an occupation strike and programmatically with systemic structural demands upon the state, the MKS was a powerful instrument indeed. It is remarkable that its origin, development, and function in 1980 has received so little attention. The MKS was the structural basis not only of the Polish August but of Solidarność itself.

18. Paris interview no. 2, Nov. 20, 1982; American interview no 8, June 10, 1984.

19. See Chapter 1.

20. Staniszkis, *Poland's Self-Limiting Revolution*.

21. In the highly stratified societies of the West, amid the political confusion that attends the glut and irrelevance of most social information, the educational task that is prerequisite to mobilizing an engaged "civil society" seems at least as great as the task that faced coastal activists in Leninist Poland in 1980. Very little is transparent in the West.

22. With respect to the movement's internal communications network, it might be added that claims made for Radio Free Europe do not explain why the shipyard strikers placed such great stress on their preconditions that the telephone blockade cease and their imprisoned couriers be released. This apparent contradiction is easily resolvable as a narrative problem. References to the struggle between the workers and the

party over the preconditions are simply passed over. As a strategic factor, the courier war does not exist in the literature.

23. Chapter 1, p. 12.

24. Interview with Bronisław Geremek (Gdańsk), Aug. 9, 1989.

25. See Chapter 2, pp. 91–92.

26. The fact that workers performed in a discernibly logical manner in the construction of Solidarność in Poland does not, of course, make a case that workers, *per se*, are either "blessed" or "saddled" with some historically endowed role—in Poland or anywhere else; nor does the evidence concerning opposition intellectuals necessarily carry explicit long-term meaning that can automatically be applied to future situations beyond this particular historical moment in Poland. One systemic reality, of course, has existed in all societies since industrialization: any nation's workers always possess, through a general shutdown of production, the political potential of forcing the existing regime to renegotiate the society's basic social contract. Beyond this, and in light of twentieth-century history, predictive generalizations seem perilous.

27. The three relevant documents are Adam Michnik, "KOR i Solidarność," *Zeszyty Historyczne*, no. 64 (1984), in which Michnik refers to the buildup to the Polish August as "the KOR epoch"; the KOR advisory of August 25, available in English in: Radio Free Europe, *Foreign Broadcast Information Service*, Aug. 26, 1980; and the "Charter of Workers' Rights," *Robotnik*, Sept. 1979, first reprinted in Raina, *Independent Social Movements in Poland*, 374–75.

The KOR myth began to appear in stray paragraphs in the world's daily press during the Polish August, made its appearance in Sunday supplements in September, and reached periodical status in October 1980. The further removed the source, the more fanciful the description of causality. In an article entitled "La Stratégie des intellectuels: Vers la solidarité" in *Le Monde Diplomatique* (October 1980), Ignacio Ramonet, drawing on KOR-connected émigré sources in Paris, reported that "the strategy of civic resistance was devised by the intellectuals of the opposition after a long analysis of the failures that had gone before. It was unquestionably they who laid down the 'general line' the resistance should take, who spread word of it to every corner of the country and who managed—no mean achievement—to win over the whole of the working class to this strategy." The "strategy" moved from idea to action by the following process: "The resistance strategy . . . is worked out mainly by three men—the philosopher Leszek Kołakowski, the historian Adam Michnik and the teaching specialist Jacek Kuroń." After citing analytical articles by the three written in 1978–79 (none of which mentioned independent trade unions or occupation strikes and interfactory strike committees as vehicles to attain independent trade unions), Ramonet concluded, "KOR organized within the working-class movement the idea of autonomous trade unions." In sum, the idea itself (even when not explicitly present) produced the action. Since 1980, such flights of fancy have gradually become somewhat tempered, further diminishing the evidential support for the KOR thesis but leaving the thesis itself intact. It exists now, without evidential support, as one of the bedrock interpretations of the Polish August sanctioned within the Polish intellectual community and among foreign Poland watchers generally. The Catholic Church plays a greater or lesser role in ways that do not disturb the essential features of the consensus concerning the intellectuals' own role. This (fundamentally cultural-ideological) way of seeing is subjected to summary interpretation in the "Critical Essay on Authorities."

28. Just as the months of September to December 1980 were critical for millions of Poles who were able to perform autonomous *private* political acts for the first time in their lives, simply by personally joining Solidarność while party functionaries looked

on helplessly, so the later months in 1981—the era of local fires and of "fire fighting"—was a time for autonomous *public* political acts. Through these two processes, civil society formed and activated itself in Poland.

In this social sense, the era of Solidrność in Poland may be seen to have contained four discernible stages: (1) the Polish August, in which popular mobilization, drawing on past experiential organizing development, was centered in the maritime provinces; (2) the autumn months of 1980, when nine million other Poles went through the experiences that some 700,000 or so workers on the coast had lived through—namely, face-to-face confrontation with the party to create local self-governing units; (3) the months of organizational innovation engendered by the Network for self-management and tactical maneuvering in 1981, dating roughly from the contentions over work-free Saturdays and the Bydgoszcz crisis in the spring through the second national congress of Solidarność in October, and (4) the twilight months of November and early December 1981, a time of heightened popular assertion amidst a public mood of national crisis and a gathering sense of desperation and doom. For the Polish people, these four stages, it seems to me, encompass the totality of the public meaning of the Solidarność experience. Because people actively participated in civil society, these rhythms characterizing high politics were personal rhythms, too.

This is the verifying meaning of an authentic civil society: instead of being alien to one another, public politics and private experiences merge. The theoretical conclusion is straightforward. The creation, maintenance, and evolution of democratic forms in modern life requires the presence in society of an organized, popularly based social formation. In its absence, civil society cannot be said to exist, and the idea of democracy, although outwardly undamaged, in fact withers to effective political control by ever-narrowing groups of elites. The extent to which the reality does or does not become publicly apparent in any society is an area of philosophy and politics that can be approached through communications theory and cultural theory. The relevance of any "theories" that might appear is directly related, of course, to the extent and relevance of prior empirical research upon which such speculations rest. Material conditions, it goes without saying, are material in undetermined ways.

29. This figure is, at best, a guess that is somewhat informed and also conservative. I have interviewed no worker activists who put the number of militants in the Lenin Shipyard much below this figure and several who put it quite a bit higher. One's own intuitive standard as to what constitutes "militancy" obviously determines each estimate.

30. The figure used in early published reports and reproduced in most subsequent accounts in "about one thousand," "upward of a thousand," and "some one thousand." A photograph taken of a gathering of strikers on this fateful weekend indicates something less than five hundred in attendance. Not all participants in the occupation strike, of course, were necessarily at the site of the photograph. Of one thing we may be certain; after the Polish August, the number of shipyard workers willing to associate themselves with this group exceeds all published figures by a considerable margin. And why not? These workers kept the hope alive when the strategic balance was tilted most ominously against the democratic movement. Without their efforts, Wałęsa's decisive speech leading to the consolidation of the occupation strike on Monday morning, Aug. 18, would not have had the setting that it possessed. Indeed, if none of them had been at the gate, there could have been no speech, as there would have been no social formation for Wałęsa to have been a spokesman for.

31. In 1989, Andrzej Gwiazda, Jan Rulewski and the "Working Group" found this form of isolation and consolation. One projects that an economic collapse can offer

new prospects for them, but only if their politics becomes relational and not merely "militant."

32. In itself, the Polish structural innovation of the Interfactory Strike Committee cannot easily make its way from idea to action in any society. Enormous prior organizing activity, over time, must precede it as an experiential training school for any militants who might dream of creating such a structure. Nevertheless, in the hands of experienced activists and with adequate means of internal communications to all participating enterprises, it can be an enormously powerful weapon of popular democratic assertion. This is the foremost strategic lesson, historically, of the Polish August.

33. The Polish August is rich in folklore, as one might expect of such a transformative historical moment. One widely circulated story concerns a one-sentence admonition Kuroń is reputed to have given to workers: "Don't burn down party committees [headquarters], found your own." As mentioned, the line even made it into Wajda's famous motion picture of the Polish August, *Man of Iron*. The story, like the "long-distance line to Borusewicz" and the "loop through Radio Free Europe" is part of the illustrative evidence of how KOR "prepared the consciousness of the workers for the strikes." Empirically, these matters have now been addressed. The methodological routes to these non sequiturs are considered in the "Critical Essay on Authorities."

Ways of Seeing: A Critical Essay on Authorities

In the aggregate, the scholarly and journalistic literature on the Polish democratic movement is a body of interpretation that exists entirely outside the intricate methodological traditions developed by the Annales School in France, the new social history in Britain and America, and the somewhat less-developed Bielefeld School in Germany. Most of the literature on Solidarność exists in a time warp. The reason is as simple as it is controlling. While Western scholars have over the past two generations been learning how to employ quantitative analysis linked to computers and oral history linked to tape recorders, producing in the process such a massive new array of evidence about social "classes" that the very definition of class itself has undergone conceptual reevaluation, "Poland watchers" have continued to work in another and older tradition. Blocked off as they have been from the base of society, they have relied on research traditions inherited from conventional political studies at the apex of society. Because the research techniques Western scholars have developed to uncover the whole of society have not been applied to the origins and development of Solidarność, the literature on the Polish social movement effectively restricts historical agency within very narrow parameters. Simply enough, causality is seen as centered in small groups of people visible at the apex of society because research on social activity elsewhere in society has not been performed. In terms of the dynamics that produced Solidarność, this circumstance has proved crippling.

It is now commonplace in the West for scholars to engage in subtle disputes with one another as to relative degrees of agency that can be ascribed to certain sectors of society at certain historical moments; indeed, illuminated as it has been by the work of social historians, the issue of agency has increasingly come to play a role in post-modern literary criticism as well. But it is striking to discover that the custom of conceptualizing the possibilities of agency very narrowly affects virtually an entire literature on a scale as large as that represented by extant studies of Solidarność in Poland.

This is not to suggest that Polish social scientists have labored under Leninist constraints by celebrating the fact that their eyes must remain closed. Scholars such as W. Morawski ("Society and the Strategy of Imposed Industrialization") have found ways to point out that Leninist systems of economic governance ("imposed industrialization") could not account for the varieties of life, ideas, and interests in a modern social system ("uncontrolled phenomena") that produced and reproduced systemic crises at the highest level of the Leninist pyramid ("deficiencies in the multilevel structure of decision making"). But to be forced for two generations to write in such an elliptical manner carries its own dangers in terms of analytical precision. A second layer of problems turns on the virtually complete absence of accessible written evidence on social activity. The only broad type of social information against which to test social interpretations emanating from the party-state resides in polling data, which are tainted in terms of salience by the power relationships surrounding the polling process itself.

The end result has been a necessary over-reliance on party sources, with scholarly achievement being measured by how well one can "read through" such self-serving materials and how skilled one is in packaging findings in acceptable terminology. This is a helpful and, in Leninist societies, an essential scholarly research and intellectual skill. But meanwhile, evolving developments in research methods for penetrating society do not become a part of the functioning equipment of the scholar, because however well one "understands" a methodological approach, it cannot be practiced. While the constraint on scholarly writing imprisons observers in Poland, the constraint on accessible evidence handcuffs Poland watchers both in Poland and elsewhere. The lacuna that persists (the evidential blank where social activity is occurring) gradually becomes unacknowledged. Scholars debate the meaning of evidence; when there is no evidence, there is no debate. Under the circumstances, the dynamics of large-scale movement building and social change hover out of reach, not only in terms of social evidence but also in terms of social thought itself.

The dismaying result is that scholars of Poland attempt to fill the lacuna by employing a number of primitive categories of behavioral and deterministic description that serve to characterize large groups of people on whose activity systematic research has not otherwise been performed. Abstract description has thus replaced social research in Poland in a way that is simply no longer possible, or remotely persuasive, in the West. Not only sociological literature, but historical scholarship, too, inevitably becomes choked with static terms of description that project static characteristics of social passivity or social insight upon large and rigidly described sectors of the population—"the workers" and "the intelligentsia" to name the two most operative examples. While the problem is systemic, it is massively exacerbated at historical moments of large-scale social transformation when highly visible popular activity, such as that engendered by the Gdańsk Interfactory Strike Committee, emerges out of prior activity that has not been researched. Sociological and historical scholarship on postwar Poland has been virtually immobilized by these reinforcing circumstances. The remarkable discovery about Polish scholarship on Solidarność is not that a sociologist such as Jadwiga Staniszkis can attempt a book-length study based almost entirely on static abstractions, but rather that so many other students of Poland share in her unsubstantiated behavioral assumptions that she is sometimes taken with a modicum of seriousness.

This circumstance points to the deeper problem underlying scholarship on Poland. Cultural assumptions (present in all modern societies but especially vivid in Poland apparently partly traceable to the *Szlachta* heritage) have produced an extremely simplified and class-influenced view of social life. Thus, an abstraction called "the intelligentsia" is routinely employed to describe people who are perceived to be both politically sophisticated and occasionally politically active, while another abstraction called "the working class" is used to describe a much larger number of people who are understood as inarticulate, authoritarian, and either politically inert or anarchic. When applied to social conduct, these highly generalized and sweeping formulations yield descriptively elaborate but socially primitive premises that, when untested by research, automatically translate into descriptively elaborate but socially primitive conclusions. In the social universe so constructed, the millions of Polish people who comprise "the working class" appear essentially as an undifferentiated social abstraction that can be "manipulated' by the party or, conceivably, have its "consciousness raised" by religious or secular intellectuals.

Though workers in different parts of Poland and in different branches of industry have had discernibly different historical experiences since the onset of the Leninist system, the custom of treating "the working class" in a distant and abstract manner has had three decisive effects: (1) it provides a way for scholars to believe that sustained research on the self-activity of workers over time continues to be unnecessary; (2) any differentiation in what is taken to be "working-class consciousness" that materializes out of these varied experiences necessarily remains undetected; (3) no countervailing evidence having been uncovered, the combination of these two factors persuades Poland watchers to believe their behavioral and deterministic categories of analysis constitute adequate means to describe social reality from afar.

These modes of description, substituting for social research, have fed into a far-reaching political assumption that has been dominant in studies of Solidarność—namely, that insurgent ideas possess a direct connection to subsequent insurgent actions. Thus, the act of writing down the idea that free trade unions are desirable under certain ideal circumstances has been seen as causal to the creation of free trade unions. In the absence of research, a kind of social telepathy seems to have connected, in unknown ways, the one to the other. Literally hundreds of articles and books on Solidarność reflect this presumption, though to have it, one must specifically pass over the entire process of sequential organizing experiences that constitute the building blocks of social movements. The results has been an empirical leap over the interior of the Baltic movement that was the source of Solidarność.

At bottom, what has dominated scholarship on Poland, whether conducted by Westerners or by Poles, is a conception of politics grounded in a conscious or unconscious presumption that intellectual attitudes are, easily enough, routinely static and can be intuited from afar—that is, without focused research on popular activity that might undermine such remote presumptions. A significant amount of scholarly literature on Poland, particularly sociological literature, is not utilized in this book for the elementary reason that it is grounded in static conceptions of consciousness unsupported by social research on the insurgent activity that generated the Polish movement. As emphasized in the Introduction of this volume, Leninist constraints on scholarly research in Poland explain this decisive lacuna surrounding the politics of popular protest in Poland throughout the period from 1948 to 1980; the overwhelming bulk of evidence of popular assertion was confined to police files. But the persistence of this research lacuna after 1980 cannot be so readily tributable solely to Leninist constraints. It is, in fact, traceable to research habits developed by Poland watchers in the earlier period. Static categories of analysis have continued to dominate social inquiry in Poland since 1980. This is evident in the work of conscientious scholars Jerzy Holzer and Stefan Nowak, to name two of the most distinguished.

Collectively, these related presumptions and analytical categories have produced an explanation of causality that is bizarre: (1) the literature of Solidarność privileges a cautious and qualified endorsement of the idea of free trade unions written by KOR in 1979 (an idea that the authors of the document themselves did not act on in 1980) over the broad-based drive by Baltic workers for free trade unions in 1970–71, an intention the workers did act on, over the objections of the intelligentsia, in 1980; and (2) though written evidence generated during the coastal mobilization of 1980 (and massively augmented by social evidence generated at the same moment) conclusively verified the comparative levels of political consciousness among workers and intellectuals at the moment of crisis, the universal failure to investigate the development of self-generated worker activity in the 35-year postwar period effectively concealed this social reality that decisively informed the politics of the Polish August. In the absence of research on the multiple prior social experiences that armed the workers of the coast to analyze events at a level beyond the capabilities of intellectuals and the episcopate in Warsaw, this transformative strategic development was left undetected and thus unexplained in the literature on Solidarność. The result is that the movement has been missed.

The utility of the extensive secondary literature on the Polish movement is thus confined to the era of legal Solidarność between September 1980 and December 1981. But given the erroneous sense of causality presumed to be operative and the static categories of analysis that continue in vogue, even the dramatic and highly visible events of this time period have been intermittently misread. Thus, when the open deliberations of the Solidarność National Congress in the fall of 1981 demonstrated the weakness of KOR influence, journalists and scholars took this as a fairly recent development. In fact, KOR could not "lose" its influence in the fall of 1981; KOR activists never possessed functioning policy-level influence on events during the long development of the Baltic movement or within the structure of Solidarność that materialized out of that development in the 1980–81 period. (This would change, of course, in 1989–90.) In 1980–81, the rather unflagging drive by KOR activists to maximize their influence, a circumstance that heightened tensions inside the movement, contrasts markedly with the unobtrusive and communitarian support given to the worker leadership by Tadeusz Mazowiecki and Bronisław Geremek. The latter two men thus became highly relevant to the movement, Mazowiecki as editor of *Tygodnik Solidarność*, Geremek as chairman of the Action Program, and both as advisors to the Solidarność National Coordinating Commission. Their emergence in 1989 at a level of national prominence is rooted in the relations of workers and intellectuals in the interior of the movement in 1980–81. Unfortunately, it is precisely the interior of the Baltic movement before 1980 that has been left unresearched to such an extent that its maturing form during the Polish August has been misread. Under the radically altered political conditions that have begun to materialize since 1989, it is problematic how long these cooperative relationships can endure, especially given the crushing economic strains suffusing Polish society.

It needs to be acknowledged at once that the two sentences used to open the preceding paragraph—concerning the secondary literature on Solidarność—will, without the slightest doubt, be taken as surprising by scholars of Poland. Indeed, the word "surprising" understates the matter; it may well be taken by some as shocking or, perhaps, incomprehensible. This is a regrettable result, but an unavoidable one. Simply enough, the judgment is a product of an encounter by a student of social movements with a literature produced by scholars who, as specialists in other fields, are not trained to focus their research on popular social activity and traditionally do not do so. Since social movements, however, are the product of sequential social activity over time, the scholarly habit of substituting perceived *attitudes* of static social groups (for example, the mindless provincialism of "the workers" or of "worker-peasants") for sustained research on the *activity* of evolving groups of historical actors (for example, workers in the maritime provinces) is fatal. In all cases where the organic historical question concerns the origin and development of the social movement itself, such a distantly frozen angle of vision is so narrow that it literally collapses human agency into prepackaged clichés.

Moreover, even though the success of the movement transformed social relations throughout society, creating new pressures not only on the state or on the Solidarność leadership but also on society itself, inherited habits of analysis have the effect of promptly redirecting almost all attention not to these multiple new social realities but rather to the newly visible struggle at the level of the state. The party as an object of research fixation is now supplemented by a new site of research, the Solidarność National Coordinating Commission. Granted that this struggle is both dramatic and germane, it is also one that is informed by the political, economic, and social pressures created by an activated popular constituency. While it therefore continues to be necessary for scholars of Solidarność Poland to inquire into such matters as the registration crisis, the Bydgoszcz affair, the July party congress, and the great national congresses of Solidarność, activity within newly constructed civil society—including its role in generating both the horizontal movement in the party and the Network in Solidarność—merits close attention because these multiple dynamics collectively shape, enlarge, and confine the outer parameters of political and social possibility. (In ways that are not "normal," social life and politics became so complex and so expansive in the Solidarność era that inherited ideological traditions of analysis were put under a strain too great to bear. The fundamentally different typologies of analysis and description developed in the study of "normal" hierarchical politics at the apex of society, as contrasted with the modes of analysis and description appropriate for the study of the "abnormal" and incipiently democratic politics generated by large-scale social movements, is discussed on pp. 450–53.

Background to Sources

The initial postwar struggle for free trade unions—the bloody episodes of 1944–47 detailed in Chapter 2—has received attention from three sources useful to this study: Jean Malara and Lucienne Rey, *La Pologne: D'une occupation à l'autre, 1944–1952* (Paris, 1962); Jaime Reynolds, "Communists, Socialists and Workers: Poland 1944–48," *Soviet Studies* 30, 4 (Oct. 1978):516–39; and Paul Barton, *Misère et révolte de l'ouvrier Polonais* (Paris, 1971).

From the moment the Polish party consolidated power in 1948, a veil of silence cloaked autonomous social activity, forcing scholars to view from afar through whatever openings they could intuit as existing in party-condoned sources. This circumstance defined scholarship on Poland for over thirty years. In 1980, however, a team of social scientists in the provincial city of Poznań, taking advantage of the social opening created by the appearance of Solidarność, conducted a hurried but highly germane investigation of the "Poznań events" of 1956. These oral interviews and narratives drawing on relevant housing, income, and other economic data that provided a context for oral evidence were published in 1981 before martial law ended such modes of scholarship later in the year: Jarosław Maciejewski and Zofia Trojanowicz, *Poznański Czerwiec* (Poznań, 1981). It is a pioneering work in Polish social history.

Another evidential opening to the social activity of protest in Poland came on the Baltic coast. In the aftermath of the victory in the Lenin Shipyard, the Gdańsk Solidarność created a historical commission to recover material bearing on events before, during and after the 1970 massacre. An immense body of carefully preserved evidence was assembled, detailing with remarkable precision the worker demands generated in late 1970 and again in 1971 by maritime enterprises, including demands from individual shop sections. The research team also acquired (from disillusioned and therefore cooperative party sources) revealing intraparty and worker-party communications during the crisis. Finally, an impressive body of social evidence was accumulated from within the ranks of protesting workers. Working with this historical team in Gdańsk was the American political scientist Roman Laba. After martial law, Professor Laba was entrusted with this archival material and subsequently managed to get it out of the country. Back in the United States, he organized this evidence and computerized portions of it that enabled him to create an analysis of worker demands differentiated by industry, crafts within industry, and comparative levels of self-assertion by broad groups of people. He generously shared this material with me. Chapter 3 of this book is drawn from many sources, but the analysis that informs the detailed happenings of December 14–18, 1970, and the subsequent demands of workers in Gdańsk, Gdynia, and Szczecin are a product of Professor Laba's collegiality in sharing his materials and his insights. This material is quite instructive. Laba's study of the complex interaction of Baltic workers with the party, the army, and the Church in 1970–71 will, I have every confidence, be acclaimed the definitive work on the subject and collectively provides relevant background for his ensuing authoritative analysis of Solidarność. His deeply researched and sophisticated account is germane not only to his fellow political scientists but also to historians, sociologists, philosophers, and anyone else concerned with social and political theory (see Laba, *The Roots of Solidarity* (Princeton, 1991). Laba's work is cited in this volume in its dissertation form, "The Roots of Solidarity: A Political Sociology of Poland's Working-Class Democratization" (Ph.D. dissertation, University of Wisconsin, 1989). Much of the interview material, correspondence, and other data collected by the Gdańsk team have been published in *Grudzień 1970* (Paris, 1986). A significant portion is now available only in the Houghton Library at Harvard, which houses a 1970 archive and a 1980 archive on the Baltic movement. One other written source offers insight into the world of the Baltic working class—Lech Wałęsa's account. The original French edition, *Un Chemin d'espoir*, is more useful since the English translation, *Way of Hope*, unaccountably omits relevant sections dealing with working-class organizing on the Baltic coast in the 1970s.

Collectively, this material contains a wealth of concrete detail surrounding the social dynamics of the two large-scale Polish insurgencies in the period from 1948 to 1979—at Poznań and on the Baltic coast—and, as such, provides the basis upon which the research agenda for this book was constructed. There exists as well a fair amount of secondary literature dealing with the apex of society from 1948 to 1980 and even more that treats the highly visible tensions between society and the state from 1980 to the present.

But to unearth the causal dynamics that lurked beneath the public surface in Poland for the years leading up to Solidarność, I have essentially been forced to employ oral investigatory techniques. Over a period of seven years, I interviewed activists and other Poles in America, France, Germany, and Poland. In the early years of martial law, when many of my sources were ex-prisoners of the regime or were in enforced exile, they agreed to interviews on the condition that their names not be publicly revealed. Indeed, some did not divulge their names. The exceptions are persons who at the time, or subsequently, gave me specific permission to identify them. These persons are visible, by name, in the notes; the others are identified by number. The subject matter, in any case, is not controversial. It has merely been unknown. As such, it represents an early exploration of that vast social space where Poles have lived for forty-five years.

Background to Research Assumptions

In terms of historical methodology, a kind of rolling intellectual dynamic is at work that has produced the decisive evidential gap between idea and action in the literature of Solidarność. In the first instance, the absence of easily recoverable social evidence on the politics of protest in Poland has encouraged the appearance of studies based on the activity of the party counterpoised against that of oppositionists who generated recoverable written material. These sources collectively produce the evidential scaffolding upon which ultimate interpretations of Solidarność have been erected. The activity of the rest of the population, unknown as it is, has been glossed over through generalized references to something called "society," something called "the mass of the Polish people," or something called "the workers." Embedded within such categories, of course, is a conscious or unconscious political perspective, even a political theory of society, as well as a series of class-derived cultural presumptions. This approach inevitably yields traditional, nondialectical "political history," even when (as in occasional articles) the subject under review is the Polish working class.

Secondly, in a gesture toward bridging the gap between idea and action, specific attributes can be affixed to the undifferentiated collage of people who have been suggested as existing but whose social activity has not, in fact, been researched. Thus, the social abstraction known as "the Polish people" can be described as "devoutly Catholic" and the active participation of some of them can be seen from afar to be a result of a "transformation of consciousness" traceable to a visit by the Polish Pope. Again, what specific component of consciousness is taken to be "transformed" is left unspecified, as is the reason why this process affected the (Catholic) maritime workers of the Baltic coast in some massively galvanizing way different from other Catholics or other workers everywhere else in Poland. Conversely, adopting a different view from afar, observers who are put off by the anti-Semitic proclivities of various high churchmen can find evidence of religious symbolism harnessed by worker activists as a fatal sign of the ideological incoherence of the worker movement. It may be observed that both views of workers, one benign and the other critical, can be reached as a function of one's prior attitude toward the Polish Church rather than through research on worker activity. In this manner, the specific details of causation are obliterated by the sheer breadth and vagueness of the categories of analysis being employed. By this process, causation at the base of society is explained in terms of actions at the top, or alternatively as a kind of mystic expression of some endemic cultural phenomenon long known to be in place. As in all views form afar, these modes of speculation serve to fill the lacuna created by the absence of research on the social formation being described.

Similarly, habits of research honed by Poland watchers in the relatively static era of party governance prior to 1980—habits that produce a routine and energized preoccupation with party activity and party-condoned evidence and opinion on nonparty Poles—can instrumentally control interpretations of the Solidarność era. As Foucault has helped to clarify, paradigms not only make knowledge possible but also serve in subtle ways to limit, constrain, and ultimately to shape the form of knowledge they facilitate. For example, in Solidarność Poland, the Network's striking exploration of the potential of self-management as an instrument of social and economic democracy in 1981 can be seen as a mere counterpoint to previous party experiments in 1945–48 and 1956–58, despite the fact that the outward trappings of party-inspired "self-management" schemes have historically been employed as a cynical organizational tool to stifle popular self-activity and thereby redomesticate workers under "Leninist norms." Thus, a scholar accustomed to working with party sources, George Kolankiewicz, views worker self-management activity in 1981 in static terms, as a "morass" that can be compared to previous party-controlled "self-management" devices and observes both party efforts and movement efforts preponderantly through elite sources in ways that drain a good deal of the social and theoretical content form the pioneering democratic initiatives of the Network. The interior dynamics of the Network are left unexplored in this traditional methodology.

Such approaches may perhaps most accurately be described as "unconsciously ideological," insofar as they privilege certain class-influenced cultural presumptions about historical causality. But another way to organize a narrative from afar is through a preemptive empowerment of formal ideology. For example, some Westerners ideologically link egalitarian objectives to activities by the party. Since the record of the Polish party from 1945 to 1980 is so disappointing in this respect, such observers look to "party reformers" as the logical sources of innovative change. The Experience and the Future group that wrote the DiP report in 1979 can thus be seen as causative to the "raising of worker consciousness"—despite the fact that the report itself tellingly reflects the power relationships between society and the state existing at the time it was written; that is, the report, while dutifully detailing numerous malfunctions in governance, is distressingly timid in the democratizing remedies offered. Nevertheless, in dramatic views from afar, Neal Ascherson in Britain, Jean-Yves Potel in France, and Adam Bromke in Canada find the DiP report notably influential in shaping worker demands—notwithstanding the fact that the worker vision of free democratic space independent of the party shattered the Leninist paradigm in ways not only beyond the conceptual range of the authors of the DiP report but also far beyond any of the specific remedies they suggested. It perhaps goes without saying that no effort is made by Ascherson, Bromke, or Potel to connect the DiP analysis, truncated as it is, to subsequent activity by Polish workers in general or Baltic workers in particular. As with other views from afar that link causality to the intelligentsia or the episcopate, no need is perceived to connect ideas to action.

But such vague inferences leave uneasy even some of the most habituated practitioners of views from afar. If causation can be suggested as emerging from the top, there remains the continuing need to suggest the existence of some sort of mechanism of receptivity at the base. Here, too, the absence of specific evidence can be surmounted through recourse to class-influenced cultural categories. Thus, the drive by workers for openness in their own internal dialogue—that is, the drive for the creation of democratic mechanisms of decision making in the Lenin Shipyard and in Solidarność itself—an be explained as a surfacing of an innate "capacity for suspicion" on the part of "the workers." The latter can be further understood, abstractly, to possess a "semantic shame" that renders "suspicion" as functionally ever present. Unforeseen ghosts of Émile Durkheim and Basil Bernstein! In this evidentially barren manner, "suspicion" explains the appearance of democratic forms, though, unfortunately, once again without research into the activity of Lenin Shipyard workers presumed to be in the grip of this social deformity. A number of studies of Solidarność listed herein are grounded in such behavioral categories of interpretation. They seem to dominate the practice of sociology among scholars of Poland.

While social historians and others experienced in researching social movements have learned to be wary of these kinds of generalized descriptive allusions, Poland watchers, closed off as they have been for over thirty years from social evidence at the base of society, seem to have remained unaware of this necessary caution, judging by the studies that have appeared on Solidarność.

Ways of Seeing

On this fundamental issue of social and political methodology, a fine distinction requires special emphasis. In all historical periods of "normal" politics—those instances where inherited forms of power function in the absence of an engaged popular movement—conventional research techniques at the apex of society can produce what appears to be a workable framework of interpretation for the simple reason that operative causal dynamics are, in fact, functioning at the elite level. Power is active and power is being analyzed. But in this sort of scholarly labor, familiarly known in the West as "political history," the absence of additional broad social evidence at the base of society tends to produce a skewed and often celebratory vision of the efficacy and rationality of politics at the apex. Even under so-called "normal" conditions of inherited politics, coherent analysis of the whole of society requires evidence drawn from the whole of society—a working hypothesis that social historians have been endeavoring to emphasize for more than a generation.

But these criteria become overwhelmingly operative in all historical situations in which a social movement is engaged in the process of transforming, or attempting to transform, the inherited parameters of political custom. Under such relatively rare circumstances, detailed evidence from the interior of the social movement is absolutely imperative if the functioning dynamics of struggle are to be revealed and if the tensions pervading all segments of society are to become contextually visible. At such a time—and 1980–81 in Poland is certainly a classic example of such a moment—research restricted to the apex of society will simply produce a systemic misreading of causality. This circumstance helps explain why the literature of Solidarność is so muddled.

The question arises, Can the depth and breadth of the evidential vacuum existing in the literature on Poland not be seen beforehand—in advance of evidential material highlighting its previous absence?

The answer goes to the heart of deep-seated cultural presumptions pervading the social environment in which intellectual and political elites move throughout the modern world—namely, the intuition that social change is routinely the product of elite theorizing and elite activity. This perception constitutes a fundamental "way of seeing" that literally colors all that is seen and provides an apparently reasonable explanation for all that is not seen. But to call this deeply ingrained presumption into question is to create a hypothesis that is psychologically difficult for a great many professionals to consider because it throws in doubt many or most of the functioning procedures governing their academic research. It challenges their worldview. If valid, the possibility that social ideas can, on a specific historical occasion, emanate from nonelite sectors of the population propels into question a vast amount of scholarly research historically designed to illuminate that familiar form of "normal" politics wherein all political argumentation occurs within received parameters of social debate in any society. To inject this hypothesis has an even more totally undermining effect upon inquiries into the kind of political universe that is created by the emergence of a large-scale social movement—on those historical occasions when inherited parameters of politics are breached or even temporarily shattered. When the political paradigm is fissured from below, "normal" modes of scholarly inquiry at the apex of society inevitably prove unworkable; they are too shallow in depth to unearth causation from below.

It is imperative to spell out this relationship of source material to scholarship, for it bears decisively on the flawed methodologies governing the listed works, which collectively constitute the received literature on Solidarność. The historical and sociological practice of engaging in the professional study of what I have described as "normal politics," including the high politics of normality at the level of the state bureaucracy, develops in its practitioners not only a number of sophisticated research and interpretive skills but also a number of assumptions and expectations as to where relevant evidence may be found. Thus, causation is normally understood to lie with the powerful and the mode of conveyance is understood to be some inherited variety of hierarchical form, ranging from effective to severely effective, unconsciously intuited as a permanently functioning "given." In contrast, in a political constellation shaped by the appearance of a large-scale insurgent movement, causation often—very often—lies with the less powerful. To approach the "era of Solidarność," then, as if it were simply another round of conventional politics, with one additional actor in the form of the social movement, offers the prospect of book-length studies in which received patterns of governance struggling in the presence of radical new forms of social interaction are subjected to inherited habits of interpretation. But since social movements, including the Polish movement, routinely respond to received hierarchical forms with counterassertion rather than conventional deference—producing in the process cognitive disjunctures in many sectors of society, specifically including the presumptions governing the perspectives of outside observers—the "politics of protest" in rarely a rewarding arena of inquiry for traditionalists uninstructed in researching social movements. It is in this decisive sense that the literature on Solidarność can be understood as conventional.

There is, of course, an abstract means of dismissing such cautionary criteria as I have here outlined. One can simply retreat to the secure high ground of received cultural tradition. Because this admonition, if considered superficially, can be taken as "privileging" nonintellectuals on one or more specific historical occasions, it can be understood less as a reasonable conclusion derived from prolonged social research than as an unreasonable example of "anti-intellectual" theorizing. To old Leninists or Trotskyists, such a broad-gauged perspective is "workerist." To Western traditionalists, it can be construed as "populist." In France, an aggrieved sector of the citizenry, mobilized for political assertion, can be seen as practicing a mindless "politics of *ressentiment*." Such emotions can, in fact, be present on specific historical occasions and need to be fully considered, but to project them abstractly as some sort of controlling *mentalité* levers a preconception into a conclusion and is pointless in the absence of corroborating social evidence. To see such class-based or ideology-based speculations as causal, without research, simply shifts analysis from one untenable ground to another. It also has the unfortunate effect of obscuring much of the concrete evidence of democratic striving within the expansive social polity created by the movement. At a time when people throughout the world are both resigned and dismayed by the rigidities of the sundry hierarchies within which they are caught, a time when functioning civil societies exist almost nowhere on the planet, it would seem especially inappropriate to pass over the instructive details of large-scale democratic activity such as were made visible by the appearance of Solidarność in Poland.

To an American social historian, there is a final dismaying element in the historiography of modern Poland—namely, the seemingly pervasive tendency (not only in the sociological and journalistic literature but also among Polish, émigré, and Western historians) to depict the Polish population holistically, even during the tumultuous period of transformation that rippled across the country from August 14 to December 1980 and, in a maturing form, thereafter. As this book emphasizes, the level of political consciousness achieved during the Polish August was not a phenomenon of the intelligentsia or the episcopate; it also was not a phenomenon of Polish workers generally. It was achieved initially by workers on the Baltic coast and, aside from themselves, was understood most easily by priests, party officials, activist intellectuals, and secret policemen who lived among the coastal workers. This dynamic was not true elsewhere in Poland. The demands put forward during the strikes of July and August by workers outside the coastal region did not, either in terms of organizational class action or political awareness, approach the level achieved on the Baltic. One could call the roll of major Polish industrial centers, including regions such as Warsaw nearest to intelligentsia activism, and not find one to which this comparison would not apply with varying degrees of starkness. Even the intense citywide strike in Lublin did not generate either organizational forms or programmatic demands of the sweep generated on the coast. Under the circumstances, to discover Polish historians such as Jerzy Holzer in

Warsaw or Krzysztof Pomian in Paris focusing on how "Polish society" or "the intelligentsia" grew disenchanted with the party not only blurs causality but ultimately obscures it. As E. P. Thompson wrote more than a generation ago, a "class" is not what it used to be or what it will become but only what it is at a given historical moment, as verified by its actions. The static terms of description routinely employed by sociologists and historians of Poland fail to envelop this inherent dynamic quality of social action that reached its culminating level of assertion on the Baltic coast, and such lifeless terminology (about presumed worker "authoritarianism" or "semantic" deficiencies) effectively conceals the political significance of the social forms that were, in fact, at work. The larger point is that the social rhythms of August on the coast subsequently coursed through all of Poland, ultimately engendering such a sense of self among the citizenry that movement activists were pressed into "fire fighting." The democratic dialogue that then ensued was, in terms of social relations, the high point of democratic achievement in Solidarność Poland. Similarly, the outer limit of democratic speculation—theoretically—was reached inside the Network. Instructive as both of these developments were, for non-Poles as well as for Poles, their essential qualities and the dynamics that produced those qualities are virtually invisible in the literature on Solidarność. Quite simply, these dynamics remain impervious to static categories of description projected from afar.

The literature on Solidarność listed here is, by and large, a product of conventional beliefs about the static dependency of (in this case Polish) workers upon secular and religious intellectuals for their capacity to perform acts of coherent social assertion. As such, this literature fails to uncover the dynamics that produced Solidarność. These authors have all passed over the detailed social evidence generated by Baltic workers that linked idea to action in such a way as to produce the transformation of life in a one-party state.

Sources on Solidarność

Solidarność, 1980–81 (Warsaw, 1983; Paris, 1984) by Jerzy Holzer is the most extensive book-length study by a Polish academic. Though Solidarność was both a product of a social movement and was itself a social movement, the author unaccountably announces in his opening pages that his study is not focused on "the social movement." Professor Holzer has written a judicious political history of the era of Solidarność that includes a chapter on how the Polish population (as distinct from the Baltic working class) learned to resist the Leninist state. The material is not without interest, but a good portion of it is tangential to the causal dynamics that produced Solidarność in August 1980 and informed its internal life thereafter. Holzer's controlling view is geopolitical.

There are gradations in views from afar. By all odds, the most persistently abstract and evidentially empty study of Solidarność is by a sociologist who was for seventy-two hours a participant in the Polish August: Jadwiga Staniszkis, *Poland's Self-Limiting Revolution* (Princeton, 1984). An analysis of her role in the Gdańsk advisory group is on pp. 228–29 and 233–35.

The definitive history of KOR is the work of veteran member of the democratic opposition and KOR activist: Jan Józef Lipski, *KOR, A History of the Workers' Defense Committee in Poland* (Berkeley, 1985). It is an intimate and authoritative report on the courageous and creative intellectuals who publicly founded a free institution in a police state. Lipski's book is a view from afar, however, when the subject is Polish workers.

The transcripts of the Gdańsk negotiations have been published in A. Kemp-Welch, *The Birth of Solidarity* (New York, 1963). This work contains a narrative section by Tadeusz Kowalik, entitled "Experts and the Working Group," which provides interesting insights into the social and political presumptions of Poland's progressive intelligentsia—ideas which continue to have wide currency in Poland and to shape the way the origins and development of Solidarność are understood. An analysis of Kowalik's viewpoint is on pp. 420, n. 34. Another variation on this central theme is embodied in the work of Adam Michnik, which is discussed at some length on pp. 321–29.

Other works by Poles that proved useful are Stanisław Starski (pseudonym), *Class Struggle in Classless Poland* (Boston, 1982); a work by a Catholic theologian, Józef Tischner, *The Spirit of Solidarity* (San Francisco, 1984); Barbara Torunczyk (ed.), *Gdańsk 1980, Oczyma Świadkow* (London, 1980); and Teresa Toranska's book of interviews with the Communist party's postwar mandarins, *"Them": Stalin's Polish Puppets* (New York, 1987). The self-serving worldviews that surface in *"Them"* demonstrate conclusively that Poland's rulers-by-grace-of-Stalin remained tone deaf, even when the music of Solidarność was playing throughout society. In the new climate of the 1990s, as police archives become accessible, Polish social scientists can begin the long-delayed task of reconstructing, through both written and oral research, the history of popular self-activity and political striving in the locales and regions of Poland since 1944.

In terms of social activity, though not with respect to private sensibility or to high politics, émigré literature is necessarily a view from afar, even when it involves collections of essays containing contributions from members of the intelligentsia remaining in Poland: Grazyna Pomian (ed.), *Polska 'Solidarności'* (Paris, 1982); Krzysztof Pomian, *Pologne: Défi à l'impossible* (Paris, 1982); Jakub Karpiński, *Countdown: The Polish Upheavals of 1956, 1968, 1970, 1976, 1980* (New York, 1982); Alexander Smolar (ed.), *La Pologne, Une Société en dissidence* (Paris, 1978). Polish workers are not present in these books as historical actors, a circumstance that encourages the authors to find causality elsewhere. The utility of the compilations of oppositional literature collected by Peter Raina is not diminished by the fact they are encased in a political perspective that seems to change with each new work: Raina, *Political Opposition in Poland, 1954–1977* (London, 1978); *Independent Social Movements in Poland* (London, 1981); and *Poland, 1981: Toward Social Renewal* (London, 1985).

Western studies of Poland, while not remotely comparable in social depth to studies of Western societies, demonstrate a remarkable diversity. Denis MacShane, a British trade unionist, provides an interesting perspective, one that intermittently penetrates to shop-floor conditions and to the interior of the Polish movement, in *Solidarity: Poland's Independent Trade Union* (Nottingham, 1981). Lawrence Weschler's reports in the *New Yorker*, later published in book form, reveal an extremely sensitive feel for civic life and politics during the era of Solidarność: Weschler, *Solidarity: Poland in the Season of Its Passion* (New York, 1982), and *The Passion of Poland: From Solidarity through the State of War* (New York, 1982).

The most widely read study of Solidarność in the West is Timothy Garton Ash's *The Polish Revolution: Solidarity* (New York, 1984). After thirty-four pages on Polish history, the appearance of opposition sects, and the 1979 visit by the new Polish Pope, plus thirty additional pages on the Polish August, Garton Ash presents a thoughtful and well-written account of high politics in the era of Solidarność. Reflecting the author's prior assumption that the movement was produced by "the great convergence" of the opposition intelligentsia, "the workers," and the Polish Pope, *The Polish Revolution* is thinnest on the interior dynamics of the social movement. In *The Polish August: The Self-Limiting Revolution*, another English journalist, Neal Ascherson maintains an impressive psychological distance from the Warsaw intelligentsia, the Catholic episcopate, and the Polish population generally, a circumstance that seems to shepherd him toward causality without people; the movement is seen as a product of "frustrated nationalism," intellectually assisted by ideas espoused in the DiP report. Daniel Singer's *Road to Gdańsk* (New York and London, 1981), though focused chiefly on the Soviet Union in the late 1970s, is a germane and at times prophetic exploration of the crisis of Leninist production and governance that underlay the Polish August.

Research problems inherent in Leninist societies have encouraged a scholarly form that is in decreasing use in American history; namely, essay collections from a variety of contributors. These collections on Poland share one characteristic: they are evidentially thinnest and most speculative in dealing with workers. For example, in Abraham Brumberg's interesting collection of contributor essays entitled *Poland: Genesis of a Revolution* (New York, 1983), Alex Pravda's thoughtful essay on Polish workers is essentially grounded in government-generated polling data of the entire population. Attitude, as distinct from activity, is thus the operative component of social life that is subjected to interpretation. Other methodological problems with respect to the static components of attitude surveys attendant upon this approach are discussed in the opening of Chapter 2, p. 398, n. 1.

Other sources on Solidarność utilized in this study include:

Adelman, J., *Communist Armies in Politics* (Boulder, 1982).

Bauman, Zygmunt, "Intellectuals in East Central Europe: Continuity and Change," *Eastern European Politics and Societies* 1, 2 (Spring 1987):162–86.

Bielasiak, Jack, and Maurice Simon (eds.), *Polish Politics: Edge of the Abyss* (New York, 1984).

Book of Lech Wałęsa, the introduction by Neal Ascherson (New York, 1982).

Brandys, Kazimierz, *Warsaw Diary* (New York, 1984).

Bromke, Adam, *The Protracted Crisis* (Oakville, Ontario, 1982).

Brzezinski, Zbigniew, *Power and Principal* (New York, 1983).

Cynkin, Thomas M., *Soviet and American Signalling in the Polish Crisis* (New York, 1988).

Goldfarb, Jeffrey C., *On Cultural Freedom* (Chicago, 1982).

Green, Peter, "The Third Round in Poland," *New Left Review* 101–2 (1977).

Hahn, Werner G., *Democracy in a Communist Party* (New York, 1987).

Harding, Neil (ed.), *The State in Socialist Society* (Albany, 1984).

Hirszowicz, M., *Coercion and Control in Communist Society* (Brighton, 1986).

Holzer, Jerzy, "Solidarność in Wonderland," in Antony Polonsky, *Polish Paradoxes* (New York, 1990).

"Kalendarium Kryzysów w PRL lata 1953–1980," *Zeszyty Historyczne* 38, 66 (1983):142–95. (the Kubiak Report)

Kaufman, Michael, *Mad Dreams and Saving Graces* (New York, 1989).

Keane, John (ed.), *Civil Society and the State* (London, 1988).

——— (ed.), *Democracy and Civil Society* (London, 1988).

Kolankiewicz, George and Paul G. Lewis, *Poland: Politics, Economics and Society* (London and New York, 1988).

Kolkowica, Roman, and Andrzej Korboński (eds.), *Soldiers, Peasants, and Bureaucrats: Civil-Military Relations in Communist and Modernizing Societies* (London, 1982).

Kowalik, Tadeusz, "Experts and the Working Group," in A. Kemp-Welch, *The Birth of Solidarity* (New York, 1983), 143–67.

Kukliński, Col. Ryszard J., "Wojna z narodem widziana od środka," *Kultura* (Ap. 1987), 3–57.

Laba, Roman, "Solidarité et les luttes ouvrieres in Pologne 1970–1980," in *Actes de la recherche en sciences sociales*, No. 61 (Mar. 1986).

———, "Worker Roots of Solidarity, *Problems of Communism* (July–Aug. 1986).

Labedz, Leopold, *Poland Under Jaruzelski* (New York, 1984).

Lewis, Moshe, *The Gorbachev Phenomenon* (Berkeley, 1988).

Lewis, P. G., "Turbulent Priest: Political Implications of the Popiełuszko Affair," *Politics* 5, 2 (1986):33–39.

MacDonald, Oliver, *The Polish August* (San Francisco, 1981).

Majowski, Władysław, *People's Poland: Patterns of Social Inequality and Conflict* (Westport, 1985).

Malanowski, Jan, *Polscy Robotnicy* (Warsaw, 1981)

Mason, David, *Public Opinion and Political Change in Poland, 1980–82* (Cambridge, 1985).

Michalski, Franek, "Conversations with the Polish Underground," in *New Politics*, 1, 2 (Winter 1987).

Michnik, Adam, "KOR i Solidarność," *Zeszyty Historyczne*, no. 64 (1984).

Michta, Andrew A., *Red Eagle: The Army in Polish Politics, 1944–1988* (Stanford, 1990)

Misztal, B. (ed.), *Poland After Solidarity* (New Brunswick, 1985).

Montias, J. Michael, "Poland: Roots of the Economic Crisis," in ACES *Bulletin* 24, 3 (Fall 1982).

Morawski, W., "Society and Strategy of Imposed Industrialization," *Polish Sociological Bulletin*, no. 4 (1980).

Naylor, Thomas, *The Gorbachev Strategy: Opening the Closed Society* (Lexington, 1988).

Norr, Henry, "Solidarity and Self-management, May–July 1981," in *Poland* Watch, no. 7 (1985).

Nove, Alec, *The Economics of Feasible Socialism* (London, 1983).

Nowak, J., "The Church in Poland," *Problems of Communism* 31, 1 (1982):1–16.

Nowak, Stefan, "Attitudes, Values and Aspirations of Polish Society," in Stefan Nowak, *Polish Society of the Time of Crisis* (Warsaw, 1984).

Ost, David, "The Transformation of Solidarity and the Future of Central Europe," *Telos*, no. 79 (Spring 1989).

Pelczyński, Zbigniew A., "The Downfall of Gomułka," in Adam Bromke and John W. Strong (eds.), *Gierek's Poland* (New York, 1973).

———, "Solidarity and the Rebirth of Civil Society in Poland, 1976–81," in John H. Keane (ed.), *Civil Society and the State* (London, 1988).

Persky, Stan, *At the Lenin Shipyard* (Vancouver, 1981).

Ploss, Sidney I., *Moscow and the Polish Crisis: An Interpretation of Soviet Policies and Intentions* (Boulder, 1986).

Polet, Robert, *The Polish Summer* (London, 1981).

Pomian, Krzysztof, "Miracle en Pologne" *Le Debut*, no. 9 (Feb. 1981):3–18.

Potel, Jean-Yves, *The Promise of Solidarity* (London, 1982).

Radio Free Europe, *Audience and Public Opinion Research Department Reports, Poland* (Munich, 1973–1980).

Radio Free Europe, Radio Liberty, *Selected Background and Situation Reports, Poland* (Munich, 1973–1980).

Robinson, William F., *August 1980, The Strikes in Poland* (Munich, 1980).

Ruane, Kevin, *The Polish Challenge* (London, 1982).

Sanford, George, *Polish Communism in Crisis* (London and New York, 1983).
———, *Military Rule in Poland* (London, 1986).
Shoup, Paul S., *The East European and Soviet Data Handbook: Political, Social and Development Indicators, 1945–1978* (New York, 1981).
Smolar, Aleksander, "Contestation intellectuelle et mouvement populaire," in *Solidarité résiste et signe* (Paris, 1984), 135–41.
———, "Jews as a Polish Problem," *Daedalus* 116, 3 (Spring 1987).
Steven, Stewart, *The Poles* (New York, 1982).
Surdyrowski, Jerzy, *Notatki Gdańskie* (London, 1982).
Taylor, John, *Five Months with Solidarity* (London, 1981).
Terry, Sarah M. (ed.), *Soviet Policy in Eastern Europe* (New Haven, 1984).
Touraine, Alain, et al., *Solidarity: Poland 1980–81* (Cambridge, London, and New York, 1983).
Ulam, Adam, *Dangerous Relations* (New York, 1983).
de Weydenthal, Jan, *The Communists of Poland*, rev. ed. (Stanford, 1986).
Woodall, Jean, *The Socialist Corporation and Technocratic Power* (Cambridge and London, 1982).
——— (ed.), *Policy and Politics in Contemporary Poland* (New York, 1982).

Oral Sources

Interviews were conducted by the author in Warsaw, Gdańsk, Gdynia, Grojec, and Gostynin in Poland and in Paris, Berlin, Washington, New York, Boston, and Durham, North Carolina, between 1982 and 1989. For the most part, interviews were conducted during martial law or enforced exile. Some interviews were for attribution: Mirosław Chojecki and Waldemar Kuczyński (Paris); Jan Lityński, Henryk Wujec, Stefan Bratkowski, Andrzej Celiński, and Krzysztof Wyszkowski (Warsaw); Henryk Lenarciak (Gdańsk), and Bronisław Geremek (interviewed in Gdańsk).

Interviews were conducted with workers from the Lenin Shipyard (24); workers and other citizens from Gdańsk and Gdynia (13); workers from Poznań (5); Solidarność activists from other cities in Poland, including Warsaw, Kraków, Łódź, Katowice, Wrocław, Szczecin, Elbląg, Słupsk, Bydgoszcz, and Radom (38). Most of these sources were interviewed in Paris, Washington, and New York.

Sources on Pre-Solidarność Poland

Adam, Jan, *Wage Control and Inflation in the Soviet Bloc Countries* (London, 1979).
Alton, Thad P., *Polish Postwar Economy* (New York, 1955).
Banas, J. *The Scapegoats: The Exodus of the Remnants of Polish Jewry* (London, 1979).
Bethell, Nicholas, *Gomułka: His Poland, and His Communism*, rev. ed. (Harmondsworth, 1972).
Blazynski, George, *Flashpoint Poland* (New York, 1979).
Bromke, Adam, and John W. Strong (eds.) *Gierek's Poland* (New York, 1973).
Brown, Archie, and Jack Grey (eds.), *Political Culture and Political Changes in Communist States* (London, 1977).
Brzezinski, Zbigniew, *The Soviet Bloc* (New York, 1961).
Chłasiński, Józef, *Społeczna Genealogia Inteligencju Polskiej* (Poznań 1946).
Cieplak, Tadeusz (ed.), *Poland Since 1956* (New York, 1972).
Cohen, Jean, *Class and Civil Society: The Limits of Marxian Critical Theory* (Amherst, 1982).

Conference Papers, *Wydarzenia Czerwcowa w Poznańio, 1956* (Poznań, 1981).
Curry, Jane (ed.), *Black Book of Polish Censorship* (New York, 1983).
Dissent in Poland: Reports and Documents in Translation, December 1975–July 1977 (Association of Polish Students and Graduates in Exile: London, 1977).
Gati, Charles (ed.), *The Politics of Modernization in Eastern Europe* (New York, 1974).
Gella, Aleksander (ed.), *The Intelligentsia and the Intellectuals* (Beverly Hills, 1976)
Harman, Chris, *Bureaucracy and Revolution in Eastern Europe* (London, 1974).
Hirszowicz, Maria, *The Bureaucratic Leviathon: A Study in the Sociology of Communism* (New York, 1980).
Hirszowicz, Maria, and Malcolm Hamilton, *Class and Inequality in Pre-Industrial Capitalist and Communist Society* (Brighton, 1987).
Hiscocks, R., *Poland: Bridge for the Abyss?* (London, 1963).
Hohmann, Hans-Hermann, M. Kaser, and Karl C. Thalheim (eds.), *The New Economic Systems of Eastern Europe* (London, 1975).
Huntington, Samuel P., and Clement H. Moore (eds.), *Authoritarian Politics in Modern Society* (New York, 1970).
Janos, Andrew C. (ed.), *Authoritarian Politics in Communist Europe* (Berkeley, 1976).
Johnson, Chalmers (ed.), *Change in Communist Systems* (Stanford, 1970).
Karol, K. S., *Visa for Poland* (London, 1959).
Kolaja, Jiri, *A Polish Factory: A Case Study of Workers' Participation in Decision Making* (Lexington, 1960).
Kołakowski, Leszek, "Hope and Hopelessness," *Survey* 17, 3 (1982).
Kolankiewicz, George, "The Polish Industrial Manual Working Class," in *Social Groups in Polish Society*, edited by David Lane and George Kolankiewicz (New York, 1973), 88–151.
———, "The New 'Awkward-Class': the Peasant-Worker in Poland," *Sociologia Review* 20, no. 1–2.
———, "Poland, 1980: The Working Class under 'Anomic Socialism,' " in *Blue-Collar Workers in Eastern Europe*, edited by J. F. Triska and Charles Gati (London, 1981), 136–56.
Konrad, George, and Ivan Szelenyi, *The Intellectuals on the Road to Class Power* (San Diego, 1979).
Korboński, Andrzej, *The Politics of Socialist Agriculture in Poland, 1945–1960* (New York, 1965).
Korboński, Stefan, *Warsaw in Chains* (London, 1959).
Kuroń, Jacek, "Reflections on a Program of Action," *Polish Review*, no. 22 (1977).
Lane, David, and George Kolankiewicz (eds.), *Social Groups in Polish Society* (London, 1973).
Lange, Oscar, *Some Problems Relating to the Polish Road to Socialism* (Warsaw, 1957).
Lepak, Keith J., *Prelude to Solidarity: Poland and the Politics of the Gierek Regime* (New York, 1988).
Lewis, Flora, *A Case History of Hope* (New York, 1958).
Mayewski, Pawel (ed.), *The Broken Mirror* (New York, 1958).
Micewski, Andrzej, *Cardinal Wyszyński: A Biography* (San Diego, 1984).
Mieczkowski, Bogdan, *Personal and Social Consumption in Eastern Europe* (New York, 1975).
Mikolajczyk, Stanisław, *The Rape of Poland* (Westport, Conn., 1973).
Narkiewicz, Olga A., *The Green Flag: Polish Populist Politics, 1967–1970* (London, 1976).
Poland, A. J., *Lenin and the End of Politics* (Berkeley, 1984).

Poland Today: The State of the Republic, introduction by Jack Bielasiak (New York, 1981). (the DiP Report)
Polonsky, Antony and Bolesław Druhia, *The Beginnings of Communist Rule in Poland* (London, 1980).
Potel, Jean-Yves, *La Memoire ouvriere 1970–1980* (Paris, 1982).
"Przyczyny, przebieg i skutki Kryzysów społecznych w dziejach PRL," reprinted in *Zeszyty Historyczne,* no. 65 (1983). (the Kubiak Report, Part II)
Rakowska-Harmstone, Teresa (ed.), *Perspectives for Change in Communist Societies* (Boulder, 1979).
Reynolds, Jamie, "Communists, Socialists and Workers, Poland, 1944–48," in *Soviet Studies* 30, 4 (Oct. 1976):516–39.
Rothschild, Joseph, *East Central Europe Between the Two World Wars* (Seattle, 1970).
Sakwa, G., "The Polish October," *The Polish Review* 23, 3 (1978):62–78.
Schneiderman, S. L., *The Warsaw Heresy* (New York, 1959).
Shapiro, Jane, and Peter J. Potichnyi (eds.), *Change and Adaptation in Soviet and East European Politics* (New York, 1976).
Simon, Maurice D., and Roger E. Kanet (eds.), *Background to Crisis: Policy and Politics in Gierek's Poland* (Boulder, 1981).
Stehle, Hansjakob, *The Independent Satellite* (New York, 1965).
Sulik, Bolesław, "Robotnicy" in *Kultura* (Oct. 1976).
Syrop, Konrad, *Spring in October* (London, 1957).
Szafar, Tadeusz, "Anti-Semitism: A Trusty Weapon," in A. Brumberg (ed.), *Poland: The Genesis of A Revolution* (New York, 1983), 109–22.
Taras, Ray, *Ideology in a Socialist State: Poland 1956–1983* (Cambridge, 1984).
Triska, Jan, and Paul M. Cocks (eds.), *Political Development in Eastern Europe* (London, 1977).
Triska, Jan, and Charles Gati (eds.), *Blue-Collar Workers in Eastern Europe* (London, 1981).
Walicki, Andrzej, *Philosophy and Romantic Nationalism* (New York, 1982).
Weit, Erwin, *Eyewitness* (London, 1973).
de Weydenthal, Jan B., "The Worker's Dilemma of Polish Politics: A Case Study," *East European Quarterly* 13, 1 (1977):95–119.
Wisniewski, Joseph, *Who's Who in Poland* (Toronto, 1981).
Wojtka, Karol Cardinal, *Signs of Contradiction* (New York, 1979).
Zinner, Paul E., *National Communism and Popular Revolt in Eastern Europe* (New York, 1956).
Zurawski, Joseph W., *Poland: The Captive Satellite* (Detroit, 1962).

Other Secondary Sources

Studies not specifically focused upon Poland but which proved especially helpful in conceptualizing this inquiry into the origins and development of Solidarność include:

Arendt, Hanna, *The Human Condition* (London, 1958).
———, *On Revolution* (London, 1973).
Billington, James H., *Fire in the Minds of Men* (New York, 1980).
Boyte, Harry, *Commonwealth: A Return to Citizen Politics* (New York, 1989).
Calhoun, Craig, *The Question of Class Struggle: Social Foundations of Popular Radicalism in the Industrial Revolution* (1982).

Gutman, Herbert, *Work, Culture and Society in Industrializing America* (New York, 1976).
Labedz, Leopold, *Revisionism: Essays on the History of Marxist Thought* (London, 1962).
Montgomery, David, *Workers Control in America* (New York, 1979).
Moore, Barrington, Jr., *Authority and Inequality under Capitalism and Socialism* (Oxford, 1987).
Pocock, J. G. A., *The Machiavellian Moment* (Princeton, 1975).
Schaar, John, *Legitimacy and the Modern State* (New Brunswick, 1981).
Scott, James, *Moral Economy of the Peasant* (New Haven, 1976).
———, *Weapons of the Weak* (New Haven, 1985).
Thompson, E. P., *The Making of the English Working Class* (London, 1963).
Williams, Raymond, *Culture and Society, 1780–1950* (1958).
Wolin, Sheldon, *Politics and Vision* (Boston, 1960).

Index